THE FACES OF MARS: BEYOND MERE TRICKS OF LIGHT AND SHADOW

BY: **Gary Leggiere**

THE
MARTIAN
REVELATION

Table Of Content

ACKNOWLEDGEMENTS OF THANKS

Dr. John Brandenburg, Richard C. Hoagland

David Gannett

NASA

JPL

ASU

MSSS

ESA

Published by

Brookscraft Publishing
A Division of Brooks Craft LLC
Info@brookscraftpublishing.com
www.brookscraftpublishing.com

Author's Contact

To book the author as a speaker at your next event or to order bulk copies of this book, please, use the email below:

marsrevealer@gmail.com

FORWARD:

By: DR. JOHN BRANDENBURG

Gary Leggiere, AKA, the Mars Revealer, has produced an epic book on the Cydonia Face controversy on Mars and the whole question of apparent Mars Archeology. Gary has been a fellow investigator with this author on the subject of Mars archeology since the late 90's. This lavishly illustrated book by Gary lays out for the reader the whole history of the human fascination with the Red Planet and its longstanding belief that it was once the home of thriving life and civilization. This culturally imbedded belief, dating from the invention of good astronomical telescopes in the 1700's, has now been validated by cutting edge technologies employed in telescopes, and orbiters and rovers sent to Mars. This historic but covert discovery, can be said now to be the unspoken driver behind our newly announced goal of placing humans on Mars within a decade. In truth, as this book reveals, Mars has been a driver of human science since the inception of the Scientific Revolution.

This book begins with the story of the discovery of an Earthlike, rather than Moonlike, environment on Mars found by early telescopes. This coincided with the time of the revolution in thought produced by Copernicus, Kepler and Newton, all of whose work focused on the next planet out from the Sun from us. These men were all attracted by its peculiar, noticeably oval, orbit. It was red, and would disappear from view for months and then swell to great brightness as it neared Earth in its orbit. Mars was thus a flashing red beacon in the sky naturally attracting the view of telescopes as they grew more advanced.

Those early telescopes first saw details on the Red Planet that attracted even more attention. Mars, unlike the Moon, had polar caps and dust storms. This fascination was amplified further in the mid 1800's by the observation of detailed surface markings, which were called "canalli" by Schiaparelli. This was followed by the monumental hypotheses of Lowell at the dawn of the 20st Century which was fueled by the success of the Suez Canal, that these markings were signs of a civilization on Mars. Though later dismissed as an illusion created by the human vision system, designed to literally 'connect dots,' Lowell's canals can also be viewed, in the light of new data, as a form of "remote viewing" since civilization they were thought to indicate, has been shown to be real. The effect of Lowell's reports in the era of 1900 were electric, however, in their time. Most impacted were the rapidly developing frontiers of human science and engineering.

Lowell's concept of Mars transformed the rocket, formerly a tactical weapon of war, or fireworks. Into a spaceship to travel to Mars. Goddard, inspired by the work of the Lowell, developed his liquid fuel rockets expressly to take humans to Mars. Launching his successively more advanced rockets from the unlikely local of Roswell New Mexico in the early 1930's, Goddard became the world's leader in rocket science. He would become the father of every rocket taking humans or their technologies to other planets.

Also inspired by Lowell, in those days of European Colonization of the less developed world, was the possibility that imperialistic aliens might attempt colonize Earth. The shocking novel "War of the Worlds", evidently has had an unspoken but profound influence on government thinking on all things cosmic. The novel's depiction of a pitiless war between the human race and invading aliens, with images of mass death and destruction unimagined until World War II, continues to haunt the human race. Mars can be understood as not only the inspiration for our space program, but also the UFO Cover-up. But as the Mars Revealer shows in this book, the human race has actually been most fortunate.

Early Mars is now understood by probe data to have been a place similar in environment to Early Earth, where life appeared soon after the formation of a liquid ocean. While the details of this process of biogenesis on Earth are unknown, it is believed by scientists to have been a natural process, similar to the equally mysterious process of biological evolution. Both processes are equally obvious from the fossil record, even if their exact mechanisms are not.

Early Earth was obviously a place of abundant liquid water, organic matter, and a magnetic field to protect against deadly cosmic radiation. These conditions lead to the early appearance of life on Earth. This process is not considered scientifically to be miraculous, but somehow part of the natural evolution of organic matter from simple to complex forms, that will occur wherever similar conditions as found on Early Earth occur. That life, a complex system of organic chemistry once it exists, will lead to biological evolution, leading in turn to something like humanity, is also considered to be a natural process. It has now been found that conditions on Early Mars were very similar to Early Earth.

Since conditions on early Mars were similar to those that spawned life of Earth, the scientific conclusion that Mars also became an abode of life of life now seems inescapable, as would the operation of biological evolution there once life began. Hence the finding of archeology on Mars, apparently left by a humanoid culture dating from a period of past Earth-like climate, is not unexpected.

As this book by the Mars Revealer establishes, by strong pictorial evidence, it is a fact that massive archeology on Mars has been found. That evidence also exists that the government is running a "Mars Cover-up", in parallel to the UFO Cover-up, is also apparent, as this book shows. The fact that a Government Cover-up at Mars is understandable, as well as annoying, can be explained somewhat by Government unease with evidence that the Mars culture may have perished due to some cataclysm, either natural or technological. However, the author of this preface strongly recommends this book as a feast for the imagination as well as the eyes. This epic book must be viewed as part of the global effort to expand human consciousness as well as civilization to the Red Planet. This is an effort begun centuries ago, now to be realized in our lifetime. So brave reader, read on!

John Brandenburg PhD

CHAPTER 1:

TRICKS OF LIGHT AND SHADOW

It was not without reason that the planet Mars is and has been chosen by humanity to undergo, and is currently undergoing a scrutiny of observing by national and international space agencies since the space age began technologically in the 1960's. It was not merely the reason that Mars was the next planet over to us in our planetary solar system neighborhood to compel the Earth's space administration's missions to put such an important focus and effort to study the 4th world counting out from the sun. It was not certainly for a reason merely of convenience of its position next to ours to test and show off our technological prowess as humanity starts to venture forth into the heavens. History shows that Mercury and Venus certainly has not received the same attentions given unto them, compared to the intense missions and reconnaissance and scientific data gathering efforts and visual data acquisition that Mars has received, more so than any other planet in our Solar system, including of our own Moon. Considering this in mind, one must wonder and ask WHY?

What is it about Mars that grabs humanity's attention to it with such importance, that it grabs the majority of our space administration's attentions to it with such importance? One must also wonder and yet ask, if somehow, ***someone in our society knew something about Mars that possibly the rest of us maybe did not know?*** In my opinion, from studying into Mars for many years, I believe that the space exploration program was started and purposed with the immediate need to rush to Mars more so than any other reason to create and advance a technologically working space exploration program agenda. But this sounds like a crazy idea one would think, yet it is certainly not without merits to consider it. It turns out that there are multiple reasons to consider this idea being a possibility. These reasons are known, but not really taught or spoken about by the scientific mainstream as being legit. These reasons are also historical as they come from pre-technologically advanced space capable exploration.

The 1st of these historical reasons comes from the most advanced capable technology to look into space and see and study its heavenly wonders as a form of science and technology that brings us all knowledge. I of course am speaking about telescopes and the telescopic observations that drew so much attention and also speculation of the public interest and which created a lot of scientific study, especially that of the planet Mars! It was Mars that captivated the public's interest the most because of the observations by astronomer's and scientists who told and shown the public of their views of Mars that no other planet that could be seen through a telescope could compare with. It was Mars that seemed to show that there was evidence of seeing a visual reason for an intelligence that was present on Mars. How? Simply put, it appears to be reasoned by its '**Tricks of Light and Shadow**'!

1: PERCIVAL LOWELL AND MARS CANALS

The most intense research and best telescope observations and public interpretation of the telescopic views of Mars was done by Percival Lowell. He became determined to study Mars and astronomy as a full-time career after reading Camille Flammarion's *La planet Mars*. He was particularly interested in the 'canals of Mars', as drawn by Italian astronomer Giovanni Schiaparelli, who was director of the Milan Observatory. The Boston geologist George Russel Agassiz noted that Lowell made the decision to begin his observations after hearing that Schiaparelli began to experience failing eyesight.

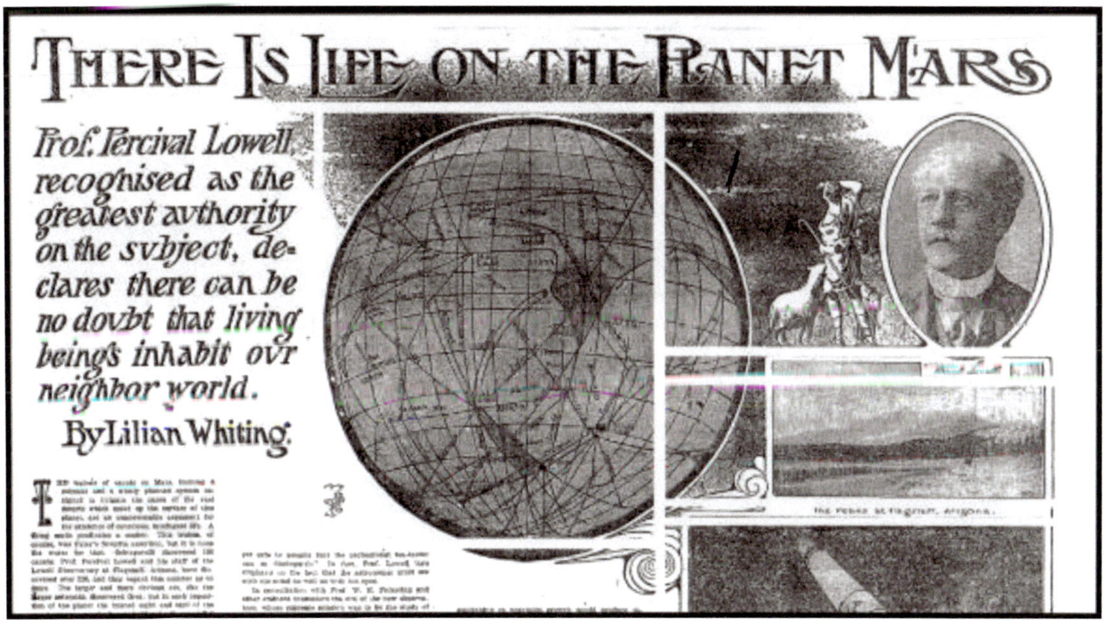
THERE IS LIFE ON THE PLANET MARS

Prof. Percival Lowell, recognised as the greatest authority on the subject, declares there can be no doubt that living beings inhabit our neighbor world.

By Lilian Whiting.

Beginning in the winter of 1893–94, using his wealth and influence, Lowell dedicated himself to the study of astronomy, founding the observatory which bears his name. He chose Flagstaff, Arizona Territory, as the home of his new observatory. At an altitude of over 2,100 meters (6,900 feet), with few cloudy nights, and far from city lights, Flagstaff was an excellent site for astronomical observations. This marked the first time an observatory had been deliberately located in a remote, elevated place for optimal seeing which included enhanced image quality, sharpness and steadiness. At his Flagstaff observatory Lowell favored the use of smaller telescopes rather than larger ones, believing that they were usually better for viewing fine planetary details. He was assisted in setting up his observatory by William Pickering, another observer of Mars who had noted the lines seen by Schiaparelli as well.

(Percival Lowell sitting at Lowell Observatory Looking at Mars)

Lowell in his day, was far and away the most influential popularizer of planetary science in America. For some fifteen years (1893 to about 1908) Lowell studied Mars extensively, making intricate drawings of the surface markings as he perceived them. Lowell published his views in three books: *Mars* (1895), *Mars and Its Canals* (1906), and *Mars As the Abode of Life* (1908). With these writings, Lowell more than anyone else popularized the long-held belief that ***these Alien Markings showed that Mars sustained intelligent life forms.***

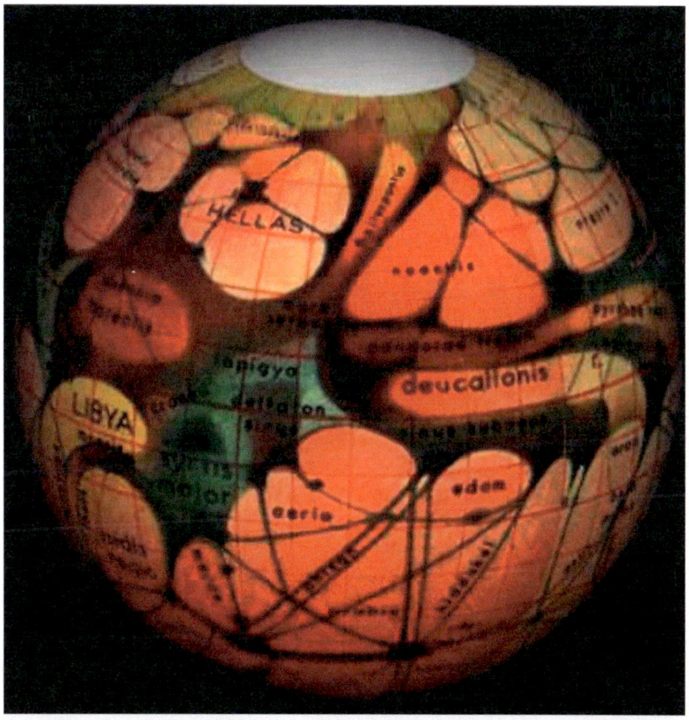

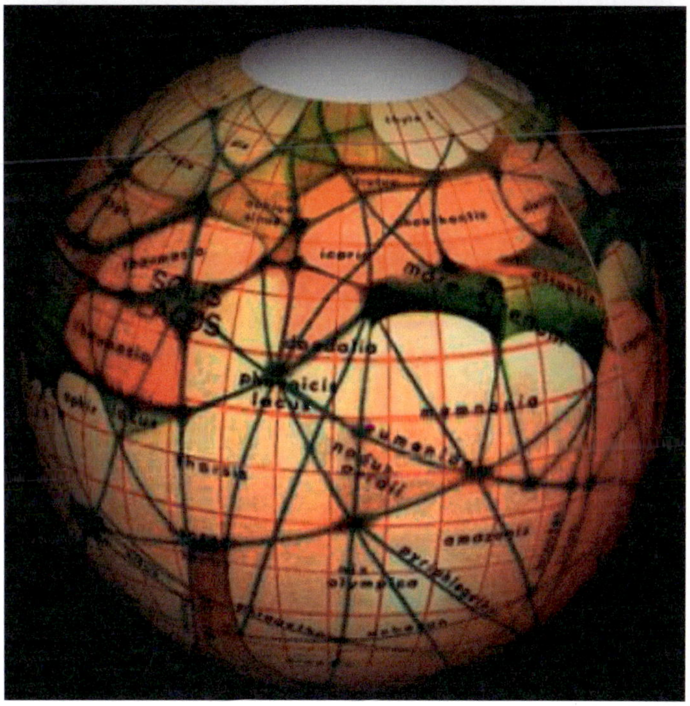

His works include a detailed description of what he termed the "non-natural features" of the planet's surface, including especially a full account of the "canals," single and double; the "oases," as he termed the dark spots at their intersections; and the varying visibility of both, depending partly on the Martian seasons. He theorized that an advanced but desperate culture had built the canals to tap Mars' polar ice caps, the last source of water on an inexorably drying planet. While this idea excited the public, the astronomical scientific community was skeptical. Many astronomers could not see these ***Alien Markings***, and few believed that they were as extensive as Lowell claimed. As a result, needless to say, that Lowell and his observatory were largely ostracized. He thought the Red Planet must once have been covered by lush greenery, but was now desiccated. The "canals" were an admirable attempt by intelligent and cooperative beings to save their home planet.

To me, this is in and of itself an important area of research science. As of today, we now call this form of scientific research looking for 'techno-signatures' of intelligence, or of intelligent life! So Percival Lowell can literally be in fact, also to add to his legacy, the father of the scientific research and study of identifying alien techno-signatures. Techno-signatures aren't just signs of life, but they are signs of intelligent life.

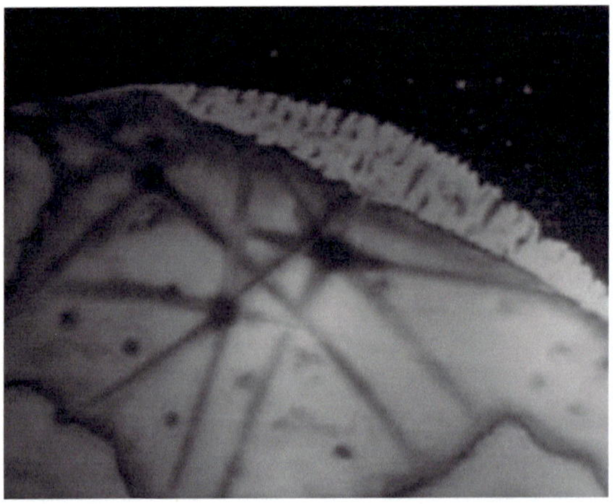

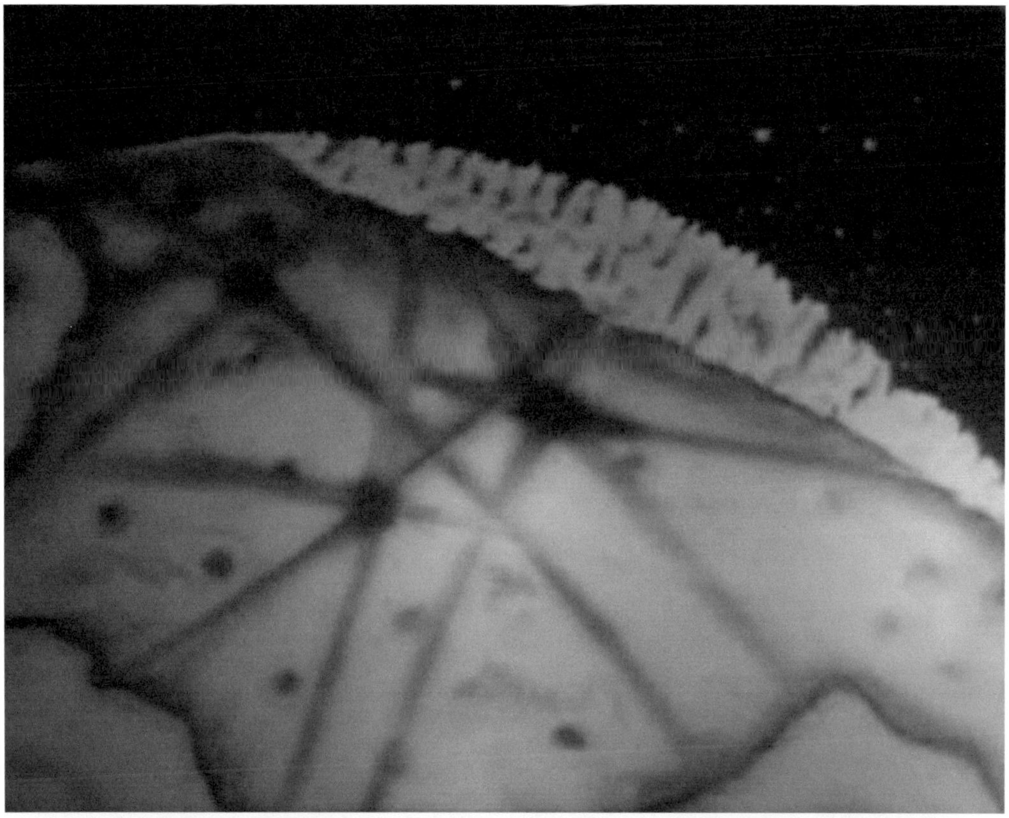

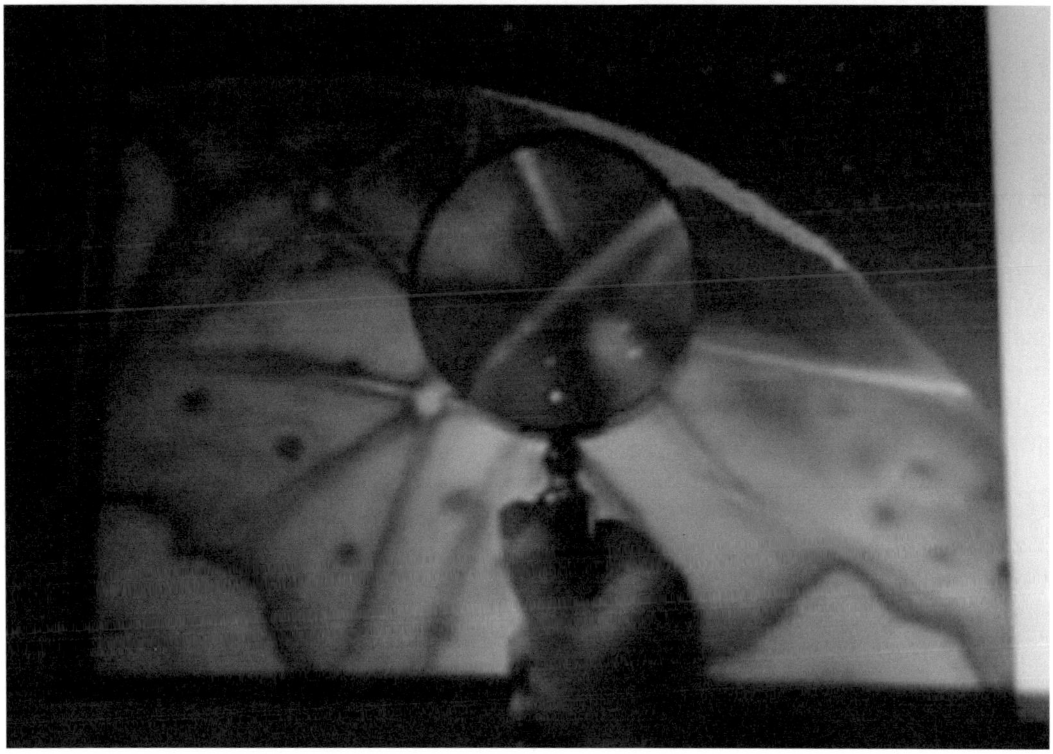

(Images seen above from RED PLANET MARS 1952 movie. Being proposed shows how the Polar Ice cap melts and water fills the canal lines, that makes the lines look shiny.)

The 2nd of these historical reasons come from yet another form of techno-signatures research; this one comes in the form of studying the frequencies of radio transmissions, also a new scientific research at the time of Lowell.

2: NIKOLA TESLA COMMUNICATION WITH MARS 1899!

Leave it to the unique mind and genius of a foreign immigrant to the United States of America, who had single handedly created controversy and started bitter corporate wars with the big name inventors and companies of that era of time for electricity and its' useful practical implications and applications to America and, even around the world. Nikola Tesla who came to America with only four cents in his pocket, successfully forced many to rethink the things they thought they knew about, to show them another and even more practical way of harnessing, transmitting, and even receiving electricity using a method unheard of, without using wires. Wireless energy and lots of it. Tesla believed that voices, pictures, and information can be carried or received wirelessly. Which is true!

Many of the corporations and scientists wanted to discredit Tesla's ideas and his inventions stating he was a mad man or that his experiments can kill people. This mostly seemed to stem from their personal feelings of jealousy and being threatened in their standing and power to the public and their livelihoods, or undermines their doctorates of knowledge, or afraid to be shown up in what they were well known for in the things they provided for the new modern America. Their main excuse to discredit Tesla was really a financial greed issue. Tesla certainly caused much need to worry for them. Yet in regards to his experiments and inventions, there were people, a select few, who took Nikola Tesla's work and his findings seriously enough to back him. At least enough for them to get what they wanted from Tesla, while at the same time, Tesla continued conducting his experiments through his financiers funds, and testing and using his inventions where he was creating them for the use of practical application.

One such discovery and research made by Tesla, was his reported discovery of a message being sent to Earth, from Mars! It is the only thing which no doubt at that time would have had the US government and scientists to want to seriously consider Tesla's reports, considering his unique way of proving how he was right regards to his inventions and applications for them. They certainly not wanted to let Tesla gain all the credit for this discovery and Tesla then show them all up once again, if they did nothing. Then once again, Tesla's name will be credited to vindication and validation to a long list of vindications and validations already mounting for this immigrated genius.

I also feel it must also have been a political decision to look at Tesla's findings not with skepticism, but with serious interest by those in government who felt they needed that information first. Especially because of the public findings and public fascinations of Percival Lowell's Mars research. The Mars fever was spreading!

Tesla's almost incredible discovery was made by him in Colorado Springs, in 1899 while he was conducting experiments in relation to wireless transmission of energy. He worked in a laboratory specially constructed. While Tesla was conducting experiments with his Magnifying Transmitter, he detected coherent signals which he determined had originated on Mars.

Nikola Tesla has had the first call of the then modern century from a neighboring planet. He declares that he has communicated with Mars, while on Pike's Peak, delving into the mysteries of the wireless transmission of electrical energy. The receiving messages was faint, but, according to Tesla, not to be mistaken.

A new voice from a planet, millions of miles removed away from the Earth, was spoken over one of the unwired telephones of the universe, and there, near the lonely mountain peak, of Pike's Peak, in the calm of night, the voice at last found a listener of another world that spoke to this world, in a language strange at first, but sure to be clearer, says Tesla, ***before the Twentieth Century has finished its course***.

Tesla also stated, ***"These were certain feeble electrical disturbances which I could barely note at times, and which by their character unmistakably showed that they were neither of solar origin nor produced by any causes known to me on the globe. What could they be?***

"I have incessantly thought of this for months, until finally I arrived at the conclusion that they must be of a planetary origin. Although I could not decipher their meaning, it was impossible for me to think of them as having been entirely accidental. The feeling is constantly growing in me that I had been the first to hear the greeting of one planet to another."

When he was asked back then to give some visible form or notation to the message he had received, he smiled emphatically and said:

"As I have already said, one of the planets in the solar system may be ahead of us in the evolution. Their means of interplanetary communication may be perfect, but we have yet to learn their sign language. It is impossible at present, even to suggest A CODE. It is enough to say at this time that a message from Mars which might be a triangle to them, would appear as some other form to us, and vice-versa. These differences can only be reconciled by time and careful study. It is wonderful enough, is it not, that a beginning has been made. As I think over it now, it seems to me that only a man absolutely stricken with blindness, and insensible to the greatness of nature, can hold that this planet is the only one inhabited by intelligent beings."

Yet here after this was reported by Tesla, it not took very long for the US government officials to be on the scene to try to play down Tesla's findings to the public perception.

Washington (D. C.), January 3. 1901— Professor S. I. Brown of the United States Naval Observatory, in discussing Nikola Tesla's announcement, says: ***"Mars may be populated. "We cannot make any positive assertions regarding that, but it will take more than a mere statement from Nikola Tesla to prove the existence of signals from the planet to us."***

Professor A. S. Skinner of the United States Naval Observatory says: ***"This is the first I have heard of Tesla's latest accomplishment. You can rest assured it is imaginative and visionary. I do not care to make any further comment on it or the possibilities of the case."***

Professor T. J. J. See of the Naval Observatory says: ***"I have read the story about this discovery. I do not care to say anything about it."***

TALKING WITH THE PLANETS
By NIKOLA TESLA

Tesla was widely criticized for his astounding claims. However yet back then, as well as now, no one could seriously dispute him, for he was a solo pioneer without peer. No one since then in that era of time of history has reported constructing a Magnifying Transmitter or otherwise replicated his experiment, so the issue remains unresolved and the mystery unsolved.

Yet all the while, I am sure the US government and its secret intense high interest to Tesla's findings, decided to take a back seat to the public perceptions and public opinion of what is real or not real on Mars, or what was being sent from Mars. If Tesla had not yet figured out yet what it was that was being communicated, there are those that wanted to find out before he did find out and he was able to tell anyone about those findings. In essence, the plan was underway to publicly control the information that seemed to be believed enough by the US government, that Tesla was supposedly able to receive such messages. They wanted to know what those radio transmissions were, and what messages they may contain, before all else may find out from Tesla, especially if no one else was to understand and act and follow up on the tip by Tesla regarding the messages being received. He would certainly once again make people look bad, this time the US government.

Unknown to most people he had a second laboratory which apparently was his main one on top of the Waldorf Astoria on the top floor and both penthouse towers. Now after Tesla retired, he maintained a laboratory in his living quarters at the Hotel New Yorker. He maintained a transmitter setup on the Waldorf; and his receiving setup, his receiving antennas and receivers which had been built by RCA under his direction, were on the New Yorker.

Two people are reported who said that they were working with Tesla, during that period during his retirement, saying ***"that he was using that equipment, and that he was talking with somebody virtually every day"***, and one of these people was emphatic: ***"it was someone off planet."*** In plain language, he was communicating with E.T.'s!

3: MILITARY INSTRUCTS NAVY TO RECEIVE SIGNALS FROM MARS 1924!

Notice that it cannot be a coincidence that Tesla had a press announcement in 1923, where he stated again that he was talking with E.T.'s off planet. This time the U.S. government became determined to figure out what this was all about. Mars was widely seen as the likeliest candidate. And while the notion of Martian radio signals was greeted by some engineers and scientists at the time to be little more than wild speculation, others believed the possibility was worth serious study. A plan was constructed to ensure radio silence worldwide, so tests can be done to capture the alien techno-signature and see if once again, Nikola Tesla was correct.

Heading the operation for the Army was Major General Charles Saltzman and Admiral Edward W. Eberlen, Chief of Naval Operations. A telegram was sent in 1924 by then Chief of Naval Operations Edward W. Eberlen instructing the United States Navy to listen for radio transmissions from the planet Mars.

The telegram read:

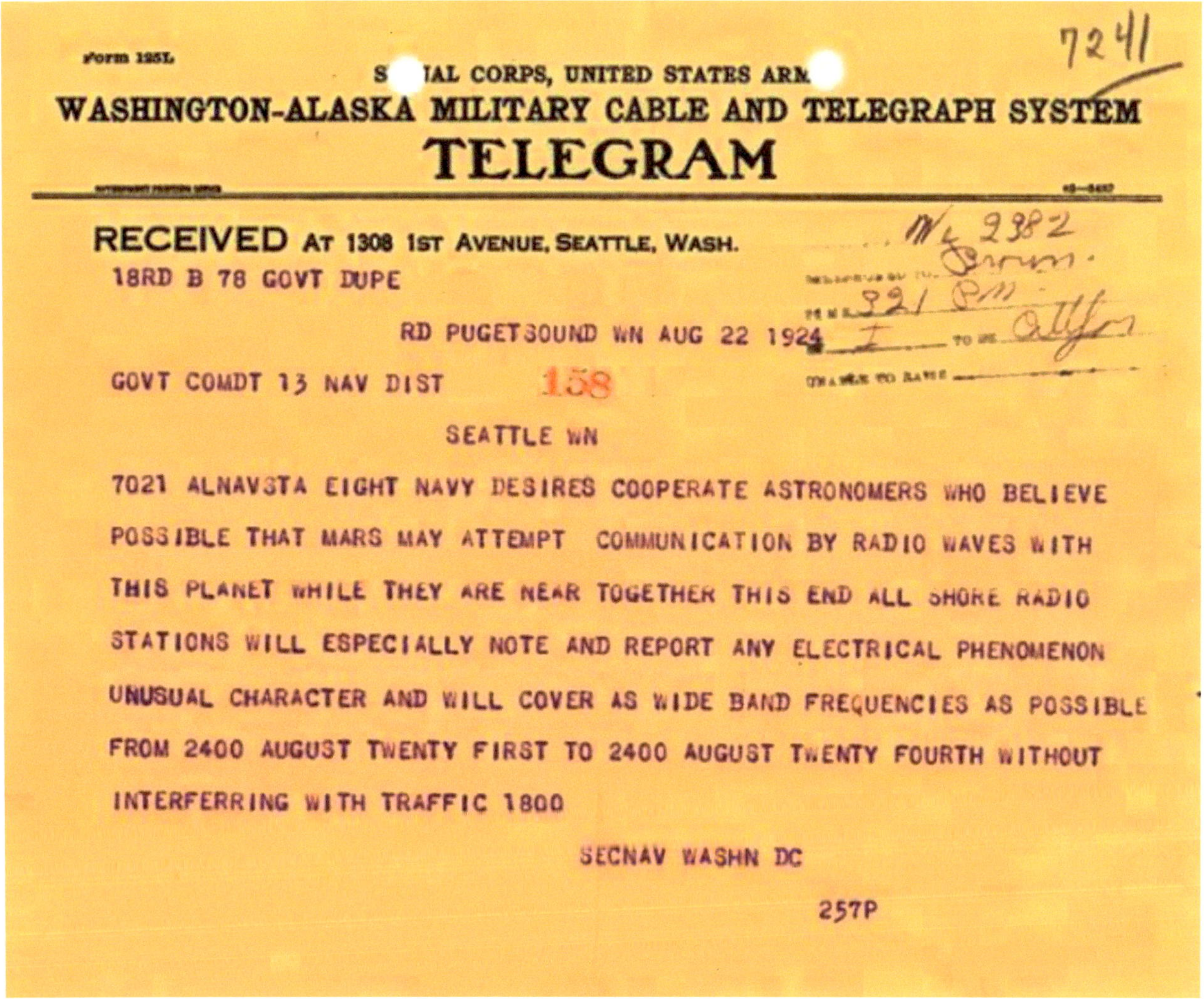

Form 125L

S IAL CORPS, UNITED STATES ARM

WASHINGTON-ALASKA MILITARY CABLE AND TELEGRAPH SYSTEM

TELEGRAM

7241

RECEIVED AT 1308 1ST AVENUE, SEATTLE, WASH.

18RD B 78 GOVT DUPE

RD PUGETSOUND WN AUG 22 1924

GOVT COMDT 13 NAV DIST 158

SEATTLE WN

7021 ALNAVSTA EIGHT NAVY DESIRES COOPERATE ASTRONOMERS WHO BELIEVE POSSIBLE THAT MARS MAY ATTEMPT COMMUNICATION BY RADIO WAVES WITH THIS PLANET WHILE THEY ARE NEAR TOGETHER THIS END ALL SHORE RADIO STATIONS WILL ESPECIALLY NOTE AND REPORT ANY ELECTRICAL PHENOMENON UNUSUAL CHARACTER AND WILL COVER AS WIDE BAND FREQUENCIES AS POSSIBLE FROM 2400 AUGUST TWENTY FIRST TO 2400 AUGUST TWENTY FOURTH WITHOUT INTERFERRING WITH TRAFFIC 1800

SECNAV WASHN DC

257P

However, it was reported also that during the three day period, during which Mars was in opposition to Earth, nothing but static was detected by Navy radio stations.

(Authors Note: A COVER UP ENSUES?! I personally do not believe this statement to be true, as we know from other sources that at the same time, who were also connected to this 'operation', it was actually reported that signals from Mars had been received. Showing very interesting results in fact.)

Which brings us all to this next historical reason example:

4: RADIO BROADCAST RECEIVED FROM MARS.... 1924!

This is a fact of historical information of which is related to the above example given. As the military of the United States had the NAVY listening for a signal from Mars, and it was not alone, as there were others of which also have been a part of a joint effort worldwide to 'listen' for a signal from Mars, and to receive that signal. The difference was, that a contradiction was seen between the findings of others who seen and shown the message was in fact received, compared to the U.S. NAVY's official statement that nothing but 'static', was detected. In my opinion, based upon my research of Mars and where it has led me, this to me shows clear evidence of possibly the first public Mars related Government 'CONSPIRACY THEORY' related to the public and to what 'THEY' really know is on Mars, that 'THEY' do not want the public to know or to see. **A Martian conspiracy theory over one hundred years old!**

This all I believe leads to and connects especially to things in regards to where we are going to be going throughout this book, as well as the purpose of this book to show. Looking back, one must ask, **"What do 'THEY' know, and when did 'THEY' know it?"**

So let's look at this very interesting and amazing story. Because this story I feel is connected to my Mars research now, in contemplating the odds of coincidence of ***coded messages being sent*** early on in our American history of the early twentieth century, of a record of a coded radio broadcast of Martians communicating to Earth and it's connections to 'The Faces Of Mars'.

In August of 1924, During the closest approach of Mars to the Earth in respect to Mars orbit, the U.S. Government ordered all radio transmitters in the country to maintain radio silence for 5 minutes on the hour, every hour for 3 days, from August 21 to August 24th in which to listen for signals during the close transit with Mars (just about thirty five million miles from Earth). This event was called "National Radio Silence Day." All American military transmitters were ordered to be shut down during this time.

The leader of the project was Dr. David Peck Todd, former head of the Amherst College astronomy department and friend of Percival Lowell. He was an enterprising scientist who had done on previous Mars oppositions, all kinds of things to try to detect life on Mars. The idea is to rise in a balloon to a height of about fifty thousand feet with the manifest purpose of overcoming the impediments of the dense air stratum to basically get a clear signal from Mars. This was done at a time and an era when people were not afraid to talk openly about the possibilities that there might be Martians and they might want to communicate, and the preferred method might be radio which at that time was not voice talk radio. It was more like Morse code radio transmissions. Yet it was very bad Morse code, because it was way down in the noise.

So during the week of Mars opposition of 1924, at radio stations, receiving stations as widely separated as Paris, London, Vancouver, and Washington D.C., a series of very powerful coded transmissions were received at all stations and recorded. And they were so loud that they blew all other stations off the air. This can be compared to listening to radio today, where loud stations or close stations would blow away weak stations. They will kind of take the air as it were, and would become the dominant source of information or the signal. It would overwhelm the fainter signal. Well man made radio signals back in the nineteen twenties were very weak, because our technology was very primitive. These signals according to the reports by these seasoned observers and radio people, electronic engineers, claimed these signals were strong enough to swamp and overwhelm the natural signals, such as thunderstorms and interference.

Not only were the natural signals overwhelmed however, but also the artificial signals, that were being generated primarily by government radio stations. And one radio station in Washington, WRC, owned by the Radio Corporation of America (RCA).

In 1924, for the night of the Mars closest orbital approach, Dr. Todd was able to arrange for the US NAVY (which is an interesting thing to think about) which the NAVY had the most powerful transmitter of information by radio, Dr. Todd got them to for five minutes of every hour go off the air and maintain radio silence during Mars closest approach. He also got RCA to take its' station off the air. So basically he got a shutdown world wide of artificial radio signals. And it was in that period of time, to have quiet times, these stations, Paris, London, Vancouver, Washington D.C. and some others, were listening and recording what they heard. This only goes to show the desire to work together to get information as well as the seriousness of cooperation applied in getting it. (It is usually what happens after getting that information which can be and is the sad thing in all this.)

To continue on, what they heard was very interesting. They heard **'A SERIES OF VERY POWERFUL CODED SIGNALS'** that appeared equally strong at all these widely separated points around the Earth! So these guys were listening to code, like Morse code accept much louder. Consider this, if you will to try to say this message was fraud, if you have an artificial transmitter, a man made transmitter on the Earth, (follow me carefully.) and let's say you are a powerful independent industrialist with his own money to have his own rig somewhere in upstate New York, and you were able to send out during these periods of five minutes of Radio Silence every hour, your own signal, to be as a kind of hoax, to hoax what was being listened for from Mars, so by others that are listening for the signals and receive it would think whatever you want them to think. Then the radio stations that were receiving the radio signals in London, Paris, Vancouver, Washington D.C., One receiving signals also in Italy as well, they would all have received different strength signals of your hoaxed broadcast during that time of radio Silence during the five minutes of every hour. Your signal would have bounced all over between the Earth surface and the ionosphere like ripples in a pond when you throw a pebble into it. Some stations would have received a fainter signal, and others a strong signal.

Yet ALL stations reported simultaneously equally powerful signals, so powerful that it literally swamped any other signal including natural background thunderstorms. That indicates that the source of the transmission is 'NOT' on planet Earth. Unless whoever was doing it, was as powerful and rich as the U.S. government and the NAVY, and had transmitters all over the world at that time? In other words, you needed a source that was illuminating the entire hemisphere of planet Earth with a frequency low enough to pass through the ionosphere and to be recorded by these widely separated ground stations receiving the broadcasts during the radio silences.

Now another thing Dr. Todd did was to convince a famed inventor, Francis Jenkins who was the inventor of basically the first primitive television, and who had developed a device, which was kind of photographic equivalent of the first fax machine, to try to record on this optical gadget which would use photographic film, slowly unwinding in a sealed box, to record what he heard that night.

On August 21st, nineteen twenty four, the Jenkins Radio-Camera was turned on. This machine produced flashes of light that were recorded on film whenever an incoming radio wave was detected. Everything that was "picked up" out of the air in about 29 hours. The film, which was thirty feet long and six inches wide, was a chemically treated film that slowly moved past a point of light modulated by electrical radio signals fed from an antenna on Dr. Todd's dirigible aimed at Mars. What was imaged and shown on black and white at fairly a regular arrangement, was dots and dashes. The "messages" running the full length of the film down its left side.

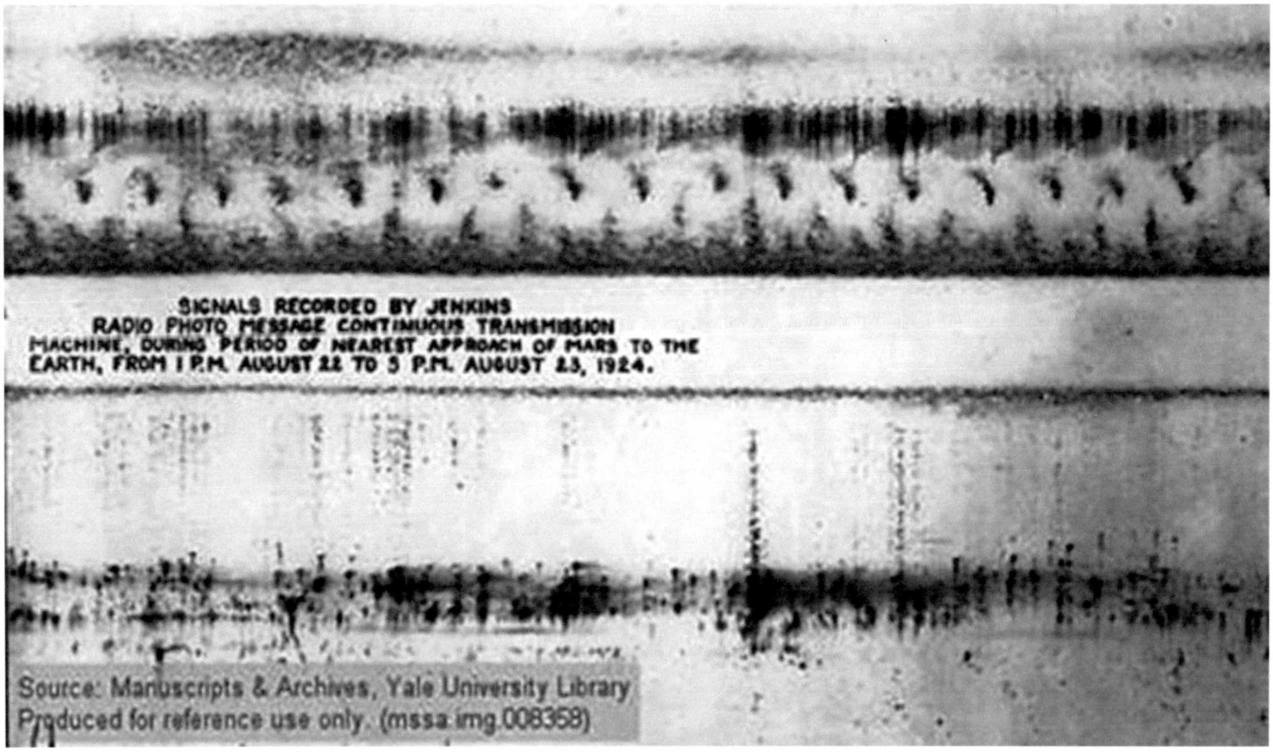

On the right side however a very strange and yet very amazing set of 'messages', repeated at regular intervals of about a half hour, was seen on the film in curiously jumbled groups, that appeared to show a fairly weird looking side profile of a crudely drawn 'HUMAN LIKE FACE'!

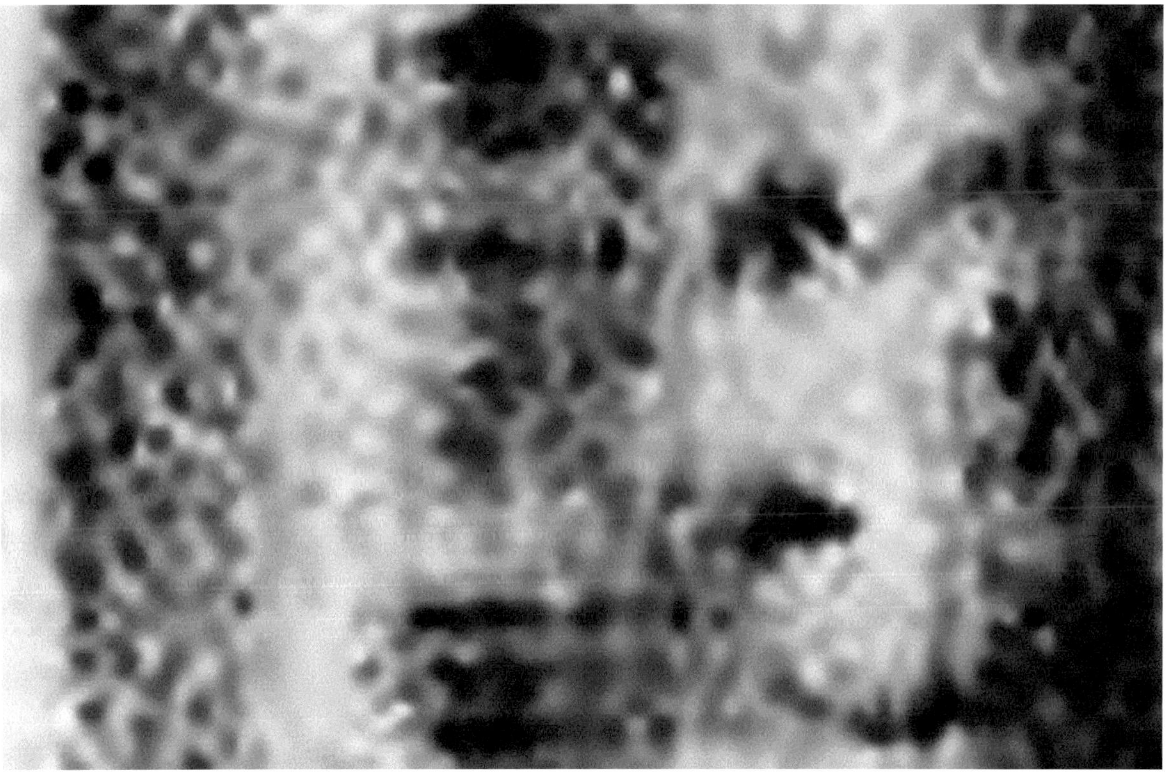

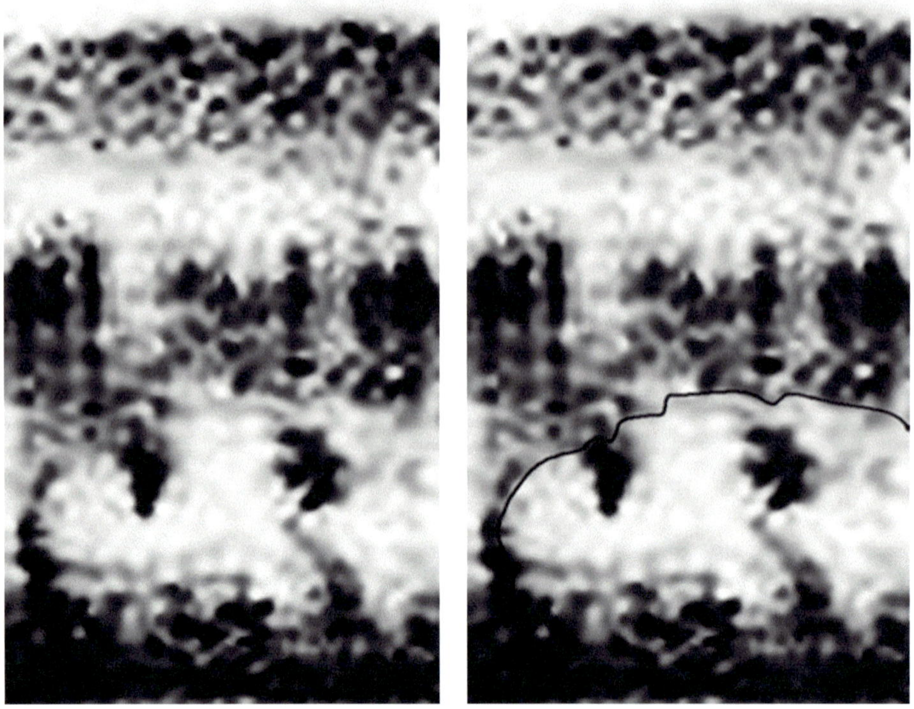

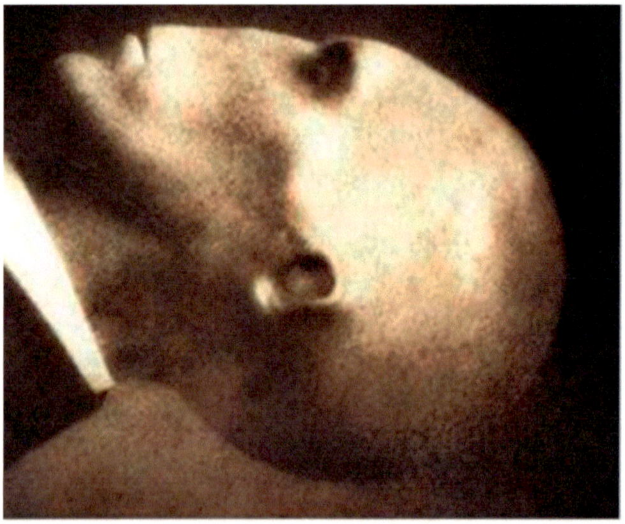

After 36 hours of recording, the film was developed, and Francis Jenkins, the machine's inventor, told a press conference that the device had received signals. These strange patterns were reported to the public five days later on August 28, 1924 in the New York Times. The development of the photographic film of the radio signals for the 29 hour period while Mars was close to Earth, deepened the mystery of the dots and dashes heard by widely separated powerful stations.

It would be worthwhile to note here at this point in the story, that the U.S. Naval Observatory worked with Amherst College's newly developed "radio photo message continuous recording machine" or "Radio-Camera" invented by Francis Jenkins. The then Chief of the Code Section in the office of the Chief Signal Officer of the Army, a world renowned code expert William F. Friedman, was standing by to translate any messages that may have come from Mars, which was some very hopeful thinking! Not only was it hypothesized that there might be life, but to consider that it was also sufficiently advanced from some single celled basic form of life, as to have a technology capable of interplanetary communication, and to also think that a code expert might be able to read anything that came in, was a real reach of imagination to consider. It seems however, it was seriously considered.

It would also be worthwhile to note here that William Friedman became chief technical consultant to the National security Agency in 1952 and two years later, he became the special assistant to the director of the National Security Agency.

Francis Jenkins, inventor of the “radio photo message continuous transmission machine” used in the experiment, doubted that the recorded radio sounds had anything to do with Mars. He declared, however, that it was the most curious phenomenon ever photographed by his apparatus.

Mr. Jenkins was reluctant at first to have part of the film reproduced, fearing that it might be regarded by the public as a hoax. He was prevailed upon, however, because of the widespread interest in the experiment, to permit its reproduction. He wished it was understood that it was from a scientific curiosity, and not belief in the existence of intelligent beings on Mars, that caused him to use his machine for this purpose. “I don't think the results have anything to do with Mars,” Mr. Jenkins said. “Quite likely the sounds recorded are the result of heterodyning, or interference of radio signals. The film shows a repetition, at intervals of about a half hour, of what appears to be a man's face. It's a freak which we can't explain.” (Author’s note: I like to add in speculation that it was probably because he was told he was not allowed to explain it to the public?)

Jenkins said he had expected no more than a blank on the film, and said he is certain the record could not be caused by a defect in operation of his apparatus. The film was the same kind used on other occasions, he asserted.

“As the film unwinds,” he explained, “an instrument passes over it from left to right 50 times every inch. An incoming sound causes a light to be flashed on the film, and this produces a black line.”

Dr. David Todd, professor emeritus of astronomy at Amherst College, who enlisted Mr. Jenkins' cooperation in the widespread “listening-in on Mars” plan which he organized, (This was the magnificent "National Radio Silence Day"), was equally at a loss to account for the freak results. He regarded the film seriously, however.

“We now have a permanent record which can be studied, and who knows, until we have studied it, just what these signals may have been? But the important thing is to have a record.”

Discussing the striking resemblance of some of the “faces” on the film to artists' and writers' conceptions of the theoretical Martians, as shown in novels and pictures, Dr. Todd said: “The Jenkins machine is perhaps the hypothetical Martians' best chance of making themselves known to Earth! If they have, as well they may, a machine that now is transmitting Earthward their 'close-up' of faces, scenes, buildings, landscapes and what not, their sunlight values having been converted into electric values before projection Earthward, all these would surely register on the weirdly unique little mechanism.”

Dr. Todd also stated “that three years before this broadcast to Earth being received, “Marconi was reported as saying he had heard signals from Mars as well! A few days before this experiment, he was quoted as saying he was too busy to listen to possible messages from Mars and that it was a ridiculous idea to do so. And no one knows what he heard the first time. With our photograph, however, it is not a question of what one man heard. It is a permanent record, which all can study."

Dr. Todd said he turned over about 12 feet of the developed film to William F. Friedman, chief of the code section in the office of the chief signal officer of the Army. Friedman, the astronomer said, will try to decipher the strange signals. The film record of the radio signals was later sent to Virginia Military Institute where ***it was confiscated by the NSA (National Security Agency).***

There is also the makings of yet another conspiracy theory about how the 'message' was kept by William F. Friedman (as he has been described as 'the father of the CIA'), and only found in his private quarters after his death! Along with that letter, was also something found written about a 'message' from Mars.....!

William Friedman died November twelfth, nineteen sixty nine, at the age of seventy-eight. ***But the Agency, concerned about 'security', confiscated the reference materials from Friedman's home. What did the reference materials show that caused security concerns? What do THEY Know?***

One must begin to wonder at this point not only about the mystery behind the reason of why this information was being taken and secreted away from public consumption, but what fascinates me is the fact that a big political push and effort was put into effect by US government and scientists alike in embarking upon what must have sounded like to the rest of the world a silly idea to begin with, to communicate with intelligences not of this world. With Martians, as if they had an understanding already that such communication was taking place, or would take place. To go about doing an extraordinary test using lots of effort, money, and coordination with people and countries around the world to receive messages from Martians with a notion of expecting a Martian communication to the Earth, seems suspicious enough to wonder about it. It was possibly one of the largest public experiments in the history of experimentation in the United States, and was also one of the earliest SETI attempts to search for intelligent extraterrestrial life, as well as early concepts of the search and study for alien techno-signatures!

I would also like to add an interesting tidbit of C.F. Jenkins. (Charles Francis) was a pioneer in early television. Only a few years after this coded Martian message photograph was taken that also showed a Face, he had opened the first television broadcasting station in the United States.

Nikola Tesla also eventually died on January 7th, in the year 1943. It was also reported that his written notebooks and papers, and designs were also scooped up by the NSA! It would have been awesome to see what was communicated between Tesla and the extraterrestrial Martians in his notes. But once again, like with Todd and Jenkins, the NSA secrets away important science data from the public knowledge. Critics, of course, immediately rushed to dismiss this astonishing new evidence as merely "a series of coincidences."

Which brings us all to this next incredible historical reason example:

5: THE FACE ON MARS.... COMIC BOOK 1958!

It was not until 34 years later, after Todd and Jenkins "radio photo message continuous transmission machine" had captured messages from Mars showing the first crudely drawn side profile of a Face, when in September 1958 where once again, there was another drawn example of showing a FACE on Mars that had captivated the public interest. This time, it came in the form of a comic book science fiction limited series of 4 comics called THE RACE FOR THE MOON! Comic Book artist Jack Kirby wrote a 5 page story of the #2 of the comic series entitled 'THE FACE ON MARS'.

This was yet another instance of a historical record where a Face on Mars is being focused upon almost 20 years before the Viking Orbiter 1 spacecraft had photographed in 1976, the **'REAL' FACE ON MARS!**

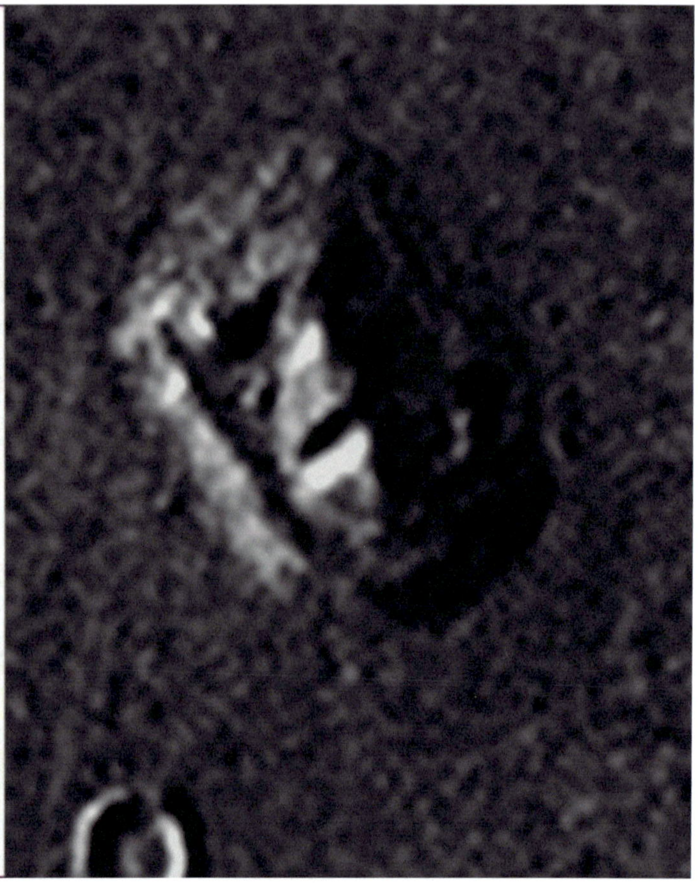

Though it was not a techno-signature, if anything in and of itself, this was a form of predictive programming? Or was it evidence of Remote Viewing programs? Certainly too many coincidences. Unless, way back then, someone also knew?! A thing to ponder. As it can indeed be possible that in all actuality and in all reality, we really could have been told about it from Nikola Tesla's radio communications and Todd and Jenkins radio photo messages. This can be a confirmation being given to us all down through time, in the form of this special comic. Which would bring us all to ask, that if this is so, then ***'HOW DID KIRBY KNOW'?***

I think we can all rule out that "someone" decades before they ever sent missions to Mars and found it, had went to Mars and found it. The only possible obvious and logical answer is that somehow, someone knew what NASA in 1976 would discover as "the Face on Mars!"

Kirby wrote this amazingly prophetic tale only eleven months after the Space Age officially began, with the surprising Russian launch of Sputnik 1, on October 4, 1957.

Critics, of course, immediately rushed to dismiss this astonishing new evidence as merely ***"a series of coincidences."*** (This ***'A Series Of Coincidences'*** connects to what book 3 of this series will detail and reveal about.)

So again, how did Jack Kirby become privy to the Greatest Secret in the History of Man, before it could be known?! The answer is again obvious, he was simply told. By "who" is the key question, isn't it? As well as "why"? Why would Jack Kirby, the King of Comics, put it in a comic book 18 years before the Viking Orbiter 1, would give us the images that proved that Kirby was right? Maybe to better contemplate an answer to this would be to look at the information he provided us in that comic? Let's see what it says.

It is almost also comical, pun intended, that we are being teased with what we see written on the 1st page of this comic book. The 1st thing that makes me laugh, is the statement: ***"Upon Reaching Red Planet, Made Discovery Of Startling Nature: Initial Study Of Object Now Under Way: Have Measured Its Dimensions, Analyze Its Substance: Present Objective Is To Find Out The Secret Of The Face On Mars!"***

I find this kind of funny because again, this is very predicting phenomena taking place here, as that is exactly what the Viking Orbiter 1 did. It also tells us a remarkable predicting phenomena right now, for us all, as it is cluing us all into regards to ***Our Present Objective***, me writing and you all reading this book, about this particular issue, pun intended, and us all to see and to understand we are being given a clue and an answer by Kirby. ***Our Present Objective Is To Find Out … 'THE SECRET',.. Of The Face On Mars!*** Which is exactly what we are doing now. Finding out about The Secret, of The Face on Mars! From Jack Kirby. He is revealing a secret to us all.

I also find it ironic that the last words spoken on the comic book page says it all clearly: "***Our Job With The Mars Expedition Is Exploration Isn't It?***" That is a clue in time to us all also, for us to explore this further, Explore what is secret knowledge of the Face On Mars, that apparently Kirby is going to spill the beans to us all about. Again, through a science fiction comic book story.

So one of the astronauts, Ben Fisher, falls through one of the eyes and into the structure of the Face, and has an incredible virtual experience. He sees a noble race of large Martian giants, living in incredible cities, suddenly attacked by alien invaders that destroy the cities, destroy the civilization, and destroy the planet. And in the last scene Brennan sees the few survivors, sending a rocket to blow the invading aliens planet up, in the middle of the asteroid belt, or what would become the asteroid belt.

If we are all supposed to seriously consider this is a secret of the Face being told unto us all by Kirby, of a known reality, then we must also assume that the added information being detailed unto us all is also secret knowledge and considered factual, about the Face on Mars, and of the past history of Mars itself!

I ADMIT IT'S A BIG SURPRISE TO FIND THE STATUE OF A MAN ON ANOTHER WORLD BESIDES EARTH! BUT, WHY THIS CLIMB?
WHAT YOU SHOULD HAVE ASKED IS, WHY ARE THE EYES OF THIS STATUE--HOLLOW?--NOW LET'S FIND OUT WHY!
THEN I TOOK A STEP FORWARD--A LONG ONE!
I COULDN'T JUDGE THE DISTANCE THAT I FELL! IT SEEMED LIKE A THOUSAND MILES--IT COULD HAVE BEEN A FEW FEET! AT ANY RATE, I LANDED SAFELY--IN A LUSH, GREEN, SUNLIT COUNTRYSIDE!
AND THERE WAS AIR--COOL, RICH AND BREATHABLE--NOT THE THIN, STRANGLING AIR OF THE MARTIAN DESERT. IT WAS A PLEASURE TO REMOVE MY OXYGEN MASK!
WELL, I DON'T KNOW WHERE I AM--OR HOW I GOT HERE. BUT IT DOESN'T SEEM UN-FRIENDLY!
"BEAUTIFUL" WAS REALLY THE WORD FOR IT! THE PLACE WAS A PARADISE! I WAS EVEN MORE ASTOUNDED WHEN I SAW THE CITY--AND THE PEOPLE!
GIANTS! MAGNIFICENT GIANTS!--WITH A CIVILIZATION BEFITTING THEIR STATURE!
2

THEN, I SAW THE OTHERS--SCREAMING INTO VIEW IN SHIPS OF A STRANGE, REPELLENT FORM!
IN SECONDS, ALL THE WONDERS I HAD WITNESSED ERUPTED IN SEARING FLAME!
I COULDN'T FIGURE OUT HOW I WAS LEFT UNHURT IN THAT TERRIBLE HOLOCAUST! BUT THAT QUESTION FLED MY THOUGHTS WHEN I SAW WHAT CAME OUT OF THOSE STRANGE MACHINES!
I REMEMBER SHOUTING MY HATRED AND EMPTYING MY PISTOL AT THEM!
HERE'S SOME OF YOUR OWN MEDICINE! GO ON, TAKE IT! TAKE IT!
THOSE THINGS ACTED AS IF THEY DIDN'T SEE ME. THEY PASSED ME BY--LOOKED THE PLACE OVER AND TOOK OFF! I WAS LEFT ALONE IN A SMASHED CITY!
ALTHOUGH I COULDN'T UNDERSTAND HOW I'D GOTTEN THERE, I SEEMED TO KNOW WHAT WAS GOING ON. IT WAS LIKE THE FACTS WERE BEING STUFFED IN MY BRAIN AND I WAS LIVING THEM!
IT'S AN INTERPLANETARY INVASION! THOSE CREATURES ARE DESTROYING ALL LIFE ON A WORLD-WIDE SCALE!

I HAD NO SENSE OF TIME THERE--YET I KNEW THAT TIME HAD PASSED--THAT I WAS COVERING YEARS IN A MATTER OF MINUTES. THAT THE MARTIANS HAD LOST THE WAR--AND THE LAST HANDFUL WERE GATHERED IN AN UNDERGROUND REFUGE!
THE MARTIANS ARE DOWN THERE! I CAN REACH THEM BY CRAWLING THROUGH THIS VENTILATION TUBE!
THEY WERE GATHERED IN A CHAMBER BEFORE A CHART SHOWING THE PLANETS OF THE SOLAR SYSTEM! A MAN WAS POINTING TO A LARGE WORLD--THE HOME OF THE ENEMY!
HE'S POINTING TO A PLANET BETWEEN MARS AND JUPITER! BUT BUT THERE ISN'T ANY SUCH WORLD!
YET, SOMETHING IN MY MIND TOLD ME THAT THE PLANET EXISTED--AND IT WAS THE MARTIAN'S INTENTION TO DESTROY IT! THE MAN AT THE CHART WAS GOING TO PILOT THE BOMB!
THEN HE WAS GONE--LANCING INTO SPACE... AND, SOMEHOW, I KNEW HE WOULD BE SUCCESSFUL. HE WOULD BLOW THE ENEMY AND HIMSELF INTO ETERNITY!
THE MARTIANS WILL COME OUT OF HIDING, NOW. BUT IT WILL BE TOO LATE FOR THEM, TOO! THE WAR AND ITS WEAPONS HAVE AFFECTED THEIR ATMOSPHERE--
THE AIR--IT--IT'S THIN--RAW--HARD TO BREATH--I--I CAN'T--
I FELT MYSELF SUFFOCATING--BLACKING OUT--FALLING--FALLING--FALLING--

THE NEXT THING I KNEW, SOMEONE WAS FORCING AN OXYGEN MASK ON MY FACE!
HE'S COMING TO! IT'S A WONDER HE'S ALIVE AT ALL!
IT WAS KOSKI BRENNAN AND LAKE! HOURS LATER WE WERE STILL REVIEWING MY RESCUE--AND MY STRANGE EXPERIENCE!
WELL, IT'S A LOT MORE COMFORTABLE ON THE ROCKET THAN THE MARTIAN DESERT!
FEEL BETTER NOW, PAL?
YOU SAY YOU FOUND ME ON THE GROUND INSIDE THE HOLLOW STATUE?
HOW YOU SURVIVED THAT FALL AND THE LOSS OF YOUR MASK IS BEYOND ME!
THAT STATUE'S GOT CONTRIVANCES WE CAN'T SEE IN THE DARK--GADGETS THAT WORK WONDERS ON THE BODY AND THE MIND!
YOU MEAN THINGS THAT MAKE YOU DREAM--LIKE THE KIND YOU WERE TELLING US ABOUT?
YES! IN THAT STATUE IS A VISUAL HISTORY OF A RACE'S HEROIC DEATH--AND THE TRIUMPH OF A SURVIVING MEMORY!
LATER, DURING OUR EXPLORATORY FLIGHT TO JUPITER, WE PASSED THE FLOATING DEBRIS OF THE ASTEROID BELT WHERE WE HAD TO MAKE REPAIRS!
IT MUST HAVE BEEN THE HOME OF MONSTERS--TO HAVE SUFFERED SUCH A FATE!
THERE'S A THEORY THAT THESE ASTEROIDS ARE PIECES OF A PLANET THAT BLEW UP BETWEEN MARS AND JUPITER!
I DIDN'T EXPLAIN IT ALL TO BRENNAN! HE AND ALL MANKIND WOULD LEARN IT, SOME-DAY--FROM THE FACE ON MARS!
5

Yet again, then now we must come to ask ourselves, HOW did Kirby know? Was all this secret information of the Face and the history of the civilization that constructed the Face, and the exploded planet coming from Tesla's communications with the Martians, or was translated from the messages received by Todd and Jenkins "radio photo message continuous transmission machine" message recordings? To me, this is the only options that makes any sense at all.

Notice also the final words of this comic book. We are being told in code speak it seems, that Kirby did NOT tell us all about it of these secrets as he given unto us all, only some of them. ***"But someday, all mankind would learn about it, from the Face on Mars!"***

So how did Kirby know? The answer is that Jack Kirby led a very interesting life. He was born of Austrian and German parents. He immigrated to New York and lived in the not too wealthy section of New York. He grew up wanting to draw comic books. He joined the army during World War 2, he was a young guy at draft age. He became a spy in Germany, in Europe, in Austria. He helped save a whole bunch of guys.

Now the only guy who was running spies during World War 2, was a guy named Wild Bill Donovan, who created something called the O.S.S., the Office of Special services, which went on to become the C.I.A., The Central Intelligence Agency. So jack Kirby had to have had access to intelligence information, from the war, from Europe which is where I think he first encountered what the Martian documents really are all about. Because we also know that from separate sources, that Wernher Von Braun and Willey Ley knew about what was waiting on Mars. And that was a key reason for why Von Braun wanted to send a manned expedition there in the middle of the 1950's, and wrote all those books with Chesley Bonestell.

It really makes me think about how much that Hollywood and the media have known, and how early they have been trying to tell the truth here ***'Between The Lines'***, to excite and inspire generations, of young kids to grow up to become engineers, and scientists to join a NASA, to make a real space program to go and really find out that everything they thought was fiction, is in fact true!

The thing that I find astonishing about the Jack Kirby comic book, is that it was at a time, where there was no possibility that anybody could not have known what was waiting on Mars, a giant human like Face. That sounds absolutely ludicrous for anyone in 1958 to propose that. And then a giant human like Face connected to an ancient civilization that was destroyed in the last gasp effort the Martians went to destroy the planet that the invaders had come from. That was not in the cards in any reasonable scientific scenario.

So this comic book was created I believe from documents which Kirby had access to about someone who knew something, of the former life and civilization on Mars. What did Kirby know, and when did he know it, and from whom? And especially Wernher Von Braun. That is the connection to the German space effort, and remember Kirby's O.S.S. activities in Germany dealing with German prisoners and prisoners of war and guys that have been captured. So that is where some believe Kirby had ran into this material in Germany. The same material that Willey Ley and Von Braun had access to.

(Wernher Von Braun with his fellow NAZI's)

So then the question arises where then did Von Braun and Willey Ley get their information from? That means you have to go into the depths of the German secret society in Germany some claim, that preserved information that had either been preserved indigenous to Germany, or had possibly been brought in by Hitler and Himmler's combing the world, for all these secret treasures and archives and notes and documents of past civilizations from all over the Earth.

Unless of course, the NAZI's also had the transmitted messages from the 1924 receiving of broadcasts that Tesla spoken about and what Todd and Jenkins were receiving and had received? Not forget that this 1924 radio silence mission to acquire radio messages from Mars was done all around the world. This is of course speculation, but this also can be highly revealing as to the possible nature of what those received messages had to say. Remember also that the NSA had secreted away Nikola Tesla's notebooks and Todd and Jenkins received messages.

What we seen in Jack Kirby's comic was in reality as was stated therein the pages themselves, was only parts of the secrets being revealed to us all. Who really knows the extent of the information from ALL the secrets that these messages may have contained and revealed? It is obvious that more information was known about this all, also from 1958, the same year Jack Kirby released THE FACE ON MARS comic book, a special study was conducted. This really seems to be more than a mere coincidence.

Which brings us all to this next incredible historical reason example:

6: BROOKINGS STUDY 1958!

The implications of a discovery of extraterrestrial life

Recent publicity given to efforts to detect extraterrestrial messages via radio telescope has popularized -- and legitimized -- speculations about the impact of such a discovery on human values.33/ It is conceivable that there is semi-intelligent life in some part of our solar system or highly intelligent life which is not technologically oriented, and many cosmologists and astronomers think it very likely that there is intelligent life in many other solar systems. While face-to-face meetings with it will not occur within the next twenty years (unless its technology is more advanced than ours, qualifying it to visit earth), artifacts left at some point in time by these life forms might possibly be

-183-

discovered through our space activities on the Moon, Mars, or Venus. If there is any contact to be made during the next twenty years it would most likely be by radio -- which would indicate that these beings had at least equaled our own technological level.

An individual's reactions to such a radio contact would in part depend on his cultural, religious, and social background, as well as on the actions of those he considered authorities and leaders, and their behavior, in turn, would in part depend on their cultural, social, and religious environment.34/

It is no coincidence as to my speculation that it is obvious that more information was known about this all that- "**The Implications Of A Discovery Of Extraterrestrial Life** was indeed legitimized by the recent publicity given to the efforts to detect extraterrestrial messages via radio telescope!" For those who don't know, Brookings is a major think tank in Washington D.C.

I believe this confirms my theory to a degree of relevance to this all in connection. But what you see written in the image about this actually comes from an official Brookings document which was commissioned by NASA when it was just born in 1958! It had literally been just made into an agency by President Eisenhower. They commissioned this study out of what they call "The Long Range Committee". Meaning they were looking long range into the future from a world perspective and knowledge of 1958-59.

For those who don't know, Brookings is a major think tank in Washington D.C. What Brookings was commissioned to do, was that NASA turned to Brookings and said "Ok we have a space program now, and we have a space agency. Tell us what impact it will have in the out years, meaning the future. How space will impact the world, our culture, the people, their jobs, the view of the universe and the view of who they were. Not just in the United States, but the Soviet Union, in Europe, in South America and Asia and all over the world." And they brought together a whole bunch of blue ribbon people from industry, military and government, business and law etc., and they took about a year to conduct this serious study.

The most interesting part of this study, is a certain section that deals with people's perceived reaction to the knowledge, to the scientific certitude knowledge that NASA could give them, that we are not alone!

One of the commissioned study participants was named Margaret Meade. She was probably the most famous anthropologist in America in the 1950's and 1960's. She would walk around with a big walking stick because she had some kind of problem and she needed assistance. So she had this hug walking stick that looked like something that Friar Tuck would use in some Robin Hood movie to knock guys off a bridge.

(MARGARET MEADE WITH HER WALKING STICK)

Being a participant in this semi-secret study, she would walk around talking about the future of the space program. She would say that from her position of an anthropologist studying so-called primitive civilizations in American Samoa in the South Pacific, her position was **that if NASA ever found anything out there in space that even hinted at ruins or intelligence, or libraries, or some previous civilization; that they should not tell us, because it would destroy us!** And that was an incredible political position that she and her colleagues took with NASA back in 1958-59.

I and others think that what happened was that this study, which recommended a whole bunch of stuff, became policy. It became the secret bible by which NASA conducted itself in all the future years when in fact NASA has found ruins on its photographs. Because part of this study predicted that NASA with our space exploration missions might find artifacts of a former civilization on the Moon, Mars, or Venus, on a future space mission!

I am assuming that it is plausible they were also considering the radio signal transmissions which was received in earlier history which was studied and potentially deciphered that may have helped them come to the predictive conclusion that NASA may discover ruins or an ancient civilization on Mars, since it was alleged that the radio transmissions originated from Mars. This probability seems likely in my opinion. Something put that idea in their minds to make such a strong prediction.

Yet this is telling us all something very important. This tells us of something which we learned over the years, which was that there is a lot of fear surrounding this subject from a lot of people. This is not intellectual. This is not studied, or reasoned academic. This is a gut fear! This is some archetypal level of perception. It almost taps into the soul of people when they take one look at this, and they run metaphorically, screaming in fear! And I and others believe that this is why we (meaning independent Mars researchers and scientists) have had such an enormous problem over the last 46 years (since the Viking Orbiters and landers) getting NASA and the mainstream science and political system to pay serious attention, to what is obviously the greatest discovery in the history of the human race; Namely, that someone had built a monument, an intelligently designed and created structure seen on another world, especially specifically on Mars, to the fact that we are not alone!

Now if you have run studies, or if you have DONE SECRET RESEARCH, or if you have done psychological profiles or in the field polling, or some other work which we are not aware of and NASA has not told the American people that they had done all this, and they get a lot of people with a reaction of fear? This is the heart of the problem!

The NASA and Brookings study was done in 1958-59, and we are now 64 years from that time as this book is being written now in December 2022. More than a half century ago, and think how morals and perception and social contracts and the way we do business with each other, think of all that has changed in our life and in our society after 64 years! So what5 Brookings basically said is that a lot of people would react to that knowledge with fear, that someone, an unknown intelligent race of beings had built a monument on another world, especially upon Mars, that we are not alone!

Now to me, this is the most important question we can ask because if the human race is really to become free, well, to me being free is being able to ask any question you want, and pursue any answer that you deem i9mportant! This is telling us something really important about why NASA has been able to get away with lying to the public about what is really out there in space for many, many, many years. The folks in Washington D.C., who basically run the lives of those who work for NASA, who pays their salary, who tell you what too do, and who to tell you that if you go beyond the edge of the paper, you will get fired; Who tell you that there are certain things that you can talk about, and certain things that you cannot talk about in public.

The people who are controlling the strings and the flow of dollars into NASA employee's pockets, the people who have guided the space program for all these many years. It is "THEY' who control what is real knowledge and mainstream science and news. Considering based upon the findings of the Brookings study. It is some of those people who need to be called out and fired! To then replace these people with real public servants who would do the agency that their employees serve with honor that they all deserve! The point is that the American people should have a right to the knowledge about this and have a voice to determine how we will proceed with this knowledge in mind!

But the most important thing here is that this is a study so pivotal, that it will effect so many people's lives even if they do not know about it! It should all be decided by public response, some form of democratic participation as opposed to another insider getting a job where there is no contact with the people of the public. At the very least, people should realize and take a step back and think 'hey wait a minute, we should have a say in who runs the agency that is going to determine the future of life on Earth, and away from Earth! Because that i9s how important NASA is everybody!

It will determine by what it does or by what it does not do, the future of everyone reading this now! And that is not only if you work for NASA; that is if you are a member of this society, if you are participating in this economy, if you are living on this planet, it is very important because what happens in space, is emphatically critical to what happens here on Earth. That all will be unfolded as time allows our collective voices to be heard and listened to, and not to be decided based upon propaganda and deception in order to hide this knowledge from us all from an outdated government think tank assessment study that asks under what circumstances will we be presented, or withheld by NASA, of the knowledge and the incredible reality that we are not alone. This is my opinion on the subject of Brookings and the policy it helped to create and to enact upon the public that obfuscates the truth and the data as well it presents to the public.

Not long after this important Brookings study was NASA was rushing spacecraft off to Mars.

Mysterious and enigmatic and a place currently now being invaded by alien space craft of various technology and capability_composed of robotic design and computer parts and programs and engineering from another world. Designed and sent from another planet, Earth. Various nations from Earth with various space probes sharing various capabilities which are composed with electric eyes and virtual noses of varied degrees and ability of varied visual and other technologies, are now invading Mars and are conducting reconnaissance missions above the surface of that planet patrolling its skies, as well as robotic rovers rolling and roving_around on its very interesting surface as of 2022.

The invasion of Mars began with Mariner 4. Launched on November 28, 1964. **Mariner 4** performed the first successful flyby of Mars. Which took hundreds of images, and returning the first pictures ever of the Martian surface, back to Earth.

The images the public did get to see depicted a cratered, seemingly dead world, which largely had changed the view of the scientific community of Life on Mars, as they were hoping to see based upon the studies of Mars many years earlier by such astronomers as Percival Lowell who with his telescope had shown his many reasons why Mars was an inhabited, yet a dying world and was covered with canals, or channels!

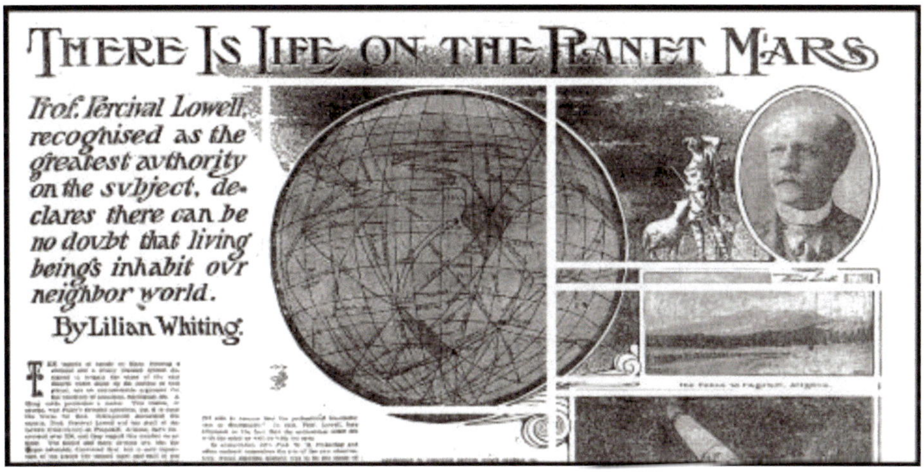

THERE IS LIFE ON THE PLANET MARS

Prof. Percival Lowell, recognised as the greatest authority on the subject, declares there can be no doubt that living beings inhabit our neighbor world.

By Lilian Whiting.

So the Mariner 4 imagery shown much to contradict the general perception of scientists, as well as the civilian populations, putting a damper on people's imaginations of the type of life that they were expecting to find there.

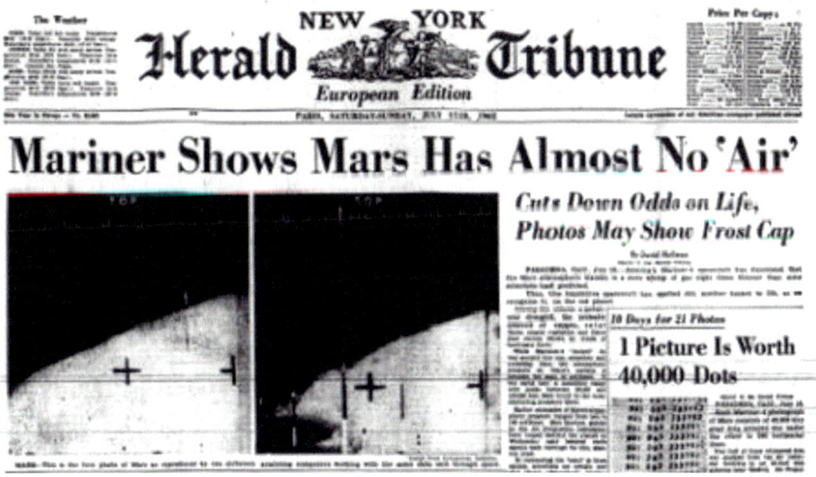

NEW YORK Herald Tribune

European Edition

Mariner Shows Mars Has Almost No 'Air'

Cuts Down Odds on Life, Photos May Show Frost Cap

10 Days for 21 Photos

1 Picture Is Worth 40,000 Dots

New York Times.

LATE CITY EDITION

NEW YORK, FRIDAY, JULY 16, 1965.

TEN CENTS

FIRST MARS PHOTO IS TRANSMITTED; MARINER SIGNALS INDICATE PLANET LACKS A LIQUID CORE LIKE EARTH'S

SENATE APPROVES HOUSING MEASURE WITH RENT SUBSIDY

Administration Forces Beat a Republican-Led Effort to Kill 'Rentcars'

OTHER DATA SENT

Sensors Find Scant Radiation Belt and Thin Atmosphere

FIRST CLOSE-UP OF MARS: Photograph made by Mariner 4 of the planet and sent

Communications with Mariner 4 were terminated on December 21, 1967. Yet this was not the end for the search for life that many people expected to see and find on Mars, nor in our imaginations that spurred our continued attempts to study Mars more closely.

Another Mariner Mars mission was created and sent to Mars. Mariner 9! Mariner 9 was launched toward Mars on May 30, 1971, and reached the planet on November 14 1971. Mariner 9 becoming the first spacecraft to orbit another planet. When Mariner 9 arrived at Mars, the planet surface was covered by the largest dust storm ever observed at anyplace. The surface was totally obscured from view. Mariner 9's computer was then reprogrammed from Earth, to delay imaging of the surface for a couple of months until the dust settled. The main surface imaging did not get underway until mid-January 1972.

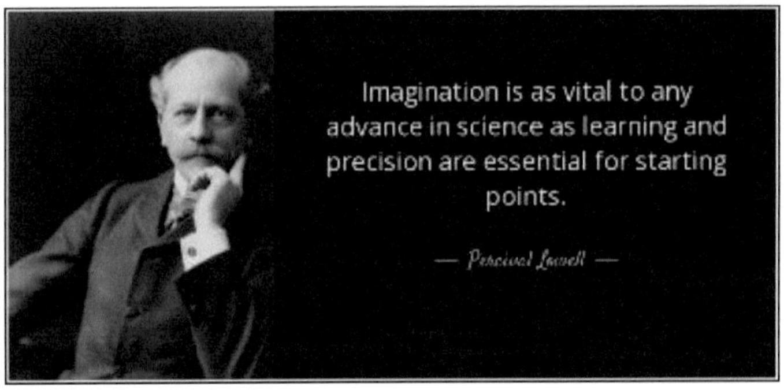

Another Mariner Mars mission was created and sent to Mars. Mariner 9! Mariner 9 was launched toward Mars on May 30, 1971, and reached the planet on November 14 1971. Mariner 9 becoming the first spacecraft to orbit another planet. When Mariner 9 arrived at Mars, the planet surface was covered by the largest dust storm ever observed at anyplace. The surface was totally obscured from view. Mariner 9's computer was then reprogrammed from Earth, to delay imaging of the surface for a couple of months until the dust settled. The main surface imaging did not get underway until mid-January 1972.

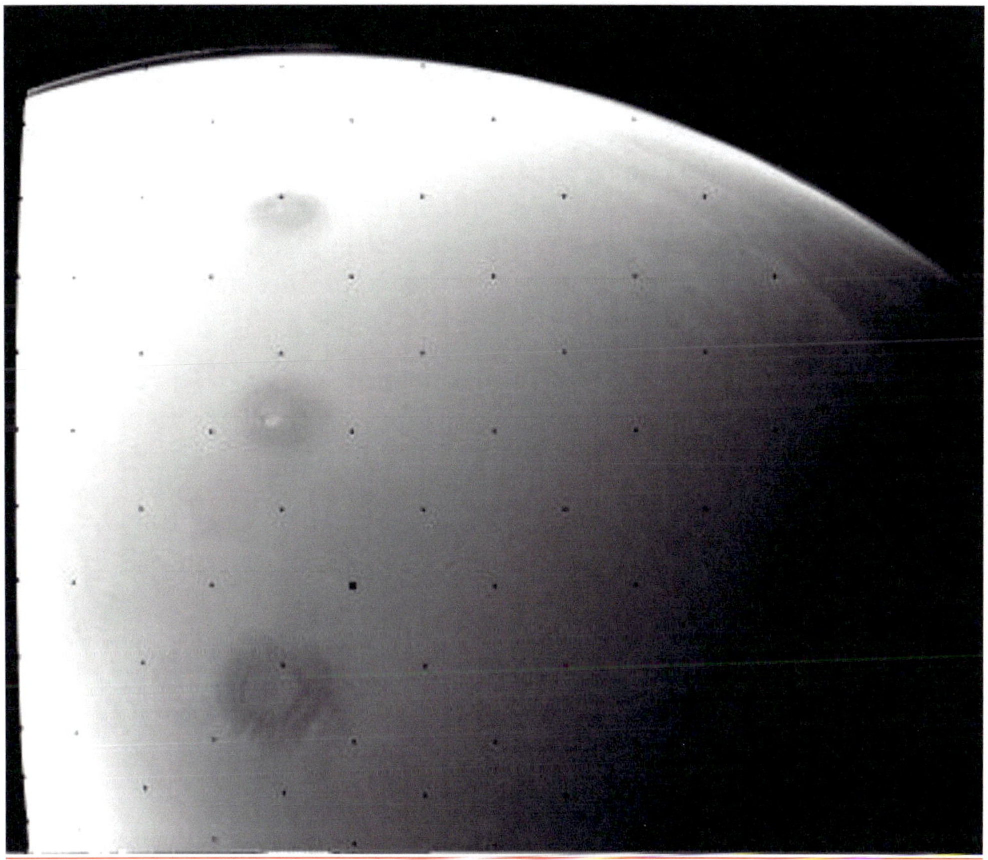

(An early image of Mars after Mariner 9's arrival showing a huge dust storm in November 1971)

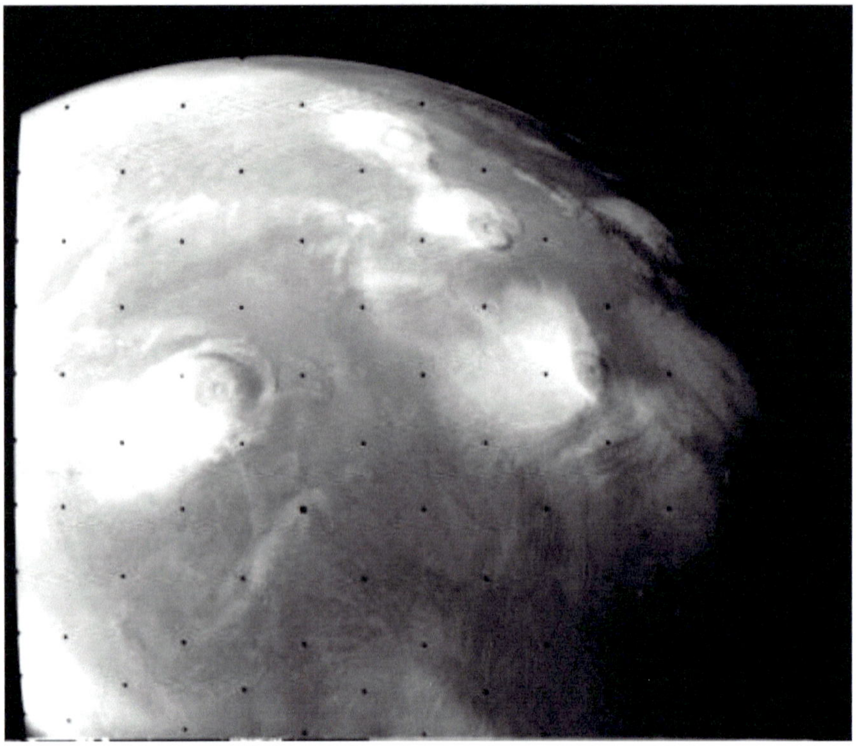

This unexpected situation made a strong case for the desirability of studying a planet from orbit rather than merely flying past. Mariner 9 had transmitted 7,329 images, covering 85% of Mars' surface, After 349 days in orbit. Whereas previous flyby missions had returned less than one thousand images covering only a small portion of the planetary surface. The enormous Valles Marineris canyon system is named after Mariner 9 in honor of its achievements. ***(However the public does not have access to all the imagery which was taken on that mission.)***

One must wonder what all was actually seen that the public did not get to see. The images we got to see revealed what appeared to be ancient river beds, craters, massive extinct volcanoes, such as Olympus Mons, the largest known volcano in the Solar System and also showing evidence of wind and water erosion and deposition, weather fronts, fogs, and more. Gone were the dreams and hopes of our imagination for the once proposed dying Martian civilization digging canals to water the planet surface.

To which brings us to the Viking Orbiter missions to Mars! A dual set of Mars probes was created and composed of two main parts: an orbiter designed to photograph the surface of Mars from orbit, and a lander, designed to study the planet from the surface. The orbiters also served as communication relays for the landers once they touched down.

Viking 1 was launched on August 20, 1975, and the second craft, *Viking 2*, was launched on September 9, 1975. *Viking 1* entered Mars orbit on June 19, 1976, with *Viking 2*, following suit on August 7.

The orbiters continued imaging and performing other scientific operations from orbit while the landers deployed instruments on the surface, after they landed. The differences between the Mariner mission imagery compared to the Viking Orbiter imagery was astounding! Much clearer images as well as closer views of Mars' surface was far superior in comparison.

(THE ABOVE IMAGE seen by the Viking lander 2 site in May, 1979. Notice the FROST on the ground, & notice the Blue skies! IMAGE BELOW: Scoop tests done on the surface on Mars in 1976 have found evidence of life.)

Yet the presence of the Lowellian intelligent Martian civilization that people were imagining and hoping to see, once again, was ever more lacking by what the Martian surface continued to show in the thousands of images that continued to come back to Earth from the Viking Orbiters, and the Viking landers.

That is, until a very controversial 'Viking Orbiter A' image came forth and came to the public attention. Image number 35A72. It was this mage that reinvigorated people's perceptions of imagining a Mars with some evidence once and for all showing finally, signs of intelligent life being found on Mars. Once again people's imaginations soared to what this means, transforming the older research and speculations into how a Martian civilization would show itself as looking like, as to what they all were expecting to find, transforming to something unexpected that we did find, A FACE staring up at you from the surface! A giant Face in fact! Jack Kirby was right after all?!

(VIKING ORBITER A Image section for 35A72. Imaged 1976. Notice the PYRAMIDS also seen in this image!)

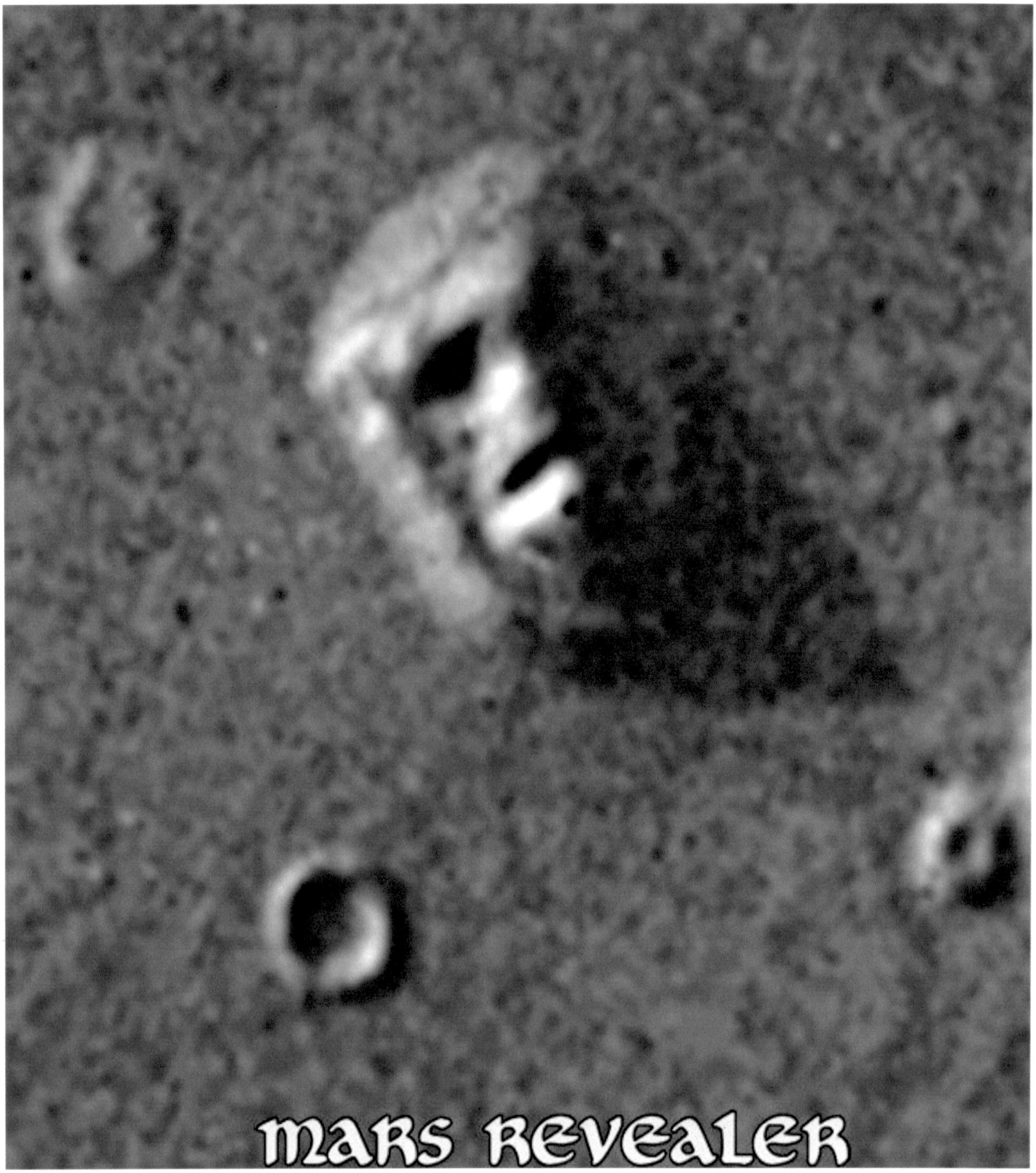

(Close up of Face from processed 35A72)

As understanding how people's imaginations were astir over this amazing image, after previous Mars missions' disappointments, losing the fervor of finding intelligent life on Mars, it seems that at least some of the NASA scientists as well thought that this Face found on Mars was also a sign to them of intelligent design, unofficially at least, because of how it was officially titled by NASA: **'HEAD'**!

At long last, it seems obvious at least unofficially, there was now being photographic evidence showing that an intelligent extraterrestrial race indeed existed on Mars all along! Despite the previous decade worth of Mars missions image data, that had saddened the ***public*** expectations of finding a world unlike that what was being hoped for of finding thriving intelligent beings fighting to keep a planet from dying, by creating canals/channels from the poles for the melting ice to irrigate the planet, to a world now where it looks like water was obviously present in the past, but no rational networks of canals designed by intelligence were seen.

Nor were there domed cities which housed and protected it's civilization from its diminishing atmosphere, so that they can live and breathe in self-contained and temperature controlled environments to keep them alive, were able to be seen.

IMAGINARY LANDSCAPE ON MARS

Instead, '*THEY*'', found this 'HEAD'!

Finally! Evidence was found that reinvigorated the public's imaginations once again, of finding any semblance of what people were theorized by Percival Lowell and mainstream science prior to the space mission age, and what they were hoping to find from the missions to Mars. Since the Mars missions beginnings, it was taught to all that any hopes or reality being seen and found as to what had previously been taught and believed of these high expectations, was now unscientific based upon the image data and had destroyed these old conceptions of finding any life at all, let alone any signs of intelligent life is being seen on Mars. At least from an official public perspective.

In fact, based upon the image data showing craters, dry waterways, and showing no signs of any present liquid water running, it was concluded as fact, that Mars was a dead planet, devoid of life, and somehow had become incorporated with the idea by NASA, that stated Mars never had any life. Therefore if something we see that somehow contradicts that idea, especially signs of intelligent life having existed upon Mars, it would just be said that it was just a fluke chance, or an anomaly that cannot be explained, accept by means of using natural explanations of course. ***Case closed!*** The mainstream scientific view was shaped and morphed by NASA's ***'policy'*** having been implemented since the Mariner 4 flyby of Mars, which has sadly taken that stand for their model of Mars they now taught to mainstream science to teach us all as scientific fact. That Mars was a dead planet, devoid of life, but now, in time, we are told now by NASA, that Mar may have once been more earthlike with a thick atmosphere, running abundance of water on the surface, but somehow was stopped in its tracks, losing its water, its once thicker atmosphere, and any life, let alone intelligent life compared to as we know it to be, was never able to take hold.

Being taught to us all that Mars had no time like our Earth, to evolve and become such a planet like our planet rife with life. No vegetation, no intelligently designed canals to water them, no animals, no birds, no civilizations. Therefore, anything that was or is seen, simply went against the model they were trying to teach as facts to the world, and using mainstream science to do it. Destroying by policy, the old public concepts and expectations of what we would find on Mars. If they can make people believe that model, although obviously more evidence was needed to make that determination, they knew most people will not care anymore, and will not ask questions if they are all taught that there is nothing to see here, so please move along!

So by using that model of 'policy' that was enacted, and taught as mainstream science, that the 'HEAD' in what it shows you to your faces, must therefore be ignored, and treated with disbelief because NASA and mainstream science teaches by 'policy', that what we see in 35A72 cannot be possible, therefore it does not exist. Despite the evidence of your own eyes and analytical prowess.

In this sense, one can understand and take it by no surprise, that instead of allowing the evidence form a Viking Orbiter image to ignite the public's imagination with desire and interest to see and learn more, and to accept this 'HEAD' as official proof that we found once and for all, evidence, of a sign of intelligent life on Mars, by the 'HEAD'. We are instead dissuaded away from investigating, and going against the mainstream model, which remember is NASA's 'policy' model, and the simpler minded which is most of the populations on Earth, are persuaded to believe in the mainstream science model, and are taught to not ask questions that goes against that mainstream model. This has been going on in public schools in the USA and around the world since the start of the space missions to Mars. It began then and continues to this very day! The only thing that makes any sense to me as to this backward type of teaching and thinking, is the 'policy' enacted model based upon Brookings. It allows 'cover' for those who are really in the seeing and knowing what is really on Mars to allow them to keep doing the real scientific work, and acknowledging what is really there, while the public is forcefully taught the 'policy' model, and keeping the truth as well as being able to hide the truth from the public.

There is much evidence in the words I am saying about this by their own 'policy' statements themselves about these issues, which stems forth from their own data to begin with. Especially regards to the officially titled 'HEAD'! When the image 35A73 was being looked at by the public for the first time (officially speaking), people were getting excited and imaginations soaring as to where our missions in space and to Mars was now going to lead us, to change the world! Speculative and intellectual conversations and ideas were being spoken about. Wanting NASA to acknowledge what their own titled release of the image stated of 35A72 was actually true.

Once again, one cannot be surprised by the 'policy' model view, the position NASA took regarding answering questions by their own scientists as well as the public of the 'HEAD' image. Instead of validating such an amazing find, verifying it's official title of the 'HEAD', it was officially told to them all that the feature we see in 35A72 was NOT REAL, but officially claiming that it was simply '***A MERE TRICK OF LIGHT AND SHADOW***'! '

Telling the public a false fact, along with truth mixed with lies! Giving the added excuses by NASA telling the public that they had taken other images of this area a few hours later, and none of them showed the feature we see, and gone was the 'HEAD'!

That may sound innocent and all ok right? It was not, as NASA had not provided any other images to back up their claims to the public, to base their words as being factual data. In fact, it was far from the truth! NASA claimed it was the only image that shown the 'HEAD' and no other images could be found that showed the enigmatic feature.

Now before I continue on, I would like to briefly talk about another finding from VIKING that also shown controversial findings regarding positive detections of life on Mars and scientists raised questions from the data gathered by that mission. This is dealing with data collected by the Viking Landers! It shows me and you all, that what I speak about regards to the 'policy' model definitely applies here as well. There can be no mistaking about it!

The Viking Landers each carried four types of biological experiments to the surface of Mars. A biology experiment whose purpose was to look for life. The Pyrolytic Release experiment (PR), the Labeled Release experiment (LR), and the Gas Exchange experiment (GEX). In addition, independent of the biology experiments, *Viking 2* carried a Gas Chromatograph/Mass Spectrometer (GCMS) that could measure the composition and abundance of organic compounds in the Martian soil.

The results were surprising and interesting: the GCMS gave a negative result; the PR gave a positive result, the GEX gave a negative result, and the LR gave a positive result. Viking scientist Dr. Gil Levin stated on ***The Martian Revelation Radio Show*** with myself as the host in 2007, 2018, 2021 that "***Our (LR) experiment was a definite positive response for life, but a lot of people have claimed that it was a false positive for a variety of reasons.***"

Dr. Gil Levin was the designated Experimenter of the Viking Labeled Release life detection experiment which landed on Mars in 1976. The experiment got positive responses at both Viking landing sites. However, a mainstream scientific consensus did not accept his results as proof of life. ***(The 'policy' model is the most obvious main reason.)***

After years of study, Dr. Levin concluded again, that the experiment had, indeed, detected life on the red planet, and he had published his conclusion despite what NASA 'policy' dictates. Some scientists still believe the results were due to living reactions. ***(One must wonder why?)***

Subsequent findings of environmental conditions on Mars and research on organisms found in extreme environments on Earth have been consistent with his claim. But yet, the mainstream 'policy' model stood strong still, and to this day it has not yet changed the findings results, nor has any attempts to continue to send any life detection experiments to Mars since 1976!

Now one must think and ask themselves, why would NASA want scientists, and mainstream science teaching us all to not follow the data and present logically the information that is staring us all right in ***THE FACE*** ?

Answer? 'POLICY'!

Let us look at 'the official' way this unfolded. The 'Face' was first noticed In July 1976 by NASA personnel Dr. Tobias Owen while searching for a possible landing site for the upcoming Viking 2 Lander, which was proposed to land originally in the area of the Cydonia region. Dr. Tobias Owen, was a member of NASA's own imaging team at the Jet Propulsion Laboratory in Pasadena, CA. Dr. Owen noticed a gigantic human like Face staring up at him from the Martian surface. Soon after its discovery, he brought it to the attention of the Viking program director Dr. Gerry Soffen. He said "Gee guess what I found? A face on Mars"!

Dr. Gerry Soffen announced that an image of an odd landform resembling a face, had been found on Mars. The 'HEAD' appeared to have an eye, nose and a mouth, and looked to be wearing a helmet. Michael Carr who was a geologist with the USGS, as well as the then head of the Viking Orbiter imaging team, immediately released the unusual image to the press. Soon after the image press release, the news media was quickly informed by NASA that when a second image was taken only a few hours later of the strange feature in the Cydonia region, the odd facial formation disappeared.

Dr. Soffen then assured the media that the facial formation was nothing more than '*A TRICK OF LIGHT AND SHODOW*'! Oddly, this second image never surfaced, and only the original high contrast picture 35A72 of the 'HEAD' was circulated to the press with it as being nothing more than a phantom novelty. From that moment forward, NASA's official position was that the Face on Mars was nothing more than an apparition of shadows and rock. Saying that the strange mesa had no resemblance to anything as being like a Face being found on Mars. ***(In essence, it was being covered up)!***

Now remember what I stated earlier, that the Cydonia area was being observed and studied for selecting a spot to land the Viking Lander 2 down in that region? That is when the Face at Cydonia was found. Well get this, as a result of releasing the 'HEAD' formation image of 35A72 to the public, NASA immediately decided not to proceed with the original plan of a Viking 2 landing within the Cydonia region. Their public reasoning was because the entire area was unsafe, even terming it with this phrase, saying that the Cydonia region was filled with too many '***NASTIES***'!

The last minute change in Viking 2 lander plans went virtually unquestioned by the mainstream media. Reaffirming at the same time with NASA's consistent stance that this facial formation was nothing more than a '*TRICK OF LIGHT AND SHADOW*"! With no surprise, the public soon lost interest. From that point on, both NASA and a dumbed down media considered '**THE FACE,... CASE CLOSED'!**

Unfortunately, the 'Face on Mars' was banished unto the sensational public pages of supermarket tabloids and the imaginative filled minds of fringe science enthusiasts and categorically placed in areas labeled as pseudo-science, as a means to dissuade anyone from taking the Face seriously or daring to investigate away from the mainstream official '**POLICY**'!

It is also ironic that these same tactics labeling and ridicule of others to help suppress real information has also been attributed to UFOlogists and UFO's in the past, which until in recent years it is now publicly and politically acknowledged by the U.S. government of their reality but with a name change. From being termed as being UFO's, Unidentified Flying Objects, to now being called Unidentified Aerial Phenomena, or UAP's, both of which consists of thinking and believing by many people and scientists, that these are objects of extraterrestrial origin. So in my opinion considering these facts and objects, it is obviously a stigma that exists to scientists and governments not just that there are objects of unknown origin or anomalies found on other worlds, that is not the problem. The problem that they are *FACED* with exists with them all is in considering the reality of intelligent extraterrestrials in the big picture of things.

CHAPTER 2:

CASE CLOSED?

NOT BY A LONG SHOT!

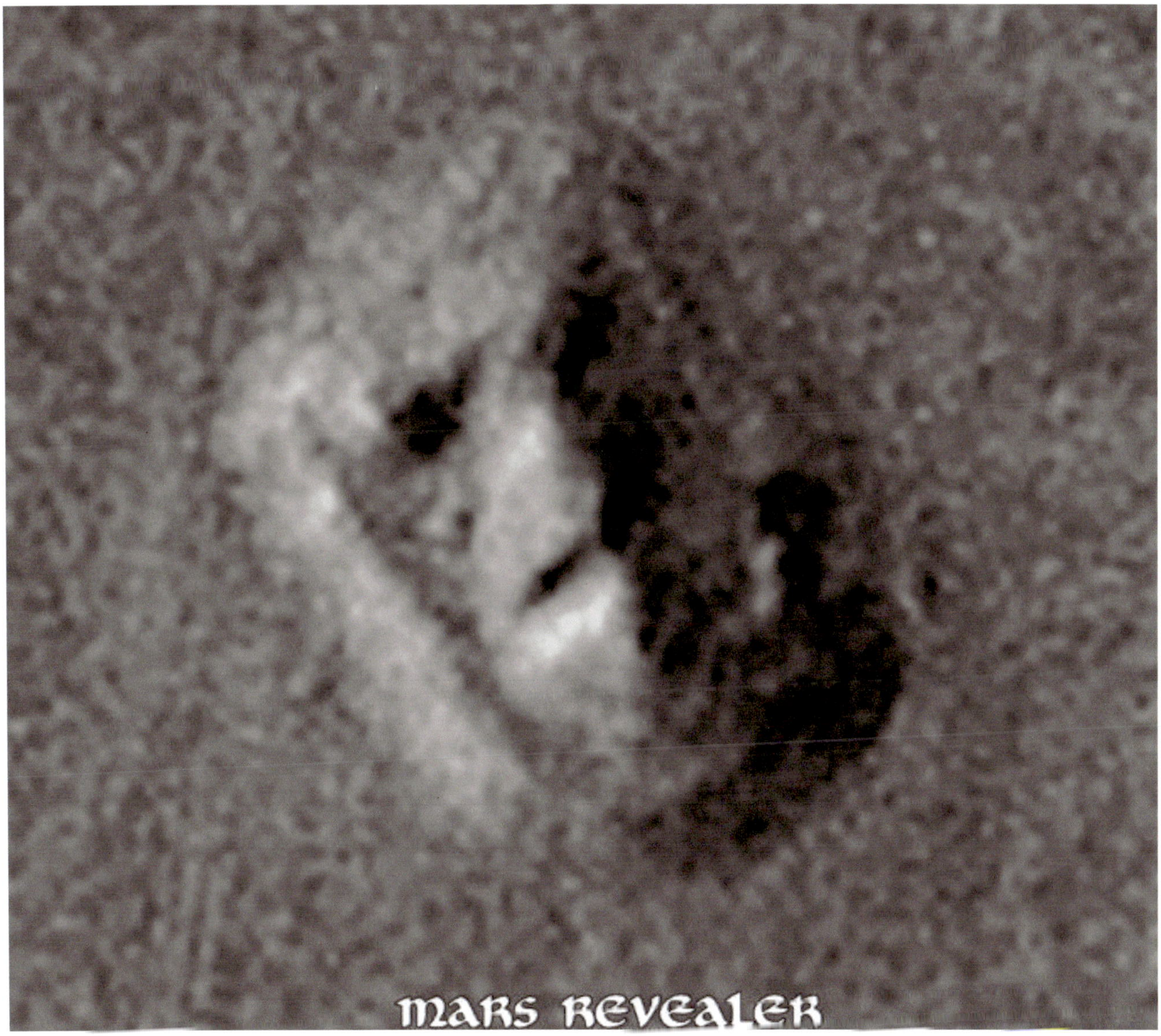

ANOTHER IMAGE WAS FOUND OF THE 'HEAD'!

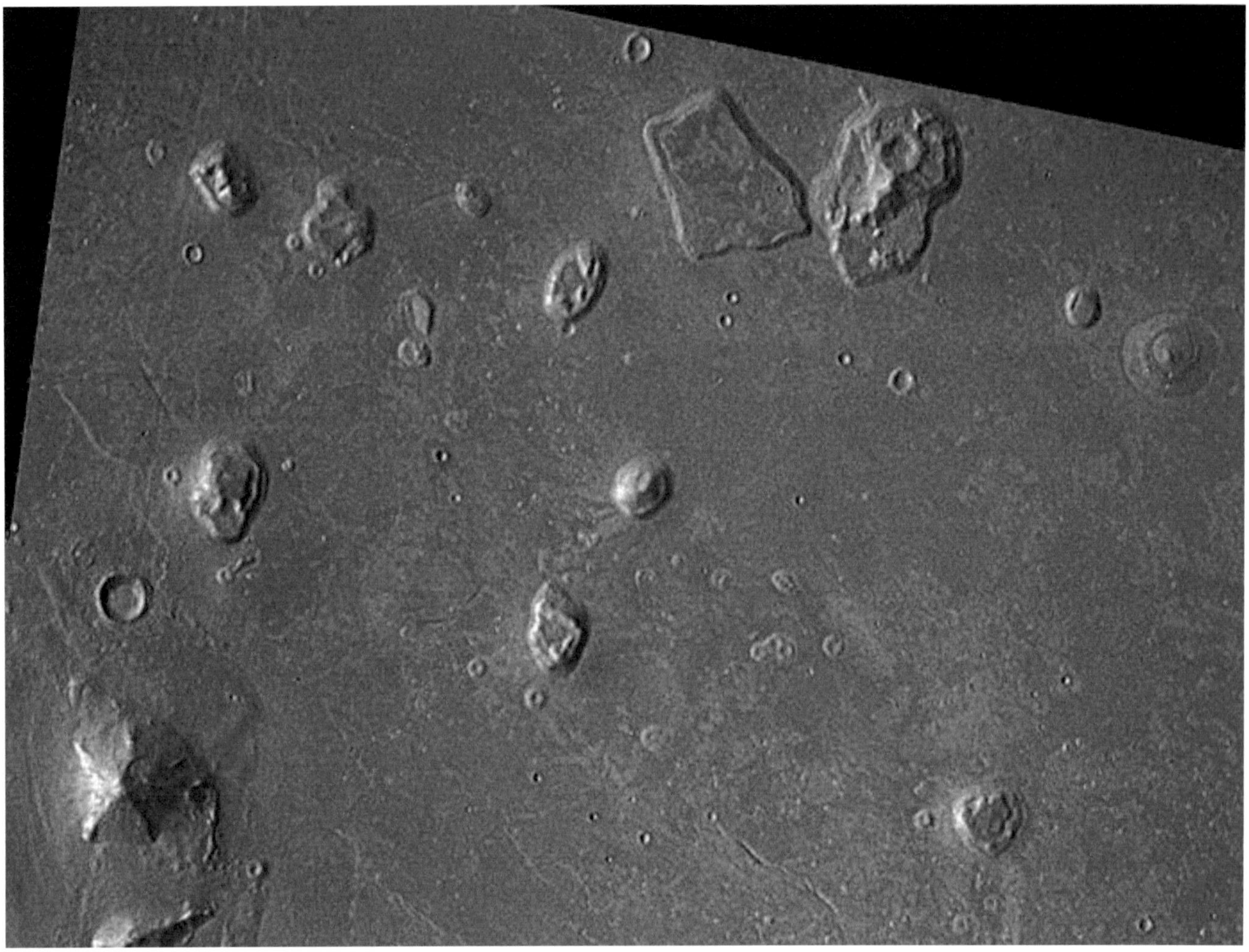

Above Image: Viking Orbiter image section of 70A13. Notice the Face top left, and the Pyramid bottom left.

Ironically, the viability of the Face on Mars being as an artificial construct was resurrected only a few years later by a couple of NASA's own employees, Vincent DiPietro and Gregory Molenaar who was working at the Goddard Space Flight Center at that time. ***Vincent and Gregory were given permission to review the entire Viking archives, including all the data tapes and printed photographs. They supposedly had everything at their disposal.***

While combing through the painstaking art of studying the vast archive data, one of the first things the pair of researchers had found was the second 'lost image' that NASA supposedly acquired only a few hours after the first image of 35A72 was taken. Although it became obvious that NASA still told a lie saying a few hours later other images were taken and the face feature disappeared, for this second image of the 'HEAD' and its image number of **70A13 was seemingly taken 35 days after 35A72!** Think about that. The new 'HEAD' image showing the Face was also apparently filed into a wrong file. Think about it like this, 35A72 was taken on its 35th orbit, the reimaged 'HEAD' of 70A13 was taken on its 70th orbit. **(The 70A13 image shows it was taken 35 days after 35A72. That is an orbit a day being imaged by Viking orbiter? As 35 plus 35, equals 70)**!

There was no way, according to the image data they had to work with, that NASA took any other images a few hours after the Face at Cydonia was seen on 35A72, that NASA had said regards the debunking of the 'HEAD' of 35A72 to the public. Again, the so called second image was taken 35 days after that news release! Thus showing NASA **lied**, or were simply keeping image data secret that was not in the vast amount of Viking data for Vincent and Gregory, or they certainly would have found it.

So this supposed second image according to the official image log shows it was acquired almost a month after 35A72, with this new image found of 70A13 had captured the Face under different lighting conditions and had illuminated the darker side of the Face as we see it in 35A72. The Face was still there, showing added details!

Image seen below taken from Viking Orbiter image 70A13

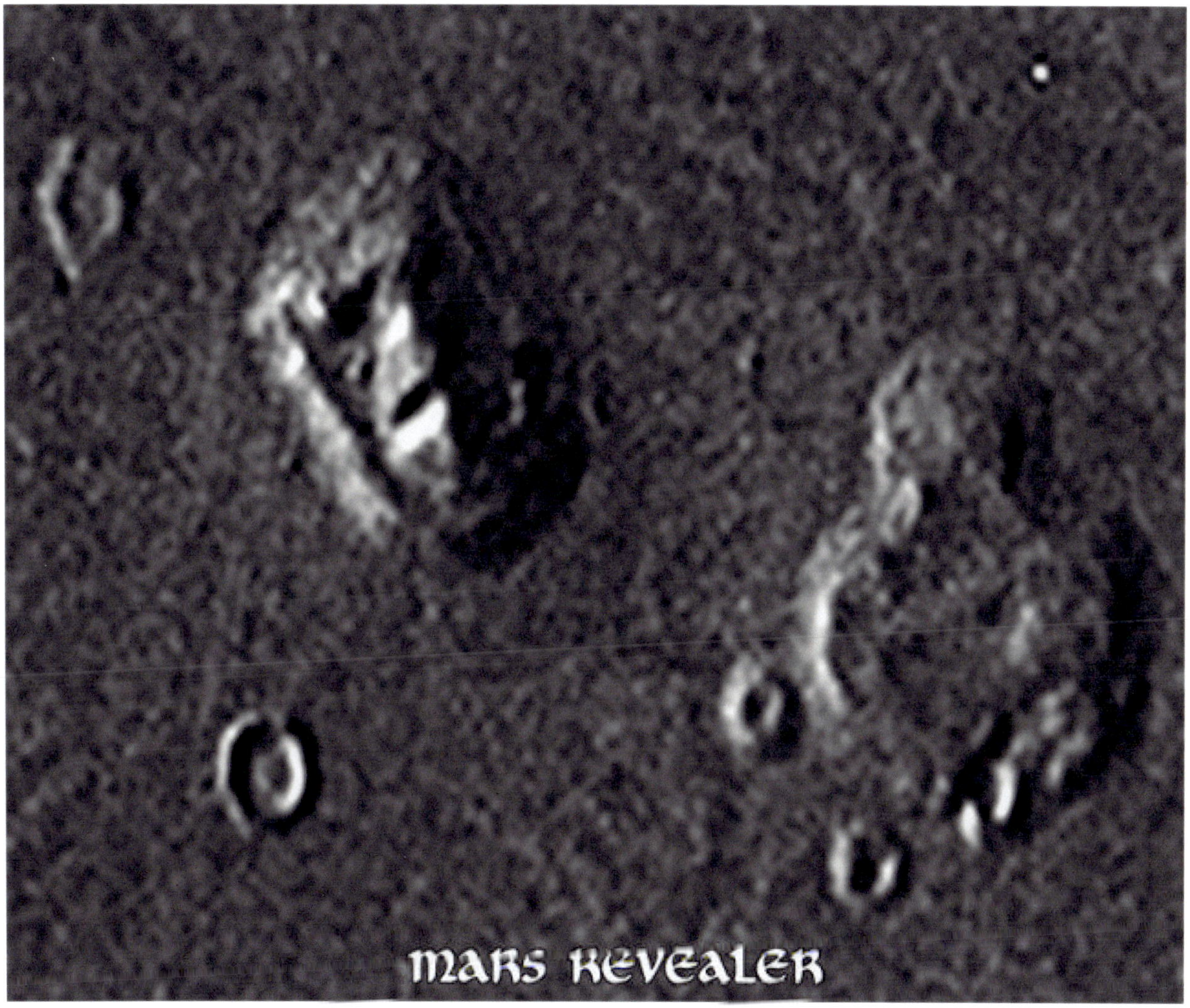

The results of their studies for the Face, and the finding of the additional Face image of 70A13, and some of its surrounding pyramidal structures, both NASA and the mainstream scientific community silently **ignored** their groundbreaking work.

(Why? The 'POLICY'!)

(Image section seen below from: 70A13)

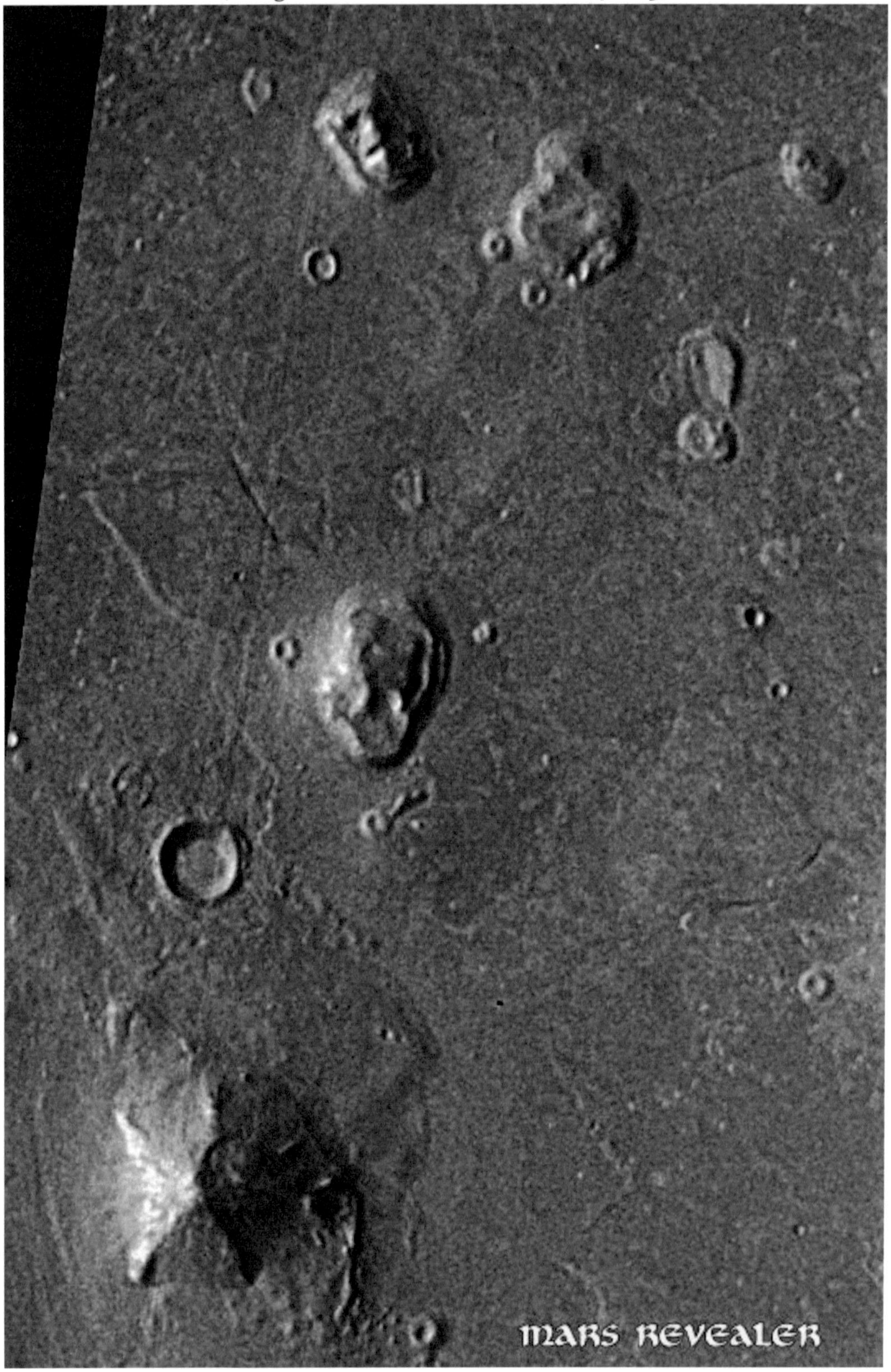

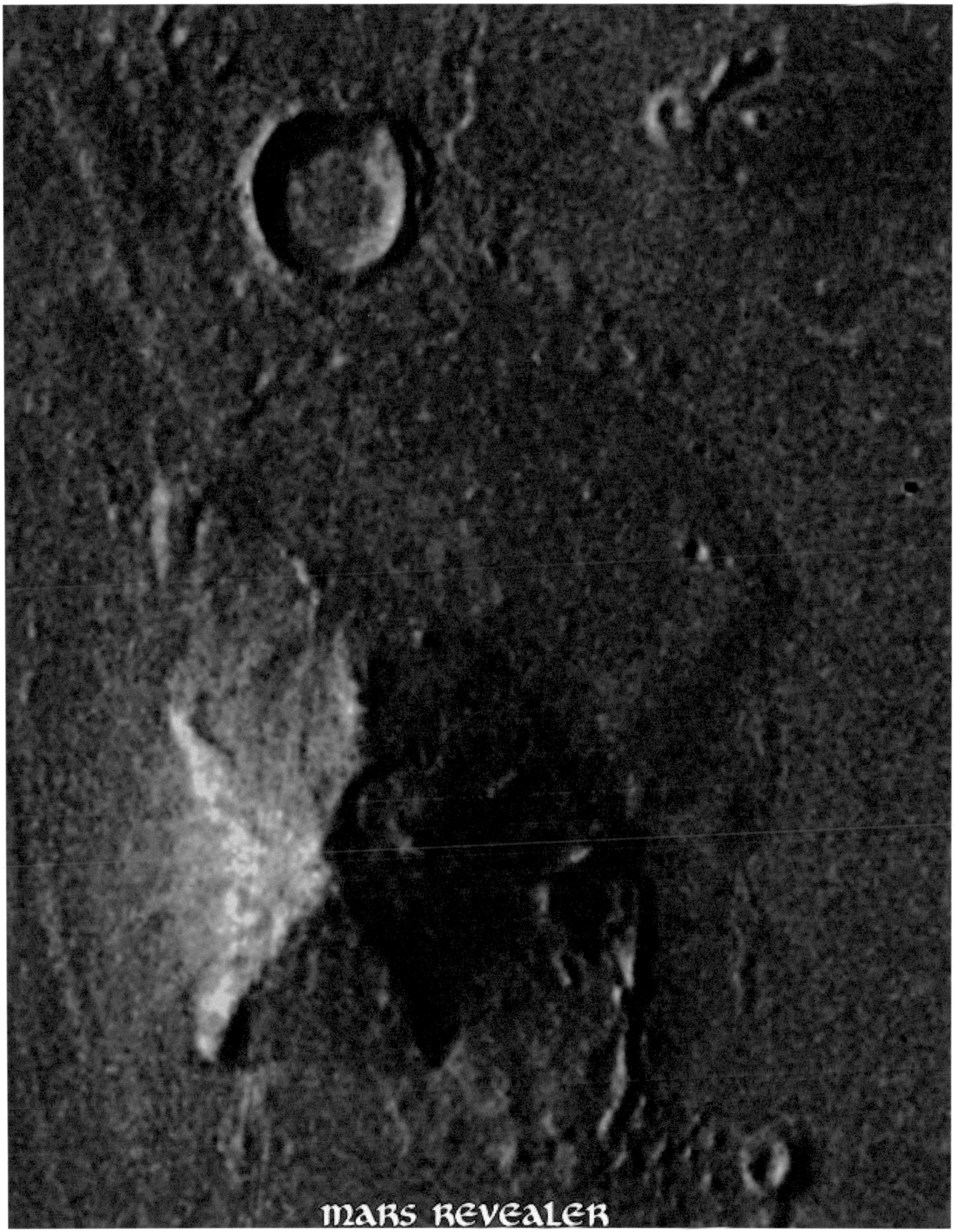

'THE D&M PYRAMID'

The immense research into these images continued on for many years by a small few and provided to the public and especially to NASA the scientific value of taking this seriously, and is even still ongoing today, despite NASA's official 'policy' position and outright lying to the public of the FACE in what it means. Added with the mainstream media's silence and bias with the intense scrutiny given unto NASA by these few researchers regards to the many things found also besides the Face at Cydonia, that screamed of intelligently design techno signatures also went ignored. The reality of the fact was and still is, by serious researchers studying these images, that this all is something that demanded continued investigation.

Instead of investigating, the NASA 'policy' driven agenda to keep this all from being taken seriously, were now using the mainstream education process to be labeling those who were researching and studying the images, and questioning against the official NASA position about the features at Cydonia were then being publicly labeled by NASA as a mere handful of kooks. A tactic of downplaying anything that is significant for the suppression of information. To try to make all of the public not believe what their own eyes tell them, and instead encouraged people to believe the public policy lies which we were all told about it instead, as if it is known facts.

Yet the research continued. Another Mars mission probe was sent in 1993, called Mars Observer, and its importance of acquiring answers for these features was pivotal. Let us just call that mission one hell of a conspiracy theory that only adds fuel to the fires of 'POLICY' enacted and a controlled shut off from the public, as the public were told the space craft was lost after it's reaching Mars. It was supposed to look more closely also to the Face in Cydonia as well as the pyramids and other interesting features. Someone it seems was not going to have that, and in some opinions, that mission went black. So one can research on the internet into the subject of Mars Observer, and its mysterious shut off, if someone wants more information on that to look into what that implicated. The main thing is the public were told no images of any kind would ever come in from Mars Observer, so the hopes of reimaging the Face at Cydonia with higher resolution were dashed.

However three years later after Mars Observer loss, the Mars Global Surveyor was next sent to Mars in November 1996.

The MGS reached Mars on September 11th, 1997. It was not until 1998 shortly after midnight on a Sunday morning (5 April 1998 12:39 AM PST), the Mars Orbiter Camera (MOC) on the Mars Global Surveyor (MGS) spacecraft successfully acquired a high resolution image of the "Face on Mars" feature in the Cydonia region. The image swath number is SP1_22003. The image was transmitted to Earth on Sunday, and retrieved from the mission computer data base on a Monday morning (April 6th 1998). The image was processed at the Malin Space Science Systems (MSSS) facility 9:15 AM and the raw image immediately transferred to the Jet Propulsion Laboratory (JPL) for release to the Internet. The image was posted on the Internet at 1:30 PM Eastern Daylight Time, April 6th.This also was the very exact day and that very image that my Mars research work had begun, I was 27, and this led me to where I am at today. (**Lots to be told there, *but that is another book***).

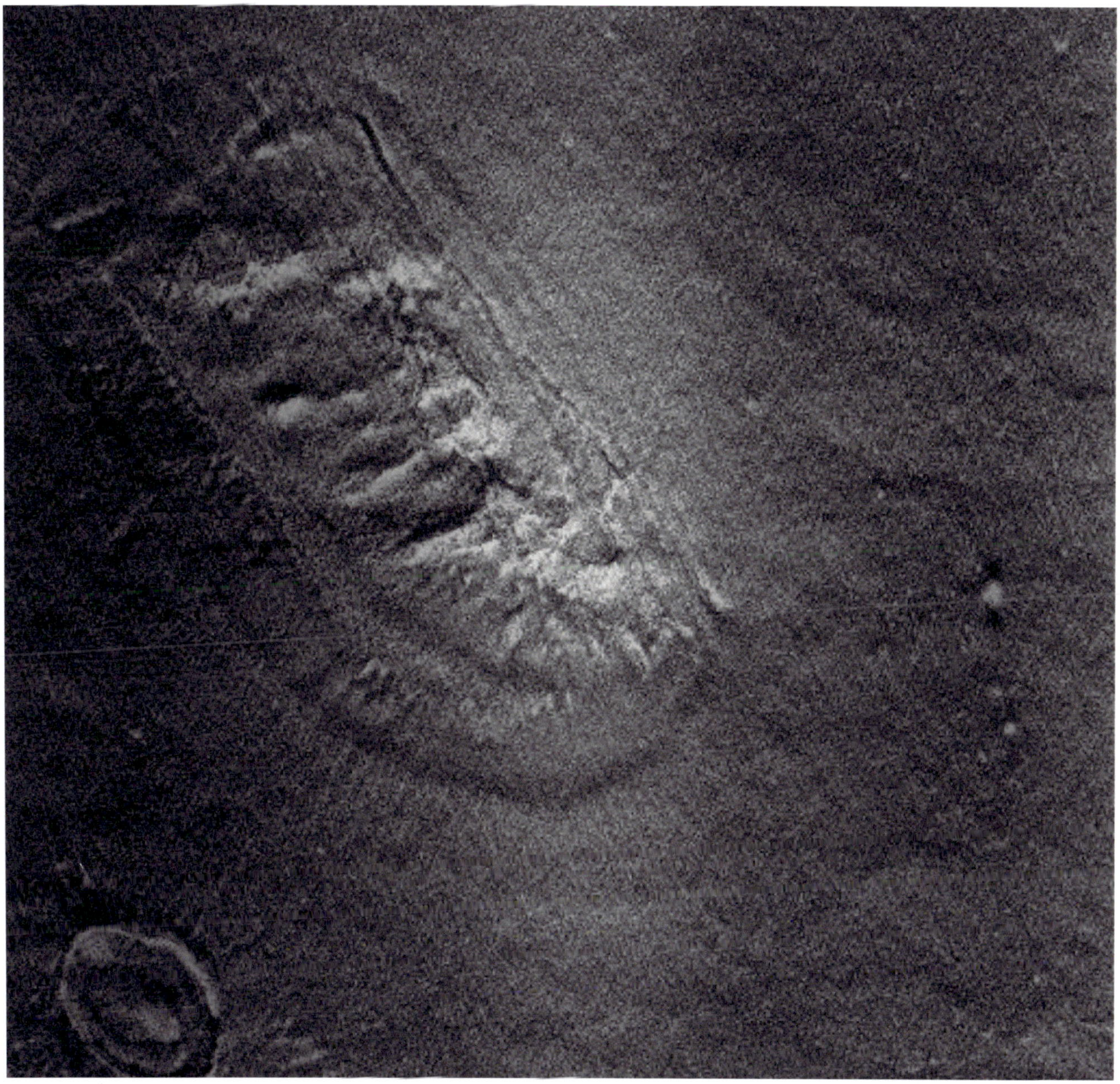

(NASA OFFICIAL 1ST PUBLIC MGS IMAGE RELEASE OF THE FACE APRIL 1998)

As exciting a time it was for everyone interested and involved with the serious research going on regards to the Viking imagery of the Face, the only data we all had to go on and base the findings thereof was from the two Viking frames of 35A72 and 70A13 taken in 1976. Waiting two decades to be actually able to get another image view, with higher resolution and better clarity. This was a main hope for the lost Mars Observer to be able to accomplish for us all. The disappointments from that loss was almost forgotten with the reinvigorated hopes of the MGS mission.

One can perceive the nail biting tensions and hopes back then, for some to be able to prove, and some be able to disprove the Face once and for all. To be able to prove NASA as being outright liars, or NASA being able to prove the researches and believers of the intelligent design techno signature for the Face, are indeed kooks after all.

After all the negative press NASA and mainstream media had applied to this subject over the many years, and how the public were treated because of their interest in learning more about this all, was finally going to be ended! Or so one would think using their common sense right? Well, think again!

It should also come to you as no surprise, that once again, controversy and conspiracy and the NASA public '**policy**' seemed to be able to continue as the norm regards even to the release of the new Face image of April 1998. It was a debacle of immense proportions and accusations flew against NASA and Mike Malin, of Malin Space Science Systems (MSSS) who built and ran the on board image camera being used to provide us all with the image data we were promised that they would provide at every opportunity as it passed over Cydonia. To take multiple images of the Face and other structures speculated about by researchers who claim the features show intelligent design techno signatures.

The main issue people had immediately regards to the new image data after over twenty years, was how 'THIS, BECOMES THAT'?

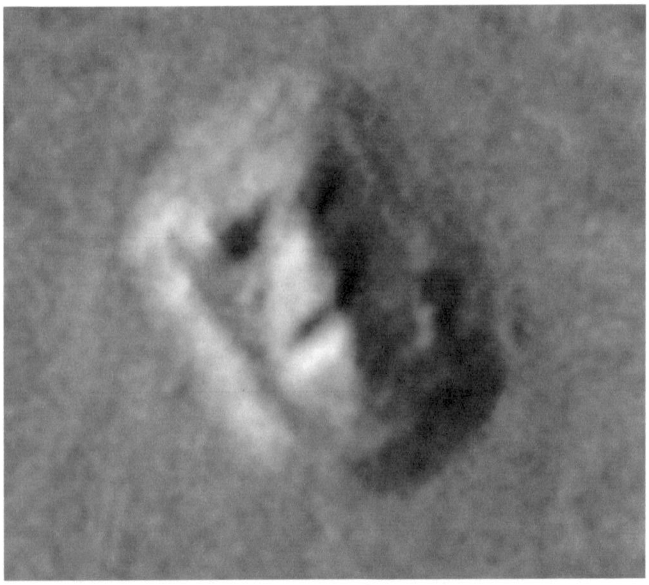

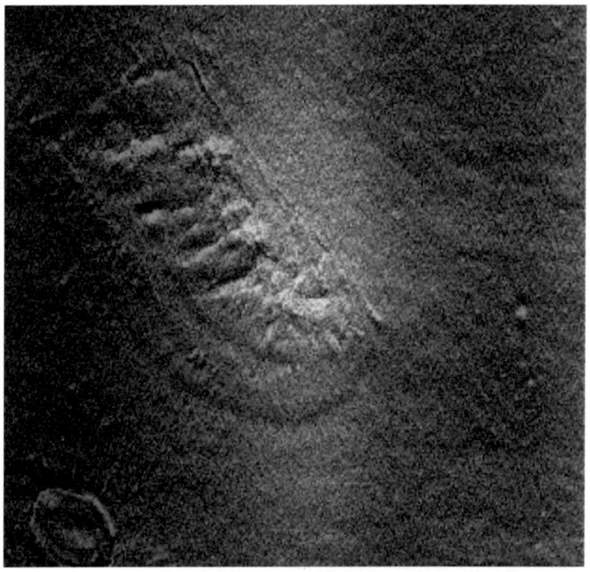

One needs to also be informed and keep in mind that before the MGS made it to Mars, at one point NASA declared that it '*WOULD NOT*' take pictures of the Face and the supposed pyramids at Cydonia, in order to conserve the battery life of the cameras! **(As if the public would not see through their fear of what we would be able to finally see?)**

So NASA wound up not going that route. After all, that would only make the 'conspiracy nuts' correct about them, I am sure they knew and understood that. A vigorous investigation has went into this all and it was public knowledge the research being done over the years previous to MGS, and what these same researchers had said about NASA in their handling and mishandling of the images years previous. The researchers alleged that NASA had things to hide and the outright refusing to reimage the Face and Cydonia would only confirm what has been declared about NASA and their 'policy' agenda to keep the Face as a 'Mere Trick of Light And Shadow', and nothing of intelligent design, and certainly not being a Face, as the reality they want you all to think is the reality as what they want to teach you what to believe it is, and you believe that. Again, against and despite what your own eyes and your own analytical prowess shows you, as well as against the scientific ethic.

TV stations, *The New York Times* and *USA Today* had all raced to cover this story. Oh how they 'Covered' it all right! Mocking the researchers and scientists and everyone else who had put over 15 years of effort by that time, and interest into the search for ancient structures on the surface of Mars which the Viking images in question indeed do appear to show. The newspapers printed the first photo released by NASA and it was a poorly processed image.

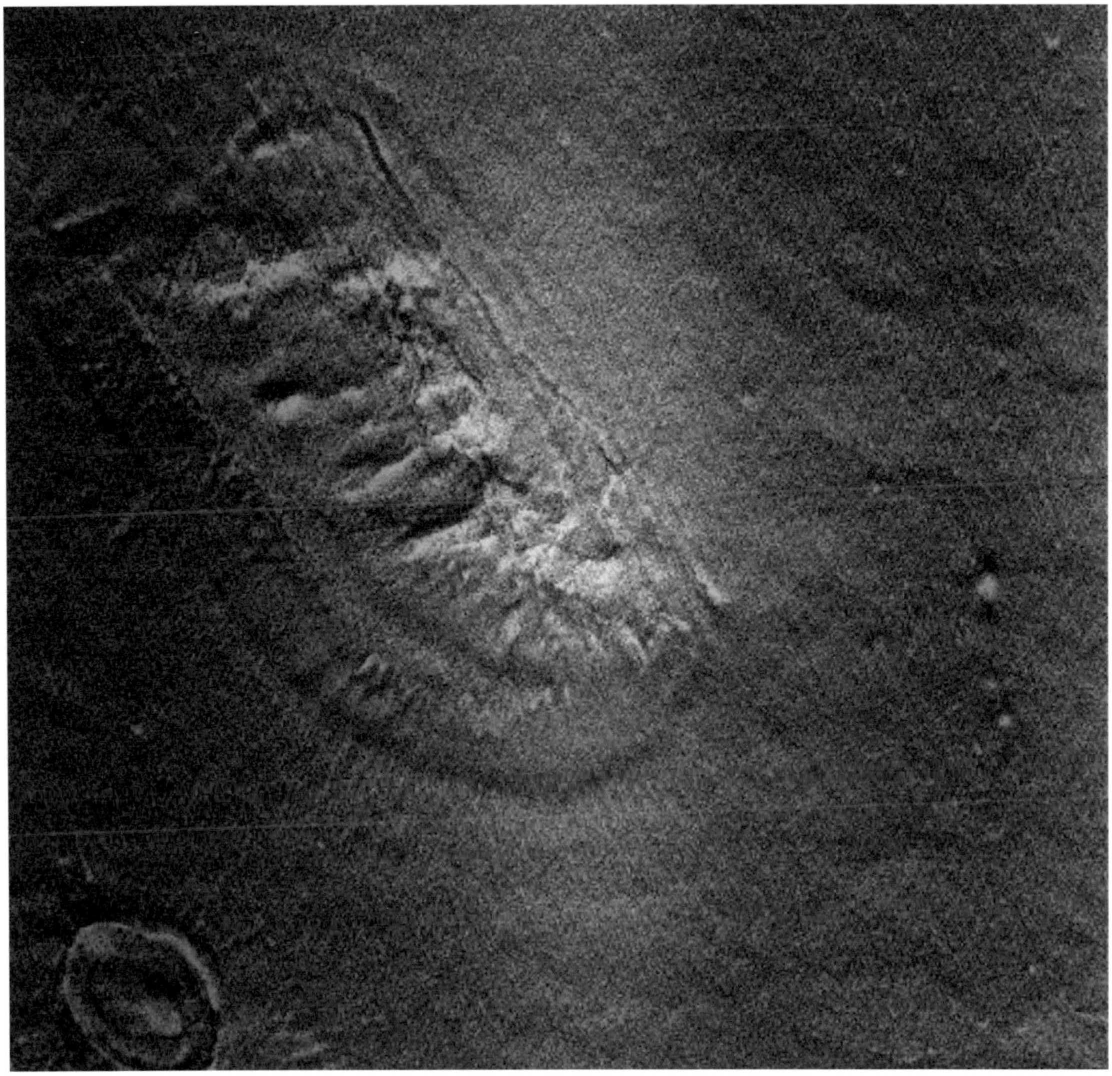

So NASA decided to image the Face, yet also decided to plan ahead of time, to create a media campaign hit-piece of 'Policy' of deniability agenda, to keep the public from seeing the Face as it really is, and to try to prevent them from accepting the independent researchers findings, by simply telling us all that the Face is not and was not a Face after all! Telling us all that the new image data taken by MGS somehow proves against intelligent design and their techno signatures!

So deciding to release the image, they decided how they can release an image, along with their official continued public 'policy' stance for the Face staying in place. By releasing the image, making it hard to even see, declaring it as proof evident that the Face was not there and was not real, and in fact was still being called a 'Mere Trick Of Light and Shadow'? Another nick name term was also then created by NASA to also mock the Face at Cydonia, as it was now also officially being called ***'THE SO-CALLED' Face on Mars***. A ridicule to the Face itself, and an insult to all those that consider it a Face can most probably be an intelligently designed techno-signature structure.

NASA cannot stop the interest for the Face on Mars, but they can certainly lie about it, and help guide the majority to not even be interested anymore about it, or consider talking about it as being of scientific value. It appears also clearly that NASA has such a hatred of the Face and a strong desire to wish that this face was never seen, let alone having to reimage it and then to reveal it to the public. The mainstream media also shares this hatred for the Face because it has been told to do so, and controlled by a 'policy' of suppression and deniability. The mainstream media are controlled news and that is no longer a secret at all today in 2022, and we who watch and believe it, are controlled to think that what the mainstream media news gives us all is the truth. In modern times again, being redundant, we now know how untrue that is!

The image looked horrible. It was even also given a comical name from talk radio show host Art Bell, who said it not looked like a Face anymore, but rather looked like his 'cat-box'! Where his cat goes to deliver his goods.

The image was all scratchy looking and not a lot of detail. The mainstream media used that image as well as NASA for their official 'Policy' agenda driven 'Hit-Piece' to try to debunk the Face as being something to be considered as being real and valid, by hoping all the people seeing it and listening to them, are all stupid enough to believe them. They were right, as far as the public majority for the most part were concerned!

It was a planned image release event to publicly again, over twenty years later, to officially use the media to denounce the Face by using an unprocessed and degraded data image, which looked twisted and out of shape. It was not an immediate obvious recognition of what we were now looking at, as looking anything like the Face we see in the Viking imagery from many years before.

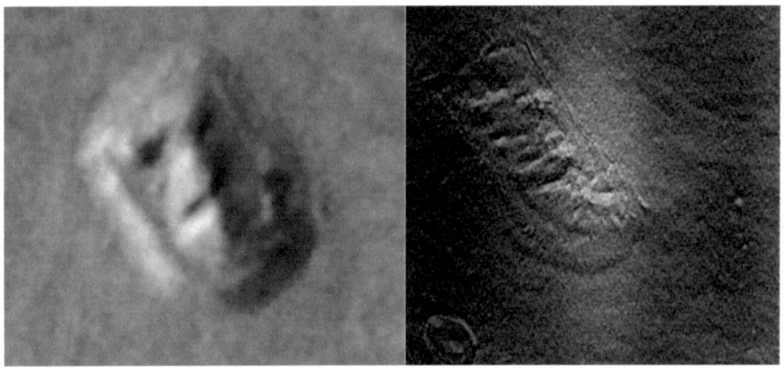

Which was their intent to begin with. To try to keep the truth covered up, that not only is the Face still there, exposing their original lies all the more from the past, but it was obvious NASA was really working hard and sweating to not show us all what is really to be seen there at Cydonia! Calling it multiple things such as a mesa, or a hill, or a mountain, to make it sound and look as natural as possible, so that you do not come to see or to accept that the Face at Cydonia is as being something real and created by an intelligence!

However, hours after the first poorly processed and un-enhanced photo was released to a ravenous media in search of debunking the story, and NASA's official position was secured to the public, a second and now, better processed and Ortho-rectified version of the same photo was released by NASA. This one was much more clear and adjusted for better viewing and elaborate, shocking and extraordinary in my opinion. The mainstream media had also not said anything at all about it, as they ran with their 'policy' controlled public hit piece horrible image earlier that day and the damage was already done. As was their plan to do. To play down the Face at Cydonia by using misconstrued image data as a means to publicly debunk the Face.

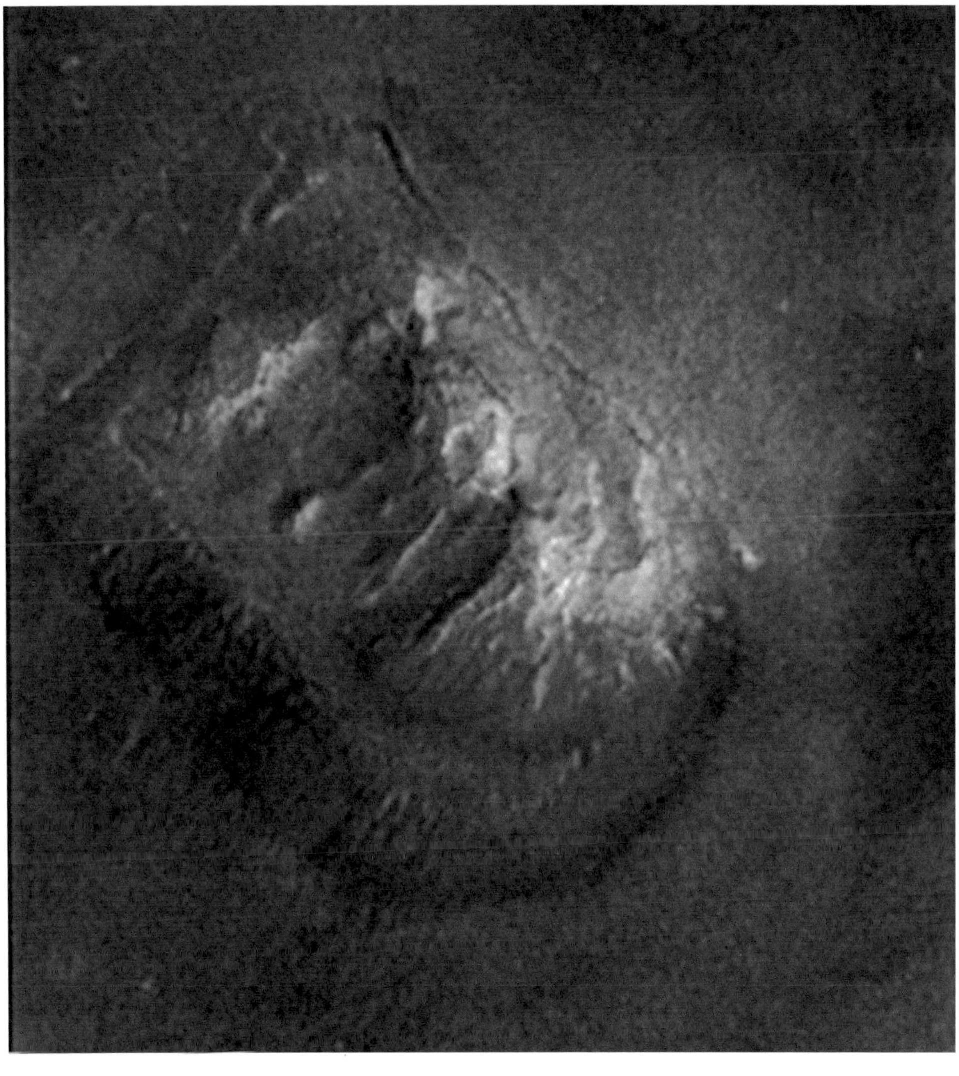

(Processed & Ortho-rectified image of the Face from MGS image SP1-22003 released April 6, 1998)

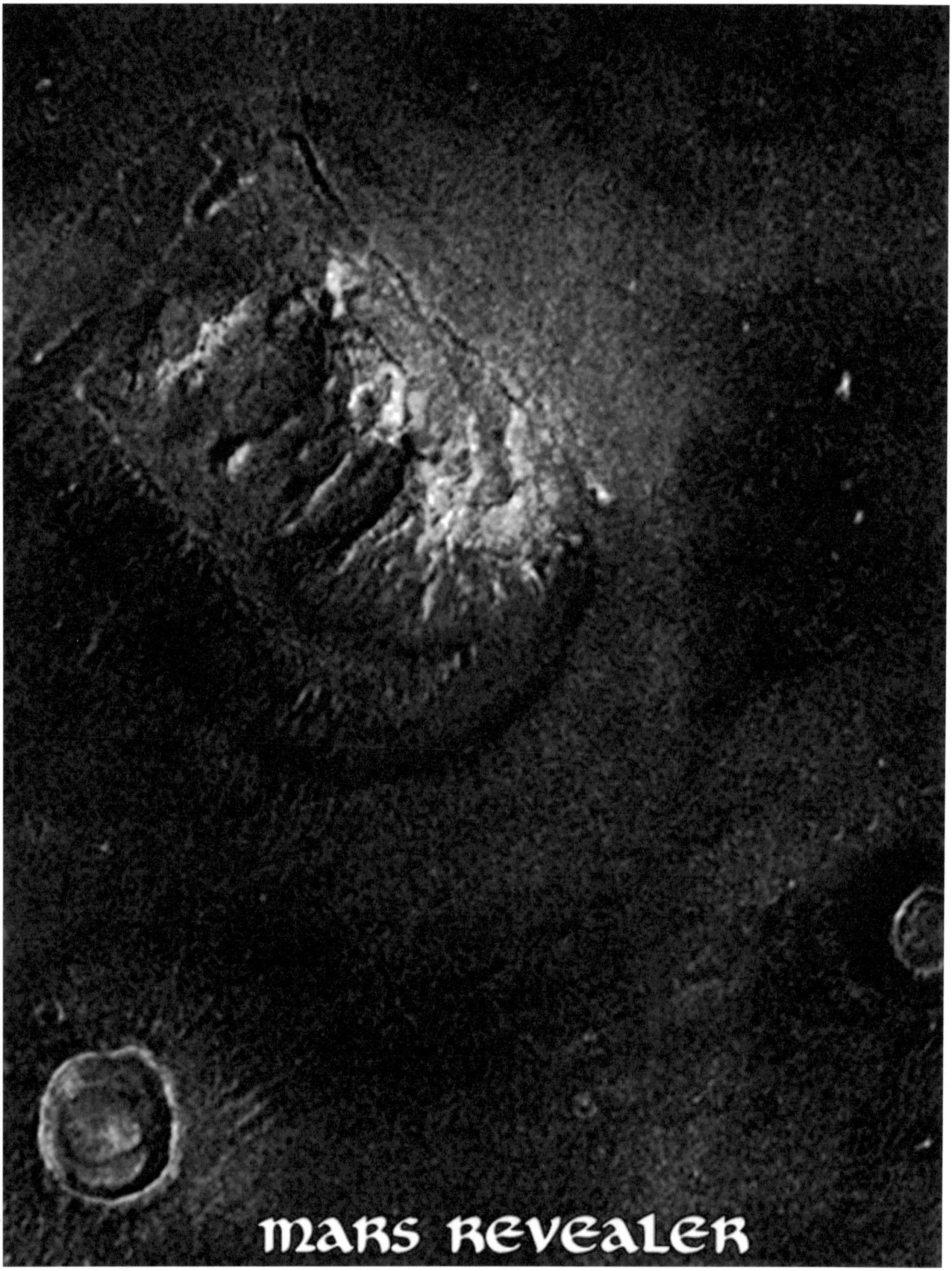
MARS REVEALER

To be fair about this Face image release, in trying to answer reasons why we received the image as we received it, it is needed to be considered that when the image for the Face at Cydonia was taken, it was during the MGS's aero braking process. The MGS was not yet correctly in an orbit to begin the mission's mapping phase, that would begin in 1999, which would allow for the ideal conditions to capture images from a position of orbit where the MGS can map the planet surface more efficiently and more accurately from a more direct overhead position to get views like we did from Viking Orbiter's view of the Face in what is seen compared to the new MGS imagery.

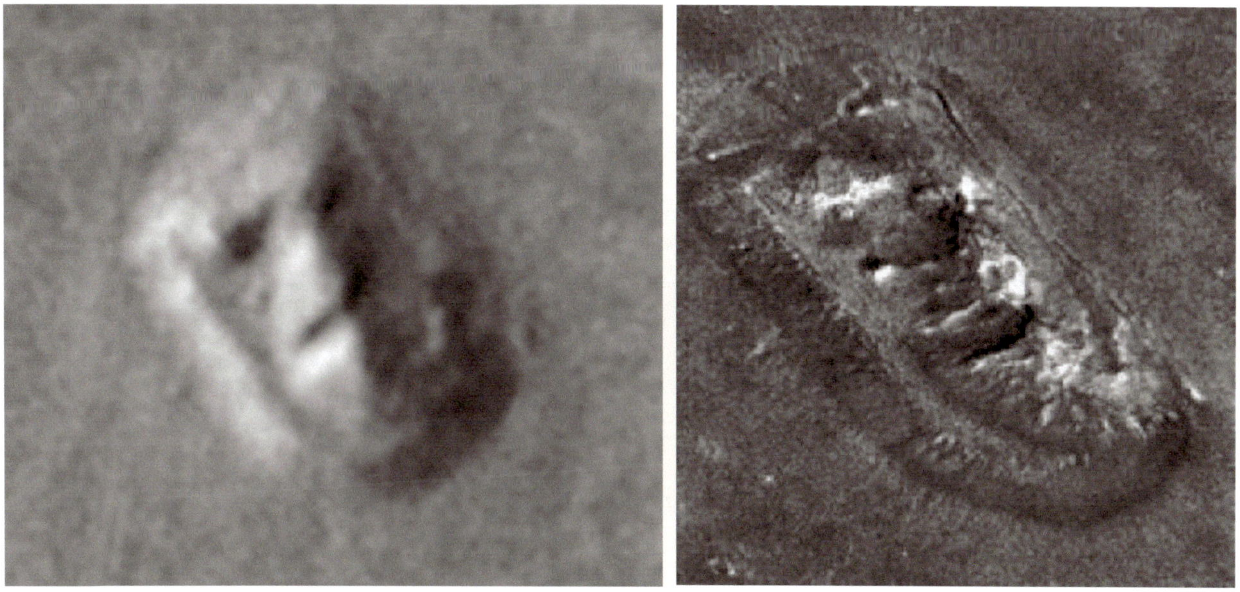

(Overhead view from Viking Orbiter 70A13 compared to the Oblique angle view from Mars Global Surveyor)

(**MSSS FACE RELEASE LINK**: https://www.msss.com/mars_images/moc/4_6_98_face_release/index.html)

To me this sounds understandable and there is actual data and facts that show us that the image of the Face we were given in April 1998, was taken from an oblique angle rather than from a more direct overhead view. (*Being seen from an oblique or side angle is called* OFF-NADIR.) A Mars researcher and image specialist involved in the serious studies of the Viking Cydonia imagery years previous, and author named Dr. Mark Carlotto PHD., pretty much summed this up in regards to the obliqueness view of the image from MGS. Here is an excerpt from Dr. Carlotto's analysis of the April 98 Face taken from:

' The New Imagery of the Face on Mars: A Preliminary Analysis '

Mark J. Carlotto (http://carlotto.us/martianenigmas/Articles/April5anal/index.shtml)

"Following the initial imagery release, JPL released a better processed and geometrically stretched version of the original image that simulates what the Face would look like from above. (*Being seen from above is called* 'NADIR'.) Ortho-rectification is a process by which an image that is acquired obliquely, is re-projected to appear as if it was acquired from directly overhead.

"Only if the terrain is flat, can orthorectification be done by simply stretching the image to compensate for the foreshortening in the direction of the observer. Because the topography of the Face is not flat, stretching the image will actually distort the internal structure of the Face, ***perhaps rendering it less face-like***."

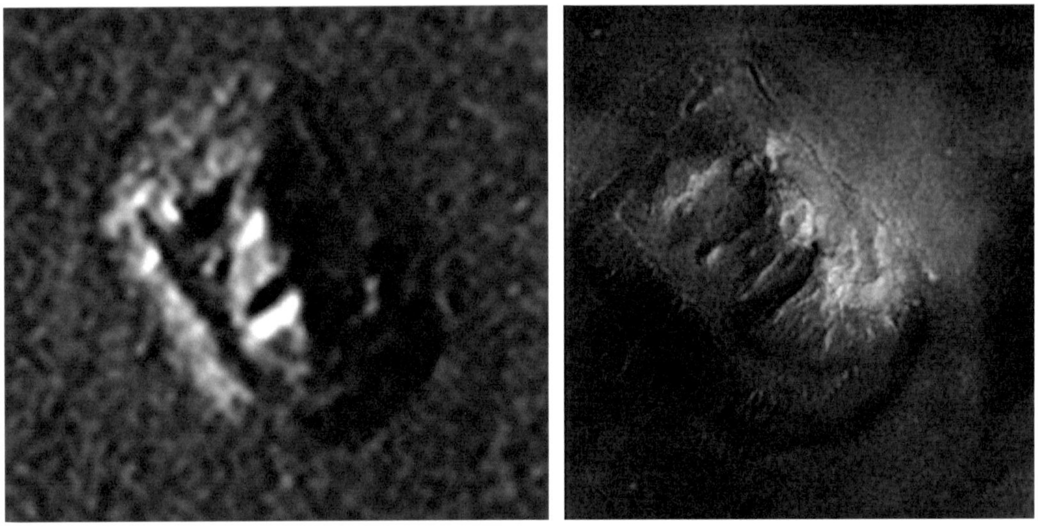

(Overhead view from Viking Orbiter 70A13 compared to the orthorectified view from MGS)

"The MGS image of the Face was taken from the west, 45 degrees off-nadir, and so much of the right side of the Face is obscured. Because it is being viewed from the left, the left side of the Face is much wider than it should be while the right side is much narrower. ***As a result it is not possible to obtain an accurate representation of its shape from the MGS image***. Fortunately the Viking images were taken overhead at low sun angles which were ideal for the use for shape from shading algorithms derived from Viking frame 70A13, and we are able to re-project it to match the Face's perspective in comparison as to what is seen from the MGS image."

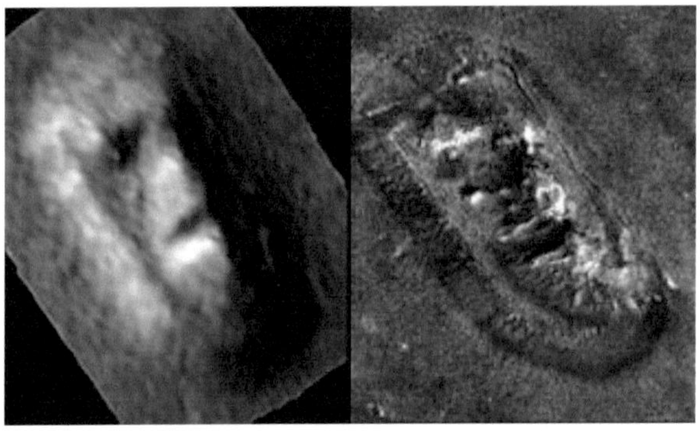

(Shape from shading algorithm perspective view derived from Viking frame 70A13 to match MGS release view)

Dr. Carlotto also stated that in the 1976 Viking images, ***the impression of a face was unmistakable. Many of the features seen in the Viking images of the Face can also be seen in the new MGS image***. The symmetry of the platform, precision of the beveled edge, linear features at the top of the head are striking. The middle of the mouth seems to come together in both views. The two bright circular features in the MGS image which look like nostrils correspond to a flat area near the tip of the nose ridge in the Viking image.

It is not clear if the left eye seen in the Viking image corresponds just to the dark circular feature on the left side of the Face in the MGS image or if perhaps the sloped surface above it also contributes to the visual effect. The crossed lines on the top of the head seen in the Viking imagery are very prominent in the new MGS image. These lines, the symmetry of the base and the precision of its beveled edge are perhaps the most striking features of this formation.

Some questions come to mind that need to still be answered that we all hoped we would have received from this 1998 Face release. First is, are there facial features on the right side of the Face? If so their presence would greatly increase the probability that this object is a face, regardless of its present condition. Second is, how does the Face appear at high resolution under the original Viking imaging geometry with the spacecraft almost directly overhead and the sun to the northwest? Third is, will NASA provide us all with that image data in future opportunities?

NASA continues their '**policy**' to not be forthcoming on the possibility of ancient Martian structures, and possible ruins on Mars' surface, and publicly entirely eliminating the theory as a possibility and labeling it instead as being foolishness. Even despite the many interesting and exciting findings some scientists and imaging experts such as Dr. Carlotto, has presented which strongly contradict NASA's official public position. ***NASA and the United States media in particular have been trying to undermine researchers and the public, and the media's job objectively it seems is to make fun of this all instead***.

NASA, this supposedly civilian space organization, is supposedly supposed to be serving the American people in detailing and revealing what we have seen and found in space, yet it has clearly NOT been forthcoming to the public about what it has found and now knows about Mars. So there was an overwhelming feeling in the public that the whole subject from NASA now could be considered and perceived as being dead and buried. '*THEY*' ' were wrong again!

Despite what the official NASA misleading statements told to us all about what it dictates about this subject, we continue to get image data that appears to prove them wrong in their official position regards to the Face. On January 31st, 2001, Malin Space Science Systems (MSSS) released a new image for the Face at Cydonia. The image number is M16-00184.

The official release page was titled: **Face-To Face with the 'Face'!**

Image release link: https://www.msss.com/mars_images/moc/01_31_01_releases/cydonia/

The official wording also states;

"This is the area where popular books, magazine articles, tabloids and other news/infotainment media have speculated that some of the hills and mesas were artificially-shaped by extraterrestrial intelligence into forms such as a pyramid, and, most publicized, a face. Owing to

this continuing interest, the Mars Orbiter Camera (MOC) team has, whenever the Mars Global Surveyor spacecraft has flown over the region, commanded the MOC to take a picture. MOC image M16-00184, which shows a portion of the famed "face" landform."

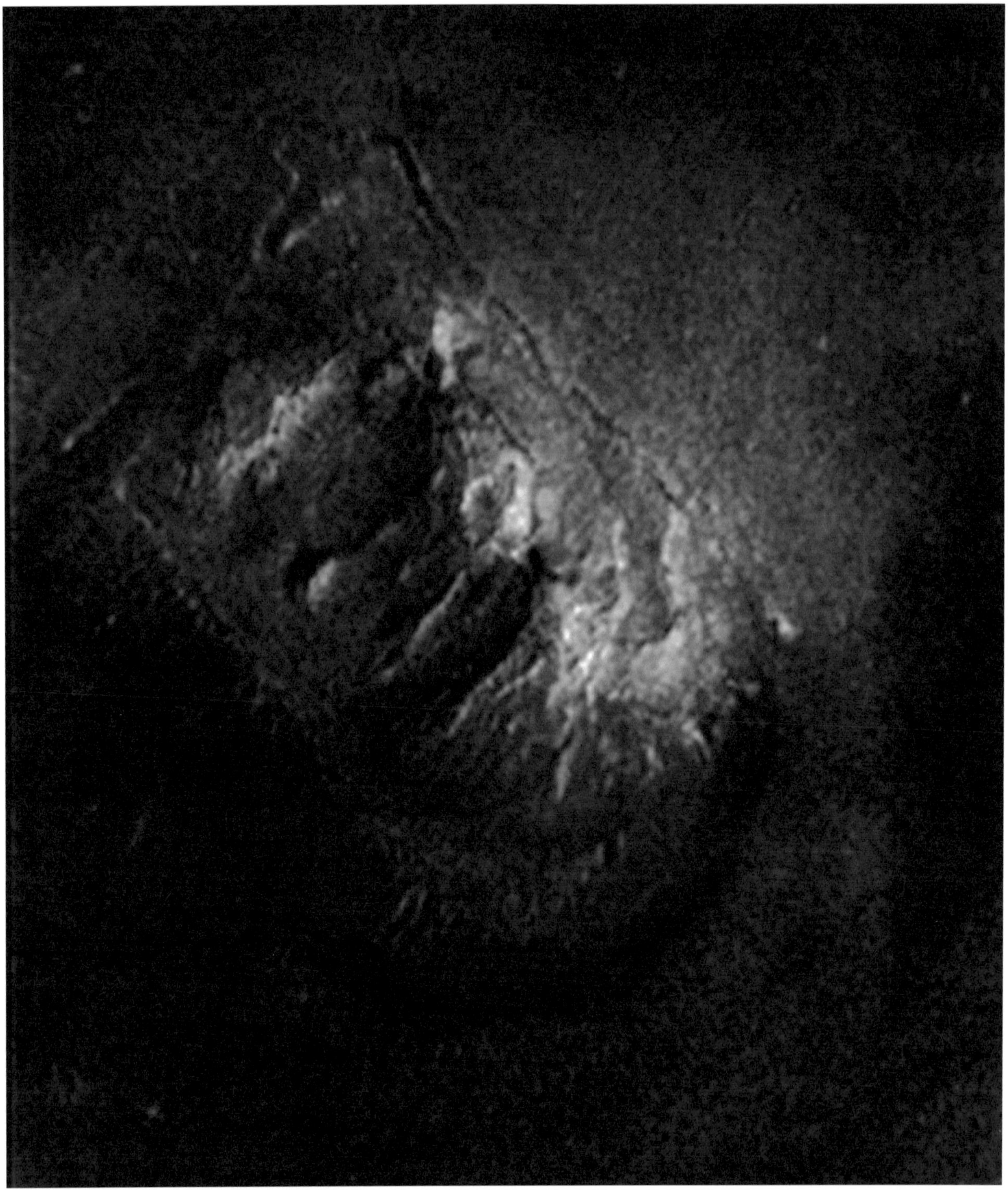

CHAPTER 3:

THE EYE OF THE BEHOLDER!

(January 31st, 2001 Face release image number: M16-00184)

MOC image M16-00184, which shows a portion of the famed "face" landform. This image release was a nice surprise for us all to see. Yet at the same time, many were not surprised with the image we received to see. I'll explain myself in what I mean here, but first I would like to point out a couple interesting facts about the release. 1st, is its 'official' title! **'Face-To-Face with the 'Face'?**

Are you kidding me? This cannot be a joke or a mere coincidence!

Remember back in July 1976, when the first Cydonia Face image was 'officially' released, of 35A72, it was given the title 'HEAD'. We were able to see how '*THEY*'', NASA, officially handled that 'HEAD' and what they officially said and labelled it as being a *Mere Trick Of Light And Shadow*, claiming it actually shown nothing at all, they had said without showing proof that it was imaged a few hours later and there was no Face or Head feature seen. It just disappeared right? By claiming it to be a mere chance also implicates that the Face is a Mere! Meaning it is an ***only one***! ***No other features like it at all***. Nothing to be seen here?!

Then in time, as we went over, the truth becomes known and a couple NASA scientists found a second image number 70A13, which was showing that the Head Face was indeed imaged again, at a much later time then NASA specified, and imaged at a different sun angle that allows more detail to be seen, and the Head was and still is, able to be seen.

Then the official April 1998 image release of the famous 'Cat-Box' image which was also titled; **Mars Orbiter Camera Views the "Face on Mars"** public hit piece, declaring a successful official NASA debunking of the Face at Cydonia, ***USING MISCONSTRUED IMAGE DATA*** through NASA's mass public media campaign, declaring once and for all, there is NO Face on Mars! That it is *NOT* a work of an intelligent techno-signature, but again, was merely tricks of light and shadow and claiming it is only a natural mountain, mesa, massif, hill, butte. I mean, which one can is it? It must be all of the above according to the official position.

What I want and to which I am focusing in on here for you, is to show you the fact that no matter what NASA officially claims over the years about The Face, it has went from OFFICALLY being called 'HEAD', to only still come to be 'OFFICIALLY' called in the 1998 image release ***'THE FACE'***? And then considering the January 31st 2001 image official release of M16-00184, **'Face-To-Face with the 'Face'?** One must see the irony here! It is like I see a double standard. Or, a CODE? Something to ponder there for certain we will also look into.

Another issue I had regards to this official release of M16-00184, **'Face-To-Face with the 'Face'**, is the way MSSS had imaged it and provided it to us all. Remember also that it was over two and a half years previous since the April 1998 Face release, which was taken during aero-braking and at an angle and not from being directly above. The Mars Global Surveyor (MGS) was now in its mapping mission phase where now any image swaths captured of the Face are most likely as being imaged from position of directly above, or pretty close to that, which would match what we see from the Viking Orbiter image views being taken from directly overhead. The 1998 face image release did not have that capability to match to that fully as to what we seen from Viking. But now it was available in high resolution!

So this face image release of M16-00184 was very important to us all, speaking in terms of independent Mars researchers, of which of whom I was and still am apart, and some mainstream scientists who were wondering if NASA would ever release another image of the Face. NASA has been pressured by independent Mars researchers like myself and others who I was affiliated with at the time, as well as again, certain few mainstream scientists as well wanted to know, and even they wondered if we would get to see what we needed to see. So instead, NASA had MSSS only give us all a 'PEEK-A-BOO' image?

NASA telling us all we are FACE TO FACE WITH THE FACE? Sounds like someone is being 2 Faced! As amazing an image it was and still is, I was not surprised how we received this image. It was indeed a deliberate PEEK-A-BOO release on their part to us all, letting us all know that what they get to see behind the scenes, we were NOT seeing. Therefore this image release was really meant to be a tease, as we are only being given a glimpse of what they get to see while we do not. The Face image release was given as looking like a Face peeking into a door or window, saying PEEK-A-BOO, HELLO! It sounds comical I agree. In their minds I am sure, it was.

To the rest of us? Well we only took for what we can get and verified some things in the data that is crucial, and have some things be confirmed. Most did not say anything in how we received the image, but I did. I felt uneasy in how they gave us this image release! This again, was intentional, and not a coincidence. It was to tell those who were demanding NASA to come clean about the Face, and what it really shows, to change their official position to where the data and the science leads them to see, and come Face To Face, with the Face!

Again it seems as if we are being led by NASA to see a code when we think about it. Giving this release image out as they did, with a window peek of the Face looking in at you as it were, from behind the dark cover that prevents us from seeing the Face as it really is, teasing us all, to show us Mars researchers demanding the truth whom NASA is openly teasing us by this release on the public surface, was also at the same time a silent confirmation under the public surface, letting us all know unofficially as a code, that '***THEY***', ***SEE IT TOO?***

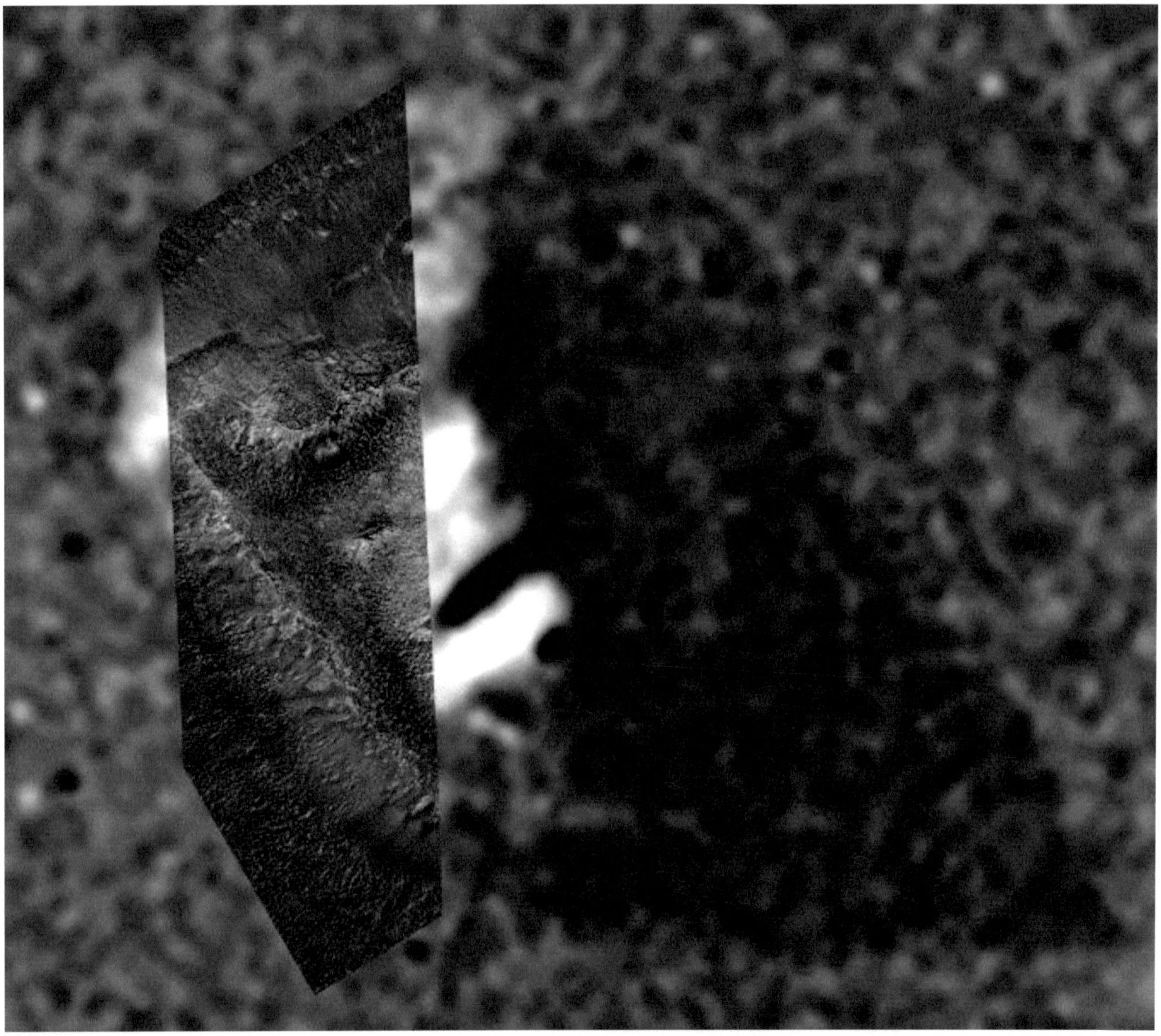

(M16-00184 Face section fused on top of Viking Orbiter image 35A72 for a comparative analysis)

Yet NASA's official position publicly not changes! How can something like this be? It is some strange things to consider indeed when seeing it all like this! After all, at this point, it is in ***The Eye of The Beholder***!

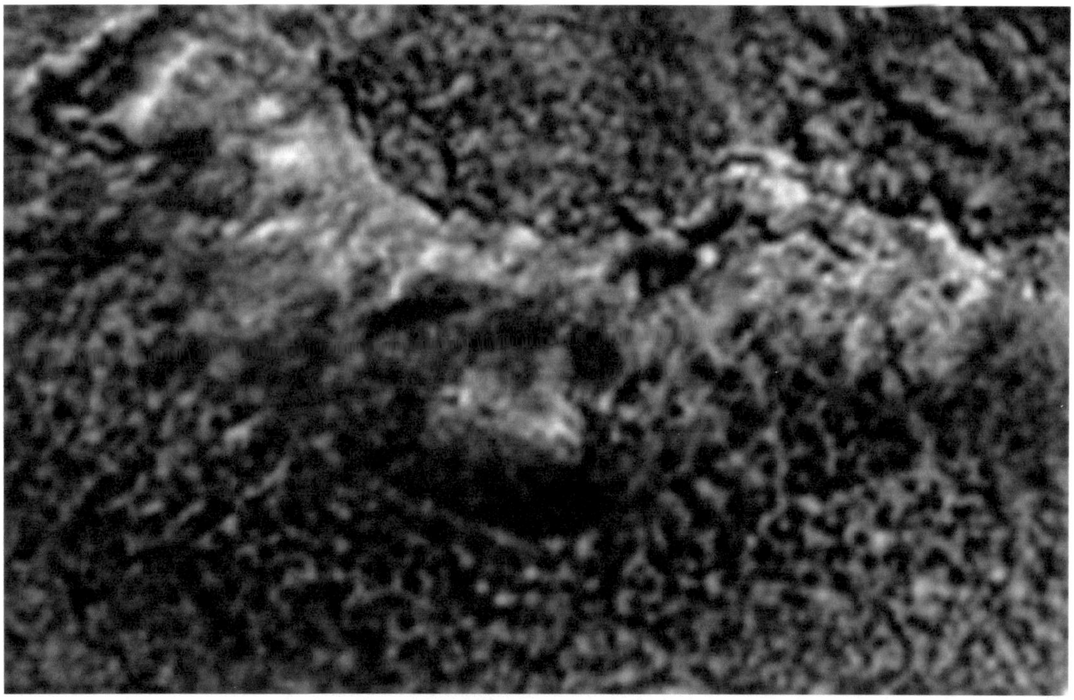

(Behold, The EYE!)

Although as beneficial to us all the Face release of 1998 really is, considering the detail that we do get to see in its processed form, this new 'Partial Face' release of **M16-00184** really gives us all a much better detailed and clearer view from an overhead viewing angle that really allows us to behold the 'EYE' of the Face! It also shows to us the extreme artistic relief all over it of which makes the Eye stand out and be able to be seen. A very strong pronouncement of itself as well, as it looks up at you from the Martian surface. It is magnificently carved with precision.

So much to see how it's artistic relief details shows us the shaping of the eye and the advanced ability of the designers to have carved this in a way that what we see depends upon the sun light angles as well as viewing angles which allows us all to see a glimpse of its intelligent techno-signature of the creators, as well as in the Face overall being a literal work of art in and of itself. In other words, The Eye, looks Intelligent itself which makes the purpose of the Face also being representing something or someone, which is intelligent. 'THEY" definitely wanted someone to notice it, and created it in such a way that it would be very hard to miss.

If I am correct in my many years in this research and in my very close study for the Face at Cydonia, I believe this Face and this artwork of the Eye is so intelligent, that if we ourselves were flying high above it, theoretically the Eye for this Face we see, would be looking at us coming, and going! It would appear as if the eye is moving, watching you as you come and go. An eye with a pupil and iris even! Only 'Intelligence' could have designed and created this artistic realistic effect. Using precision art sculpting concepts and using the knowledge how the sun can be used to create the living animation intelligence effect for the Face which by now I am certain NASA and other space agencies must have to also acknowledge amongst themselves. If I can see and reason this, by what I can see in this data alone, so can them! After all, they get to look down and see a lot more than we do or can!

Once you see the art, and see how it makes the eye itself, as well as its incredible art sculpted surroundings, that make up the entire Face itself, you can then see what I claim is certainly plausible. You can see, or almost see where the pupil and iris is and would be depending upon your viewing conditions from more than one position.

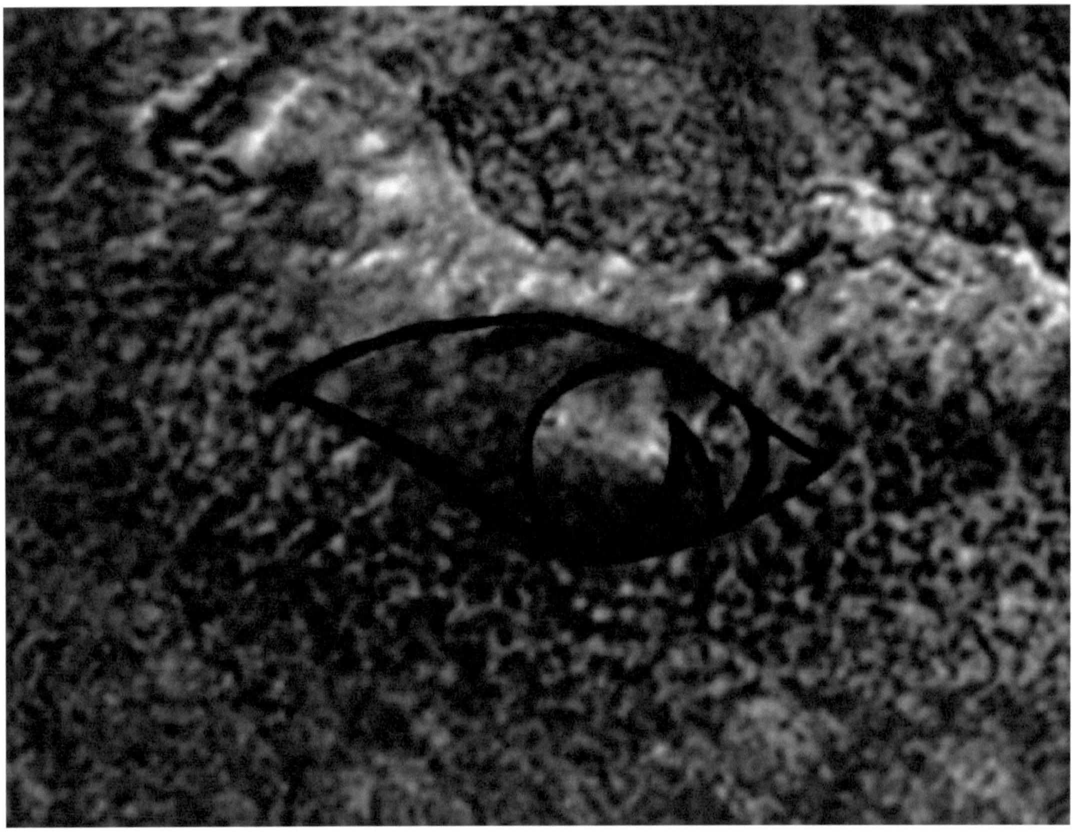

There is detailed relief there to be seen, and is certainly not a coincidence and certainly shows itself with purpose and intelligence, that clearly shows us all now, that NASA indeed really was being coy to us all when they said the Face at Cydonia was a ***'Mere Trick Of Light And Shadow'***. Again, that statement possibly being like a code? Telling us a half truth? To word the official statement in a way that they will not really be lying? It depends upon what face of understanding you see and take it as meaning. Yet pushing and adhering to the surface lie about the Face's official 'Policy' driven statement, and using it instead as an under the surface code that is unofficially claiming the Face as a work of intelligent design? Looking at it as if it is a code, we can see those words also match to what we see in the image data that implicate that artistic reality. NASA uses those words itself instead on the public surface as a means to claim it was just a ***TRICK OF LIGHT, AND SHADOW***. Which remember, that was also in their official announcement of 35A72 to which they also never provided proof of their claims for that being the reality. Now they have this visual proof, as do we now. You have seen the reality that they lied officially in 1976 and despite the new image data that was acquired, NASA never changed or corrected their official position.

So to deliberately use those very words, a Mere Trick of Light and Shadow? It was used to lie and to mock us all telling us all officially, that there is nothing at all to see there, because it is all natural, a mere trick of light and shadow. Move along! The public will be trained and forced propaganda to not believe that they are seeing intelligent design on Mars, No! They told us all we only see a Trick of Light and Shadow?! Let us look deeper into the 'EYE' of this face of understanding, that we were possibly being given hidden coded information.

A *TRICK*? (Which it was, wasn't it?)

(From Free online Dictionary:

1.

a. An act or procedure intended to achieve an end by deceptive or fraudulent means.

b. A mischievous action; a prank: *likes to play tricks on the other students in the dorm.*

c. A stupid, disgraceful, or childish act: *Don't let the kids pull any tricks while we're gone.*

2.
a. A peculiar trait or characteristic; a mannerism: *"Mimicry is the trick by which a moth or other defenseless insect comes to look like a wasp" (Marston Bates).*
b. A peculiar event with unexpected, often deceptive results: *"One of history's cruelest tricks is to take words that sounded good at the time and make them sound pretty stupid"*
c. A deceptive or illusive appearance; an illusion: *This painting plays tricks on the eyes.* *(Notice hint of work of art representing the Face of Cydonia itself, &/or the picture that is being painted for us in seeing it this way figuring out this coded information?)*

3.
a. A special skill; a knack: *Is there a trick to getting this window to stay up?* *(Notice hint of window is opened to allow us to see the face looking through?)*
b. A convention or specialized skill peculiar to a particular field of activity: *learned the tricks of the wine making trade.* *(NO WINE BEFORE IT'S TIME???!!!)*

A Trick of Light and Shadow takes on a whole new meaning at this point doesn't it? We were given a pre-scripted official announcement in 1976 that was worded carefully and correctly back then, to think about now how we are to see this truth in these words. Just not meant for at that time in 1976! The official NASA announcement we were told, was a Trick to be seen and understood down through time?!

1.

a. An act or procedure intended to achieve an end by deceptive or fraudulent means.

b. A mischievous action; a prank: *likes to play tricks on the other students in the dorm.*

c. A stupid, disgraceful, or childish act: *Don't let the kids pull any tricks while we're gone*

It certainly was ***deceptive and mischievous***, on NASA's part and certainly ***they pulled a prank on us all*** in that official position announcement. ***They like to play tricks on other scientists in the dorm.*** How correctly fitting. The official announcement also certainly was ***stupid***, ***disgraceful and childish***, to lie to us all to keep this truth from us all, and/or to unofficially tell us the truth meant for a later time, when the window is opened and we can see this face of understanding looking in at us? After all, we are told a TRICK is also ***an act or procedure intended to achieve an end by deceptive or fraudulent means***.

It was certainly also a peculiar event with unexpected and deceptive results, that I am sure you would agree with that. An Elusive appearance, with a deceptive illusion which was also indeed their official statement, an illusion! A Trick to try to get the window which was opened for us, to allow you to see and face the truth, to stay up!

"One of history's cruelest tricks is to take words that sounded good at the time and make them sound pretty stupid"

Yes, this trick all speaks for itself to show you how you see the Light behind all that Shadow. NASA sounded pretty stupid in their original official comments! It certainly sounded good to them back then didn't it? Their unofficial coded 'painting' plays tricks on the EYES!! To be seen and solved now?! This is also not a coincidence, and I firmly think and believe that this is what it was, and is, a code! A code to which we needed to face, and is now being seen and connected down through time! *THE WINDOW IS STAYING OPEN NOW!?*

This Coded official announcement is and was a puzzle for us all to see the truth behind their shadowy cover. So when the truth is found and NASA confronted on it, down through time, they can say look, we did not lie, not exactly, it is not our fault your too stupid you could not see the Light showing the Face behind the open window Shadowy cover act of our trick.

I say this because it actually gives yielding results to that very official comment that indeed does seem to match and also make sense to how they themselves laid this out, and to how I am laying this out now for you all to see through their open window given to us to see the face of it all. I should not be able to make anything out of this if it were nothing at all. A coincidence you may think? Possibly I guess. This aspect does definitely put a bizarre twist to what you all may have expected to read and see here, but we must see how this face unfolds. I say this also because it is like a 2 sided coin. Again NASA seems to be 2 FACED here. Meaning dual sided! It not yields information only for that act of deception of the ***Trick of Light and Shadow*** on their part seen on the surface of their official position, but it also yields to us all valuable intelligence information being given to us all at the same time, about the Face at Cydonia itself, through a window of time! Information given unto us of their official statement back in 1976 about the Face's constructed design that is also true in and of itself about the Face at Cydonia as being an intelligently created techno-signature as well. This cannot be a coincidence!

The art of telling it, without telling it applies here?! The unofficial back story as it were. If what I am saying is true, it would also make sense that what we were told indeed can be attributed to the Face also. Remember looking at the official statement in this Trick of Light, this was a preplanned scripted public response. An official denial and yet, an unofficial coded confirmation of information about the Face as well, that the new imagery confirms their old statements from 1976! TIME, had to go by to allow the window of *time* to be opened, for us all to see its Face and to keep that window open? Hmmmm, maybe also meaning, that window?,... *OF COMMUNICATION*?!

So it seems we are being told in this face of understanding, that the Face at Cydonia was designed with the intelligent imprint using a Trick of Light and Shadow, meaning fused in the Face's intelligent design in the features that we see in the image data of the Face. In the coded unofficial side of this, we are being led to see unofficially the Light of the fact that *'THEY' SEE IT ALSO!* Intelligence information telling us all how the Face at Cydonia was apparently designed, by using these chosen words calling it 'Merely a Trick of Light and Shadow. So in this Light, what I came to the conclusion of must be true, as in fact, it coincidentally confirms my theory of the Face at Cydonia after studying it closely since April of 1998! This also shows us that ***'THEY'***, NASA, had already came to that conclusion when they made this public official statement in 1976. Was it only their Viking Orbiter images presented to us all publicly at that time, that told them with certainty what is now clearly seen today in much higher resolution? That the Face at Cydonia is designed by the use of Tricks of Light and Shadow! It just makes me wonder, because years later, we get higher resolution images that show the artistically sculpted relief that make up the Face's constructed design to be more determinative of that fact in how it is designed. How did NASA know for sure by the less resolution imagery of the Viking Orbiters? To tell us as a fact about the Face back in 1976, that is seen true now today, in a public official statement telling us the Face was a mere ***Trick of Light and Shadow*** back then. It just makes me wonder. This is NOT a coincidence!

Again if I am able to determine this, then so can 'THEY', in how it seems obvious to me by what I see, is that the creators of the Face designed this face with a mastery of art and sculpture, that has capability to do this no less on a massive scale, utilizing the sun's light and using also shadow to fuse within its design. Certainly a major fingerprint of an intelligent techno-signature this is! Considering the fact also that means that the creators of the Face at Cydonia as well, were using their understanding of the sun lighting and viewing angles and lighting angles and points of position, in determining how to create and carve this Face, as for representing their intelligence in and of itself. While standing for vast ages of time with or without their past or extant presence to have this Face do what it was designed to do, by using their masterful Tricks of Light and Shadow, to create the illusion of their intelligence watching you as you fly by overhead, with its eye following you as you go, letting you know and see, their presence. As if to let you know and to see, that the intelligence seen there, ***can and does see you as well***?

So my *CRAZY* theory seems sound regards to the official 1976 Face on Mars 'policy' driven specially worded announcement being a possible code meant for us '***Down Through Time***.' As it does seem to show now down through time, consistency in the Light of how we came to see and understand the coded official Light and Shadow Tricks act that NASA pulled regards to the 1976 public response. It is a theory but a plausible one, as we are able to make a model of it, using it itself, to glean information from 2 sides of a coin as it were. This also tells us and should show us all also, that what I described to you in how I see the Face is designed as, 'THEY' (NASA), also must have actually knew it as well, before their pre-planned public 'policy' announcement for the 'HEAD" in July 1976.

To deliberately 'code' truthful information as a short and simple excuse that is not so simple in this Trick of Light and Shadow is it? Using the words to describe the Face as being 'Merely Tricks of Light and Shadow' which on the surface acts as a debunking of the Face being natural in the public 'EYE', and not a work of intelligent design. Yet on the other side of the coin, we get an unofficial way of them telling us 'THEY' see it to. But if one was to see this back then, and one who maybe would have asked about this 'anomaly' in their wording, they certainly would never ever admit it to you.

Only down through time, when this theory can be spoken about now, looked at, and possibly being figured out and connected, can we meet with added data given to us all down through time, given to us all by NASA! To see how the future would present itself of this potential fact. To put the pieces together if and when seen, that will show indeed, that NASA in a twilight zone sort of way, have told us all the truth, that only down through time, we can only figure out and try to piece it together ourselves to be able to see it, as you will never get an official recognition of this theory that sounds so crazy, yet is all so very real!

Therefore, in order to test this theory, we again have consistency and connections to it by NASA's own continued wording, it is wise then to see and to study then what is said in the future image releases for the Face at Cydonia. The best example is the partial face release view of **M16-00184.** By its very wording as well as being a visual representation which is to be acting as symbolic, to directly show and connect to this theory. In essence, it is part of the main reason in my opinion, why we were deliberately given this partial Face release as how we were given it, and how they deliberately imaged it, to be given unto us all at this point in time in the end of January 2001. Yes, I am saying that Face image swath was deliberately taken like that as NASA gave it to us for this very purpose. To Come ***FACE TO FACE, WITH THE FACE***! This was a tease release. Something more, in my mind, was coming!

CHAPTER 4:

UNMASKING,... THE FACE OF MARS!

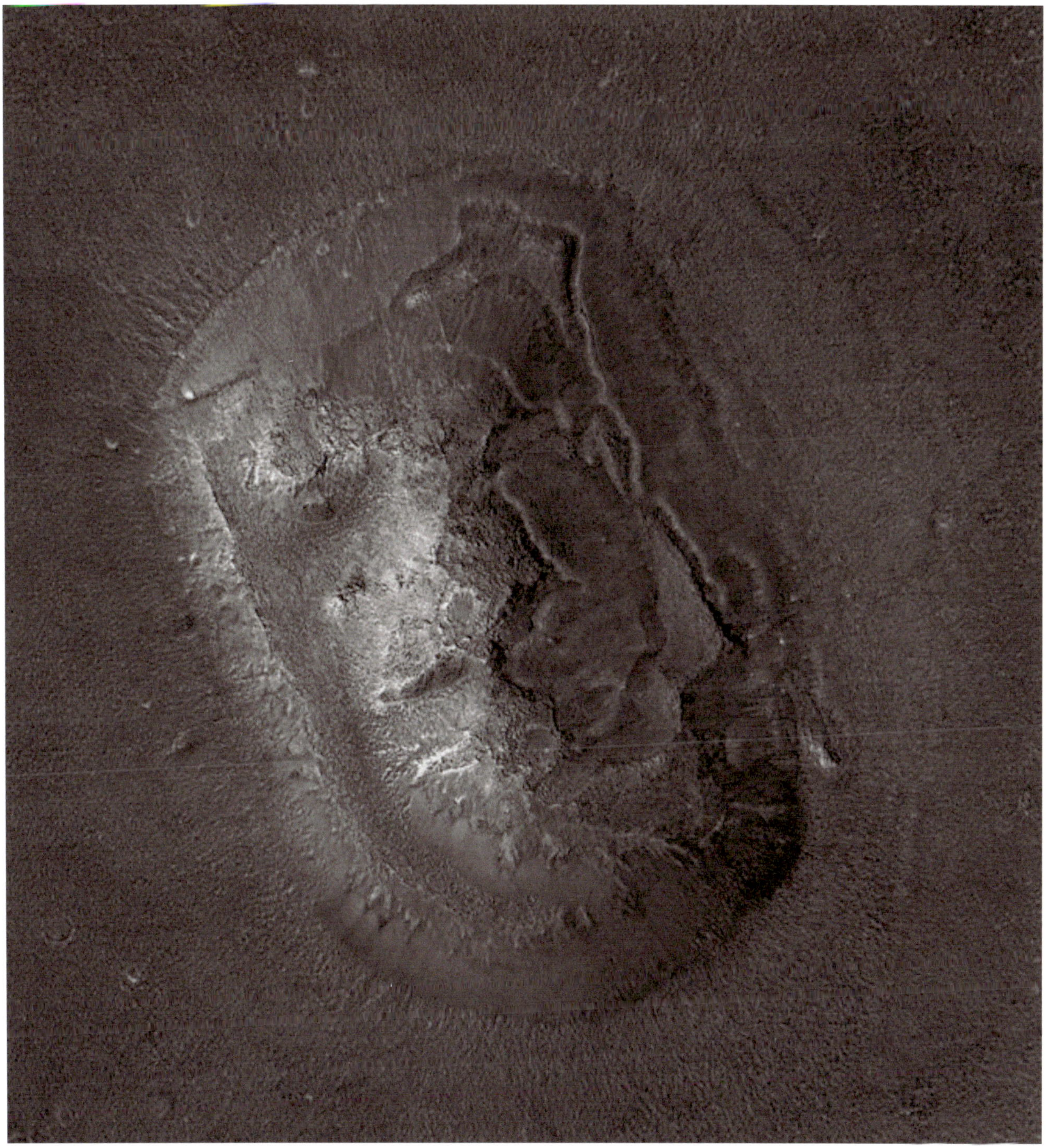

(MSSS FULL ON IMAGE RELEASE E03-00824 APRIL 21, 2001)

Only now down through time, I can now attribute it and connect it to the working theory of their 'coded' official statements as being 2 Faced, consisting of an official side, and an unofficial side. Letting us all get a glimpse of the 'EYE' staring at us all in the Face, peeking in at you in the window, while the 'coded' title NASA gave it was confirming and validating by calling it **'Face-To-Face with the 'Face'!** Officially titling it as if for the purpose of mocking us all who think and see that the Face is of intelligent design. Making it to appear as if NASA are sticking to the official public 'policy' position and yet unofficially purposed in coding information confirming to us all that the Face is a Face, and thus being a feature of intelligent design, because it looks like a Face, even to NASA! Admitting it without admitting it.

My 'crazy' theory of a coded 2 Faced NASA 'policy' of official position is of which on the surface on one face of understanding, we see clearly that NASA is purposely dis-informing the public about the seriousness and the reality of the Face on Mars in its official public dealings, as well as it's important image data. Officially playing it down to the public as something unreal and not worthy for anyone to look at or speak about, let alone needing to acknowledge it to the public mainstream in any scientific way. It is easier to use scripted words and public policy, than to use that image data to scientifically debunk the Face seen in their image data.

On the other face of understanding of the 2 Faced 'policy' looking at this in the mindset of the unofficial side of NASA's coded intelligence information where it seems in fact, NASA are in fact validating to us all that the Face at Cydonia is in fact of intelligent design. It is a feature that has not disappeared, as it is obvious that the feature in question, we can all see just keeps showing up in NASA's imagery. They knew that the public was not satisfied by their official public announcements in 1976 and onward over the years that should have closed this issue down, as was their official position and intentions were set out to do. At least again, as seen on the surface officially.

Yet on this 2 faced side of the unofficial coded 'policy' driven information, if we all just consider this crazy mind set, we can really see that NASA were telling us all from the outset that the Face at Cydonia indeed was a construct designed as a face, and that it was intelligently created with high intelligence and incredible capability by using mere tricks of light and shadow to help design and construct it. Validating to us all in this 'crazy' unofficial 2 faced NASA public announcements for the Face, which *they know* they cannot even hide this information and image data, yet officially on the surface, it seems that is their true purpose and intent to continue to do. To HIDE IT!

It is obvious to me that NASA knew the public was not fully bought into their official side of this story, denying and debunking the Face with scripted wordings .When the first announcement in 1976 was scripted and fed through a 'policy' of public dialog as a means to deliberately lie about the feature, and play it down by calling it a Mere Trick of Light and Shadow. Only now, Down Through Time, we can see what that was really about.

The evidence grows for this as a real possibility as I have laid out already the basics of how to see this from a 2 faced coded perspective. It is not a coincidence! We also can see how NASA keeps officially despite the official 'policy' treatment for the feature, keeps on creating 'titles' to their continued public announcements which despite their official position, they keep acknowledging the feature as '***THE FACE***' in their public announcements!

The April 1998 image release officially calling it '**MARS ORBITER CAMERA VIEWS** '***THE FACE ON MARS***'.

The January 31 2001 partial Face image release calling it **'FACE TO FACE, WITH '*THE FACE*'**!

I found this to be very hypocritical for NASA to keep calling something that *'THEY"* officially not want anyone to consider this feature as an intelligent design techno-signature, and wanted it to simply disappear, and then just to give an official title for the feature that was trying to be hid and covered up, with a title called *'THE FACE ON MARS'*? To draw attention of the people to it even more as time progresses, just to lead the people to see how they officially handled it? To expose themselves as deliberately lying and trying to cover this up? It not makes any sense. Unless, we consider the 2 faced 'policy' of coded information we are laying out here and learning to see.

Once again as time went by after the January 31, 2001 partial face 'PEEK A BOO' **'FACE TO FACE, WITH '*THE FACE*'** tease image of M16-00184, we were once again given yet another official Face image release only 4 months later, on May 24, 2001once again titled: **'Highest-Resolution View of "Face on Mars"** The image appears to have been taken during April 2001, ***and was held for release until May 24th 2001***. One may ask why? We will try to see that answer later.

'Highest-Resolution View of "Face on Mars"

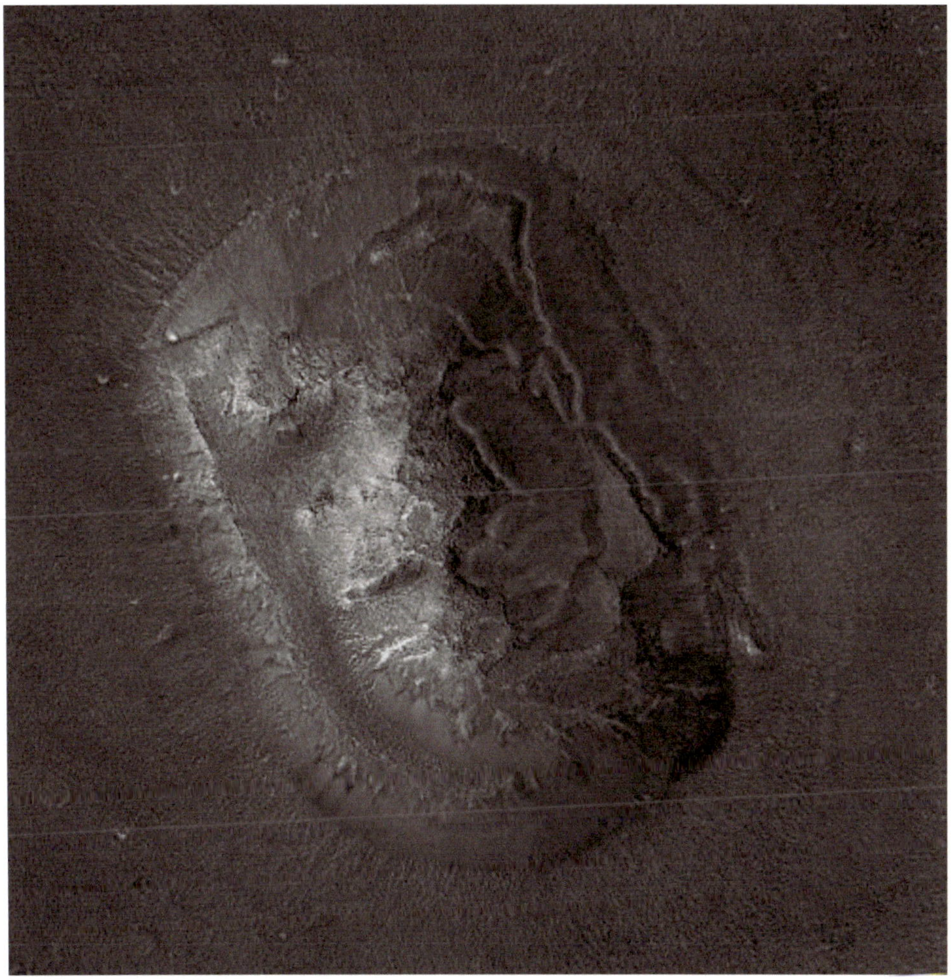

(MAY 24, 2001 MSSS FACE IMAGE RELEASE E03-00824)
https://mars.jpl.nasa.gov/mgs/msss/camera/images/moc_5_24_01/face/index.html

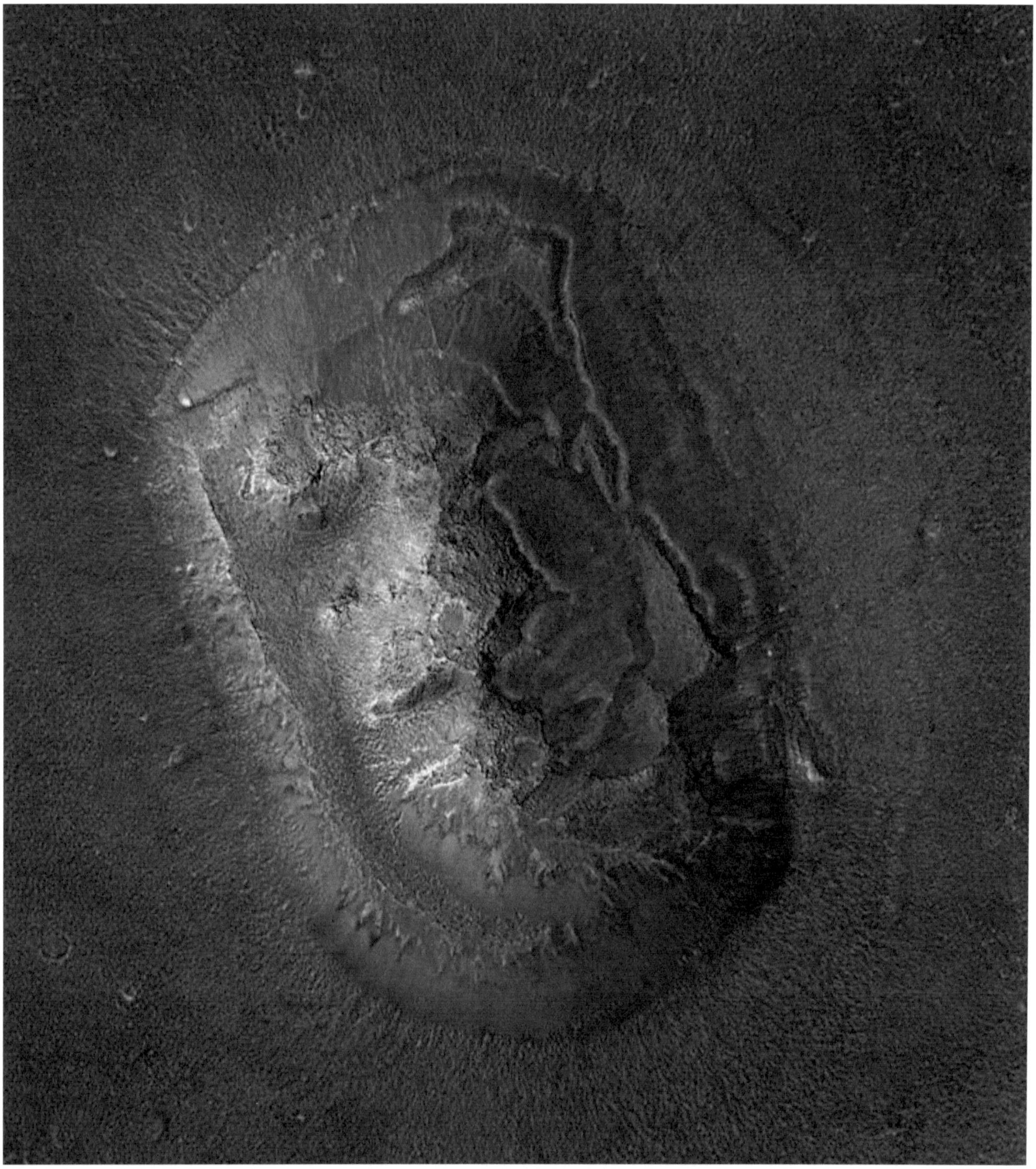

It is at this point where I would like to raise a scientific fact. For scientists to come to conclusions to base a theory as a scientific fact, it is well known that ***scientists usually require 3 sources,*** to base a claim of scientific fact to present as a scientific hypothesis as a working model of fact. So in this face of understanding, we now have 3 sources of which to base on the hypothesis of my 'crazy' theory of a NASA 2 Faced coded policy. Let us all look at them all together shall we?

1: The April 1998 image release officially calling it: 'MARS ORBITER CAMERA VIEWS '*THE FACE ON MARS*'.

2: The January 31 2001 partial Face image release calling it: 'FACE TO FACE, WITH '*THE FACE*'!

3: The May 24, 2001 Face image release titled: 'HIGHEST RESOLUTION VIEW OF "*FACE ON MARS*"!

So basing these facts on using science, I am proposing that this NASA 2 Faced 'policy' of coded public announcement scripts are being as an unofficial validation that my crazy theory here as being of a scientiffic fact! A working model, which I have just shown. Impossible as it should be and as 'crazy' as that theory sounds, there should be nothing at all. This in what your reading should not and would not even exist at all. Yet here we are, with a scientiffic model that is consistent because we not only have 3 sources of their coded public 'policy' scripting that allows us to source it as scientiffic fact.

But we must also consider something else just as amazing and also using the scientiffic model. As we also now have 3 images of 'THE FACE'. Since the original 35A72 public image release with the coded public 'policy' driven announcement telling us all 'THEY', NASA, called it 'HEAD' first, and then officially stated that it was a Mere Trick of Light and Shadow! That 'THEY', NASA, said a few hours after that image was taken, 'THEY' somehow took others, that shown the 'HEAD' feature was disappeared.They told us all It was gone! Yet remember 'THEY' never shown any proof of this obvious lie. Only proof exists that other images were aquired and given to us all that clearly show, the feature is not only still there, but it still looks like a face!

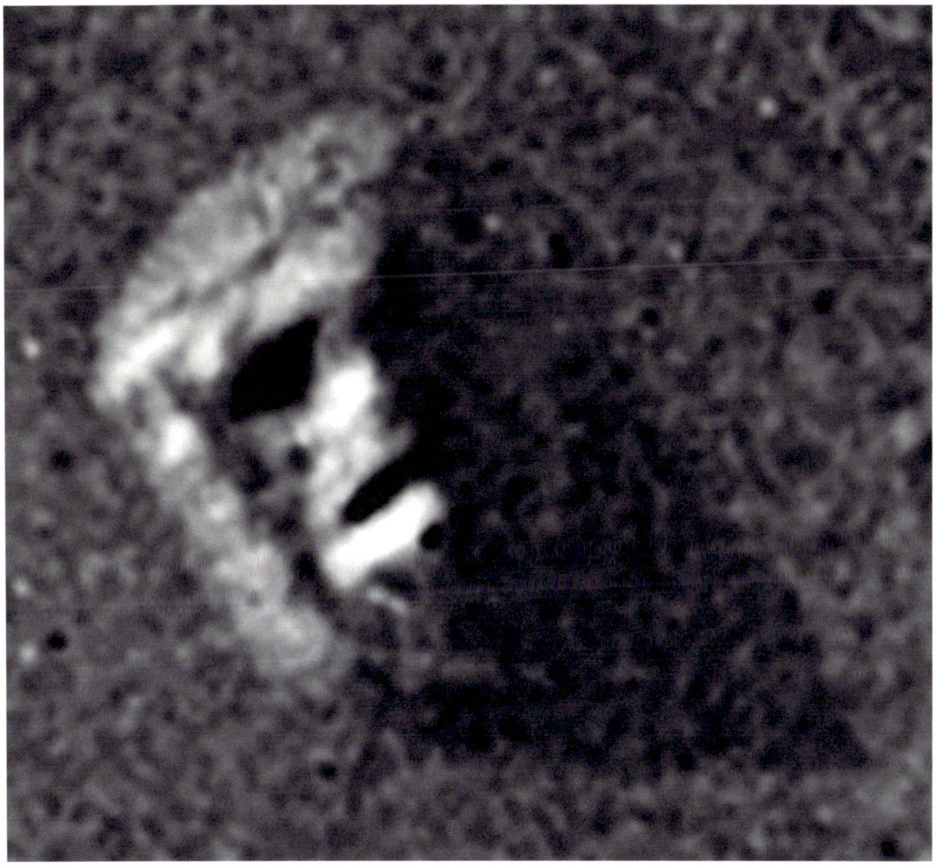

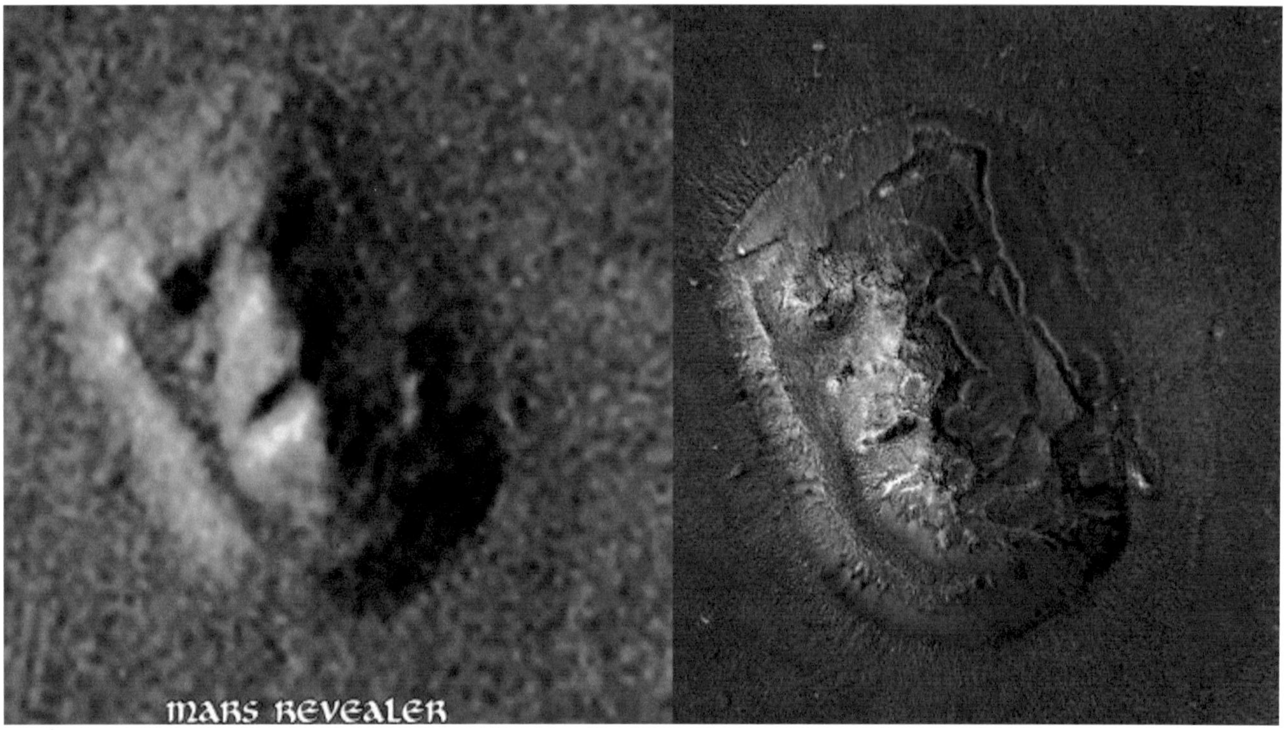

The release of **E03-00824** again, was on **May 24, 2001**. On the release page from MSSS, it ***states that on April 8, 2001, was the first opportunity since April 1998 arose to turn the spacecraft and point the MOC at the popular "Face on Mars" feature. The Viking Orbiter 'images' acquired in 1976 showed that one of thousands of buttes, mesas, ridges, and knobs*** in the transition zone between the cratered uplands of western Arabia Terra and the low, northern plains of Mars, ***looked somewhat like a human face***. The ***feature was subsequently popularized as a potential 'alien artifact' in books, tabloids, radio talk shows, television, and even a major motion picture. Given the popularity of this landform, a new high resolution view was targeted***. The resulting image has a resolution of about 6.6 feet per pixel. If present on Mars, objects the size of a typical passenger jet 747 airplane would be distinguishable in an image of this scale.

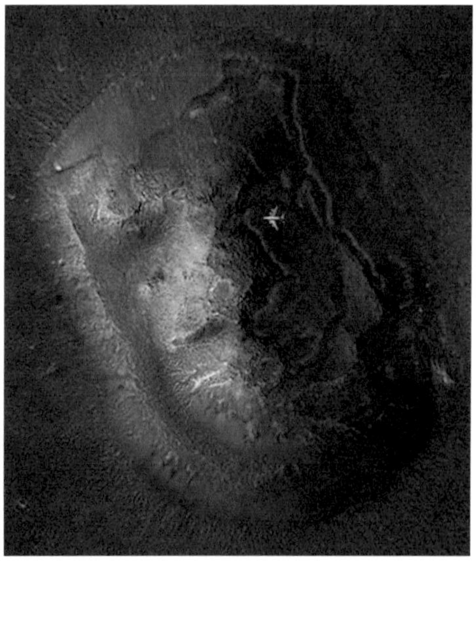

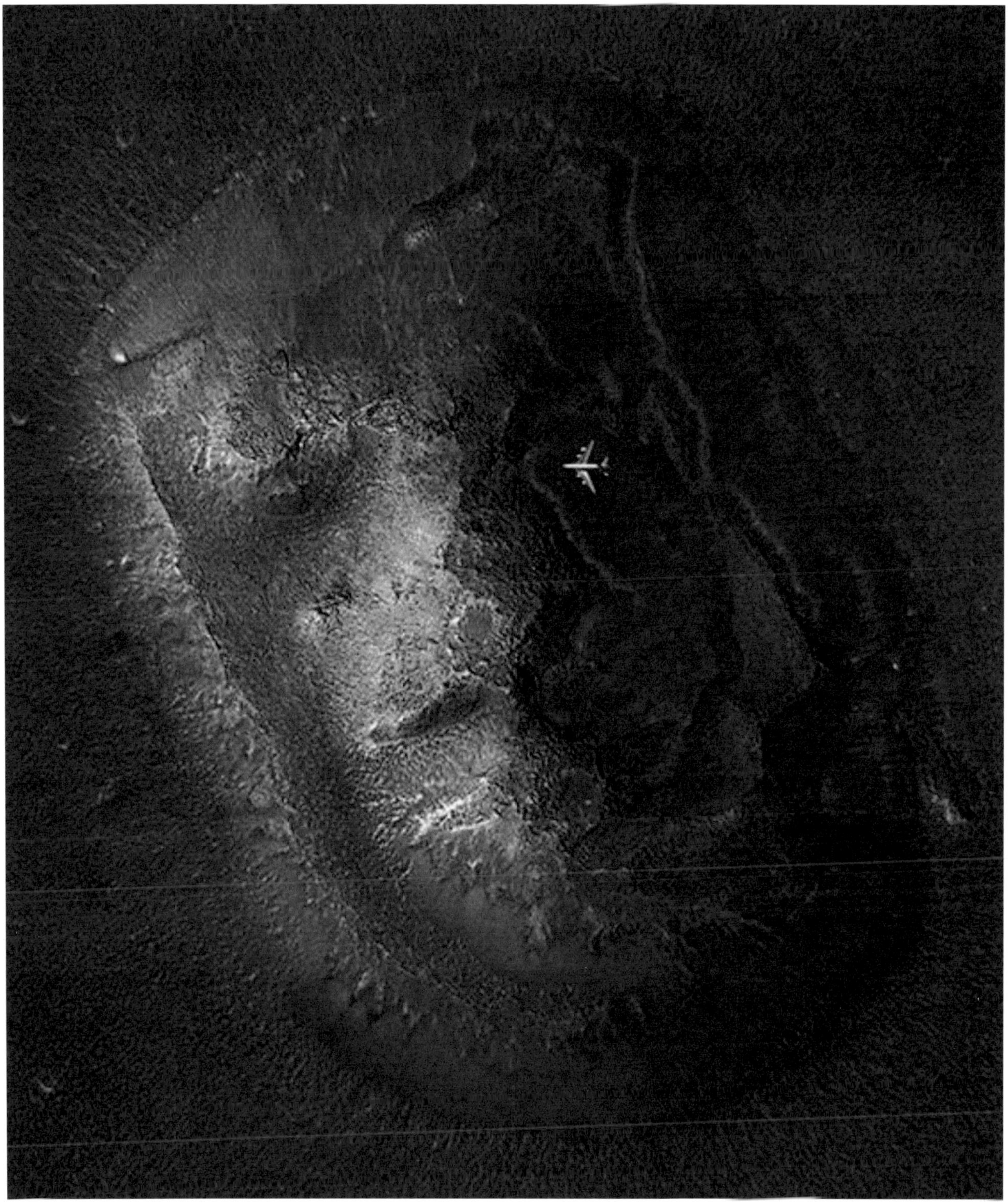

I have taken it upon myself to create an illustration to show how large the Face is compared to a 747 airplane and what it would look like as it is flying above it. Keeping in mind also that the Face is 6,067 feet across! The size of the airplane seen is 231 feet in length and the view is accurate in how it would look, if looking down and seeing it fly over the Face. I can only imagine the price for a window seat for that view, flying over the Eyes of the Face!

I would also like to point out another lie put forth in some of the wording from the E03-00824 full face image release page. Notice we are told that **on April 8, 2001, was the first opportunity since April 1998 arose to turn the spacecraft and point the MOC at the popular "Face on Mars" feature**. Again, NASA thinks or puts out the subliminal idea that we are all stupid. It seems NASA has forgotten the fact just 3 months previous, on January 31, 2001, we were just given the partial face release view of M16-00184. So it seems obvious to me that there was an opportunity to image the Face since April 1998. The only problem is how we were given the image of the Face for that obvious opportunity which was there on January 31st 2001. Instead of the full face view we are getting now in April 2001, only a couple months previous there was the opportunity to do the same, but NASA didn't take it. The opportunity was obviously there, but the camera eye was shifted from a full face view to that of a partial view. This also tells us that we exposed yet another lie in the words into the full face image release of E03-00824.

In the image release from January 31, 2001, it was stated also that the Mars Orbiter Camera (MOC) team has, 'whenever' the Mars Global Surveyor spacecraft has flown over the region of Cydonia, commanded the MOC to take a picture. This tells us all that the opportunity exists more often than not in taking images of the Face. This also shows us all that I was seemingly correct in my assessment at the end of chapter 3, where I had said that the image release for the partial view of the Face in image release of **'Face to Face With 'The Face'** of M16-00184 was a tease release. Where I said that something, in my mind, was coming. This full Face view image of E03-00824 is what I knew now is the something which was coming. This is also something that tells me and should tell us all that this full Face release of **April 8, 2001,** was indeed a pre-planned event.

I like you all to see the illustration comparative of the Face at Cydonia and the 747 airliner again. Realize that the Face is far larger than that of the airplane. The 747 is 231 feet in length. The wingspan is

261 feet in width. The Face is 1.63 miles in length, by just over 1.15 miles in width. Its height is 1500 feet high. This Face is not a small feature, nor one which would simply vanish away on its own.

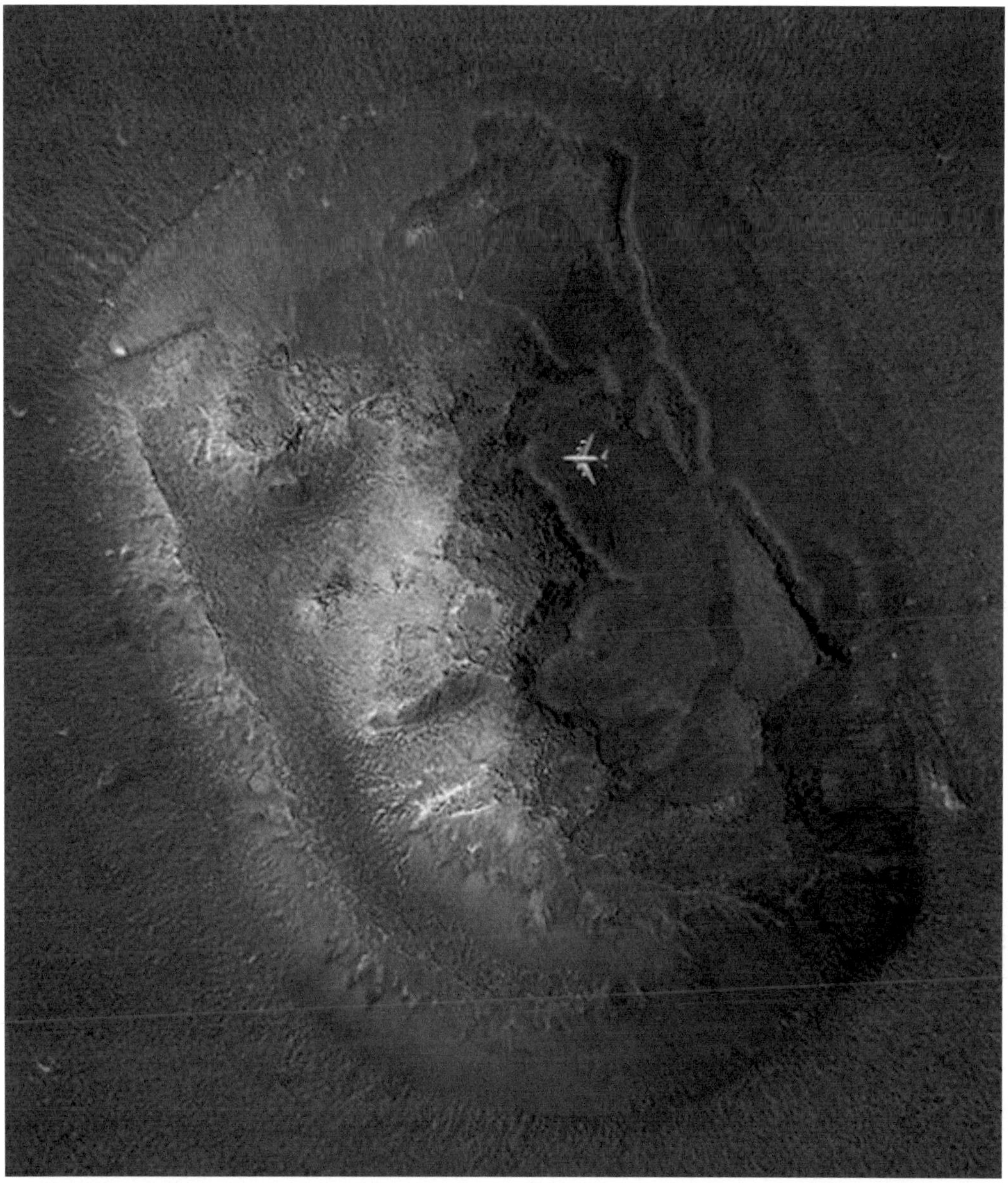

One would notice by the wording of the Face image release on **May 24, 2001**, it officially suggests a natural reasoning because we were told that this Face feature was just '***one of thousands of buttes, mesas, ridges, and knobs***'. In other words, the subtlety of the wording shows deliberate reasoning to have the intention once again publicly, to diminish the obvious importance and implications that this Face demands to be addressed with scientific seriousness, and not with the unscientific contempt that has been given unto it all these years. Forever staining what NASA in the minds of the people is supposed to represent. To look for life or past life and signs of life. The only advantage we have is that we were given this image data. We just have to study it for all of ourselves to determine the weight and the veracity of NASA's 'policy' driven public wordings, regards to the reality of the findings from those who study the data. Findings of which not matches the official policy driven agenda to try to keep making this

Face appear at least in the public minds, as being nothing more than a natural feature. No more trick of Light and Shadow? Maybe one cannot see as deeply into the MSSS Face release wording as I can and do, but there is yet more data to the public regards to this Face image release that supports my theory as being a fact!

Along with this MSSS Face image release, there was also a public official NASA news press hit piece put out for this MSSS Face release that shown a desperate attempt to really go out of its way to try to make you think it is a natural formation. It is also not so coincidental that this NASA news press hit piece was released 1 day before the MSSS Full Face image release. Why? Simply to steer the public away from seeing what NASA did not want you to see the next following day being released from MSSS (Malin Space Science Systems).

The title of the NASA news press release was extremely revealing for its intent and purpose.
Unmasking The Face On Mars: New high-resolution images and 3D altimetry from NASA's Mars Global Surveyor spacecraft reveal the Face on Mars for what it really is: *A MESA*!

I agree, it is a very long title. This also will take some time to look into this NASA official 'policy' driven news hit piece. As it shows right from the outset given to us all in the title where the public was to be lead, on an official fantasy filled hike up to the top of a MOLA hilled mesa!

The internet link for the '**Unmasking the Face on Mars**':

https://science.nasa.gov/science-news/science-at-nasa/2001/ast24may_1

If you go to that link on the internet and read it, you will indeed see that it was obviously another biased public dismaissal of the Face being artificial that obviously took some time to put together. So I want to look at the press statement in detail of what it says.

The official Unmasking The Face hit piece story continues:
May 24, 2001 -- Twenty five years ago something funny happened around Mars. NASA's Viking 1 spacecraft was circling the planet, snapping photos of possible landing sites for its sister ship Viking 2, when it spotted the shadowy likeness of a human face. An enormous head nearly two miles from end to end seemed to be staring back at the cameras from a region of the Red Planet called Cydonia.

Notice here in the opening storyline, it is detailing the history of the Face being found. Notice also how it was worded. '***The Shadowy Likeness Of A Human Face. An enormous head nearly two miles from end to end seemed to be staring back at the cameras from a region of the Red Planet called Cydonia***'.

This is actually a step up in their public comments, and has upgraded 'information' since the days from a mere trick of light and shadow. Notice the words do in fact acknowledge we are still yet being given coded information as well. NASA still officially states it as being a Face, despite the cover story which attempts to play it down on the surface claiming it instead as a natural feature. Now we are also given new coded information from the unofficial side of this deliberate writing. It is being reaffirmed to us all that we are being told that the Face, is '***AN ENOMROUS HEAD STARING BACK AT THE CAMERA, FROM A REGION CALLED CYDONIA***'!

Therefore in the sense it was hinting also to what I claim about the Face, that the eye of the Face would be watching you coming and going. This new information tells us all that NASA, noticed and noted also, that the Face is indeed ***AN ENORMOUS HEAD, STARING BACK AT THE CAMERA***! There can be no denying that the words are there written in their own 2 Faced 'policy' driven news hit piece.

One must admit that looking at it all in the Light I am trying to reflect to you, that it always appears as if all these 'policy' driven official wordings over the years, do indeed show contradictions to itself. In this light of understanding, in how I am trying to present this, as a code to be deciphered. we should be able to see that as strange as it all sounds, it actually all makes weird sense. Just entertain the thought if anything else.

But let us continue looking at the wording from the official NASA press release, as it continues to yield us all information again. We are looking at things in a way where we can look at the relevant subjects, and piece it all together.

The official Unmasking The Face hit piece story continues:

There must have been a degree of surprise among mission controllers back at the Jet Propulsion Lab when the face appeared on their monitors. But the sensation was short lived. Scientists figured it was just another Martian mesa, common enough around Cydonia, '*only this one had unusual shadows that made it look like an Egyptian Pharaoh*'. The original caption for 35A72 Face release in July 1976 noted a "huge rock formation ... which resembles a human head ... formed by shadows giving the illusion of eyes, nose and mouth."

Ok so we are told that when the Face appeared on the monitors at the JPL, in 1976, **'there must have been a degree of surprise among the mission controllers'.** One must wonder about the surprise they had. That surprise maybe being how the issue of the Face at Cydonia was taken over and being withheld from the public about it's reality, and and leaving the mission controllers surprised by what NASA were saying how to deal with this Head instead? To their surprise, NASA made it clear no one was or is to claim that the Face is an intelligent structure?

As nutty as this all sounds, just follow me here, because we are being told also that **THE SENSATION among the mission controllers WAS SHORT LIVED**! Which sensations? If this is indeed a NASA 2 Faced 'coded' unoffficial information we are being given, then we are being told that **the sensation, of the mission controllers** seeing the Face **on their monitors at the JPL, looking up at the camera**, revealing to the mission controllers they are seeing intelligence and intelligent design, was short lived!

Why? Because that 'sensation' among the mission controllers was officially killed fast. Therefore, what or who represents the **degree of surprise**, who killed their sensations among all the misson controllers of seeing the Face for what it literally appeard to be to them? An intelligently created techno-signature structure. Did they get the 3rd Degree about it, and that really surprised them? Being it was NASA who enforced this demand? It does makes sense in all this.

It is only by and because of the 'official' wording in which this theory is able to even be seen, so it is not something I am just making up. I am just putting forth the data and breaking it down with you. I was as surprised as you are regarding considering these possibilities!

So after the mission controllers original sensations of the Face were 'short lived', we are all being told we are to believe **the scientists figured it was just another Martian mesa**? I will only say that is a great leap backward from the sensations the mission controllers originally had, in seeing the intelligence and the artificiality of the Face staring literally at the camera watching the Viking Orbiter as it passed overhead.

It is at this point where it seems we now know what '**THE DEGREE OF SURPRISE**' stands also for and represents, the WHO that killed the mission controllers sensations. **The SCIENTISTS! The DEGREE!** Get it? Scientists have degrees. It is and was 'THEY', THE SCIENTISTS, who had THE DEGREES, that SURPRISED the mission controllers who had denied their own eyes in what was so obvious and easy to see, and adding to the mission controllers degree of surprise to only hear and see that instead of the scientists validating the obvious, they instead officially claimed that they figured that the Face was an ordinary mesa instead.

I want to point out to you all to notice by their official words, NASA exposes themselves again to the original lie from 1976 that after the 35A72 image of the Head was taken, NASA said that a few hours later another image was taken and the feature was gone. Notice that lie never gets addressed by NASA again! Their own words itself also proves the feature is obviously still there. This tells us all that there must have been infighting between the mission controllers and the 'degreed scientists' that surprised them, in what was to be wanted to be told to the public. Which instead of the NASA scientists telling us the truth, the NASA scientists instead tried to outright lie to cover it up, and NASA also continues to mislead us all to the reality of the Face at Cydonia to this very day, in regards to the public about what 'THEY', the mission controllers, and the NASA scientists, had really found back then.

We are told again in the official NASA press release for **Unmasking the Face on Mars**, where their own official words about the Face release in 1976 state:

"This one had unusual shadows that made it look like an Egyptian Pharaoh. Which resembles a human head ... formed by shadows giving the illusion of eyes, nose and mouth. The authors reasoned it would be a good way to engage the public and attract attention to Mars."

(I agree, I see in my mind the authors, had reasoned it would be a good way to engage the public, via their creepy official and unofficial 2 faced 'policy' public information agenda script, by showing and telling the public the truth! Therefore now, in time, after this becomes seen, is meant to then attract the public attention to Mars now, after we Unmask its code of the Face on Mars down through time, to reveal this truth. Even truths about themselves and truths about how the mission controllers and the public were handled regards to this Face back then, because only now, as we '**Unmask the Face on Mars**', and put all these pieces together can the light of the truth be seen down through time.)

It was not the misssion controllers whose surprises were short lived that publicly changed their minds about what they knewthat they were seeing, as they did not change their minds at all. It was indeed a Face. That is why they were surprised by the degreed scientists officials. The mission controllers figuring the obvious, and then the handling of what they officially claimed to the public as being named HEAD, had then switched to what the degreed scientists wanted to say the Face really is. Just another Martian mesa! The mission controllers were astounded by it all in other words. As it was also made the official NASA position that the Face is a mesa.

The mission controllers sensations for the Face looking up at the camera in what they were easily able to see and to ascertain, was that they all knew that they were seeing intelligent design. So then we are told that it was the scientists, and not the mission controllers themselves, who are the ones who made all the mission controllers sensations about the Face, to be short lived. To kill their sensations, to destroy their positions as well, if the mission controllers spoke out in defence of themselves and defence of artificiality of the Face. One may ask then why wouldn't anyone of the mission controllers speak out against the official degreed NASA scientists position and declaration that they obviously knew was incorrect? I am sure that not one of them wished to lose their jobs. Not to also mention any papers that needed to be signed in order for the mission controllers to do their jobs, that prevents them from saying anything at all officially or otherwise.

The scientists **figured it was to be officially publicly known as being just another Martian mesa.** We are indeed **UNMASKING,... THE FACE ON MARS**!

It seems I was right in also claiming that evidence would show if my crazy theory is correct, about there being a NASA 2 Faced official and unofficial 'policy' of given information, because it seems this very image release and official wording was very connecting to what we are doing literally right now, which is Unmasking The Face of Mars. We are Unmasking the unofficial side of the official story of the Face on Mars. Only in this light, can we see the dual sided nature of their deliberate wording that contradicts their official positions on the surface. Again why keep contradicting their own position, while at the same time keep officially calling the Face on Mars, THE FACE ON MARS?

I think this 'policy' driven code of THE FACE ON MARS, is now UNMASKED! As was it's true intentions?!

Therefore this is all just being a type of intelligent code to which was scripted by the 'authors' for the official NASA press releases, and intended for it to be seen and figured out one day. But not for at that time, and then later in time intended to be seen and to be figured out and to be pieced together? Again, as crazy this all sounds, this actually makes sense. It is actually what we are all doing now isn't it? Unmasking the code of **The Face Of Mars**? One must admit that it would and does answer for the contradictions in their official claims and misleadings and NASA's own 2 faced nature about them all. Which also all connect to each other in the same manner reaffirming itself. With each official release yielding new information, over time, to be seen and put all together like how we are doing now. By digging deep into their script the secret authors scripted for NASA's official press releases.

WHY? Well we were told that **the authors 'reasoned' it would be a good way to engage the public and attract attention to Mars,** once this subtle code is seen at a later date, and understood and most importantly, revealed.

The official Unmasking the Face hit piece story continues:

"Some people think the Face is *bona fide* evidence of life on Mars -- evidence that NASA would rather hide, say conspiracy theorists. Meanwhile, defenders of the NASA budget wish there *was* an ancient civilization on Mars."

This is interesting, it seems that we are being given information of who it was that was responsible for the NASA scientists figuring that the Face was just another Martian mesa. We are told that **THE DEFENDERS OF THE NASA BUDGET '*WISH*', THERE WAS AN ANCIENT CIVILIZATION ON MARS.**

We are being clued into the fact that the '**Defenders of the NASA Budget**' are the ones who made the NASA scientists officially claim that the Face was just another Martian mesa. Telling us all that the NASA Budget Defenders **wish** that **there was an ancient civilization on Mars**, because they now see and know that they are 'faced' with one being there now!? THEY', the Defenders of the NASA Budget, apparently not wishes to see it officially, and therefore not really wants any of us all either to see it. Yet we are being given coded information. It seems someone, inside NASA, wants us to see and to know this information as well.

Therefore **THAT**, is **the evidence** that **conspiracy theorists** (*obviously me in this light of understanding*) say that NASA scientists and obviously the defenders of their budget, would rather hide this all. **To MASK THE FACE OF MARS**! Not that there **was** an ancient civilization found on Mars that **they** are trying to hide from us all, but that there currently is an ancient civilization found on Mars that is still there now? Of course this fact would bother them so much, that **they** would wish that there **was an ancient civilization on Mars**. It is comical in a way.

It was a reality that the mission controllers had seen the obvious when the Face was first imaged in 1976, telling us all that they were surprised by the degreed NASA scientists denying and officially claiming that the Face was just another Martian mesa. We now know why, that it was because of the NASA budget defenders, which obviously must be seen as being only **the U.S. government**, who wanted it that way!

The official Unmasking The Face hit piece story continues:

Although few scientists believed the Face was an alien artifact, photographing Cydonia became a priority for NASA when Mars Global Surveyor (MGS) arrived at the Red Planet in Sept. 1997, eighteen long years after the Viking missions ended. "We felt this was important to taxpayers," explained Jim Garvin, chief scientist for NASA's Mars Exploration Program.

"We photographed the Face as soon as we could get a good shot at it." .....

"Revealing ... a natural landform. There was no alien monument after all."

It is interesting here to point out that we are yet being given more intelligence in the wording. We are being told that although how the Face was officially handled and labeled, by the NASA scientists claiming that it was just another normal Martian Mesa, we are being told now behind the scenes that a **few scientists believed the Face was an alien artifact!** So not all of the NASA scientists bought into the claiming of the Face at Cydonia was just as being a normal Martian mesa. Well at least that is some good news right?

We already know why though that the opinions not counted for the scientists who seen the Face for what it really is, an alien artifact! The NASA budget Defenders. The United States government! The Face we are told, **became a priority for NASA!? NASA *FELT THIS WAS IMPORTANT TO TAXPAYERS***!? *To Mask the Face*! Yet it is the taxpayers who pay the NASA Budget Defenders, representing the taxpayers whose interest in the Face demand answers and images, are who make the NASA scientists officially want you all to think that the Face is just another normal Martian mesa! While to them, the NASA Budget Defenders, it was a priority to them to photograph the Face at Cydonia, because they see what is really there and not want you all to know that the Face is a priority to them! But the higher priority regarding the public, to them was to again, *MASK THE FACE,* because of the taxpayers continued interest despite NASA's official positions on the matter! They want this subject put to bed!

NASA officially wants you to think and believe that from the images they have taken, that the Face at Cydonia is not showing an artificial alien artifact, but instead is revealing a natural landform!

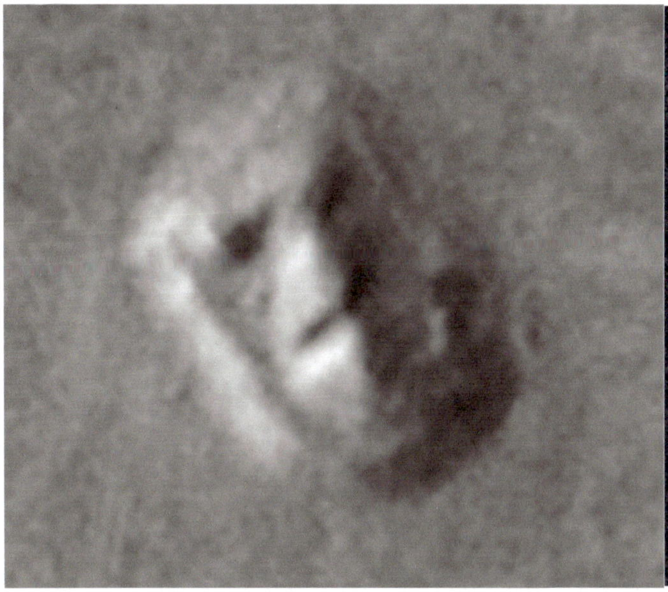

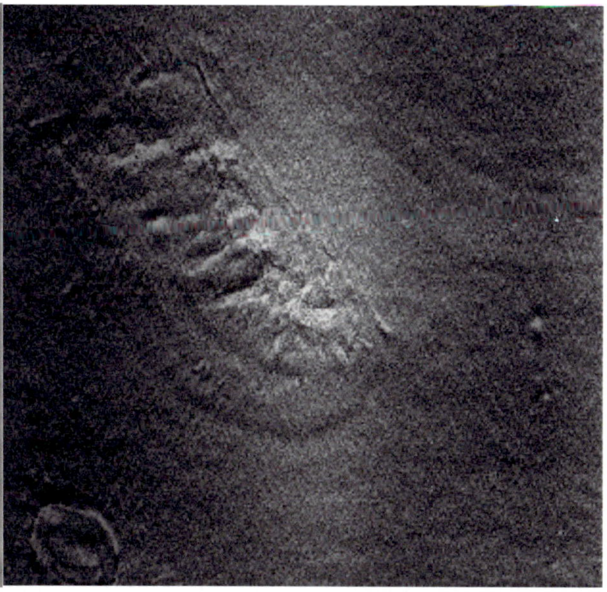

The official Unmasking the Face hit piece story continues:

"But not everyone was satisfied. The Face on Mars is located at 41 degrees north Martian latitude where it was winter in April '98 -- a cloudy time of year on the Red Planet. The camera on board MGS had to peer through wispy clouds to see the Face. Perhaps, said skeptics, alien markings were hidden by haze."

Mission controllers prepared to look again. "It's not easy to target Cydonia," says Garvin. "In fact, it's hard work." Mars Global Surveyor is a mapping spacecraft that normally looks straight down and scans the planet like a fax machine in narrow 2.5 km-wide strips. "We just don't pass over the Face very often," he noted.

Nevertheless, on April 8, 2001 -- a cloudless summer day in Cydonia -- Mars Global Surveyor drew close enough for a second look. "We had to roll the spacecraft 25 degrees to center the Face in the field of view," said Garvin. "Malin's team captured an extraordinary photo using the camera's absolute maximum resolution." Each pixel in the 2001 image spans 1.56 meters, compared to 43 meters per pixel in the best 1976 Viking photo."

This is also correct. *Not everyone was satisfied* by the official NASA proclamation of 'policy' and hidden agenda. In other words, their propaganda. We the People put pressure upon NASA to re-photograph the Face. There is no mistaking that fact. Notice that it was the official NASA excuse which was the reason to take the image and provide it to us. Yet it really was to cover for the fact that the Face at Cydonia was really a priority to the US government to try to prevent the Face being seen by the public. It also is possible that when the April 1998 Face image release of **SP1-22003** was taken, it was done as a priority of

NASA and the NASA budget Defenders, the U.S. Government, to present the image data taken of the Face to present misconstrued image data to a mainstream public to try to debunk the Face a to try to fool us all and try to convince us the image shown a natural landform? In essence once again trying to say that once again, like the official statements for the Viking image 35A72, that the Face was not being seen when taken a few hours later that there was no face being seen. Yet NASA never shown that image, and just wanted us all to believe them. Telling us all that there was NO an alien monument after all? Yet as we already went over, Vince DiPietro and Greg Molenaar found the PROOF image, 70A13, found in a misplaced file, and shows that it was taken months after the 35A72 image. We also seen that yes, the Face was still seen there. Obviously proving NASA had lied and had made official statements without showing the data or the science that show as proof that the Face was not there anymore. To me, NASA did more to make the Face harder to see in the official public image data releases since the Viking Orbiter images of 1976 and of April 1998, and January 31st, and **tried to hide its alien markings** and claim that ***the Face Was NO Alien Monument After All***. Talk about adding insult to injury!

Their public excuse for the horrible presented April 1998 image of **SP1-22003,** was done so *with an intent* to *NOT provide us all with an accurate image* of the Face to begin with. Notice we are told specifically that ***"The camera on board MGS had to peer through wispy clouds to see the Face!"*** That is because NASA's policy given by the U.S. Government had wanted it to be that way to begin with. To look through the Mars Orbiter Camera (MOC), until they can acquire an image that makes it hard to see the Face. The reason for this was that obviously each time they seen the Face to take an official public release image, NASA and the U.S Government still seen the Face! So their hopes to finally debunk this Face structure tabloid icon to the public and to save FACE was by using image data showing that the Face is not really there. It was determined, that the only hope they had to try again to unscientifically debunk the Face and to convince the public that the feature is natural, was by getting an image taken while the Face is viewed at an oblique angle (off Nadir) view. Talk about continued desperate attempts to make the Face disappear by taking another image, to try to prove NASA not lied to us back in 1976 about their public position of taking another Viking Orbiter image a few hours after 35A72 was taken and saying an image was taken, and it shown and proved that the Face was not there and was no longer able to be seen!

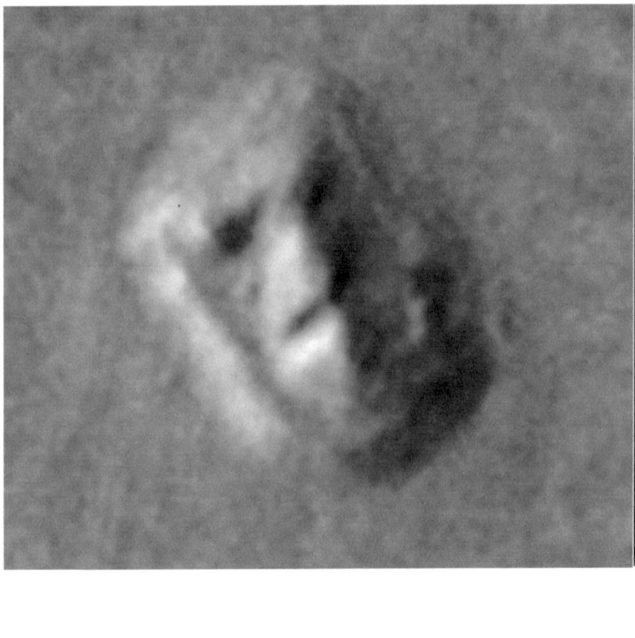

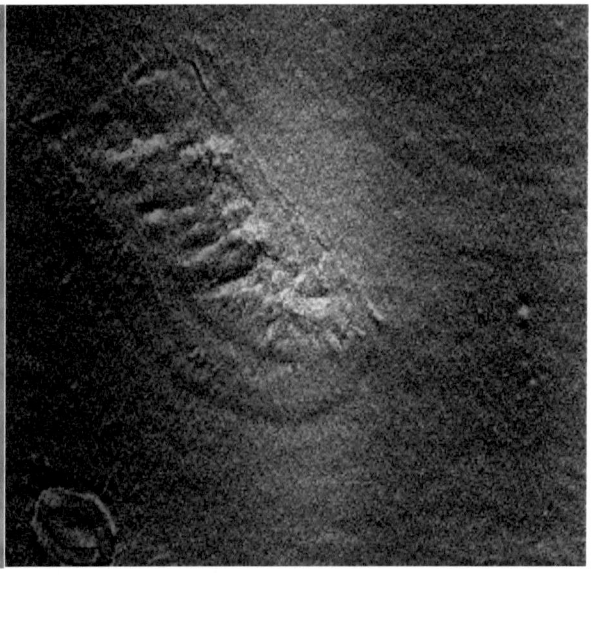

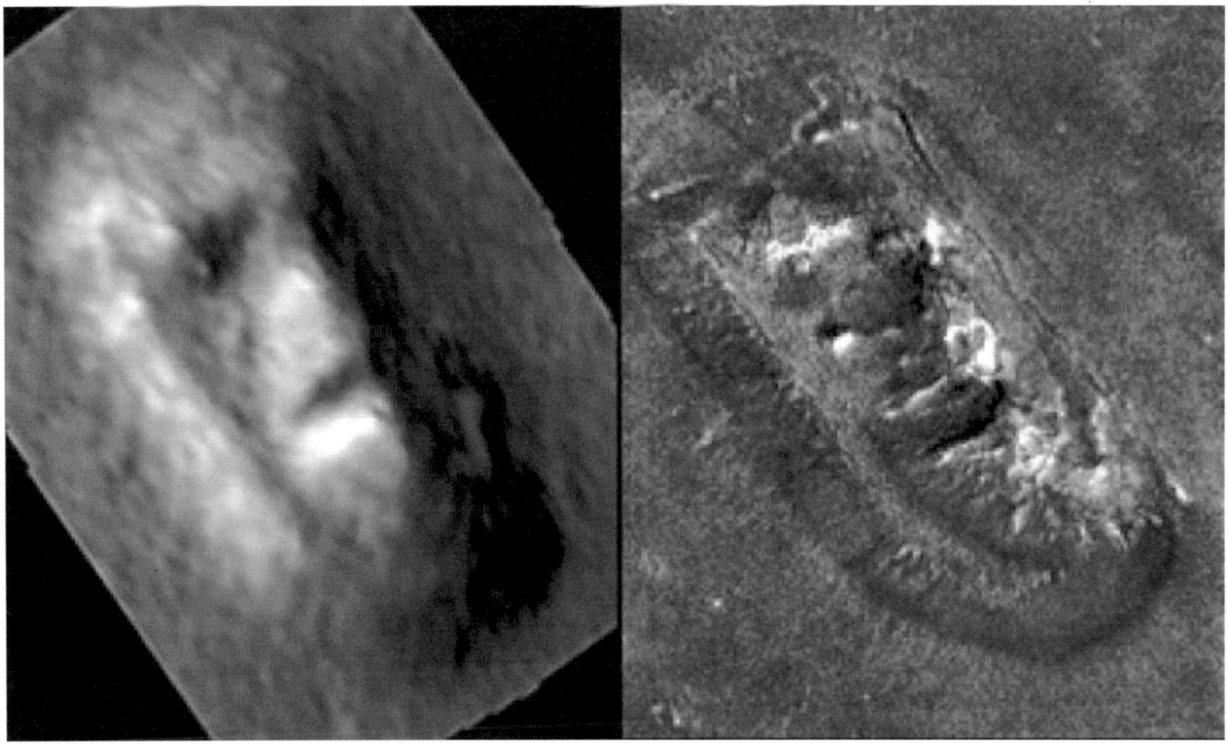

(Off nadir view with Ortho-Rectified comparison by Dr. Mark Carlotto)

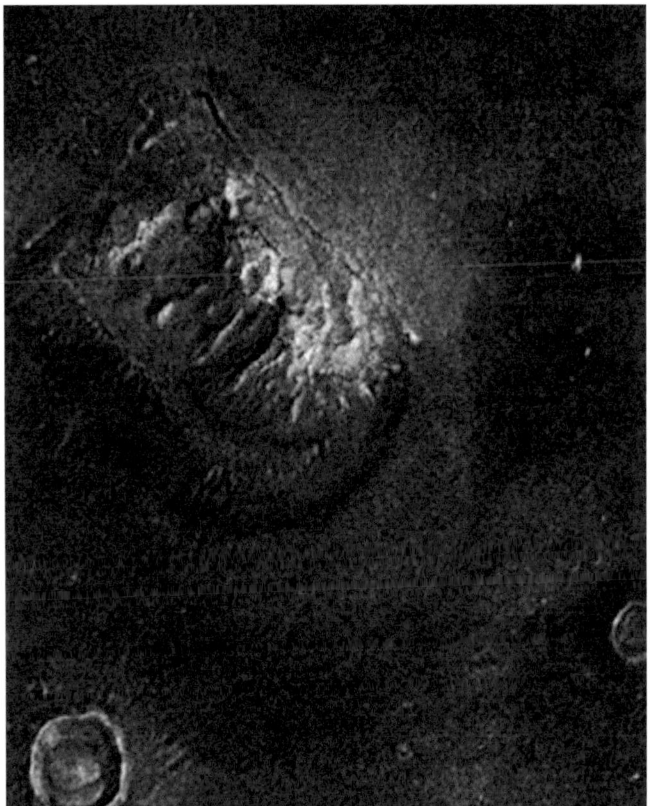

It is rather comical on the surface of their official story statement and also revealing in code, that all I am laying out for you here is true. Notice that we are also told: ***"Perhaps, said skeptics, alien markings were hidden by haze."*** Let us look closer to this statement, as it is a coded riddle. That allows us to yield information. Looking deeply into it is obviously needed. What is important here, is that we have another source, (*as scientists usually require three sources*) we have consistency so far of what I am laying out here for you, and is therefore, scientifically plausible despite this sounding so crazy. That there is a hidden code in all these public statements by NASA over the years of the Face at Cydonia. That allows us to come to see that what I am trying to show here now, is accurate and also, the reality! The truth coming as from an uneasily seen and in an UN-OFFICIAL 2 faced sides of position of this information given unto the public. Let's again, look at this official statement:

"The camera on board MGS had to peer through wispy clouds to see the Face. Perhaps, said skeptics, alien markings were hidden by haze."

After considering all that we have read and went over and are being led to discover, we should all see the picture that is being painted about the official and un-official sides to this bizarre story. After considering that NASA's official statements about the Face image release of April 1998, was being explained to us all under the surface of this coded information explaining that it was and is the priority of the U.S. Budget Defenders, to continue the policy of hiding the Face from the public, and prioritizing NASA to image the Face and make a public release using deceptive actions to try to end the public interest for the Face in unscientific ways. This was and is a censoring policy priority. *The Camera Had To Peer Through Wispy Clouds To See The Face.*

Did you catch it? ***THE CAMERA... ON BOARD MGS, HAD TO PEER THROUGH WISSPY CLOUDS, TO IMAGE THE FACE!!!!*** That was a priority order given to NASA, by the U.S. Budget Defenders! We are being passed secret information given unto us all in an un-official way, via its coded scripts! Therefore it is funny, once we see it in this way, what is said next is a clarifying and vindicating statement:

"Perhaps, said skeptics, alien markings were hidden by haze." Incredible! We are being tested to make sure we see its true meaning of information we are being told. We are to think now, considering what we are being steered to see, is a question being posed upon us all, that we can answer with a certainty. We are asked ***PERHAPS?.... SAID SKEPTICS, ALIEN MARKINGS??,... WERE HIDDEN BY HAZE?*** Yes, the wispy clouds representing the haze part of the statement, making it hard to see efficiently to really be able to determine what the Face is. You may be the sceptics considering that by NASA prioritizing imaging a public release image to debunk the Face by taking an image at an oblique angle, taken through wispy clouds in order to hide ***what***?? ***ALIEN MARKINGS***! There is so much information being given unto us all in this statement alone! We are being told in this 2 Faced coded official statement, ***by James Garvin***, that NASA and the U/S. Budget Defenders also acknowledge that their priority policy was and is to hide the Face and the fact and reality is its ***ALIEN MARKINGS***?! This in itself is a disclosure point of perspective if we consider this 2 faced code seriously. Let's look what it says next to end this test to see it and understand we are solving its riddle of disclosure:

PERHAPS?.... SAID SKEPTICS, ALIEN MARKINGS??,... WERE HIDDEN BY HAZE?

INCREDIBLE! So ***PERHAPS***? The ***ALIEN MARKINGS***, ***WERE HIDDEN BY HAZE***? Yes as stated in the official statement we see and understand that the U.S. NASA Budget Defenders prioritized NASA to take an image to use as an official desperate attempt to continue hiding the Face and it's reality from us all, by taking it at an oblique angle, and through wispy clouds as a haze, to do what? ***SAID SCEPTICS***, *WHICH IS YOU*, **TO *HIDE ALIEN MARKINGS*** ! You're being tested to see and understand the message that is trying to be told and shown here to you in this coded script! YOU are making the conscious opinion as you read this as to what the answer is, even as it is being given unto you, as scripted to do! I can also prove this all to you be providing the definition of: **HAZE**! Which also provides more information attributing and connecting to this all, and making sense to what I see and am showing you all here now. Let us look at the definition of:

haze [1]

(hāz)
n.
1.
a. Atmospheric moisture, dust, smoke, and vapor that diminishes visibility.
b. A partially opaque covering: *Let the polish dry to a haze before buffing it.*
2. A vague or confused state of mind.
intr.v. **hazed**, **haz·ing**, **haz·es**
To become misty or hazy; blur.

haze [2]

(hāz)
tr.v. **hazed**, **haz·ing**, **haz·es**
1. To persecute or harass with meaningless, difficult, or humiliating tasks.
2. To initiate, as into a college fraternity, by exacting humiliating performances from or playing rough practical joke supon.

So when we consider all we were told, and broken down and figured out? When we see that it says: ***PERHAPS,... SAID SKEPTICS, ALIEN MARKINGS??,... WERE HIDDEN BY HAZE?*** We can see that the added information of haze indeed does clarify itself, and to us with this information being true. Especially in regard to the second part of the HAZE definition!

1: To persecute or harass with meaningless, difficult, or humiliating tasks.

(To me this represents affirmation of what we were told. That NASA had to prioritize taking the image at an angle, through a haze of wispy clouds, which is true, to try to keep the Alien Markings HIDDEN from the public. Telling us all that this was to continue the persecution of those who wanted to see and believe the Face was and is an *Alien* created structure with *Alien Markings*! Telling us also that this was done ***to HARASS WITH MEANINGLESS, DIFFICULT, OR HUMILIATING TASKS?*** Thee attempts were to be *meaningless* to science and to reach any conclusion regards to the public of the official April 1998 public release and official statements to debunk the Face. Telling us the task was difficult, TO HIDE THE FACE AND ITS *ALIEN MARKINGS*, and to try to fool the public. Telling us also, confirming information that all thee desperate attempts by NASA and the NASA U.S. Budget Defenders are really humiliating tasks, taken by THEM, to continue to try to hide the Face with its *Alien Markings*, and continued censoring policy of trying to lie to and fool the public! Humiliating themselves by these task done to continue the LIE that the Face was imaged a few hours after 35A72, and saying it disappeared. Trying to continue to SAVE FACE, while yet continuing on with their humiliating attempts to continue to lie again to u all saying that the Face is not an alien creation that has ALIEN MARKINGS that they want to keep hidden from us all. Do I get at least a B+, in breaking this information down and presenting it to you?

2:

To initiate, as into a college fraternity, by exacting humiliating performances from or playing rough practical jokes upon.

However, there is yet more information being presented to us all also to consider to add onto the information we connected to see, *Through* ***A Vague Or Confused state Of Mind*** which we all have now. The second part of the continued definition of Haze, hints to us that this process of what we are doing to try to see the hidden meanings of this code possibly? *Is meaning as a means as* ***To Initiate****?* ***As Into A College Fraternity****?* It seems we are being told that we are being initiated into some form of *College Fraternity*, of which we are learning how to SEE this hidden language communication that is shared by some form of Fraternity?! You must admit it, that it is interesting to see things this way as we are considering something that sounds so crazy. It seems to be explaining itself, knowing somehow, that we are trying to reason with these ideas, as if to help make it seem plausible and factual. (***Just make note, that I have shivers running down my spine as I write this now***.)

Yet still more information is to be gleaned in this information being given to us. Now it seems as if this script is affirming the process upon which this **HAZE** was done unto us. We are being told what we already knew all along! That these public Face releases and NASA's desperate attempts priority of the NASA U.S. Budget Defenders to debunk and hide the Face along with its Alien Markings, (which all is designed and carved with the inclusions of using *Tricks of Light and Shadow*.) By ***Exacting Humiliating Performances*** of their deceptive acts and actions upon we the people, and/**or from** ***Playing Rough Practical Jokes Upon*** we the people, as if it were our faults that they had to humiliate themselves in their desperation to suppress the Face at Cydonia from the public and scientific mainstream of its reality of there being a Face with Alien Markings covering it!

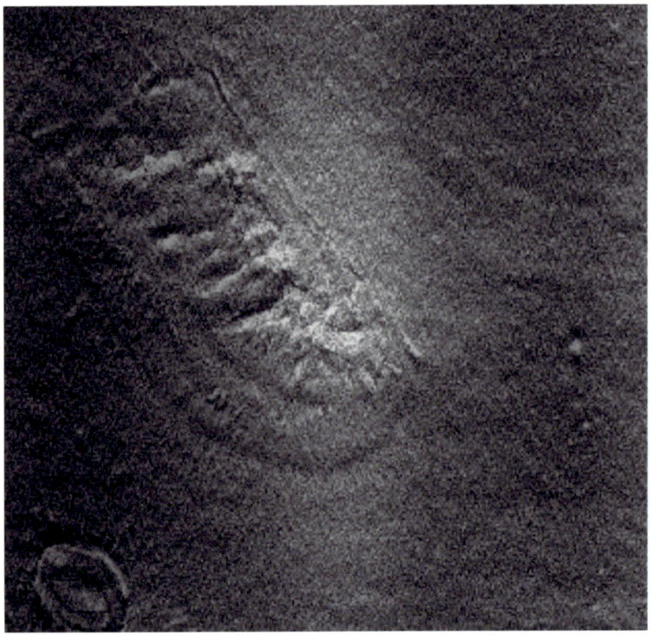

Mission controllers prepared to look again. "It's not easy to target Cydonia," says Garvin. "In fact, it's hard work." Mars Global Surveyor is a mapping spacecraft that normally looks straight down and scans the planet like a fax machine in narrow 2.5 km-wide strips. "We just don't pass over the Face very often," he noted.

Nevertheless, on April 8, 2001 -- a cloudless summer day in Cydonia -- Mars Global Surveyor drew close enough for a second look. "We had to roll the spacecraft 25 degrees to center the Face in the field of view," said Garvin. "Malin's team captured an extraordinary photo using the camera's absolute maximum resolution." Each pixel in the 2001 image spans 1.56 meters, compared to 43 meters per pixel in the best 1976 Viking photo."

As we go back to ***The official Unmasking The Face*** hit piece's official statements, we are then told **Mission controllers prepared to look again. "It's not easy to target Cydonia. In fact it's hard work. We just don't pass over the Face very often."** The excuse it seems is being given to us as to why we have not received any images of the Face prior to this new Face public release for the full Face view of image **E03-00824** on April 8th 2001, and also why we did not receive thi Full fac view after it was taken, ***was because it was held from the public, and released unto the public as soon as this official NASA hit piece 'was ready' to publish it a month after the image was taken***. NASA sat on the image, for the time which was needed for the official 'policy' to figure out how to handle the public release of the data and for them to put their official controlled scripted and coded spin on it at the same time.

I am guessing that by Jim Garvin's excuse trying to make excuses for why we not received an imagee of the Face sooner was because how hard it is, and not so often for the Face to be able to be imaged, but I am betting they actually were able to take more images, and never shown the public, but the US government were seeing the images that we never get to see. As the images were being gathered and were being seen, the Face was obviously still being seen, and so they had to fulfill public demand and make a public release of this new full Face image of **E03-00824.** Although being forced to provide the image, they had to come up with a story to which they can try to put this subject to rest in the eyes of the public who were not so gullible as compared to those who believe in their lies about the Face, because to many, they figure why would NASA lie? Despite even what their own eyes can see in the image data, but would never admit to! Their jobs were at stake! The public Face image release was now done to continue a 'policy' to attempt to silence those in the public and certain scientists, who still see and claim that the the Face really is an alien artifact. Despite the fact of the US government's priority in the Face, which was and still is to keep this truth from the public, and their priority was and still is, to keep this Face a secret.

So it is no surprise to see the ultimate goal in the NASA **Unmasking The Face** on Mars hit piece was to provide a fake history for the Face as well as misleading you all to try to and to accept to see what they propagandize and want you to see the Face really is, which is their public official position is that the Face is just an ordinary Martian mesa. Not for what it really shows, which is an alien artifact, and the U.S. government wants NASA to steer you all to think how they want you to think, by trying to convince you that it is something else, and you are to think of it is as being all natural! Again, despite your own eyes and analytical prowess, and despite what the image data itself, actually shows you all to *YOUR FACE!*

The official Unmasking the Face hit piece story continues:

"*What the picture actually shows*" is the Martian equivalent of a butte or mesa -- landforms common around the American West. "It reminds me most of Middle Butte in the Snake River Plain of Idaho," says Garvin. "That's a lava dome that takes the form of an isolated mesa about the same height as the Face on Mars."

This is interesting. Jim Garvin chief scientist for NASA's Mars Exploration Program stated what the Face ***Actually Shows*** is equivalent of a butte or mesa, landforms common in the American west?! Jim Garvin is trying to manipulate you all to make you think that you see the Face as a natural landform, and wants you to compare the Face with an Earthly object, which is a butte or mesa. Trying to make you think of the Face with a pre-programmed bias and propaganda trying to make you think of Earth and Earthen features. That reasoning also that the surface of Mars is like Earth or the same as Earth. Mars IS NOT Earth! The Face also is NOT an Earthlike feature! The impression he wants to imply upon you all is to discredit what you can see with your own eyes in the Martian feature, by looking instead, at a feature seen on Earth as a means to somehow prove that the Face at Cydonia is not a Face with Alien Markings, but A Mesa. Not only that, they want to also convince you that it is not just a mesa now, but it is also now a being called ***LAVA DOME*?!** Claiming it to be the same thing? Claiming that is what the Face really is?! That makes no sense at all.

Let us look more closely to Jim Garvin's claims. I mean Jim Garvin must know what he is speaking about. After all, again he was the chief scientist for NASA's Mars Exploration Program. Let us compare for ourselves, if they indeed look alike shall we?

(Middle Butte, Idaho)

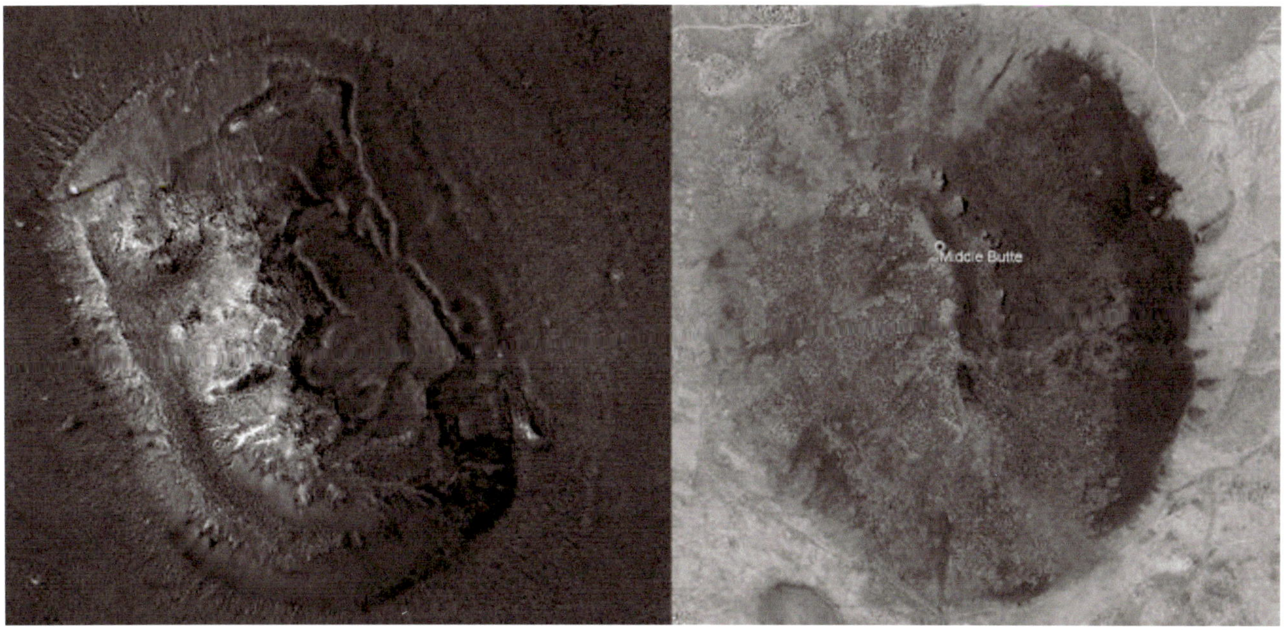

(Cydonia Face image release of E03-00824 compared with Middle Butte, Idaho)

The only thing this proves in reality as a comparison is that the Face at Cydonia is a little bigger than the Middle mesa butte in Idaho. In fact the NASA hit piece states the Face is as tall as Middle butte, which is 800 feet high. The Face at Cydonia is more like 1500 feet high! That is an enormous 700 more feet above the height of Middle Butte. It also looks ***NOTHING*** at all like the Face, nor does the Earthly mesa features even compare to what the Face at Cydonia shows us in its unearthly features of **Alien Markings**! ***It simply is not the same thing, nor can be compared as Jim Garvin tries to explain the Face as being something factually natural.*** The Middle Butte looks completely natural and without purpose or showing any sign or hint of intelligent design or purpose. The Face at Cydonia does not look natural at all, and even has the purposed look of intelligent design complete with its artistic **Alien Markings**!

So how can this account for what we see as the Face at Cydonia? The simple fact is, it does not account, nor does it count at all in trying to compare the Middle Butte with the Face at Cydonia! One can easily see in comparative imagery that they are both so different, and are not able to be compared to each other, like night does not compare at all to day. Therefore NASA's official position again, was simply to keep us all in the dark, and continue to mislead us all in the public with a 'policy' driven agenda to keep the Face at Cydonia from an official position of it actually being an intelligently designed construct, to being that of a natural coincidence of which is being reasoned at as being something very natural instead, **A MESA, AND A LAVA DOME?!**

The official 'Unmasking the Face' hit piece story continues:

"Cydonia is littered with mesas like the Face, but the others don't look like human heads and they've attracted little popular attention. Garvin and other members of the MGS Science Team have studied them carefully,...."

It is still incredible that the offfical wording continues providing us all with unofficial information in what is seen on the surface of what all is said of their official coded script. On the surface of it all, it is obviously being put forth as a deliberate continued official denial of the Face for anything of reason of it as being beyond natural. The propaganda and focus is manipulated to us all instead on by using the Earth, with features as a natural means to try to explain for what we all can see on the Martian surface. In this case it is compared to the Face at Cydonia. NASA knows most people were not expecting NASA to see and speak of intelligent design being found. Especially considering their offical statements and lies about the Face since 1976, and the dessperate attempts to contiue to try to hide the Face and its ***Alien Markings*** from us all, but obviously could not succesfully pull that off. Nor could they just continue to avoid the public with having to take images of the Face. So NASA's best atttempt to decive us all, was and sill is by comparing it to a natural feature here on Earth, caliming that their visual and unscientific comparison is some form of scientiffic evidence proving that is the case?! Talk about comparing apples to oranges!

We are then told that **Garvin and other members of the MGS science team have studied the many other mesas like the Face, that is littered in Cydonia?!** Acknowledging officially now, that only ***one of them, looks like a human face***. This is just another affirmation that what we have seen and learned behind the scenes as it were in their coded official-unofficial informations being given, that yes they still also see a Face, but they still cannot affirm to anything other than a natural explanation. Even when compared, as we had done, one can see how stupid that comparison is. As that is no comparison at all let alone being anything presented as being real proof that the Face at Cydonia is nothing but being completely natural! He also stated another official lie in regards to the **many mesas littering Cydonia**. We were told, "**Garvin and others members of the MGS science team have studied the many other mesas that is littered in Cydonia**. Telling us all that the other mesa's, ***attracted little popular attention?*** I will expose this obvious official lie in this book with only 2 Viking Orbiter images, ***35A72***, and ***70A13!*** Yes the same very images that started this all. As can be seen, there are many other interesting features in the images that is just as worthy of close study around the Face, although they do not look like the Face.

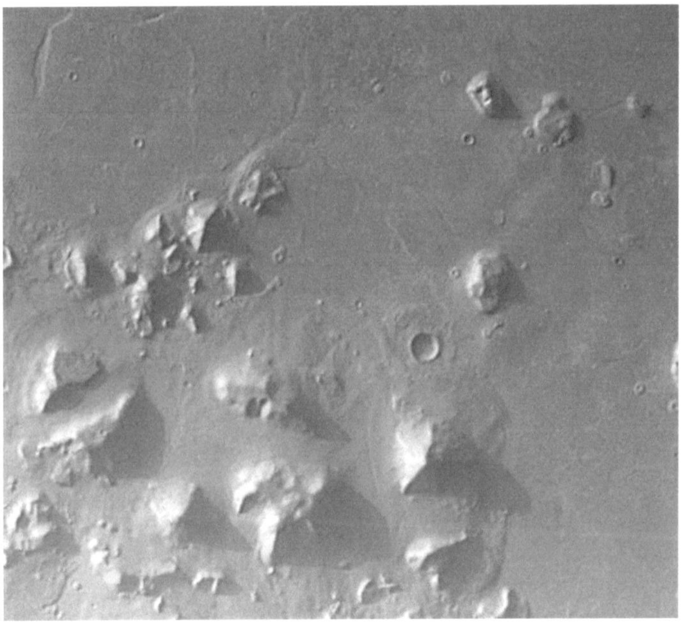

We are told that Garvin and other members of the MGS Science Team have studied them carefully,...." As anyone can see in this image from 35A72, we can see there is no other human like faces being seen. But to claim **'Little Popular Attention'** to the other **'mesa's'** in the area is another obvious lie! Nor can these be claimed as also being other Ordinary Martian Mesas! We see in both of these images what appear to be intelligently designed mountain sized miles wide pyramids among all that natural nature. ***These features have indeed become poularized attractions.*** NASA just does not share the same popular attentions toward them as many else do and have over the years, including myself! Just southwest to the Face as seen in the above image we can see what is called The D&M Pyramid, named after the initials of the of the 2 scientists that discovered it, Gregory Mollenaar and Vince DiPietro. These are the same 2 scientists who had found Viking Orbiter image 70A13, that shown the Face in better sun lighting to reveal it's details that 35A72 had shadowed, and outright exposed NASA lied about the Face.

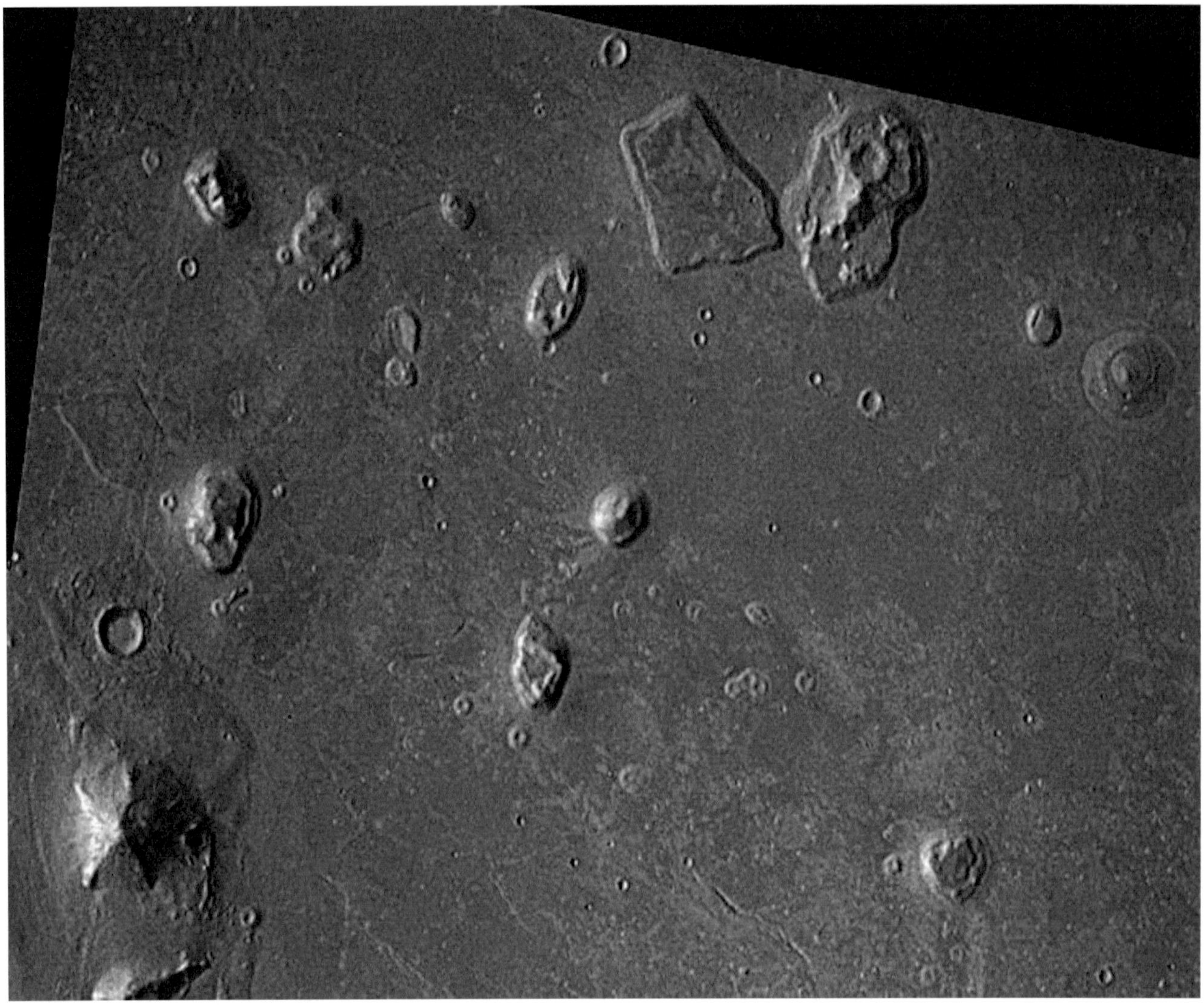

Let us look closer again to the 'D and M' Pyramid, and also look closer at some of these 'NASA so called' *Little Popular Attractive* mesa's, that has also garnered a lot of ***Popular Attention*** among Mars surface image researchers. Just remember also we are told that **'Jim Garvin and the MGS science Team have 'Studied these Other features/mesa's Carefully'!** (*A hint of sarcasm exists in that statement.*)

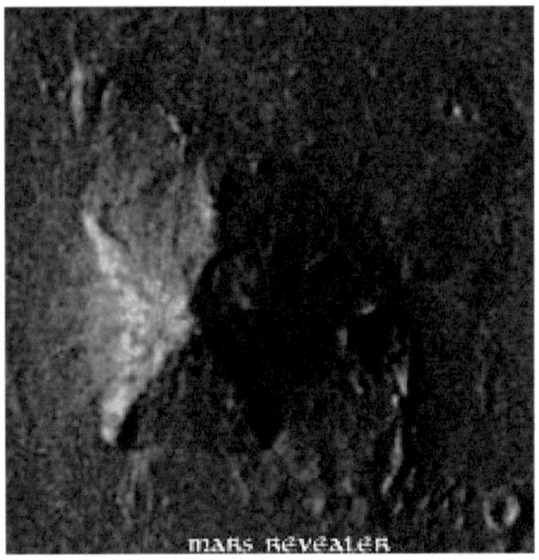

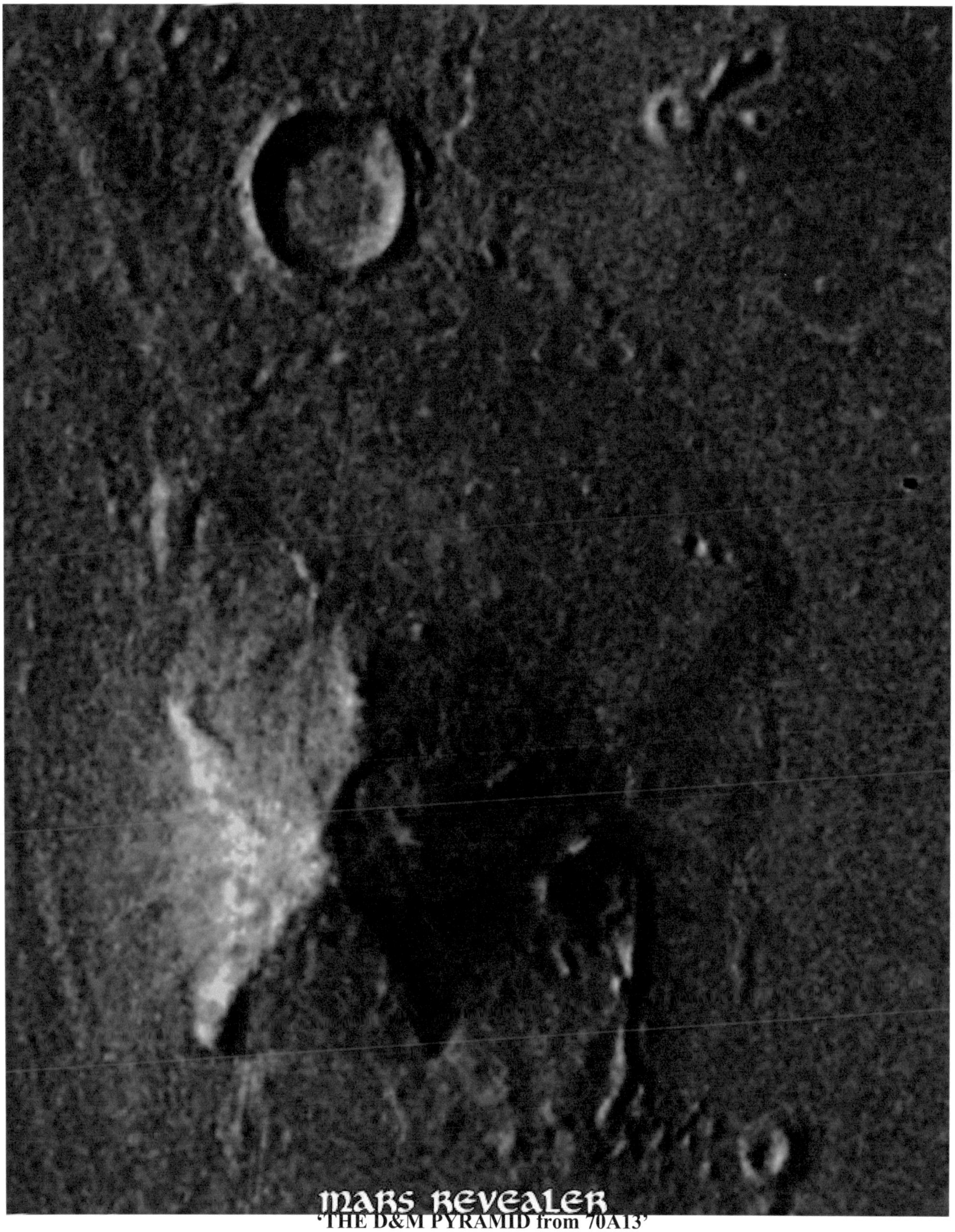

'THE D&M PYRAMID from 70A13'

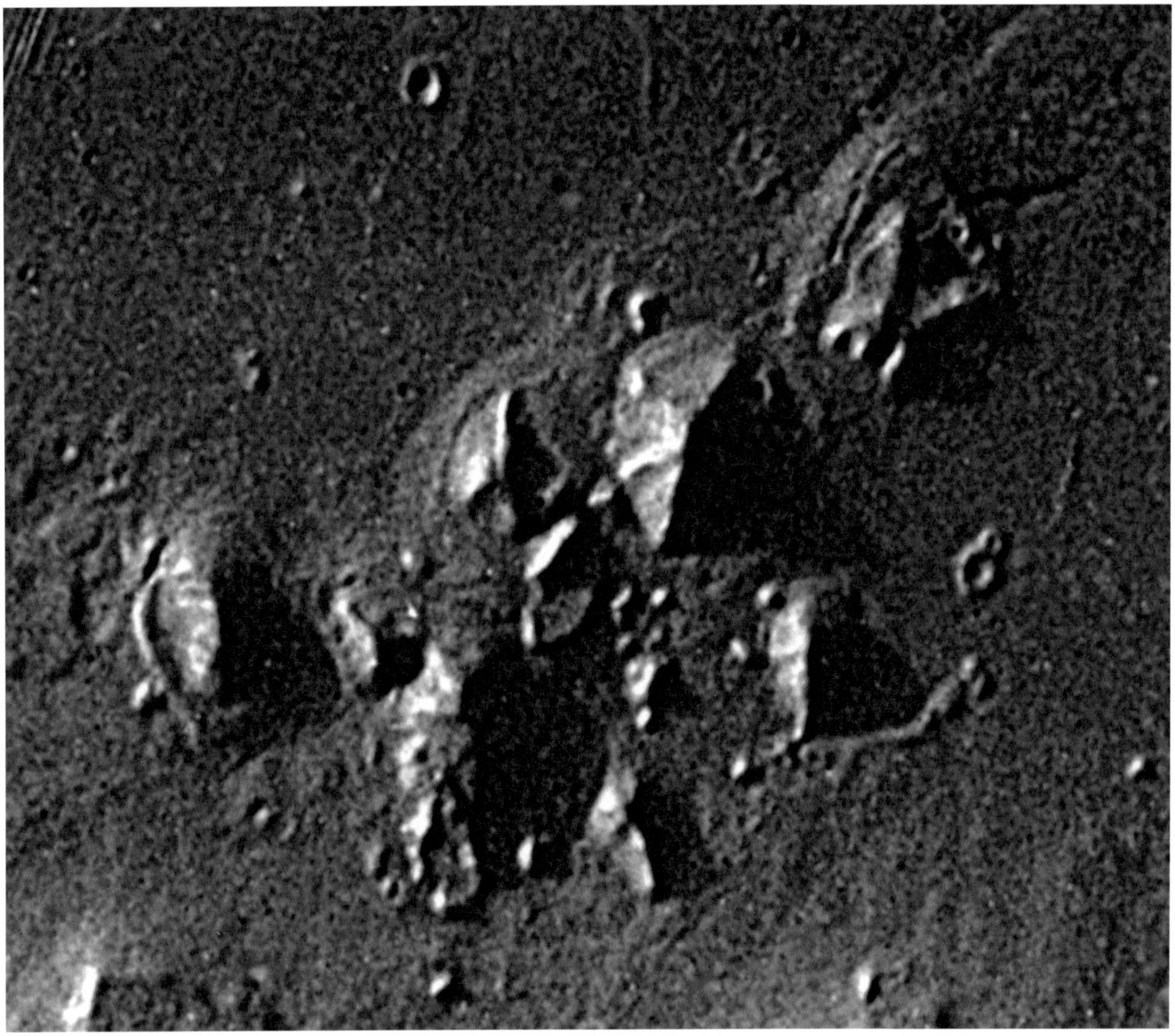

('THE CITY' FROM 35A72)

(These attractive Popular Attentions will be addressed in another book I am doing, and showing modern imagery detailing them and their Attractive Alien Markings in high resolution!)

Once again, we can see some of the other so called **Ordinary Martian Mesas** around the area of the Face. As we were told, there are no other face like formations seen nearby. However the surrounding so called ordinary Martian mesas do also have attractive popularity because these features not look odinarily natural neither. For instance once again, we can see the amazingly intelligently designed 'D&M Pyramid' and 'The City' in better detail. Yet we are being told that **Garvin and the MGS science team had studied them carefully**? Of course that I what we were told! They certainly would not officially tell you the results of their findings of their study gone into them by Garvin and the MGS Science Team, if they most certainly want to officially keep the Face of Cydonia in the eyes of the public knowledge, as being something natural by any and all means, and also it seems, almost at all costs! The question that remains and is ever present, is ***WHY?***

The official Unmasking the Face hit piece story continues:

"Cydonia is littered with mesas like the Face, but the others don't look like human heads and they've attracted little popular attention. Garvin and other members of the MGS Science Team have studied them carefully, however, using a laser altimeter called "MOLA" on board Mars Global Surveyor,
MOLA can measure the heights of things with a vertical precision of 20 to 30 cm (its horizontal resolution is 150m).

"We took hundreds of altitude measurements of the mesa-like features around Cydonia," says Garvin, "including the Face. The height of the Face, its volume and aspect ratio -- all of its dimensions, in fact -- are similar to the other mesas. It's not exotic in any way. The laser altimetry data are perhaps 'even more convincing' than overhead photos that the Face is natural. 3D elevation maps reveal the formation from any angle, unaltered by lights and shadow. There are no eyes, no nose, and no mouth!

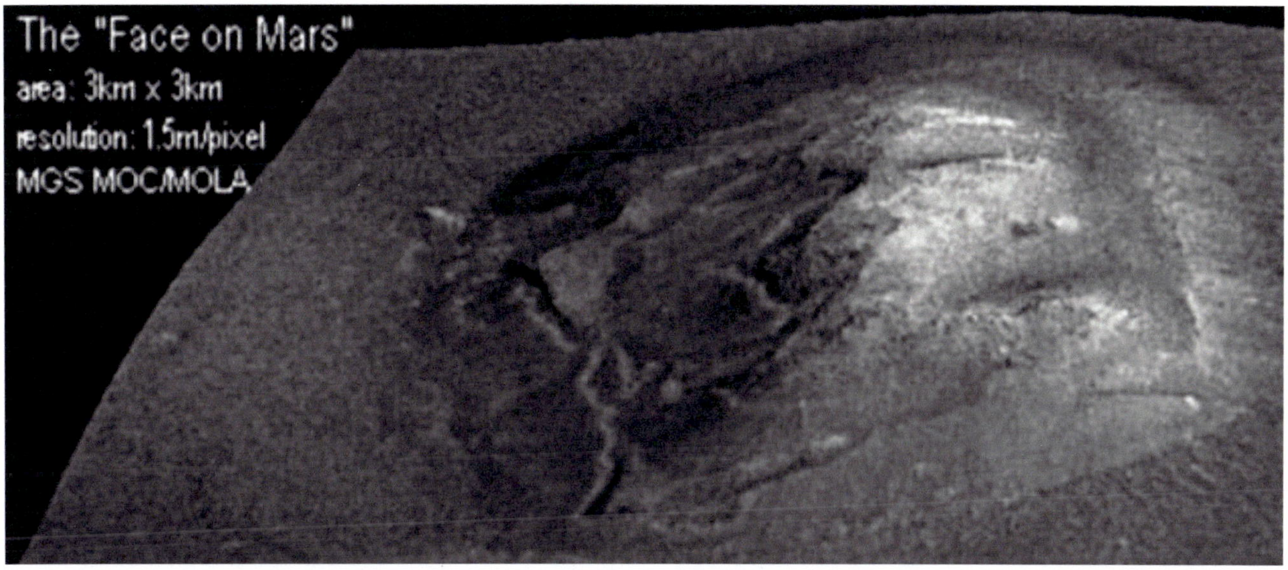

As we can see in the image shown above, a 3-D perspective view of the Face on Mars landform produced by Jim Garvin (NASA) and Jim Frawley (Herring Bay Geophysics) from the latest MOC image (April 8, 2001), to the untrained eye and to those who are ***not familiar*** wih the subject, can think that they see validation in the words spoken by Garvin saying **"It's NOT exotic in any way."**

To me it seems Garvin is trying really hard to convince us, or maybe even convince himself and others that what we see looking like a face, is not really a face at all. Garvin adding also that in the image referenced above, it is as if he is speaking with glee almost, claiming that **"There are no eyes, no nose, and no mouth!"** At this point in time, NASA is frustrated apparently. To which the lengths NASA would go to continue a deception upon the public because of a feature that so drives them to confrontational awareness of the fact that the Face of Cydonia, is in fact REAL! The fact of the matter of this discussion is how NASA and MSSS released this with the continuing official position of denial, by obfuscating the facts. In this case showing visual representations they had construed to match their official position. Which is really easy to see at this point if your following along with how we looked at this all so far to consider. To see it, all you have to do is untwist their words, as well as their visual representations, to be able to see it clearly enough to stretch it out a little, and determine that what I am saying is in fact true!

The best way to get into the science and the facts stated about this part of NASA's hit piece for The Face, it is important to focus in on the MOLA data for it is the main only scientific excuse it seems by which NASA and Garvin claim is that **'The laser altimetry data are perhaps 'even more convincing' than overhead photos that the Face is natural. 3D elevation maps reveal the formation from any angle, unaltered by lights and shadow. There are no eyes, no nose, and no mouth!**

The best way to do this I believe is to include a very detailed respondant article put forth by Richard C. Hoagland on The Enterprise Mission from 2001, to test and contradict Garvin's and NASA's position which is titled:

"HOW TO MAKE A MOUNTAIN... OUT OF A MOLA HILL".

(Richard contributed this article to publish here in this book for you all).

When the new image of the Face on Mars was released in May 2001, we had a right to expect (and hope) that it would be given a fair airing by NASA after nearly 20 years (at that time) of controversy. Instead, as we predicted, we all watched as NASA went into a relentless spin cycle on the new image, trying very hard to suppress any idea in the public's mind that it might not be what NASA has always said it was -- "a funny looking hill and a Trick Of Light and Shadow." The linchpin of this NASA public relations fusillade was the (supposed) analysis of laser altimetry data from the Mars Global Surveyor's MOLA (Mars Orbiter Laser Altimeter) instrument, which according to NASA "**showed there were no eyes, no nose, no mouth**!"

NASA's reliance on this data to "prove" anything about the Face is dubious scientific reasoning, at best. We have also called into question the fact that the images presented in the NASA hit piece (represented by NASA as being generated from MOLA) were both shown *flipped* – presumably, to reduce the chances of the public recognizing *anything* resembling a Face in them (below).

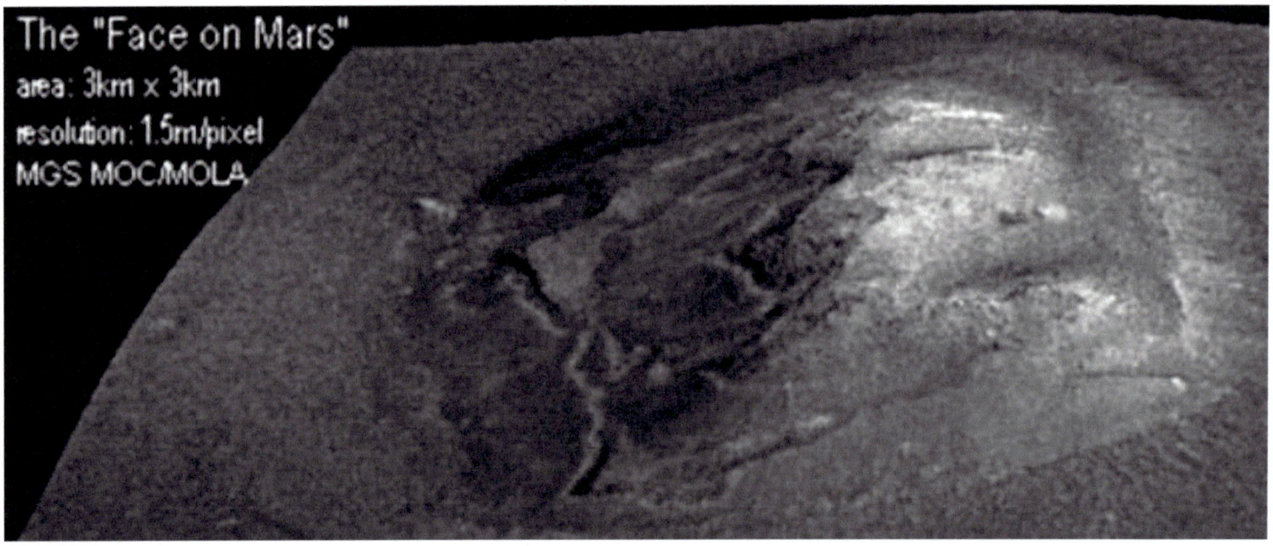

(The "Face on Mars" as presented by NASA)

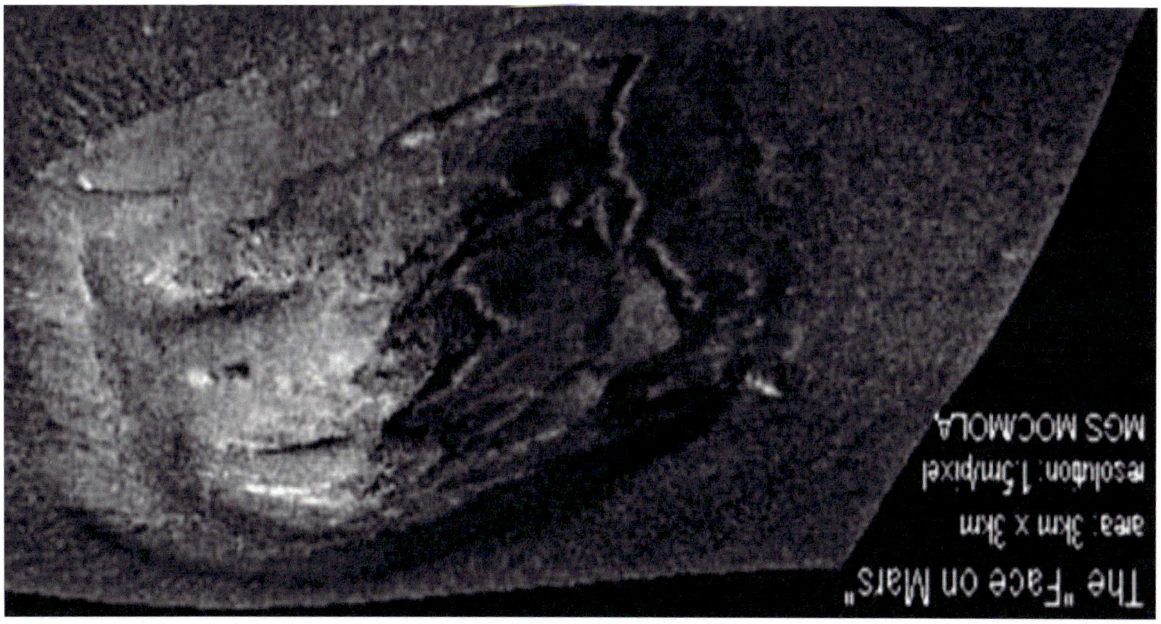

(The same image, rotated into its normally seen orientation and horizontally compressed to its proper proportions...)

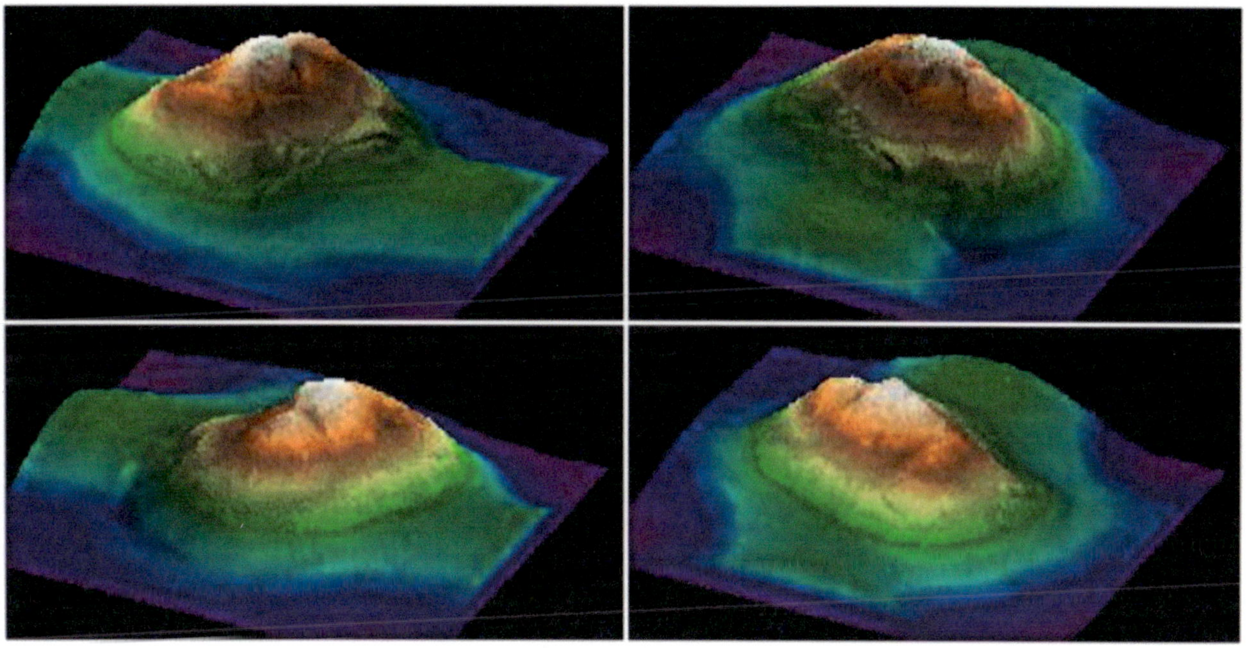

Above: Four views of the "Face on Mars." Jim Garvin (NASA) and Jim Frawley (Herring Bay Geophysics) combined MGS laser altimeter data with high resolution MOC images to produce these 3D images in which color represents height. The peak of the mesa-like landform rises 800 feet above the surrounding plain.

(... and the Colorized "MOLA" data as presented by NASA ...)

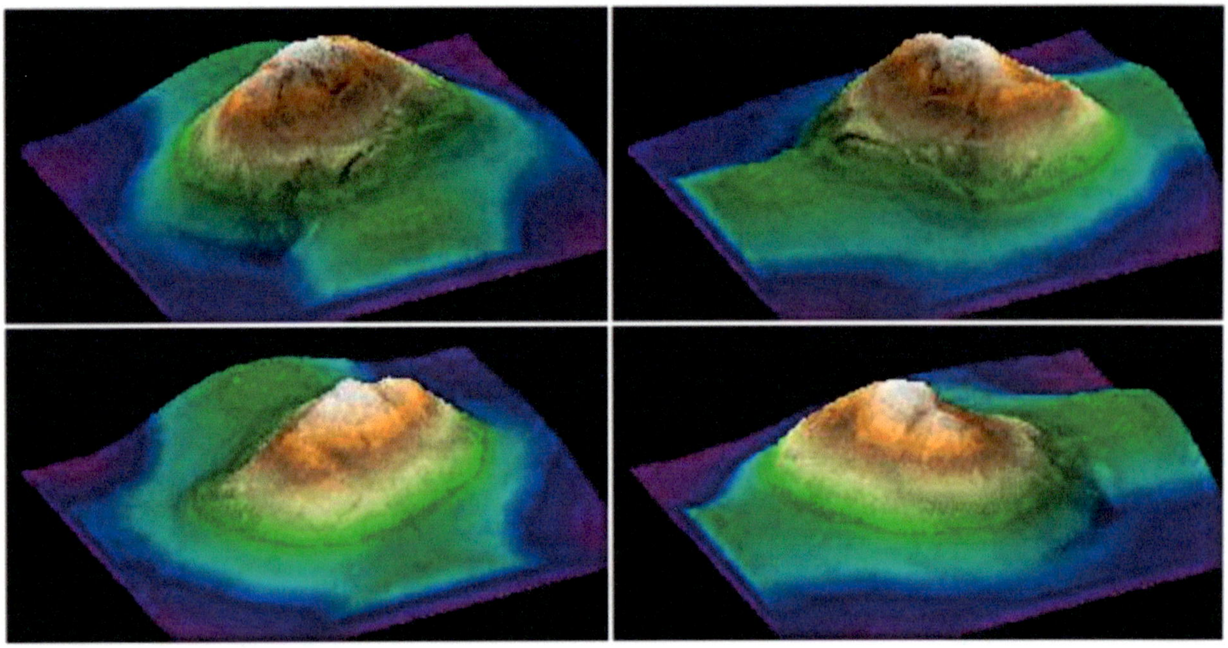

Above: Four views of the "Face on Mars." Jim Garvin (NASA) and Jim Frawley (Herring Bay Geophysics) combined MGS laser altimeter data with high resolution MOC images to produce these 3D images in which color represents height. The peak of the mesa-like landform rises 800 feet above the surrounding plain.

(... and flipped to show its proper orientation.)

But the biggest problem with passing MOLA data off as a viable representation of the Face's true shape has always been the resolution limits of the instrument. Sadly, some individuals have been so blindly loyal in their belief that NASA would "*never* hide anything" (despite their problems with getting the pictures right-side-up, above), that they have actually accused us of misrepresenting the capabilities of the MOLA instrument, in order "to further our contention that the NASA spin was unscientific and logically flawed." Well, we are here to show them (and you) that not only is NASA logic and "analysis" categorically *wrong* -- but that it is *so* wrong it can only be the result of ***a deliberate attempt to mislead the public about the true nature of the Face!***

Because it can produce neat, colorized views like the one shown here (*on the next page*), most people seem to assume that MOLA is a *high resolution* instrument. Some of our critics have even gone so far as to suggest that since MOLA has a "vertical" resolution of "only a few centimeters," it is in fact a "better" and more accurate visual instrument than the MGS camera. In fact, it is *nothing of the kind* -- and this sort of ignorance among the lay public and the press is precisely what NASA depends on to keep the Face from being seen it its true context (and if that is not enough, as we have seen, they are not above flipping it upside down and backwards even in their "MOLA" presentations to enhance this deception – (*seen on previous page*).

MGS - MOLA

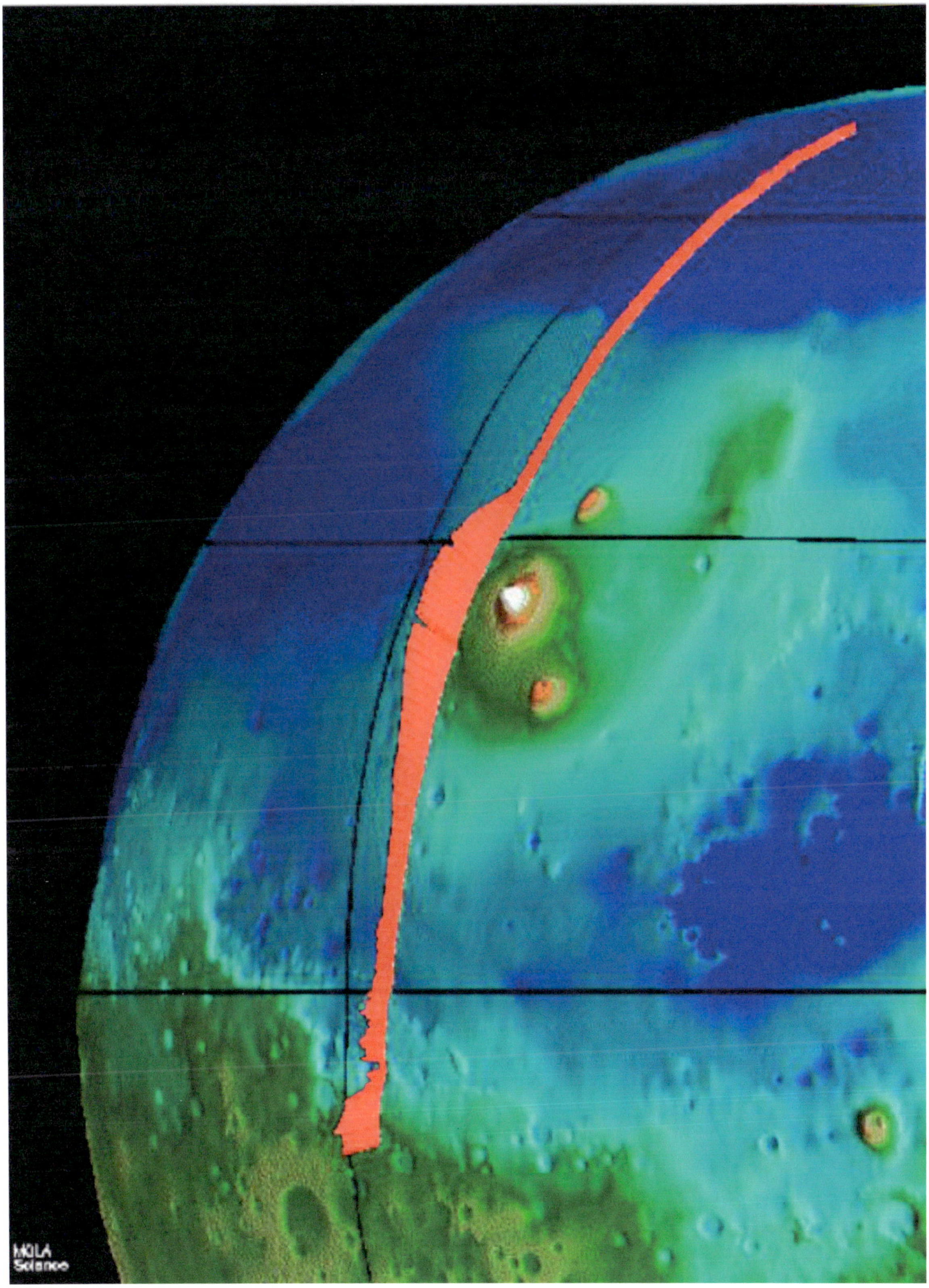

What the MOLA really does is send out a series of pulses (10 per second) of laser light that "bounce" off the Martian surface and are reflected back to the instrument's receiver. This results in a single, circular "pixel" (or picture element) of data that is some 160 meters (over 500 *feet*) in diameter. Since the spacecraft is traveling at about 3.3 km/per second, the next "dot" is some 330 meters down track (since MGS is in a roughly polar orbit), leaving a gap, or blank spot in the image track about 170 meters (about 600 feet) wide. While the spacing is greater at the equator and less at the poles, it still requires multiple scans to accurately define any object visually. MOLA has been operating continuously since MGS went into the primary science phase, except when it was turned off during solar conjunction. So what the MOLA has produced is one continuous string of data, consisting of a series of 160 diameter spots, with 170-meter gaps in between them -- winding around the planet like a ball of twine for over two years. Sounds kind of cool, doesn't it?

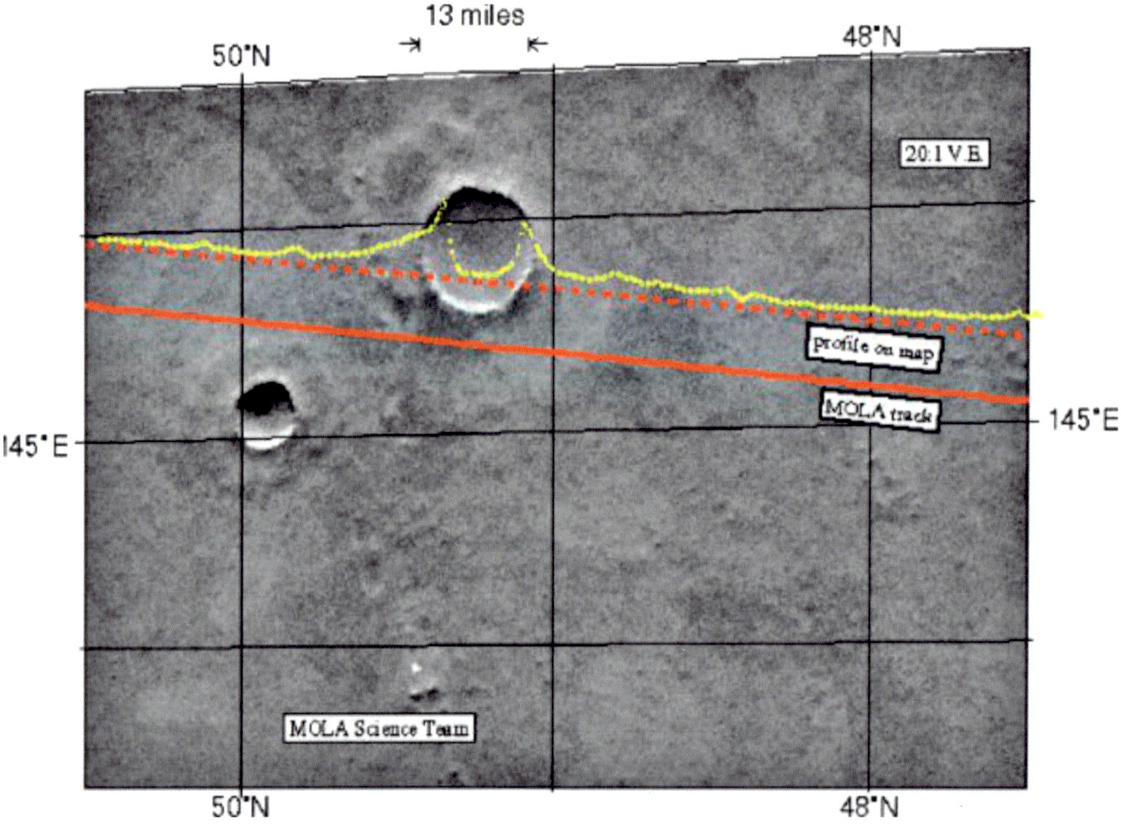

Well it is. But what it is *not,* is very specific or accurate on the scale of a mile or so. In other words, it is certainly *incapable* of "imaging" any individual object as small as The Face. Nor is it anywhere near the spatial resolution of the MGS camera, even at the latter's worst. In the example above, from a MOLA press conference, you can see just how coarse the instrument "resolution" is. The example used above is a *13-mile* wide crater, and the yellow dotted line represents the MOLA data track across it. As you can see, it is a series of dots, not one continuous scan, and it is only on a very narrow (160 meter) track -- *missing well over 99% of the crater.*

Some people have been confused by the stated "vertical" resolution of the MOLA. One particular critic, seems to be utterly incapable of grasping just what the 20-30 *centimeter* "vertical" resolution of the instrument actually means. He has even gone so far as to suggest that because of that, the MOLA instrument is "better" for resolving features on the Martian surface than the MOC camera. As you will see, this is simply laughable. Didn't it occur to this genius that if that was the case, they wouldn't have even bothered to *put a camera* on the spacecraft – thereby saving perhaps a *hundred* million dollars, over the course of the entire mission?

Let's go back to the MOLA instrument and see if our critic is right.

Within that 160-meter diameter "dot" that we keep talking about, the MOLA can discern almost no detail. Its 10 quick pulses hit the ground in the area in question and return the timing data to the instrument. MOLA then takes the *average* altitude of the spacecraft above the ground within that 160-meter "pixel" and assigns a "value" to that pixel based on the average. As a result, *every bit of detail within that pixel is reduced to a single point, a single value:* the average spacecraft altitude above the ground being "pinged." All of the individual stuff within that 160-meter circle is completely lost. Now, that average value for the area in question *is* accurate to within about 1 meter with respect to the distance below the spacecraft, but that's it.

So just how big is "160 meters?" Just how much is missed by this (as our critics would have you believe) "precision" instrument? A lot. "160 meters" works out to about 525 feet -- which is a *huge "pixel" diameter*.

It is, within about 5 feet, the diameter of the Tacoma Dome arena near Seattle (*shown on previous page at 1 meter per pixel resolution, 50% better than the MOC camera*). Within its area, this sporting arena can hold upwards of 23,000 people, not to mention the playing field, the facilities, locker rooms, concession stands, press facilities, plumbing, miles of wiring and enough concrete to build a 50 mile long highway. And MOLA would miss it all! To MOLA, it would just be one big blob. A dot. It *would* be able to give a very close estimate of the average distance the roof of the dome was from the MGS spacecraft, but that is it. It could discern no *details* about the object whatsoever.

The argument has been made that "160-meter resolution" really isn't *that* bad, that it is "only" 3 times worse than Viking. But remember, "160 meters-per-pixel," vs. Viking's "50 meters-per-pixel," is a *150%* difference. And when you consider the scale of the Face itself, it becomes obvious just how much crucial detail is missing from these "precision" MOLA scans.

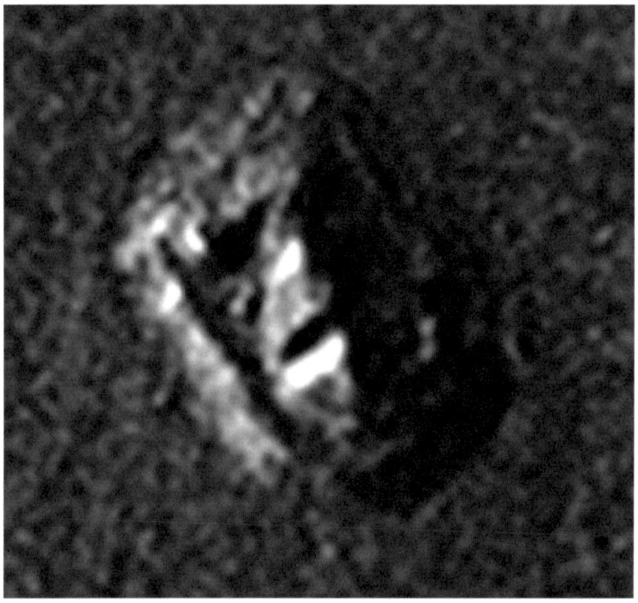

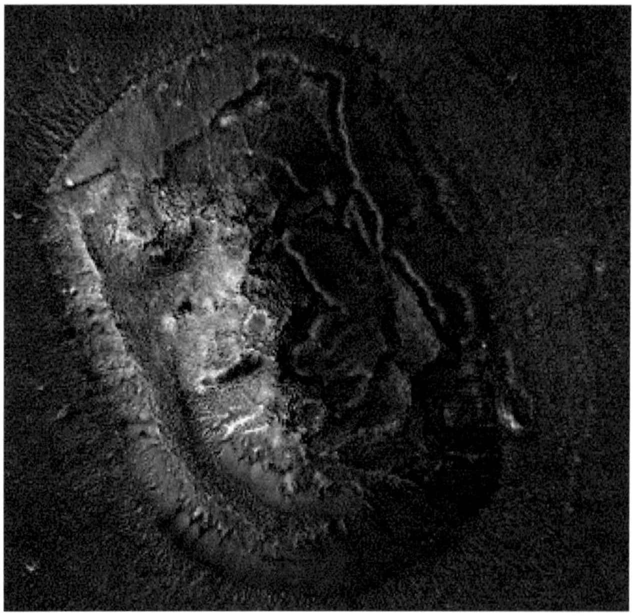

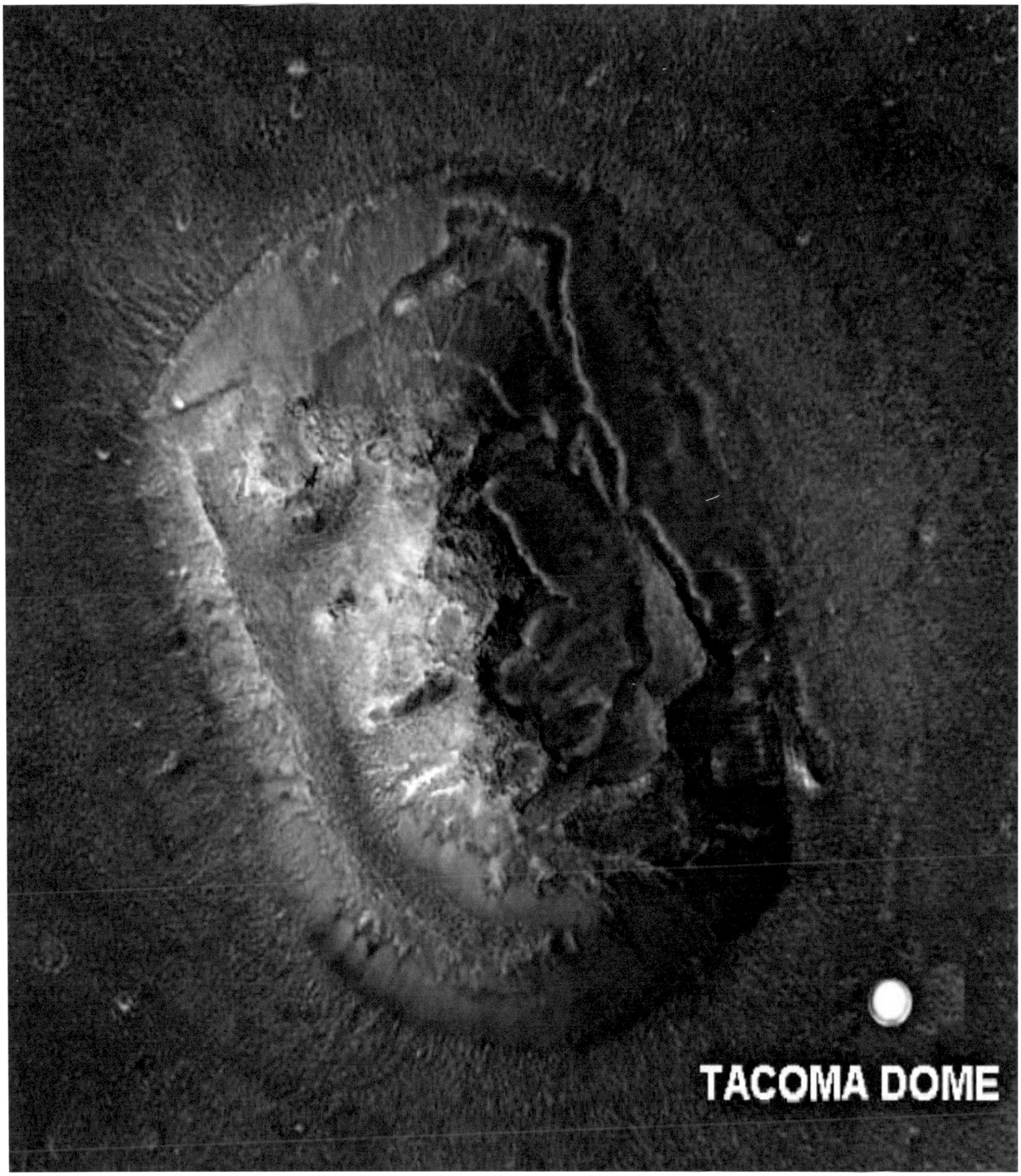

To give you some idea of the scale involved here, I have placed the Tacoma Dome -- approximately to scale -- next to the Face on Mars. As you can see, the pixel size of the MOLA is so large that a feature like the eyeball in the eye socket (which is about the same size as the Dome) would be *completely missed*, assuming that by some miracle the MOLA scan actually ran across the feature in the first place.

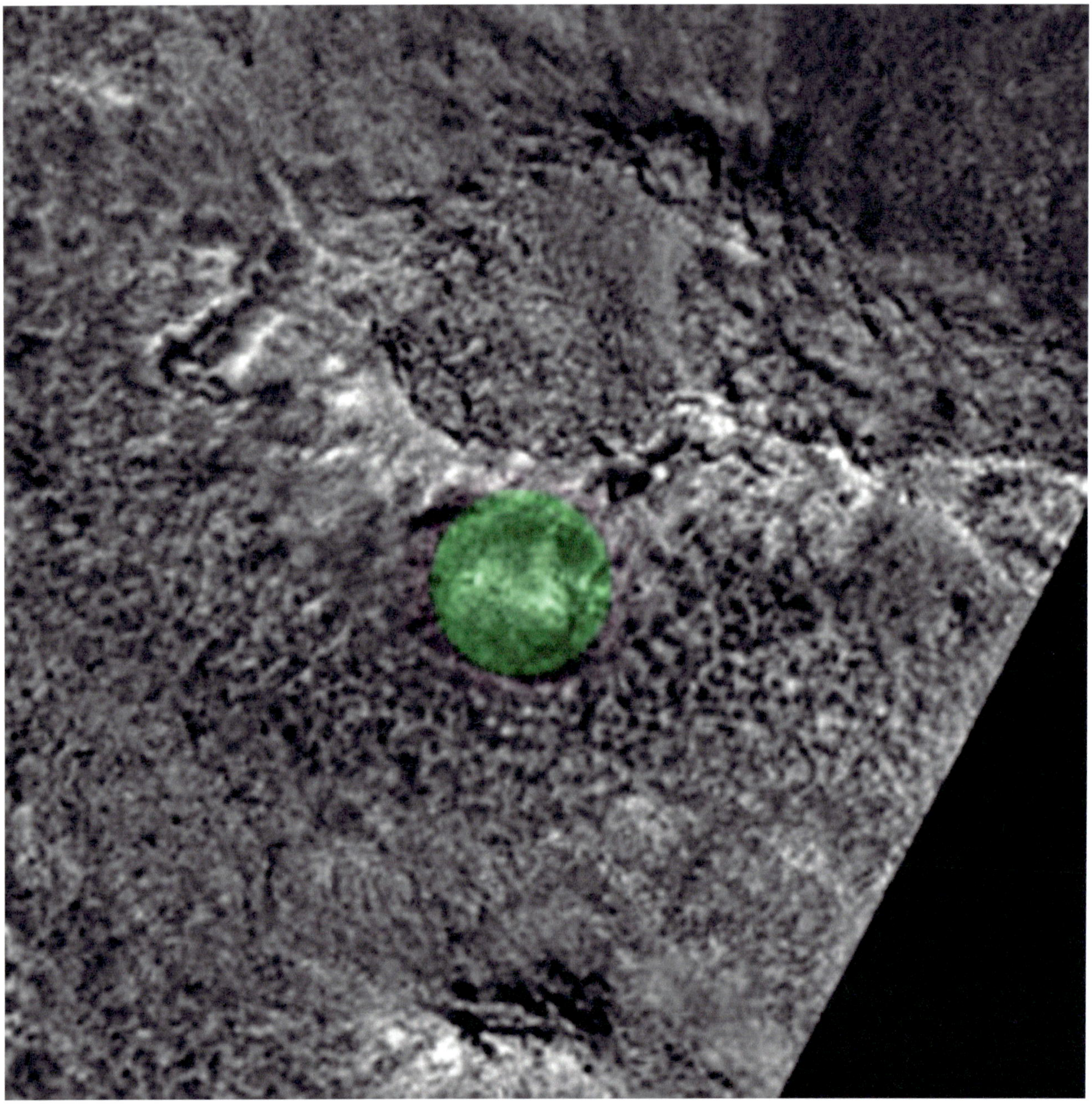

By contrast, the MGS camera, at its maximum resolution of 1.5 meters per pixel, would "see" an area thousands of pixels "square" in this same 160-meter circular space. Objects as small as passenger cars could be made out without enhancement. And each of those pixels has a specific color value assigned to it from 256 available shades of gray. Enhancement processes can use these color values to bring out even more detail, effectively increasing the spatial resolution (under certain conditions and assumptions) even further.

So to argue that there are "no eyes, no nose, no mouth!", based on such a crude instrument (MOLA) is not only scientifically absurd -- it is scientifically *dishonest*. The simple truth is that MOLA is incapable of resolving a feature as small as the (Tacoma Dome sized!) eyeball. And the old-fashioned MOC camera, with its 1.5-meter dimensional resolution, and a "mere" 256 shades of gray-scale resolution, is *thousands of times* more accurate.

Which brings us to the next problem with NASA (and our critics) using MOLA data to "debunk" the Face.

There is, again, a general misunderstanding about just how the MOLA works. Because planets are so large, and individual features like the Face are so small by comparison, the chances of MOLA actually tracking *directly* across the Face in the course of its two-year "nominal" Mission were very small indeed. Our critic friend made up this grid pattern (below) to illustrate what *he* thought was the coverage pattern of the MOLA. He assumed that MOLA, like some kind of "scanning camera," completely blanketed the Face in a tight grid-like pattern. In fact, once again, this is *completely wrong* -- but his grid overlay is useful for a couple of reasons.

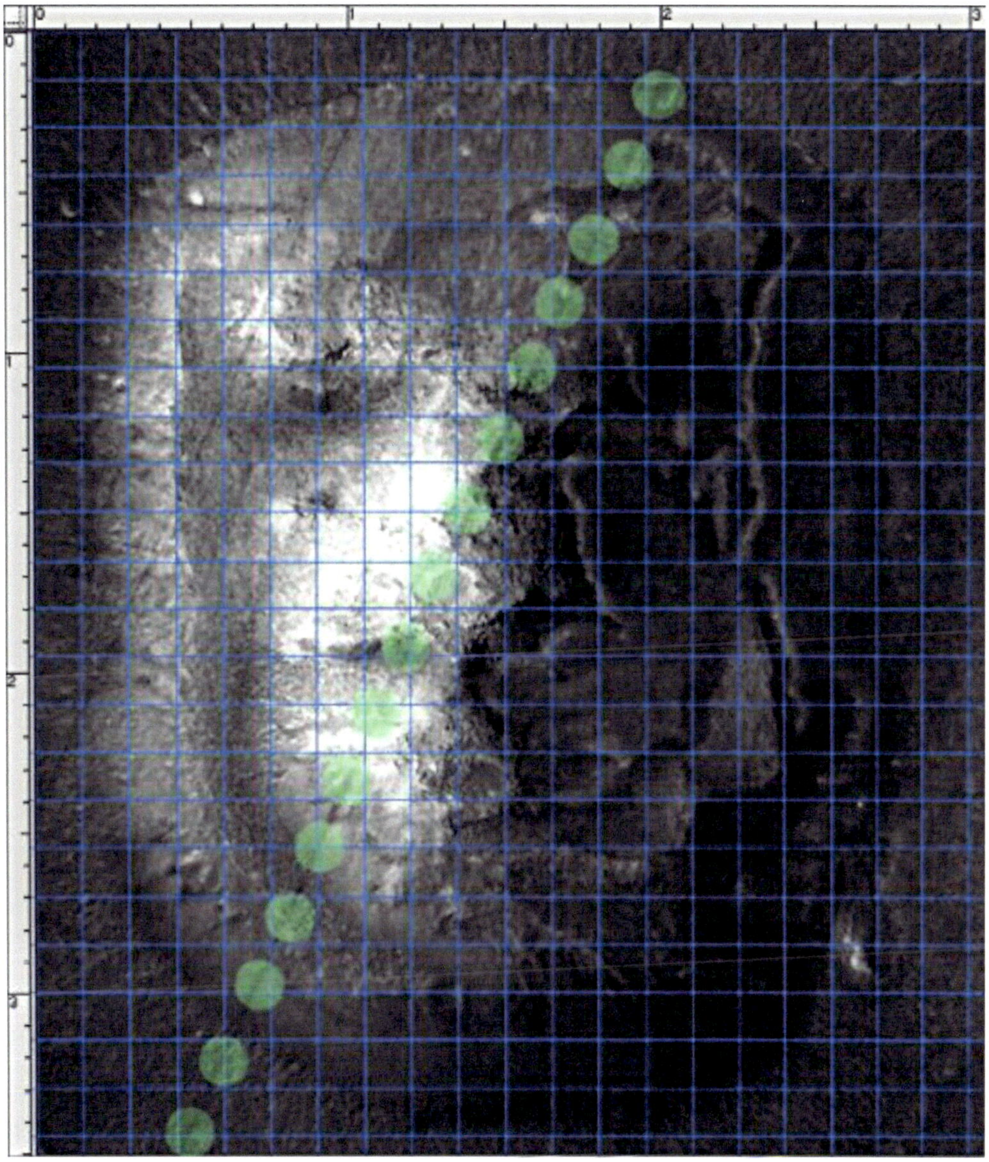

For starters, the image that he used still had visible CCD "pixel dropout" lines from the latest unprocessed MGS image of the Face. These less sensitive pixels, of the CCD "line camera" that makes up the heart of the MOC itself, represent the actual *geodetic track* around Mars that the MGS took over the Face, as it acquired this new image. In an ideal circumstance, MOLA would have tracked down *right across the center of the Face* along those darker "scan lines" (which are offset from true North by about 5°). I have added the green dots to represent the actual data pixels as they would have been acquired by the MOLA in an ideal "centerline" scan -- roughly every 160 meters (his grid is about 150 meters on a side), with the 170-meter gaps between the "pulses."

As you can see, this is *far different* than the idealized notion (represented by his grid) that there were *hundreds* of MOLA data points taken across the Face. At best, there could only be between *15-20 points* -- as you clearly can see from this example.

But wait a minute, why couldn't MOLA have made *multiple* passes across the Face, and gathered enough data to accurately measure the height of the *entire* object in its mission around Mars? After all, hasn't MGS been in orbit for *years*? Well, yes it has. But that has only amounted to about 10,000 orbits since MGS began the Mission Mapping Phase in March, 1998. This might seem like a lot of orbits, but since Mars is such an enormous place (with a surface equal to the land area of *all* the continents on Earth *combined*), it means that MOLA has only covered the planet sufficiently to date to leave *1.5-mile gaps* between the "twine" (and, at the equator). At the latitude of the Face (41° North), the distance between tracks is somewhat less -- probably about 0.80 miles. Since the Face is only about 1.2 miles wide, it is highly unlikely that any subsequent parallel tracks actually scanned across the formation more than *twice*. This is because, since there has only been one direct *overhead* MOC shot of the Face released by Dr. Malin -- *the one taken in May 2001* -- there can almost certainly be no more than two simultaneous samplings of MOLA data taken across the Face in the course of the entire mapping mission [because the first MGS image, taken in April 1998, and the latest one (above), were taken "off nadir," so MOLA was not used]. And remember, that May 2001 example did not track accurately across *the middle* of the Face (see below), but was offset to one side (*my example on the previous page is an ideal circumstance, merely to illustrate a point*).

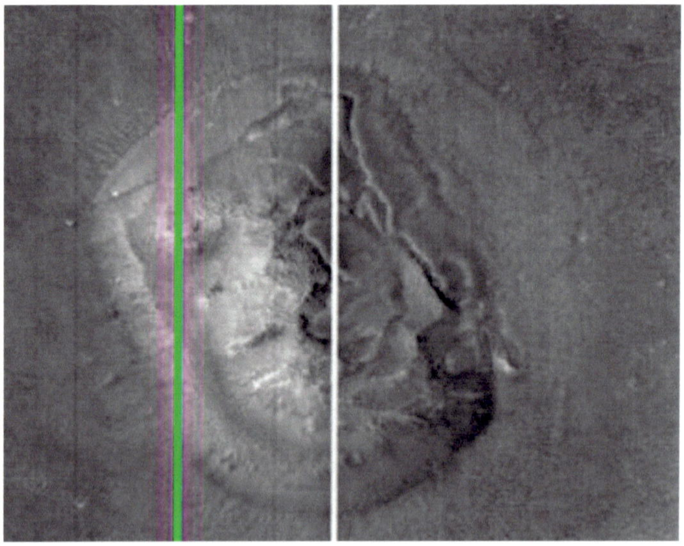

In fact, we can test all this rather easily.

If we assume that the unprocessed version of the latest off-nadir (~25-degree) Face image (above) has *not* been cropped, then the (bore sighted with the camera) MOLA scan (*if* the instrument was actually turned on) would have been pretty much right down the center of that frame. When you draw a simple line down the center of the unprocessed frame, as I have done (above), it becomes obvious that the best MOLA track (the white line) would have been off to the Cliff side, and clearly would have missed the tip of the "nose" -- which is the highest Facial point. The green line represents our estimate of the actual MOLA track of the May 2001 imaging opportunity, which would be theoretically right down the center of *that* frame. This notion -- that the MOLA scan NASA used in its (mis-) representations of the Face, missed the Nose *completely* -- is further reinforced by the claim made in the NASA hit piece: that the Face is "only" 800 feet tall. Previous estimates, made from reliable methods like comparative stereo images and measuring trigonometric shadow lengths, have shown that the Face is actually some 1,500 feet high at the nose tip. This discrepancy can be easily accounted for when you see that the MOLA scan that NASA actually used had to have tracked to the side of Face's highest point (the Nose) -- *completely missing* the tallest feature on the Face! Unfortunately, our estimates cannot as yet be verified – as, for some reason, NASA has chosen *not* to release even the raw MOLA data they obtained during the May 2001 image pass.

But Jim Frawley, the contract scientist who is credited (along with NASA's Jim Garvin) as having created the "MOLA" image used in the NASA hit piece, admitted as much in an e-mail. When asked directly if there were only *two* MOLA passes over the Face, he responded "Your [sic] right. I found just two."

So, that's *two* passes: each a series of 15 to 20 dots, *160 meters* in diameter, with absolutely *no* discernable detail about the "Tacoma Dome-sized" areas that MOLA scanned. How could NASA, from this meager data, decide that the Face was "800 feet" in height, and generate the supposed "3-D mesh" to create their now infamous "MOLA image" for the hit piece? How could they further decide, from just *two* scans that missed all these *crucial facial features*, that there was, as Garvin is quoted as emphatically stating, **"No Eyes, no Nose, and no Mouth!"**

They couldn't.

The fact is, there is *no way* for Garvin and Frawley to have created the "images" presented in the NASA "hit piece" from the available MOLA data. Further, it is equally impossible for them to have made *any kind* of accurate determination regarding the fine scale ("Tacoma Dome-sized ") features -- like the "eyeball" strikingly visible in the January 31st 2001 partial Face image and in the April 8th, 2001 second detailed image.

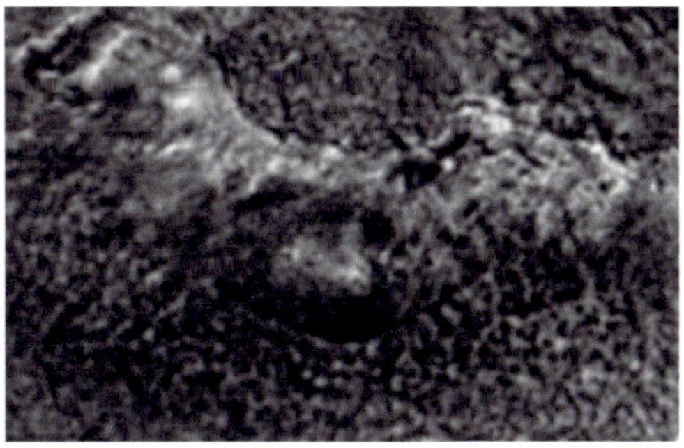

(EYE from JANUARY 31ST 2001 M16-00184 Face section)

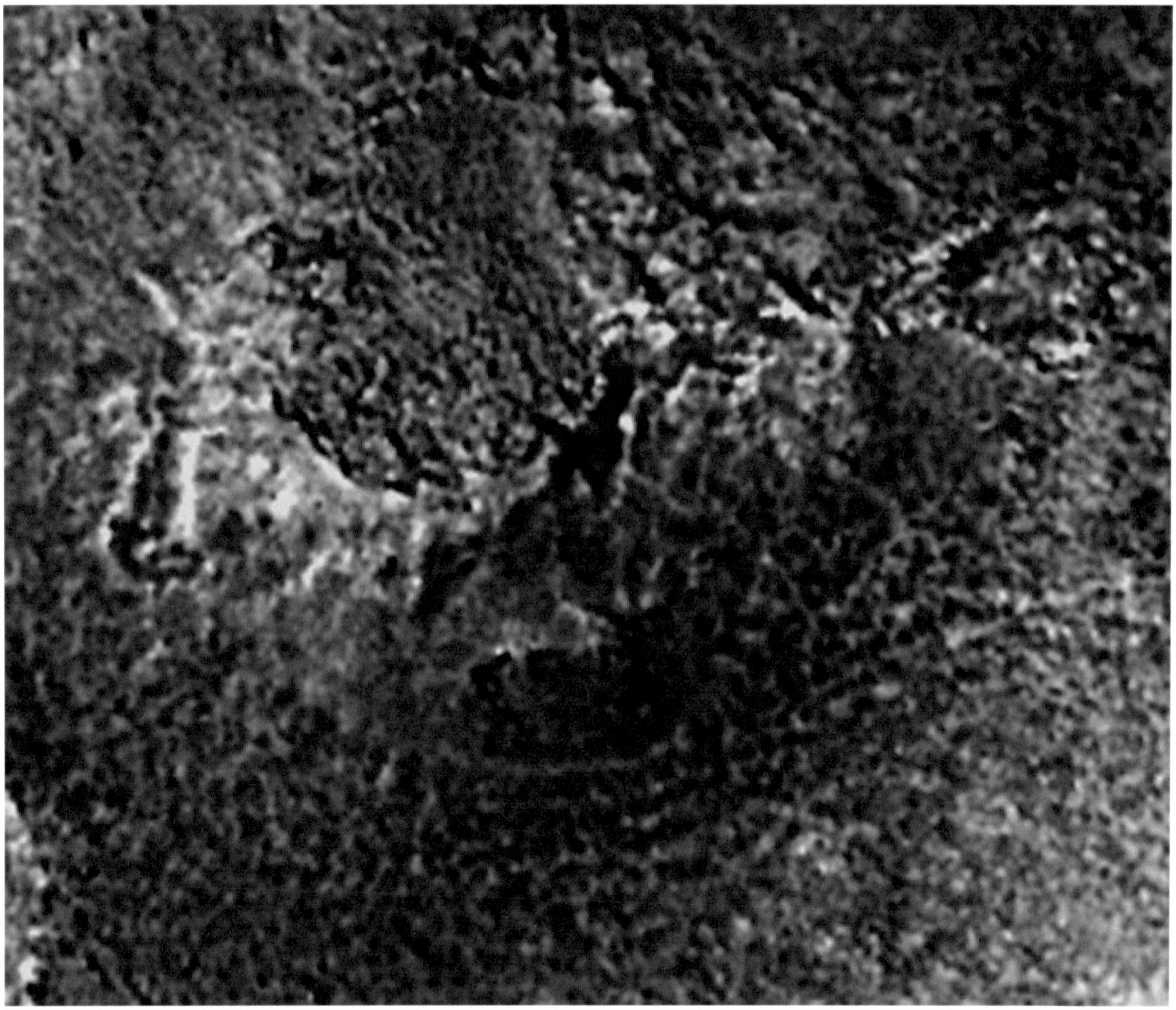

(EYE from April 8th, 2001 Face image release of E03-00824)

These facts are in stark contrast to how the data was portrayed in the NASA hit piece (*which was reprinted and treated uncritically on Space .com*). This article argues that the Face is according to Garvin, "**not exotic in any way.**" Of course it isn't, at the MOLA resolution!

NASA faked it -- pure and simple!

The image they are passing off as "MOLA-generated data" is nothing more than a deliberately "de-resed" version of the MOC image itself! Once again, confronted with this information, Frawley admitted to the truth. "You're right on this too. Image is 99% MOC. It's made with an 'inverse imaging' program I wrote some time ago. MOLA is used for constraints." "Constraints" simply means that he used the available MOLA data to make sure he had the height-to-width proportions correct when he made his shape-from-shading-image. And in reality, it's more like the image is 99.99% MOC ....

But why quibble? The key point here is that NASA has made *outright false claims* about the image they presented to the press and public as specific MOLA data, and compounded that lie by pretending that the instrument could determine more than it actually could. When that wasn't enough, *they flipped the image upside down and stretched it to ensure it was totally distorted.* To be fair to Frawley, *he simply produced the image he was requested to produce. He had no control over how Garvin and the NASA hierarchy used and distorted that information to serve their own partisan political purposes.*

And make no mistake, this article a well as what was seen in this book so far, *was all about "politics."* As we have shown, there was ***no science*** in Garvin's MOLA claim at all! In fact, like his other statement comparing the Face to Middle Butte Mesa (and then not even producing an image of the mesa to support his claim), Garvin has been shown to be lacking either the intelligence or the integrity necessary to carry out his duties as director of NASA's Mars projects. He is either ignorant of the capabilities of his own instruments, or was engaging in a ***deliberate deception***. Either way, the reality is that the MOLA claims are not only *false*, but they were calculatingly designed to "scotch this thing for good," as one unnamed JPL "scientist" put it after the 1998 "Catbox" fiasco.

Tempting as it may be for some of our critics to fall back on the "honest but stupid" model, Garvin cannot escape judgment here by claiming mere "stupidity." Surely he was well aware of the limitations of the MOLA instrument when he made his fallacious claims, just as he was aware the image was actually a deliberately degenerated MOC image. ***This shows that NASA is willing to go to considerable lengths to suppress interest in the Face* -- even when forced to release images *they'd obviously prefer to hide from the public*.**

But the tough question for our critics now, and coming back again to the question asked before going over this scientific rebuttal article, is "**WHY**"? Why, if there is *nothing* to this forty six year old controversy of "a Face on Mars," if the Face is so *obviously* a "natural and common geological formation," does NASA *need* to go to such extraordinary and Machiavellian lengths to "prove" its point? Why do they *need* to make unsupportable comparisons and outright (and easily proven) *fallacious claims* about their own instruments ... in an increasingly desperate and obvious effort to suppress *all interest* in a "natural" feature? Why do they *need* to withhold a full-on image of the Face for *two months* -- while they prepare their distorted "hit pieces," rather than just release the image without comment ... for the press and public to decide?

Why *indeed*?!...

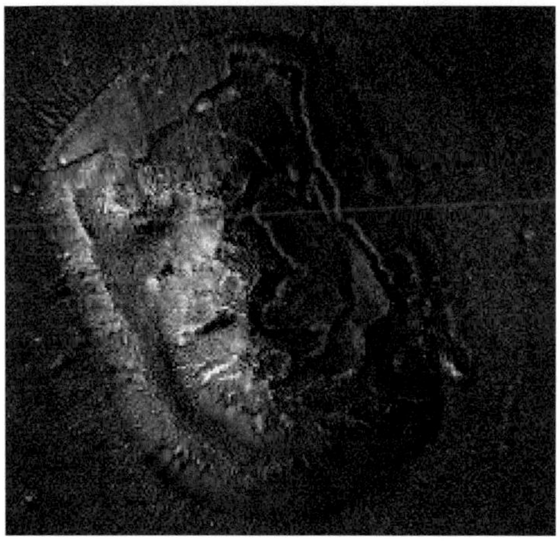

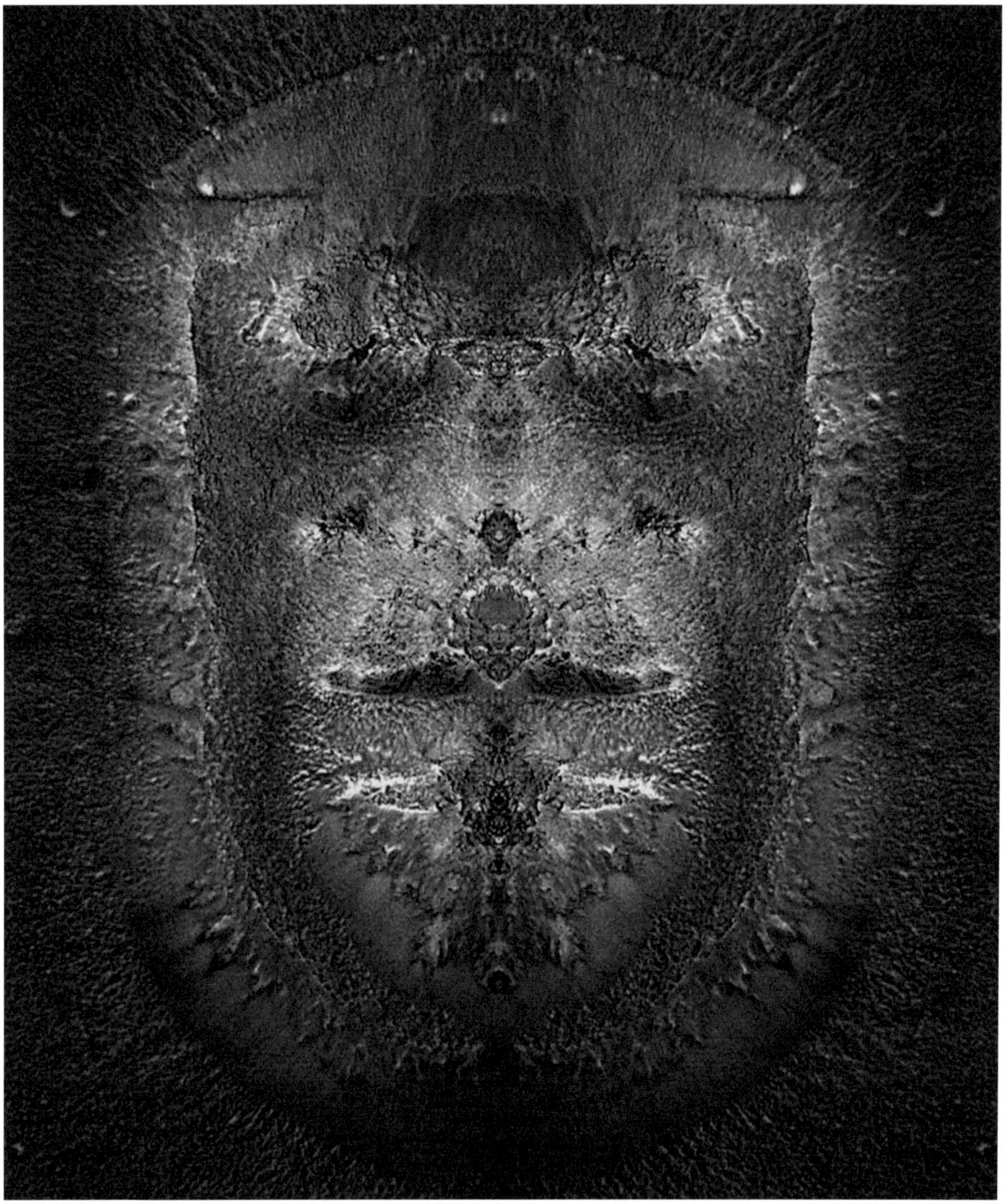

"The answer in fact, is staring you right in the Face."

"Maybe it's time to actually look."

It is intriiguing to also remember Garvin's words in the NASA hit piece had astonishingly acknowledged that the **Viking view of this Cydonia enigma bore a strong resemblance to "an Egyptian Pharaoh."**

*(**Article and scientific forenssic information and some article images provided by Richard C. Hoagland.**)*

Yet, considering the trained eye and those in knowledge of this subject and feature, all of a sudden the feature now looks quite different to those who never really looked hard enough at the data which was given to us all. The feature changes drastically from what NASA tried to portray as scientific fact when NASA never presented science.

Which brings us all once again to ask ourselves again at this point,... **WHY???**

NASA must know that someone would notice how the images were released and being released, and would also work with the images to 'fix' the unscientific NASA public perspectives, to show how NASA deliberately misconstrued the image data, and had outright lied to the public which obviously NASA knows makes them look bad. Fixing not just the images being released, but also to their scripted official wordings. So WHY?

I am believing and stating that if we look at this all in the perspective as I have laid out in these chapters, of us getting coded information and truths hidden in plain sight from NASA's official scripted choices of wording used regards to the Face in downplaying and trying to debunk it, then this all once again, kind of makes sense that someone, despite the official policy given to NASA by the U.S. Government in dealing with the Face and with the surrounding nearby other mesa's implications of them all being intelligently designed techno-signatures, was also trying to provide a harder to see dual story of events and facts.

At least we have consistencies here worthy to consider by how we are piecing this all together throughout the years, **IT IS OBVIOUS, THAT 'SOMEONE', WHILE ADHERING TO THE POLICY TO PUBLICLY DENY THE REALITY OF THE FACE, HAD ALSO SCRIPTED THESE OFFICIAL STATEMENTS CAREFULLY!** Not making it obvious, but *hard to deny once found*, that there is an underlying storyline which allows facts and truth to be clearly seen of NASA's ***UN-OFFICIAL*** positions regarding the Face. The only way to SEE it, is by considering the official code wording of their statements is just a cover story, which obviously it is, as Richard Hoagland had also uncovered and revealed for us all. A cover story that changes the picture perspective in our minds as to what we think we are seeing on one plain, (*the surface*), which we expect and naturally think and believe it is the truth. I mean who would look into their scripted official words so deeply to even think that there is something else to glean there in those words? (*Me of course*). I mean, it is CRAZY right? Or is it?

What we see, and what we would see if we indeed figure it out, will make us think about it, which is what we are doing here now. A wise man once said that "*You cannot find something if you are not looking for it*!" But how can this be true, if this is as I described it so far in this weird twilight zone of a study and investigation? In the least, as I always say, you will all be entertained if anything else. But I am trying to help you SEE what was meant to be SEEN! If this all was nothing to see at all, as NASA officially claims officially about the Face, we would NOT even be here now discussing about this and all these possibilities correct? It goes hand in hand s it were. There would be nothing at all to discuss as there would literally be nothing to see here at all! Yet there is plenty of things we have seen, showing great reason to look deeply not only into the images of the Face itself released to us all, but also of the official scripted and well thought out comments released to us all about it. It changes the perspective not just of the images for us, but of what is really trying to be told to us all where officially it cannot be stated.

Their official coded statements through TIME, is what was and *WHEN* this is and was meant to be seen and figured out! It seems NASA wants YOU to figure it all out and to SEE it also?! This makes this so much more bizarre! But as I said, at least to this point, you all should in the least be entertained. I admit, it seems hard to believe.

That if we figure this all out, then we can ascertain the unofficial position where we are told or clued into of there being a controlled policy in place by the U.S Government unto NASA that distorts the truth to the public, yet also hidden on the surface with code to piece together that an UN-OFFICIAL REALITY INDEED EXITS THERE AS WELL!.

NASA will never admit to it of course. They as well as others, and maybe you even, would say you're.... **MAD!**

Now we come to a point in the NASA HIT PIECE, that I believe we are given something REAL from them showing us all, between the lines, even on top of the surface, that we can see something which NASA really wants us to see, but shows us all that it is information and image data showing us all that the Face of Cydonia is something else as projected to us all officially. We never seen anything like this.

"The laser altimetry data are perhaps even more convincing than overhead photos that the Face is natural. 3D elevation maps reveal the formation from any angle, unaltered by lights and shadow. There are no eyes, no nose, and no mouth!"

Now since we all went over that you may be thinking, by Richard Hoagland's accurate scientific breakdown and revealing the truth NASA lied to us all, that there is nothing at all left to see in those coded scripted words, then YOU,... ARE MISTAKEN!

As far as I have seen and known throughout all these years since bginning my independent research with Mars and Cydonia and since the Face image release in April 1998, I have seeen no Mars researcher, mainstream or independent, nor especially anyone at NASA to have mentioned about this high quality information, or which has even noticed it apparently. I did see it, and I sat upon this information for many years because I was unsure of the relevance, until I pieced it all together, and became...'MAD', as I said.

Because as I see it, even on the surface as we see it, as what is shown and what is being said, we can see it's accuracy in both plains. It all depends upon how you SEE it all, to figure it out. Then one may see it seems NASA officials have indeed given us all a 'PEEK' of the truth, in an un-official manner as we now see it to be, in oder to SEE it, as we have not seen it ever like this. (*This is NOT necessarily true, but that is another book to reveal*).

For instance let us look what we were told again. Like the Face as we come to see and learn, is a dual Face with 2 different sides just as intelligently designed and purposed as its connected opposite side, while sharing their own separate information to glean from it, so indeed, is the script of what is said. Being 2 Faced is an analogy here. Having 2 sides, dual meanings. Think carefully and with your mind's eye.

"The laser altimetry data are perhaps even more convincing than overhead photos that the Face is natural. 3D elevation maps reveal the formation from any angle, unaltered by lights and shadow. There are no eyes, no nose, and no mouth!"

I firmly belive and trust I am accurate in what I am trying to sshow you all I worthy to consider is true. Let us look at the words:

"The laser altimetry data are perhaps even more convincing than overhead photos...!"

Here is what I glean from this also. The laser altimetry data are perhaps even more convincing than OVERHEAD PHOTOS! The laser altimetry data is the information we gleaned in this code. Telling us that this code information, is **EVEN MORE CONVINCING**, than the **OVER HEAD PHOTOS???** Yes, the Over Head photos, the cover story! As I mentioned about. We are being told that this coded information we are lasering in on now here, the information being given here is information being attributed as being laser altimetry data that is Even More Convincing than Overhead Photos? Yes, more convincing of the images in your mind that the scripted words that puts photos in your mind that they try to paint, by their cover story. The cover story we are being told is less convincing than the story under the surface photos in your mind you are seeing now of this reality behind their words. Because we are told that the laser altimetry data are perhaps even more convincing than OVERHEAD PHOTOS that the Face is natural! Their official storyline is not as convincing as compared to what I am laying out here now for you all to see in those same words. Yes indeed! What we see in coded information being seen and put together here now, under the surface is more convincing than the overhead photos (*the picture they paint to the public minds using their official wording*), that the Face is natural.

It is comical it seems. The code seems to be reaffirming itself?! Telling us all, as if by prediction somehow, that they (the scripters of thee official statements) can see or which sees us looking at this code, telling us all that *they know* their cover story is not convincing to you all, the public, while telling you all also at the same time, that this underlying storyline of information being decoded as it were to you *NOW*, is also more convincing than what we see on the surface in their words trying to officially state and convince us all, that The Face IS Natural! I am laughing hard, because it is true. The code is acknowledging itself, and acknowledging its opposing other side position of... THE FACE OF MARS!

This is not a coincidence. As we were lead here since the official statements of Viking, until this point of where we are at now. We would not be here now, considering or seeing this, if there was no substance to bring us all here now.

"The laser altimetry data are perhaps even more convincing than overhead photos that the Face is natural!"

In this statement alone we are being given un-official truth. The truth as I have just laid out for you all to SEE. We are being clarified of my theory as being vindicated by this code as it were.

This statement tells us information of truth. A hidden disclosure to be seen down through... TIME! As it is in fact the truth!

Remember to SEE this, we have to untwist their words, to match that what we are being told in comparison to the image releases. The COVER of the truth tells us that: **"The laser altimetry data are perhaps even more convincing than overhead photos that the Face is natural. 3D elevation maps reveal the formation from any angle, unaltered by lights and shadow. There are no eyes, no nose, and no mouth!"**

Thanks again to Richard C. Hoagland, for providing a scientific rebuttal for this statement NASA scripted and provided on the surface that shows NASA lied to us all in that statement, and shows their overall statements and attempts to deceive us all, was not as convincing to us all, as this all here now is.

However, in code? The 3-D elevation maps represent this book, in what I have been laying out to reveal this underlying storyline. This code revealed *the formation*, of *itself*, from any angle (point of view or from both sides), unaltered by *lights and shadow*! Thus, something invisible! ***We are told There Are No Eyes, No Nose, and No Mouth?!*** Indeed, because what is speaking to us, is NOT a person, but the code itself. The scripted Maps, which is guiding us and telling us information *down through Time*, that is true!

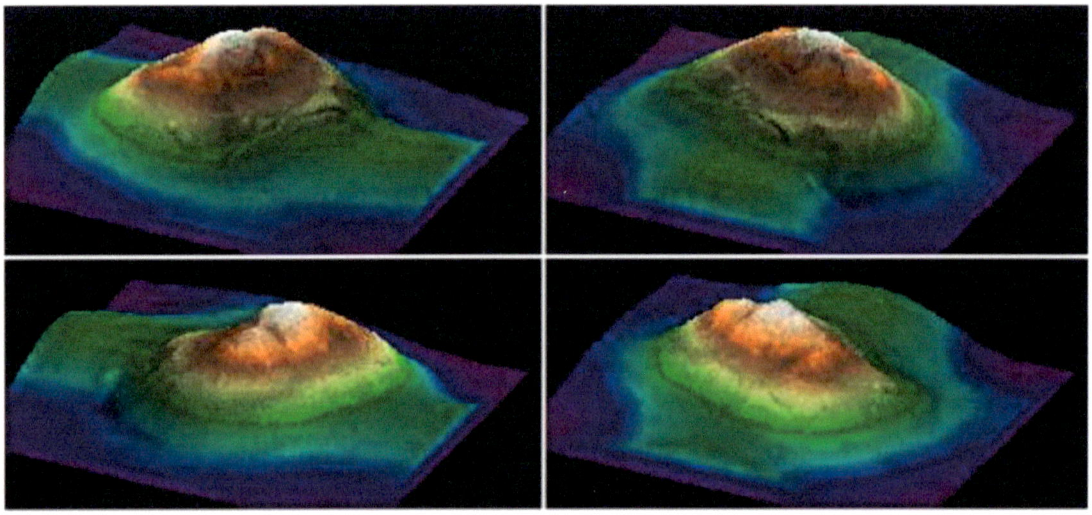

Above: Four views of the "Face on Mars." Jim Garvin (NASA) and Jim Frawley (Herring Bay Geophysics) combined MGS laser altimeter data with high resolution MOC images to produce these 3D images in which color represents height. The peak of the mesa-like landform rises 800 feet above the surrounding plain.

Once again, to SEE this decoded statement hidden within the surface of official deception, we have to untwist their words again, so therefore we must untwist the image once again that is presented to us on the surface, we now notice that the Face of Cydonia as represented here is 'flipped' over horizontally. Like the statements are flipped. So we must now un-flip it to SEE the ***right side*** of things in the correct perspective, and find out what it is we are being clued into here of what we supposed to see, that has not been seen in overhead photos. The clues lay in the official statement itself. We have to see if it is true, or false. A test as it were?

If we fail the test? We NOT-SEE the intelligence of the coded messages. We miss the truth because we did NOT-SEE the intelligence information being given to us un-officially.

If we pass the test? We DO SEE IT! (The intelligence of the coded messages.)

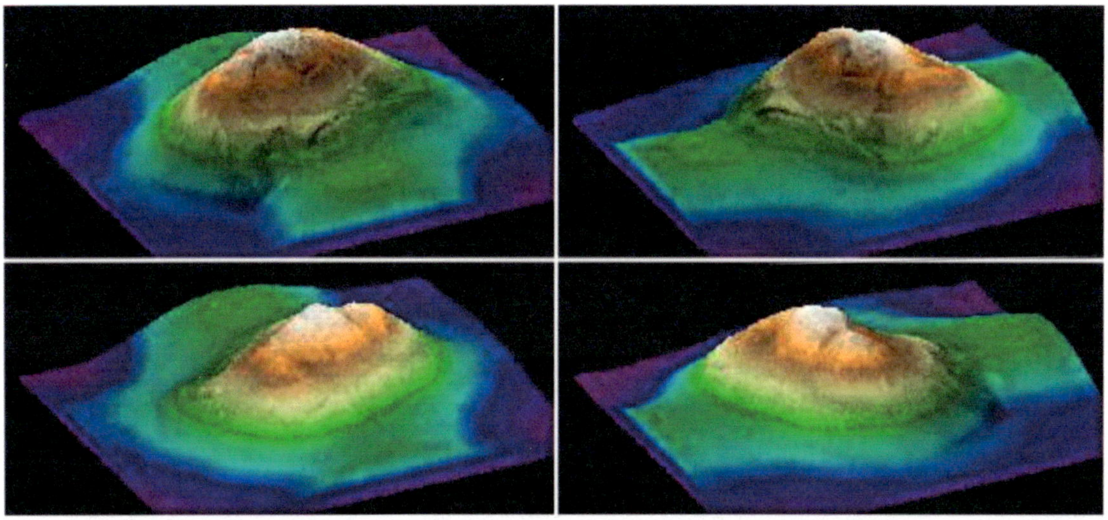

Above: Four views of the "Face on Mars." Jim Garvin (NASA) and Jim Frawley (Herring Bay Geophysics) combined MOS laser altimeter data with high resolution MOC images to produce these 3D images in which color represents height. The peak of the mesa-like landform rises 800 feet above the surrounding plain.

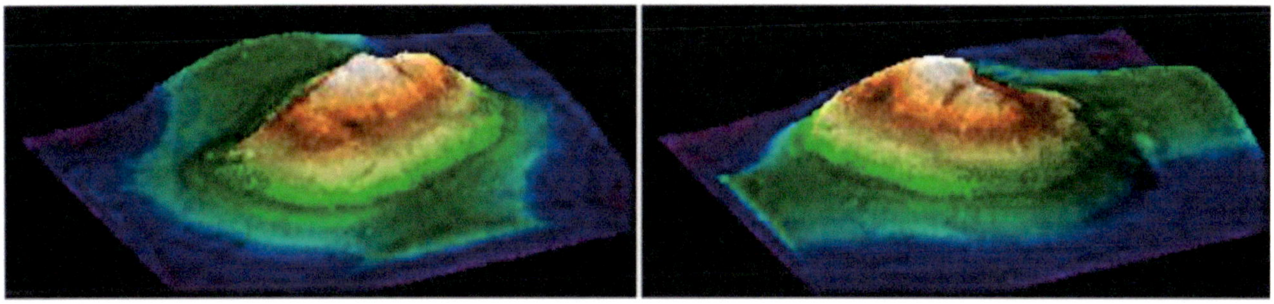

Considering this truth in this Face (of understanding), we can see clearly that it changes the overall perspective of this situation when considering this truth in what we are being told unofficially behind their official position!

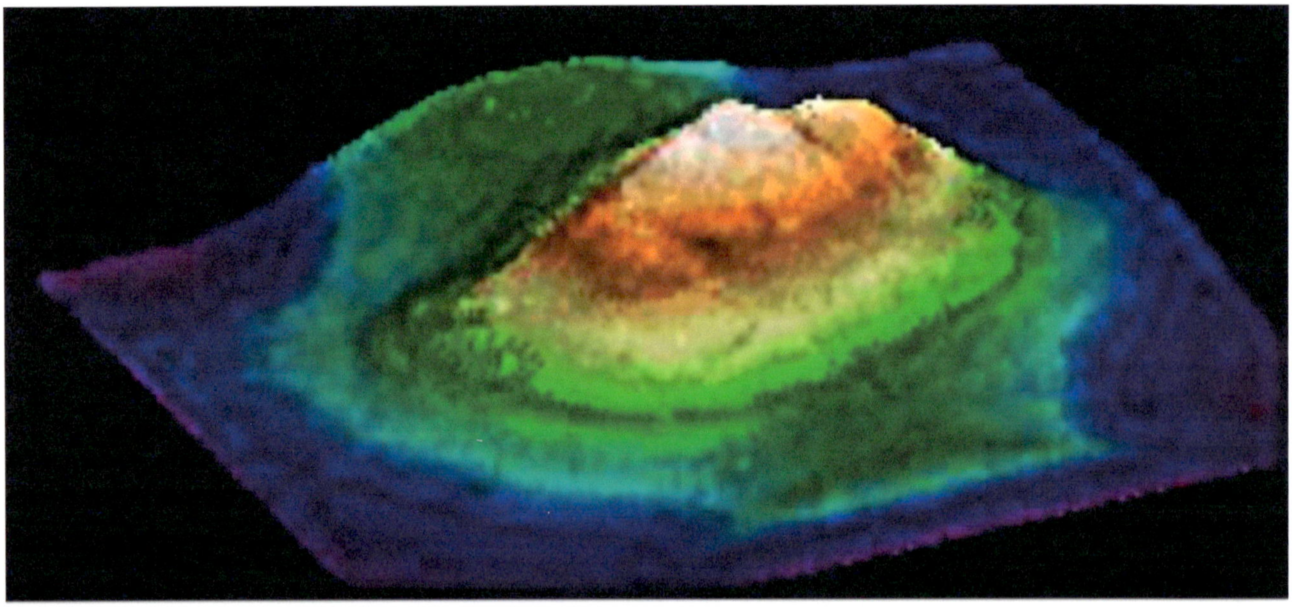

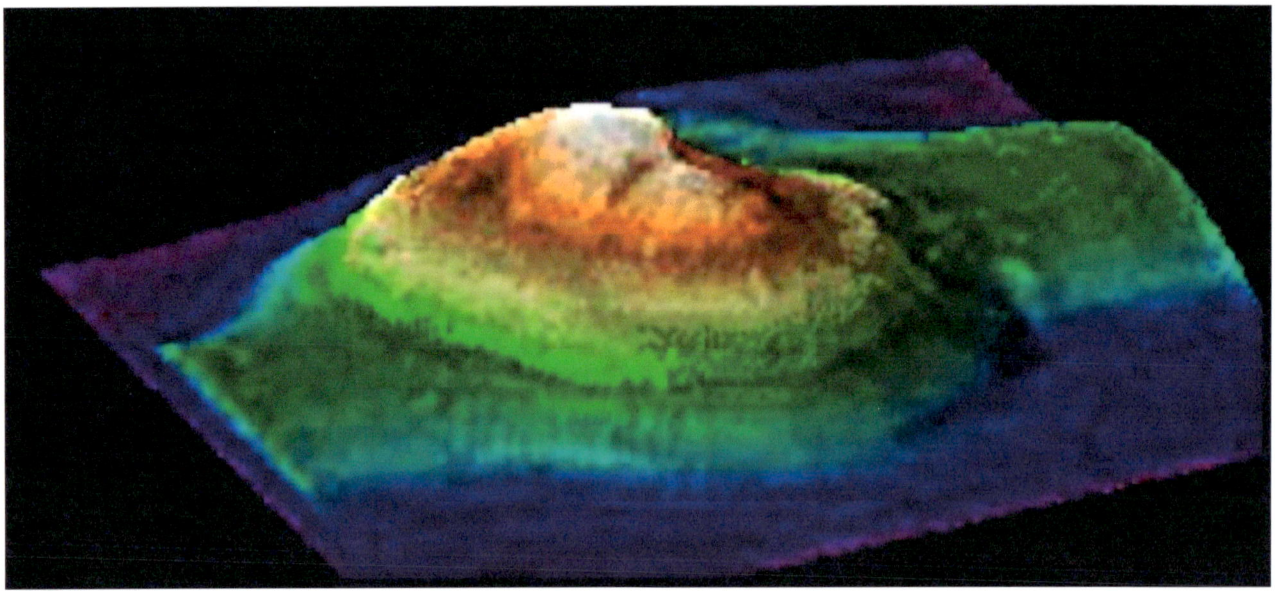

Yet another interesting fact is when we look at both image examples for the MOLA which was given to us all, no one at all noticed or mentioned the fact that even these images, which should be consistent with each other for the shape of the Face, both show different features! This leaves us yet again, with a question. Which image view is the real view and shape of the Face? **Or, is neither of these images accurate?** How can we find out?

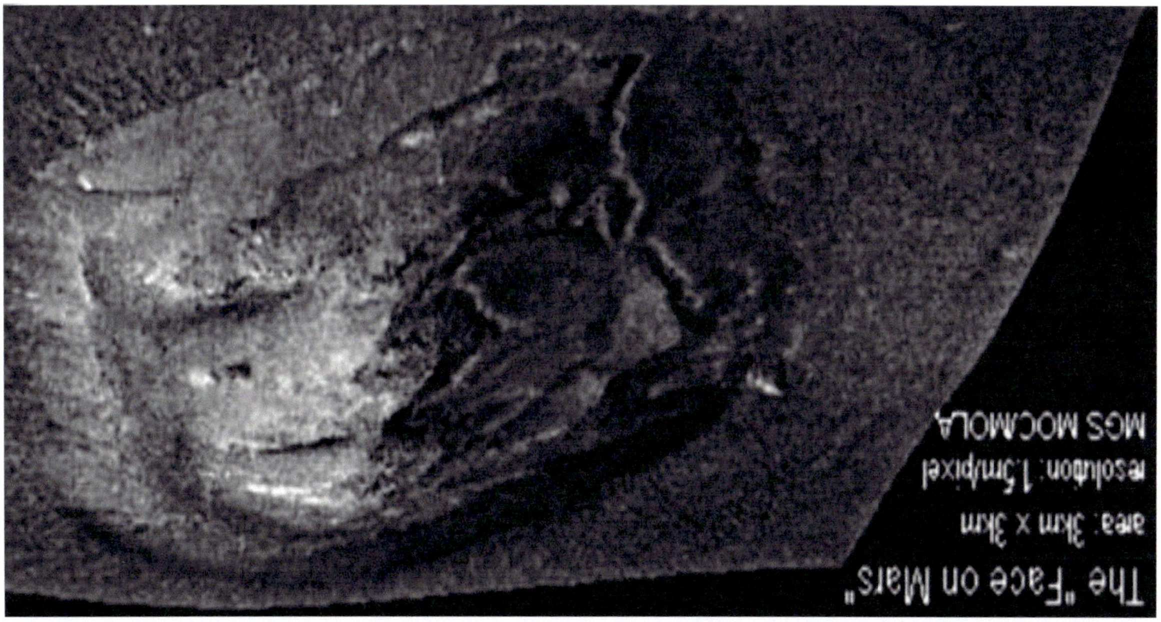

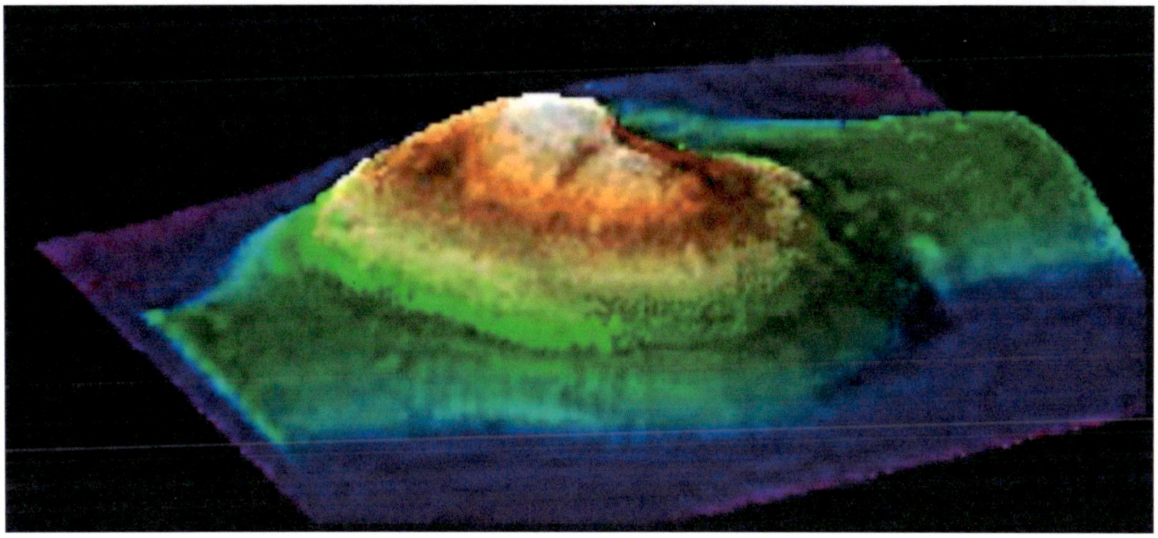

"Perhaps the best way to unravel the mystery would be to send a geologist to investigate."

We have dug through truths protective layers!

To Reveal the Tricks Of Light and Shadow' of the dual tracks of the coded script!

WE HAVE UN-MASKED,... 'THE FACE OF MARS'!

Unmasking the Face on Mars wasn't easy!

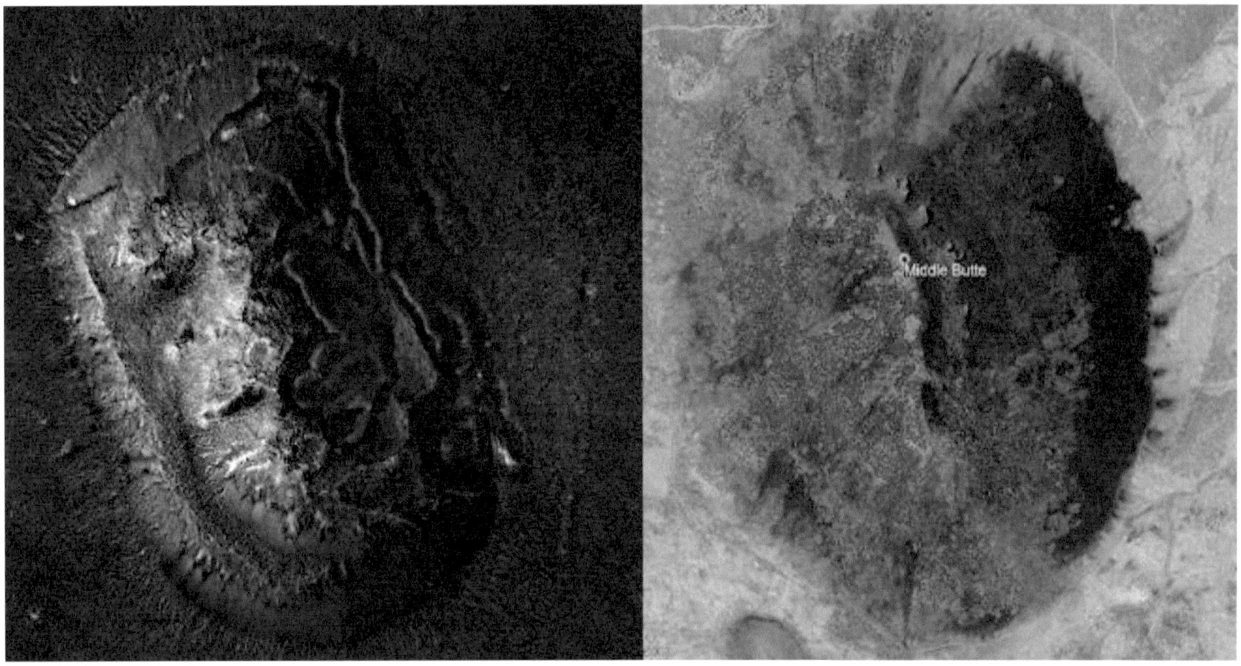
Middle Butte

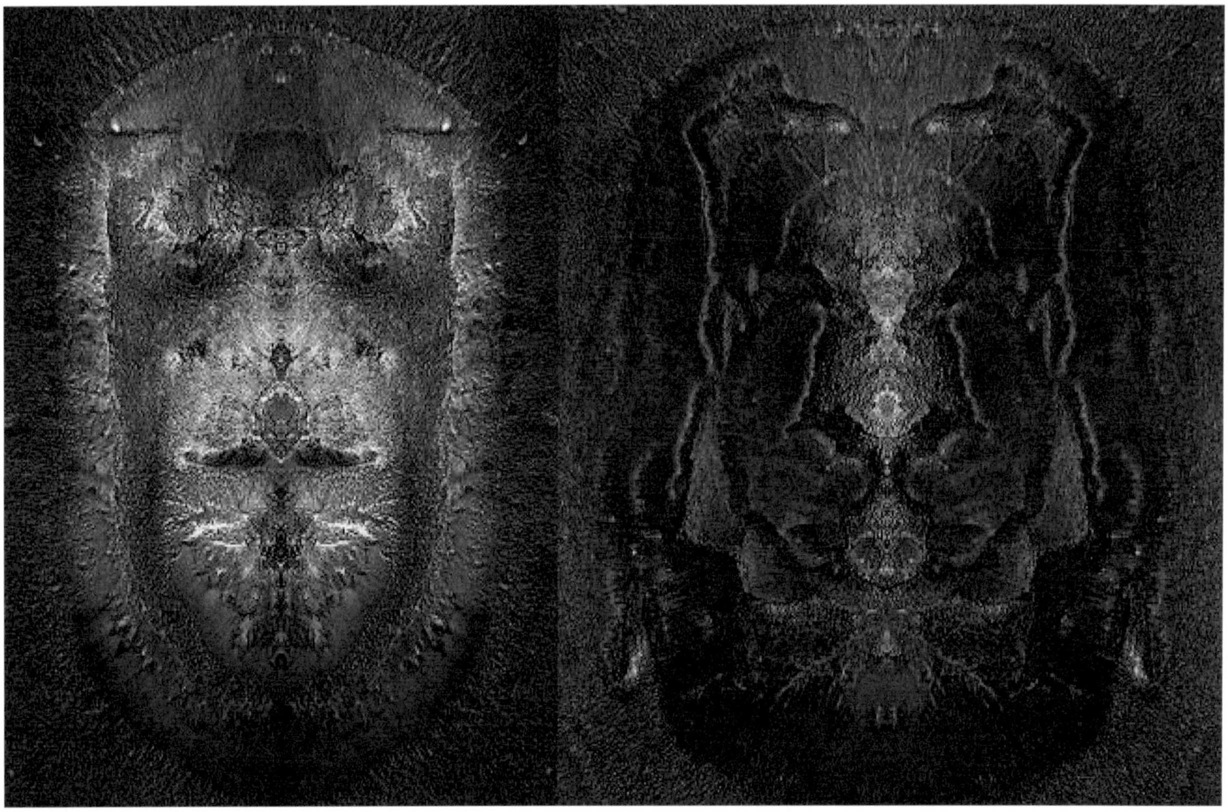

CHAPTER 5:
CAN THIS BE TRUE?

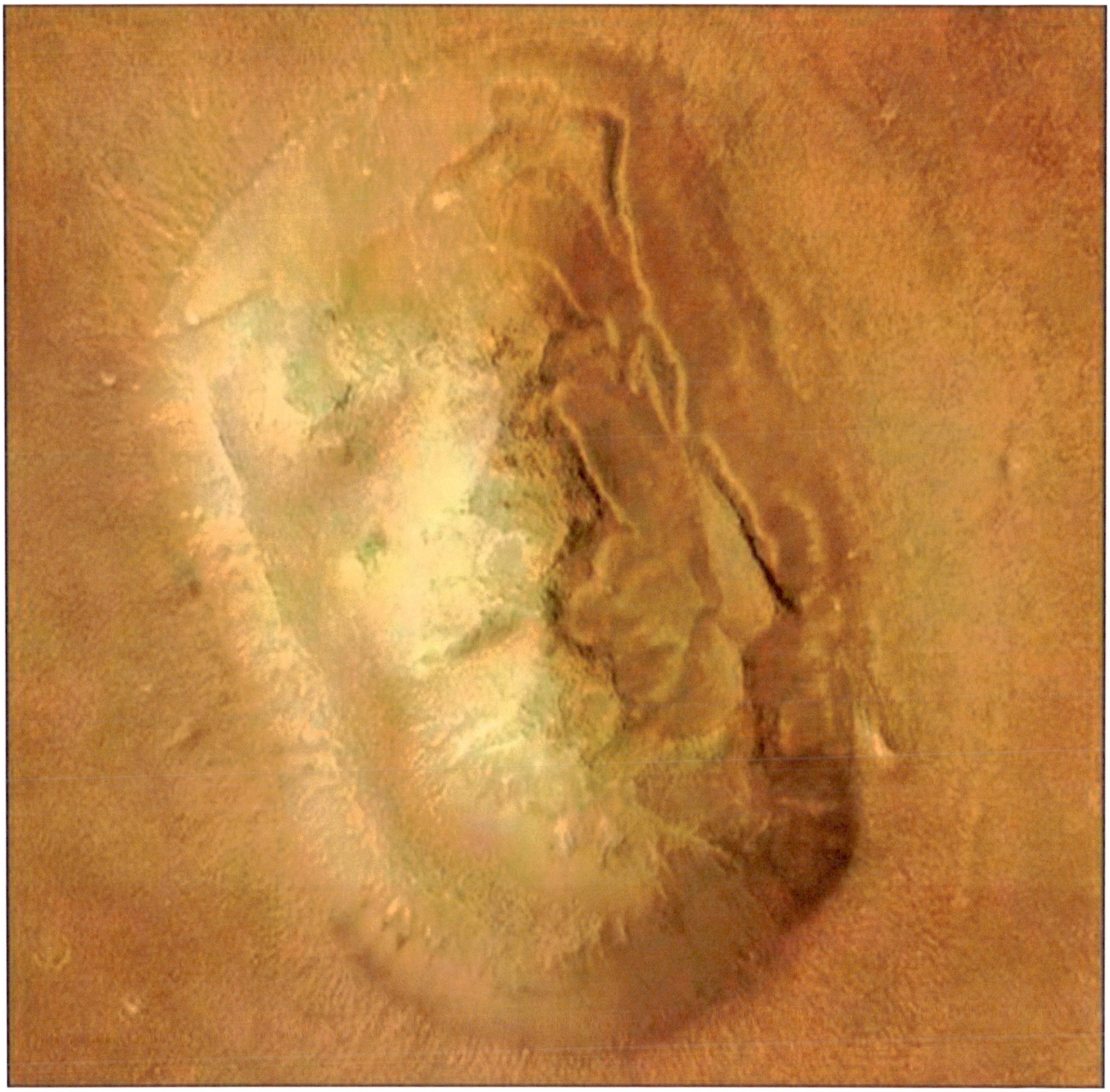

- **Released** 21/09/2006 12:23 pm
- **Copyright** ESA/DLR/FU Berlin (G. Neukum), MOC (Malin Space Science Systems)
- **Description**

21 September 2006

Finally, on the 21st day of September of 2006, a special and final 'HIT PIECE' press release was given out five years since the last hit piece of '*Unmasking the Face on Mars*'. This new hit piece was a semi-collaborative effort between Malin Space Science Systems (MSSS) Mars Global Surveyor, and the European Space Agency (ESA) Mars Express, that was supposedly focusing in on the Face at Cydonia by showing comparative results of imagery from MSSS on behalf of NASA and imagery shown to us from ESA. These images however only provided us with having more questions rather than getting any real answers yet as to what we are to see as the reality of the shaping for the Face. There was st ill issues about that Face in what was supposed to be answered by now at this point in time.

There were aspects about these amazing comparisons to which did not make any sense at all when visually comparing and studying them all together. Instead of results being given and shown that should show a replication of results? We instead get comparative image data that show instead a contradiction of results to compare between what the two space agencies separate mission imagery shows us, thus it was no comparison at all in my opinion. The results were not able to be replicated by what each side had provided and had to show to the public. I wondered and asked myself and others back then of why they even bothered? Although I really appreciated the new image data from both sides at the same time.

The way the Hit Piece press release was worded also, one can easily tell that the bias for considering the Face to be an artificial structure or designing, was immediately shown with words only, to obviously to be played down as usual, so no surprise there. I only figured that by now, from seeing and experiencing this all of their press releases over the years, why would I or anyone else for that matter, even consider what NASA says about it of being any truth? I had to study the feature intently myself to determine what my eyes were showing me, and that I was and still am seeing a Face! Despite what is officially said.

As with all previous Hit Piece press releases, the theme for the press release was to discard the notion of artificiality with the notion of the Face being a face, and forced as it were into all our minds that it is again, just a normal natural feature. It seems by NASA and any other official agency thinks by repeating something not true enough over and over, that somehow we will all think it is true, despite the evidence of our own eyes and analytical prowess by what all the image data shows. So we got nothing new there, and we were not surprised of that continuing official unscientific theme. With what that article was stating, it was propaganda of what the Face is NOT, in my opinion on behalf of NASA's position that there is nothing to see here at all but a natural mountain. The article was not science based but opinion based. Especially by those who not want to say what they really see, or not know how to explain what they really see! That speaks loudly in my opinion! The data speaks for itself!

The Hit Piece press release started off by showing us an image of a 3-D perspective view of the Face at Cydonia, which was almost kind of hard to tell what it was at first. We will get into that soon. It was a perspective image view of the Famous **MOC2-283**, or better also known as image number **E03-00824** taken in April 2001. And then released to the public on May 24th 2001.

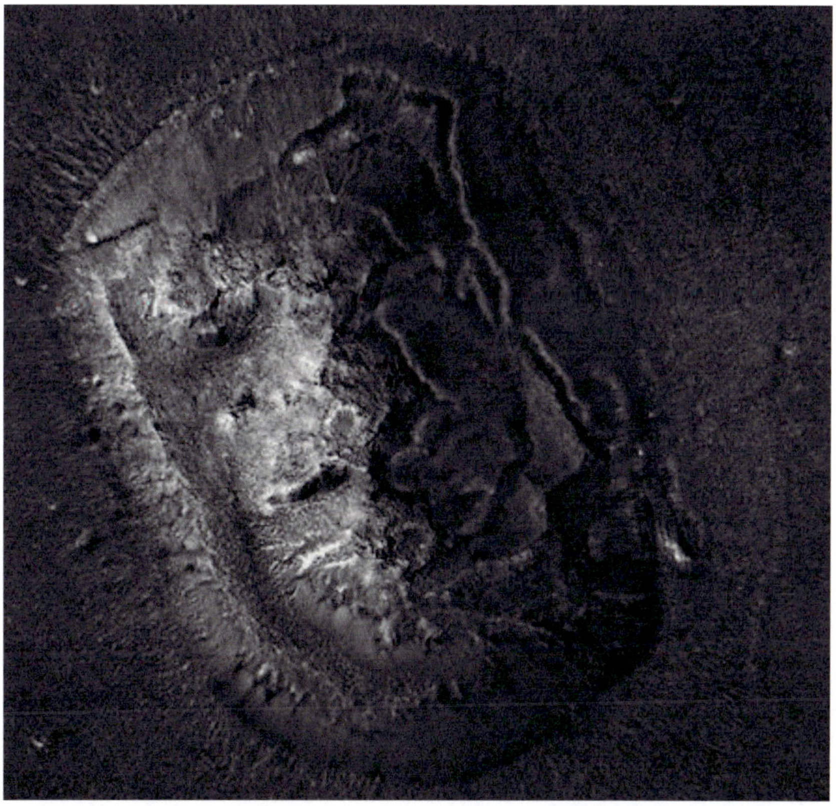

This image however was different than what was originally released to us all previously, and no one had noticed or cared it seems, but me who had made a fuss over it, that the image given to us for this new 'HIT PIECE' was amazing. As the image was not seen in black and white, but was now given and **seen in color**!

That is the first immediate fact I noticed and noted well publicly, again, that no one seemed to notice or really even care about. There was a debate that arisen which also no one else followed up with, was regards to the color source. We will get into that soon too!

The second immediate fact I noticed and to which everyone else at that time were also making note of, is what the Face perspective 3-D image was also showing, too much of everyone's surprise and even shock. It appears to show that the Face had all of a sudden received a huge 'BUMP' to the forehead, between the eyes of the Face. It looked so surreal to all of us whom researched into this Face at Cydonia, to see that it looked like as if the Face had been traumatized as if being HIT on the head with a bat or some object to cause it to swell up like that between the eyes.

I had joked publicly about it stating that this was proof of this so called press release being really yet another 'HIT PIECE', because again, it seems as if the Face was hit and given a huge bump upon it, to be seen in the perspective view. This 'BUMP' in my opinion was a distraction attempt that took away all the importance of the fact that this image was released to us all in color. No one cared surprisingly, or had focused on that fact, but instead it was the traumatic bump that caused sensation and conversation, as well as hot debate. Creating distraction amongst all the independent Mars researchers to only focus on what ***was obviously falsified data*** that was being published as an official presentation of what was supposed to be accurate scientific image data to the public.

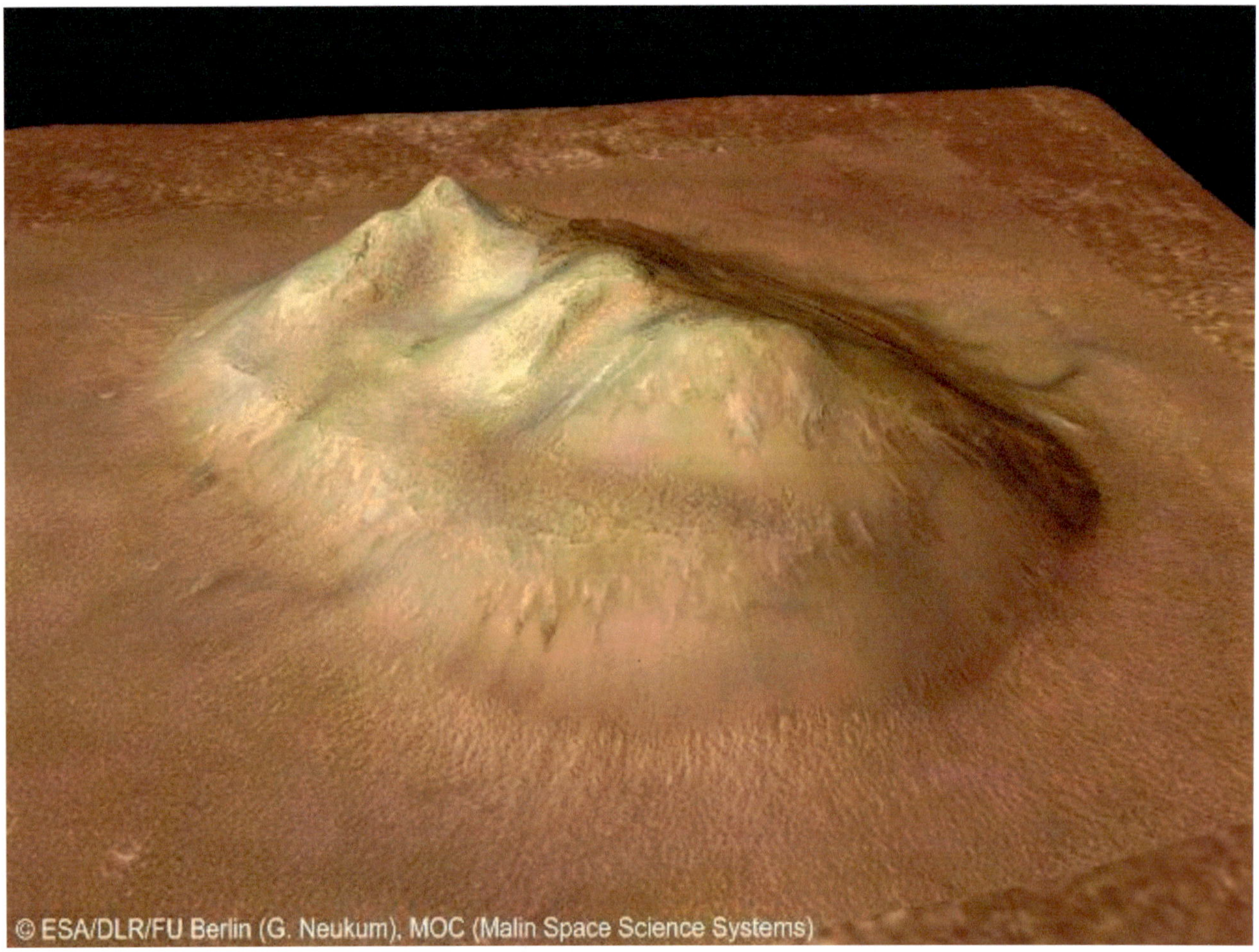

It had appeared to me that it was a proof given to us all by Mike Malin of MSSS (Malin Space Science Systems), on behalf of NASA to us all, to which seemed to say propaganda about the 'naturalness' of the Face feature almost as if MSSS were trying to one last ditch effort to officially make us believe it, yet we know and already learned to see that they do not believe it themselves. They had to HIT the HEAD between the eyes, because in my opinion, they could not kill the idea that the Face is not natural! The image data shows this even to them! So they had to once again, manipulate and twist the truth! Mike Malin, in my opinion hated and hates the Face in what it shows and represents. Those who know Malin and know his history regards the imaging of the Face and Cydonia in general can understand my opinion. Especially because of those who share a strong interest of the Face as being an important scientific techno-signature artifact worthy of close study, which NASA and Mike Malin is not ever willing to admit.

Therefore this final NASA 'Hit Piece' with NASA's official position has always remained the same; Down play the Face at any and all public announcements possible, including news media sources, and at all costs, to keep the official NASA public position for this Face as being of a natural origin alone! Even when separate image data is coming forth publicly from another international player in space exploration and missions to Mars, that show that again, the image data is questionable from NASA's official point of view and contradicting. Not in words being spoken, but by image data that is provided that shows NASA's deception on their part for failure in bringing forth accurate science and accurate image data.

To me, it was very hard to see the collaboration of ESA at all in this article, in regards to official positions on the matter. ESA's Mars Express image data speaks for itself. It appeared as if ESA was not involved in the official written position at all. They just seemed to parrot the position NASA propagandizes. I figured this was because they were counting on NASA funding and support and participation into their own Mars and space ambitions and endeavors. The article also posted links to previous press NASA releases put forth by MSSS/NASA on the Face, which we already went over in previous chapters. So at this point in time, it is no surprise yet another 'Hit Piece' was pushed forth. This time NASA and MSSS, were acting as it seemed with frustration and tried to control the mainstream and public perception and despite what ESA had to show with the release of their amazing image data views of the Face at Cydonia. With ESA remaining silent other than parroting and detailing the acquisition of images, and with MSSS having the dominant control of the wording and the **controlled narrative** for the Hit Piece.

Yet it is ESA which is seen inside this article Hit Piece, which shows clearly a new chapter to be written in history and in this book, of showing the Face in terms of coming from an image source other than NASA, and actually shows what it still actually looks like and represents, A FACE!. Although ESA has spoken parroted words of NASA's official position in the article, they had provided us all with image data instead, that we can compare with MSSS/NASA's imagery of the Face that we are able to clearly see some discrepancies that do not match up in comparison of the two sides' image data to compare. THAT is science!

Note on images: The color scenes were derived from three HRSC-color channels. The perspective views have been calculated from the digital terrain model derived from the stereo channels. The 3D anaglyph images were derived from the stereo and nadir channels. Image resolution has been decreased for use on the internet.

Released 21/09/2006 12:29pm

Copyright ESA/DLR/FU Berlin (G. Neukum)

(OFFICIAL ESA STATEMENT):

A perspective view showing the so-called 'Face on Mars' located in the Cydonia region. The image shows a remnant massif thought to have formed via landslides and an early form of debris apron formation. The massif is characterized by a western wall that has moved downslope as a coherent mass. The massif became famous as the 'Face on Mars' in a photo taken on 25 July 1976 by the American Viking 1 Orbiter.

ESA's Mars Express has obtained images of the Cydonia region, site of the famous 'Face on Mars.' The High Resolution Stereo Camera photos include some of the most spectacular views of the Red Planet ever. After multiple attempts to image the Cydonia region from April 2004 until July 2006 were frustrated by altitude and atmospheric dust and haze, the High Resolution Stereo Camera (HRSC) on board Mars Express finally obtained, on 22 July, the HRSC finally met success during orbit 3253 over the Cydonia region at the best possible resolution of approximately 13.7 meters per pixel. Being also seen in 3-D, that show the famous 'face' on Mars *in unprecedented detail*.

"These images of the Cydonia region on Mars are truly spectacular," said Dr. Agustin Chicarro, ESA Mars Express Project Scientist. "They not only provide a *completely fresh and detailed view* of an area famous to fans of space myths worldwide, but also provide an impressive close-up over an area *of great interest for planetary geologists*, and show once more the high capability of the Mars Express camera."

Title 'Face on Mars' in Cydonia region, perspective

Released 21/09/2006 12:29 pm

Copyright ESA/DLR/FU Berlin (G. Neukum), CC BY-SA 3.0 IGO

Description A perspective view showing the so-called 'Face on Mars' located in Cydonia region.

It is these perspective views provided by ESA seen above, that speak for themselves and ESA not has to say anything to contradict what NASA's or MSSS's official image and position to the public is. The language in their (ESA's) statement speaks for itself in description of their images that tell it all. THE TRUTH! Notice that they stated their image was ***"a completely fresh and detailed view, of an area of great interest, for planetary geologists,*** **and** ***show once more the high capability of the Mars Express camera."***

That statement alone says so much truth as the image indeed was a fresh view of the Face as coming from any source, other than NASA. Notice though we are told it ***was a completely fresh and detailed view!*** Yes, the devil is in the details it has been said throughout history. The details seen in these amazing images by ESA shows some amazing details that shows that the Face is far from being a natural formation as NASA is so frustrated with in trying to convince the world is natural and a coincidence. In fact the ESA statement also said that this fresh detailed view was **over an area *of great interest for planetary geologists!*** Telling us all that they too claim this area is an area of ***GREAT INTEREST to Planetary Geologists***. Meaning THEM?! It makes one wonder if what I am saying is indeed so. It seems to be an admittance, of the Face and surrounding details are detailed! Possibly mocking NASA which seem to make the details harder to see? Because they stated ***"once more the high capability of the Mars Express camera."*** Noting the fact that the Mars Express camera has a High Capability, (to see details?), whereas NASA's camera only has High Resolution? Hinting that it seems NASA's camera although had high resolution, but NASA's camera NOT has a High Capability compared to Mars Express camera? To do what? The being able to image the details? Or rather as we earlier learned, the ***Alien Markings?*** It does and should make one wonder and think about it some.

Is this some confirmation to us all by ESA of what all we seen and learned from NASA's official and unofficial scripted coded statements? This really makes me wonder and think hard. Because to me, I see a validation and a vindication of what I tried to lay out what is really seen within NASA's previous official statements for the Face, as crazy as it sounds or appears. NASA provided nothing new regards to their words and image provided to their official position for this hit piece, unless we consider the manipulation of image data to the perspective view, as well as the added color data no one wanted to focus on.

A person can easily see that in ESA's Mars Express perspective views, that there is also NO bump on the forehead between the eyes of the Face, as provided by MSSS on behalf of NASA!

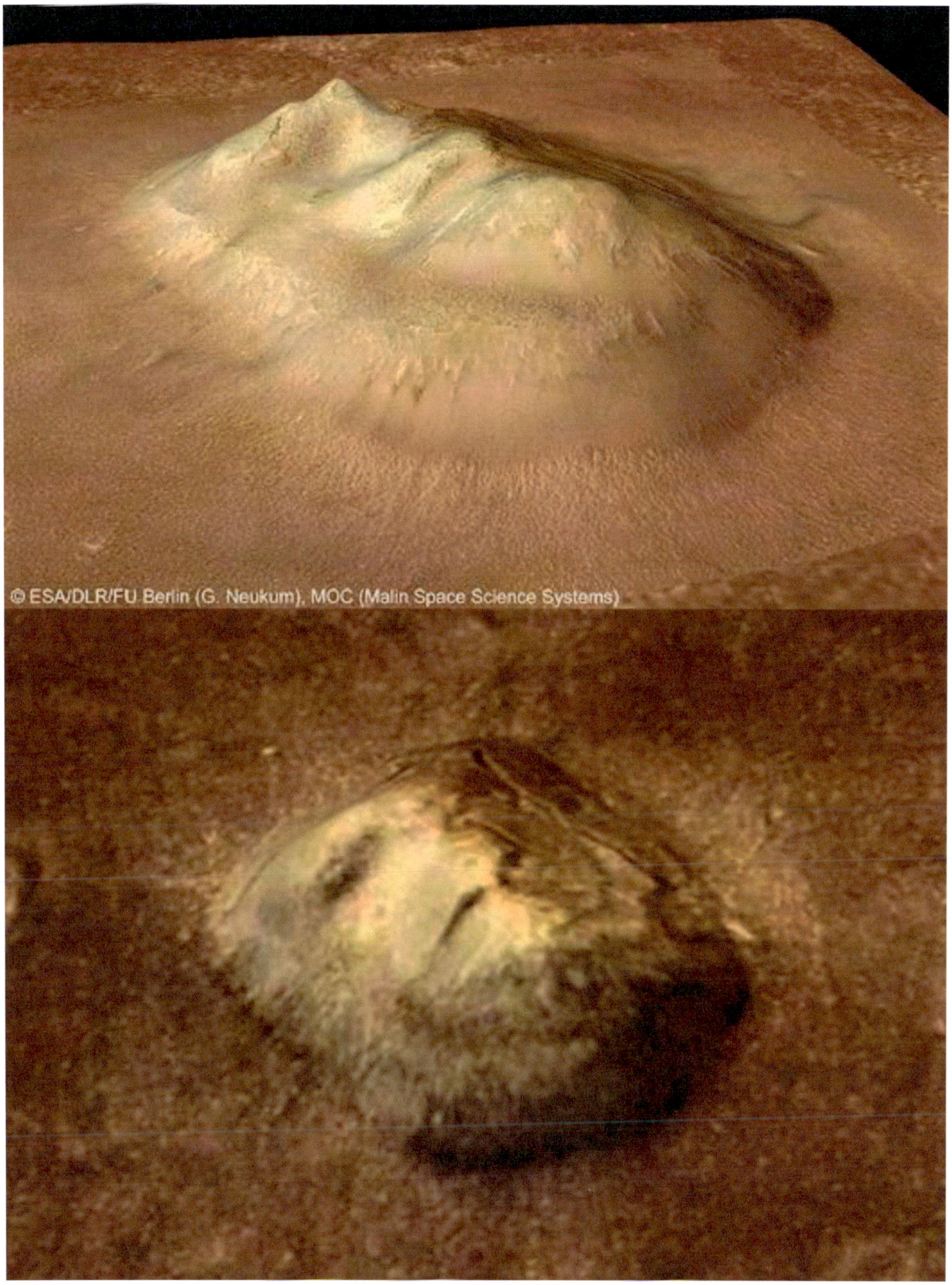

(Comparison test of MSSS and ESA image data clearly detailed shows NO BUMP on the forehead of the Face!)

To add caliber and candor to what ESA's image data shows in its perspective views as being more honest than NASA'S and MSSS's regards to the traumatic bump of the Face being manipulated and added by Malin himself, manipulating the truth of the image data to present to us all an image that is not accurate in comparison as seen in ESA's Mars Express detailed perspective imagery! To add comparative scientific analysis to validity of NASA's or ESA's image comparisons, we also have Viking orbiter image data from years previously, that Dr. Mark Carlotto had rendered with his shape from shadow algorithm to analyze the size and detailed shaping of the Face, which has been processed and rendered into video perspective views of the Face.

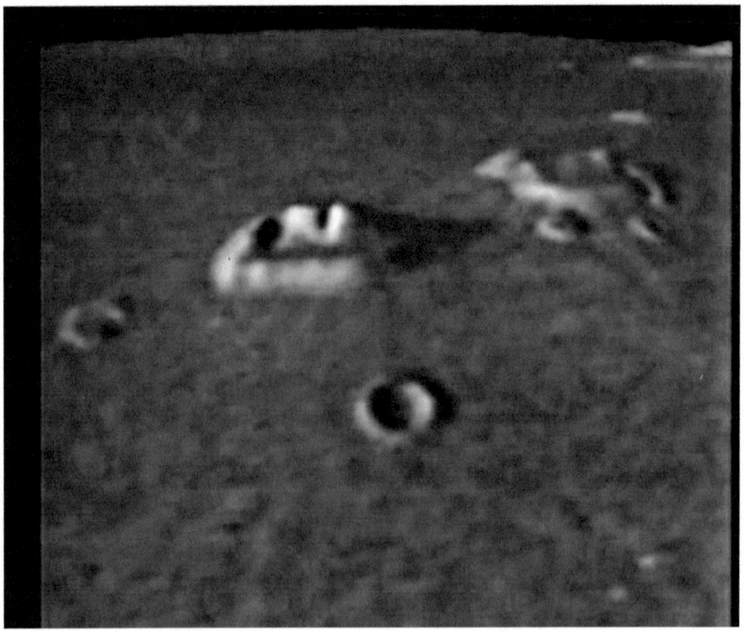

(This gets shown and explained in Dr. Mark Carlotto's book : THE CASE FOR THE FACE)

Notice again here, there is NO Bump on the forehead between the eyes of the Face!

Yet again, by comparing all these image examples, let us see once again NASA's perspective views from ***Unmasking The Face On Mars*** provided by MSSS. Notice that they do not match to what we see from NASA's Viking image data. Remember, we already know that these MOLA image views is MOLA crapola propaganda and not accurate in appearance nor are they even scientifically accurate, let alone visually comparable. So, how can NASA say they successfully were able to prove anything against any artificiality therefore so called successfully debunking it?

Their unscientific case of vocal and visual propaganda falls apart with merely using their own image data to prove their words as being false, because their case is only based upon their own words only to begin with, certainly not by using the image data only in and of itself to make their case.

At this point, all there is to say and ask is a simple question, **"Will the real shape of the Face, please stand up?"**

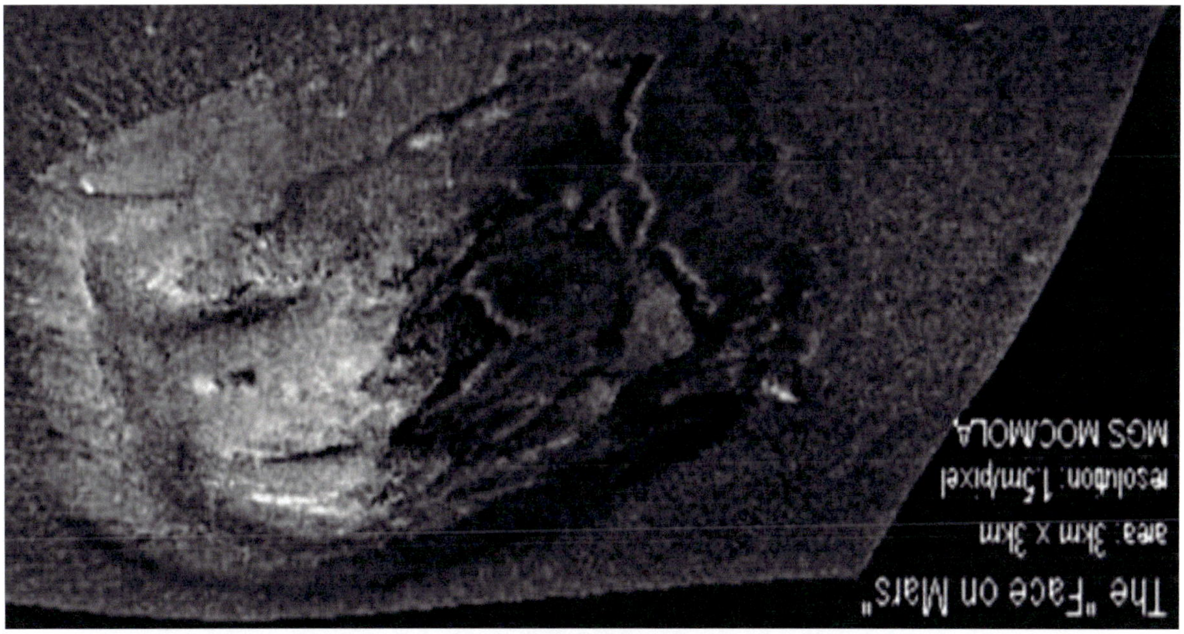

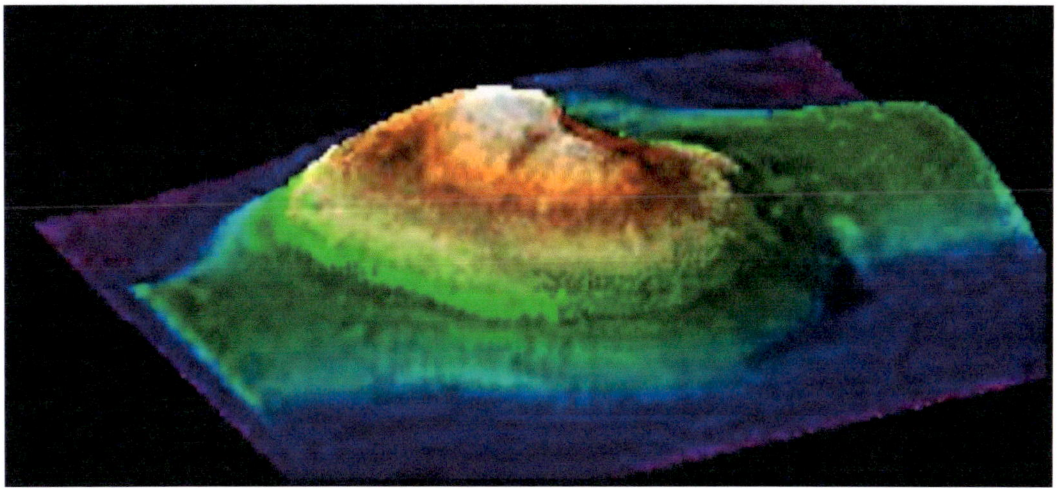

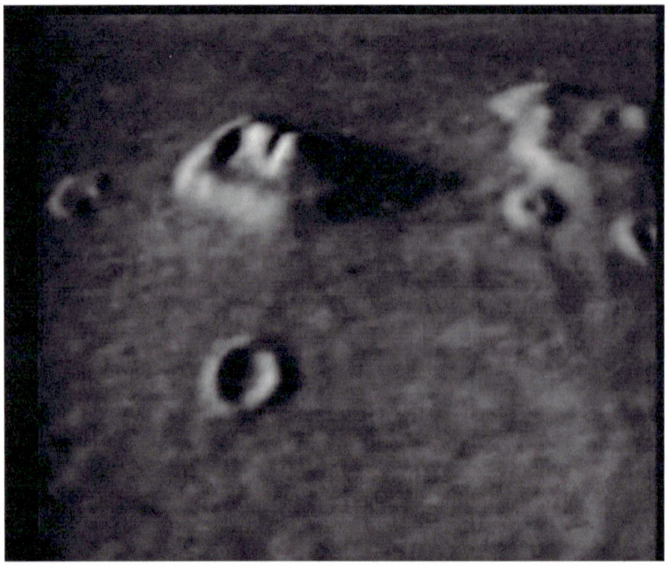

© ESA/DLR/FU Berlin (G. Neukum), MOC (Malin Space Science Systems)

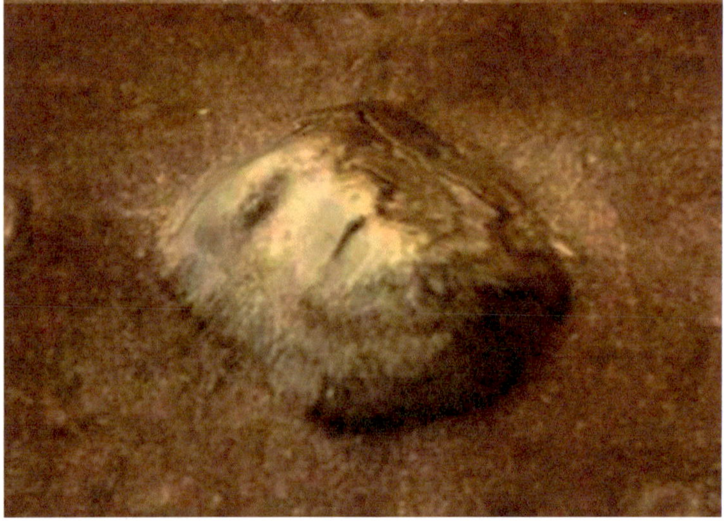

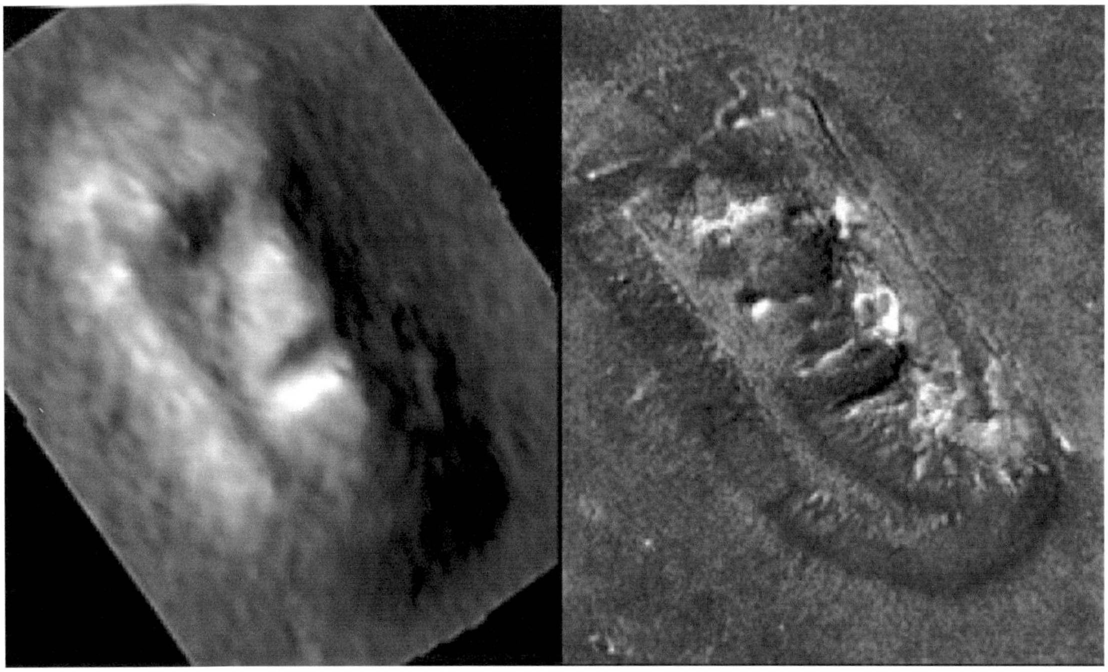

It becomes very obvious and 'detailed' to us all, that *ALL* of NASA's images that they officially released and made claims with, do not even match up with each other to compare for their own public narrative of the Face, let alone be an accurate comparison to what Mars Express provides in perspective views, and also what Viking provides in perspective view, as shown by Dr. Mark Carlotto. It is simply ridiculous for anyone to think that this collaborative ESA and NASA Hit Piece release perspective view image provided from NASA's side is genuine and accurate to the reality of what the Face is and how it is shaped. The Mars Express release of the Face and Dr. Mark Carlotto's Viking imagery shape from shading examples do not seem very consistent at all regards to the shaping of the Face.

The most important aspect these comparisons show, is that there is no trauma bump on any other of the various NASA image releases of the Face before or after this perspective image view was given, nor is there any hint of the bump seen on any of the Mars Express image data given for this hit piece. At this point in time as this is being written, one must wonder and ask whether or not the communist Chinese will release their images taken of Cydonia and the Face that we will be able to add here to also compare? Highly unlikely! (***We will be focusing on more perspective views of the Face in a later chapter in this book.***)

In my opinion, this special collaborative release was one which NASA may have been reluctant or even unhappy about having to participate in this press release effort. It is also obvious to me again that ESA parroted the official NASA position of the Face as being natural, while at the same time showing detailed imagery that openly contradicts what NASA released and says about the Face. It is obvious also that ESA and NASA must have had a preplanned effort in contract for such a public propaganda release for the Face. I am making hypothetical points here of course, trying to answer for the fact we have contradicting image data between the two agencies, which is what really counts, while the official NASA position is upheld by ESA regards to what they say to the public about the Face being a natural feature. It not makes any sense to me. I mean why then do we have the visual contradictions?

So I am thinking that possibly, NASA may have been a bit upset with ESA? Maybe for the space agency going forth with their public release for the Face, after maybe not presenting it in a way that NASA may have wanted them to release or to not release their image data of the Face? Maybe despite their contract to do a public release on this together honestly to show the detailed HRSC (High Resolution Stereo Camera) of the Face, and showing it also with color? NASA and MSSS rather, seems to have really gone out of their way, while on an international collaborative and cooperative contract, to act unjustly and unprofessionally, by presenting false manipulated image data to the public that does not match to what ESA's Mars Express image data shows ***detailed*** of all its ***Alien Markings***? In the very same public release no less!

Obviously the data provided to us by ESA of the 3-D perspective views of the Face from Mars Express shows clearly that the MSSS Mars Global Surveyor perspective view is in error regards to the accurate Face shaping details. Or is MSSS not so subtly telling ESA they are mistaken? Either it is NASA's image data is correct, and Mars Express image data is incorrect? Or the Mars Express image data is correct, and the NASA's image data is incorrect. We are deliberately left on the edge of a cliff, by not getting the definitive answers or even the definitive image data of which that was supposed to be part of a definitive final public release and official position from two international space agencies in successfully debunking the Face and proving it is after all, a natural surface feature.

I believe again, that this Hit Piece was done to provide us with also a distraction to focus on, trying to have all focus in on what the true shaping of the Face is NOT, instead of us all asking more questions to the other obvious important questions that this data from both sides also provides us with. Which is ***THE COLOR DATA*** between the two space agencies views provided to us all.

These also are different from each other. Meaning that the color data seen on and as the official MSSS release ***is not the same color data*** seen on and as the ESA color image data. Meaning that the ESA color data was NOT used to overlay on top of the MSSS image.

Why is this important? This is important because MSSS had **NEVER *released a color version for image MOC2-283***, or better also known as image number **E03-00824** taken in April 2001, and then released to the public on May 24th 2001. Yet we now know that indeed, by this public release that MSSS had obviously withheld it from public release until this September 21st 2006, special 'Hit Piece' press release was given out. Even then, ***there was NO full-on nadir* (directly above) E03-00824 *color view also released with this HIT PIECE to be downloaded at any time, then or now from MSSS, or from NASA! Why?***

Only a high-resolution false perspective color view provided that had its image data falsely manipulated to continue to mislead us of the true shaping of the Face! Yet I ask, ***does that mean that the colors of the Face is also untrue?***

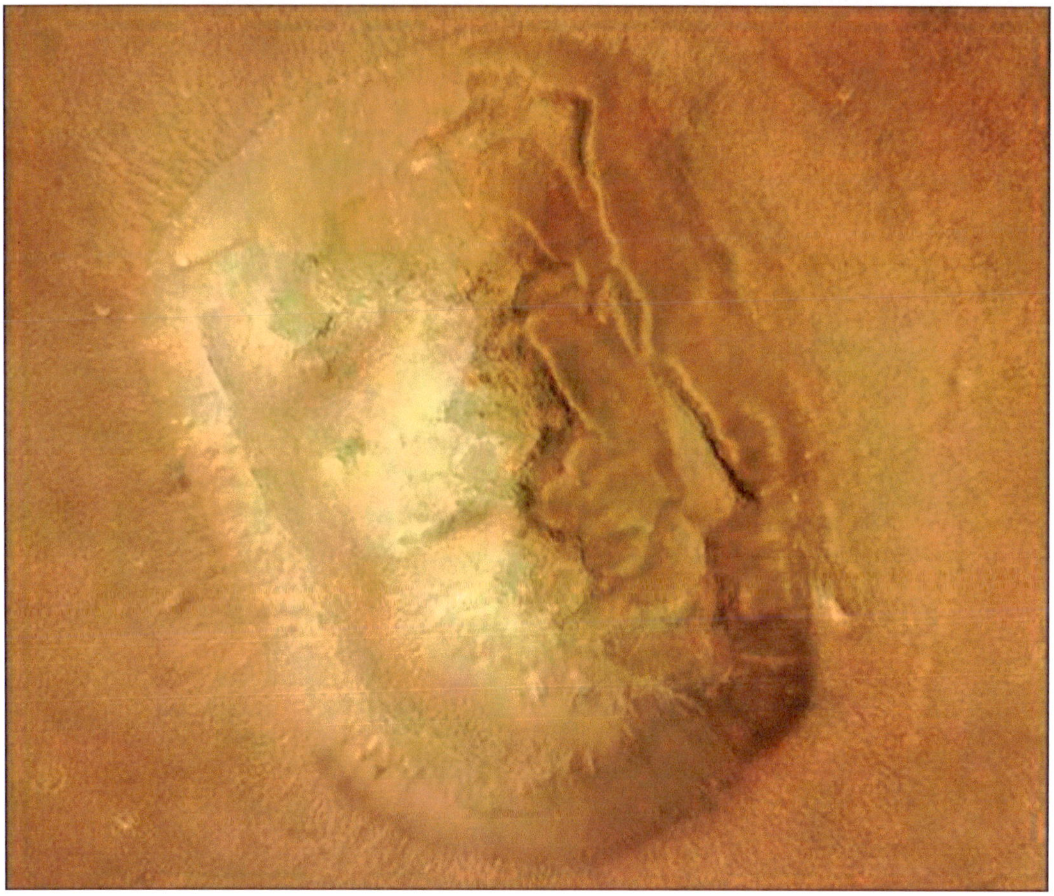

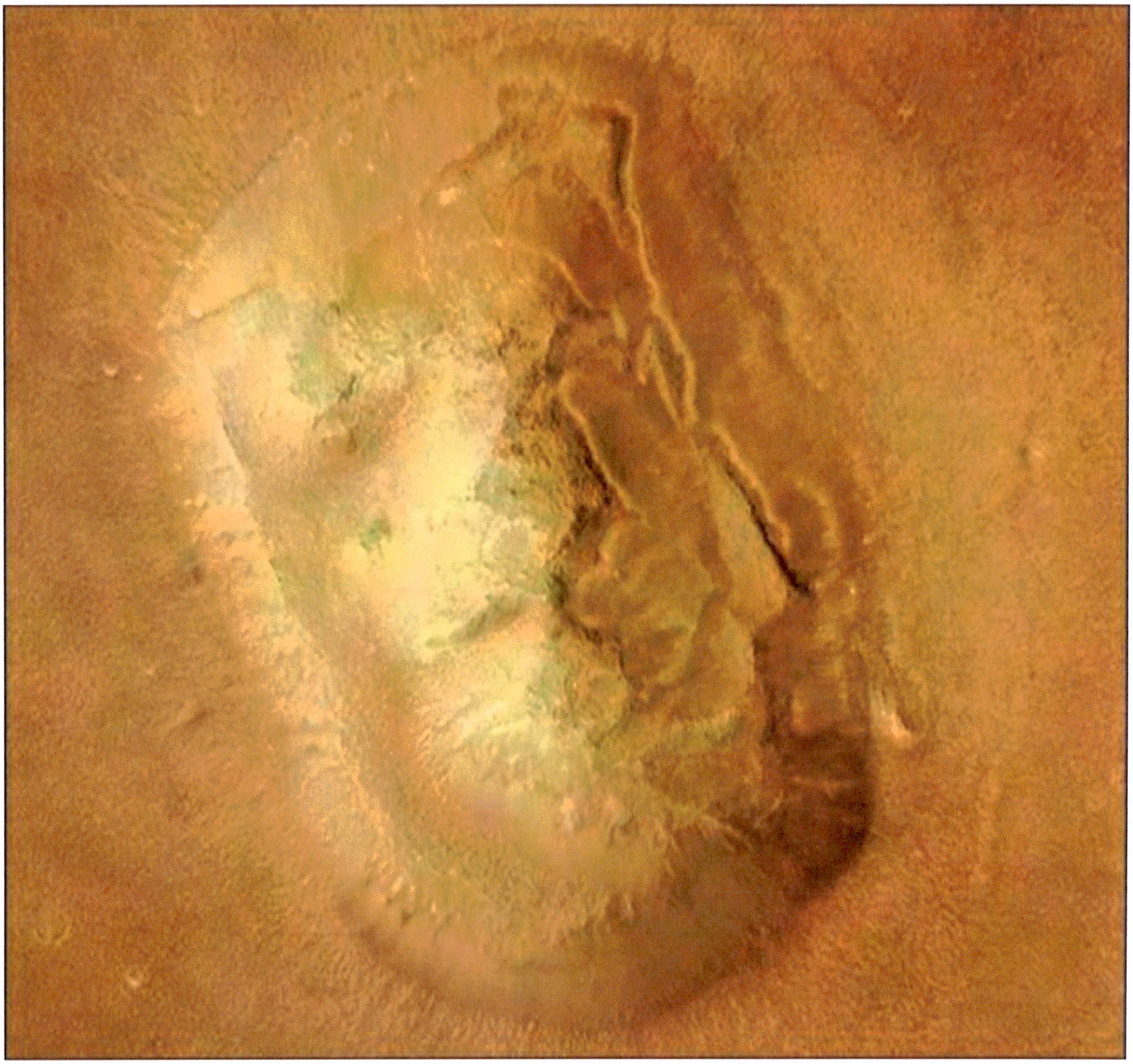

One may be wondering then how is it that we have this image view seen above? (*We will get into that later*). The focus here at the moment is the amazing color image data, because there is much more to be seen from the color data that shows also an unnatural anomalous look to it and the color data fits in more with the anomalous features that can be seen of the Face's relief itself.

As regards to what all the images show upon their release at face value, pun intended, someone may not think twice about studying the color data, like myself. Someone may think there is nothing at all to be studied or worth being studied. **I highly disagree with that mindset.** So it is my job to show you that there is much more to see than what is readily seen in the image data, again, than what is readily seen at face value. A continued obfuscation to mislead and hide image data that actually resides therein the data.

To explain what I mean, I will show you comparative image views that changes the perspective (pun intended) and value in what is given to us from MSSS and ESA.

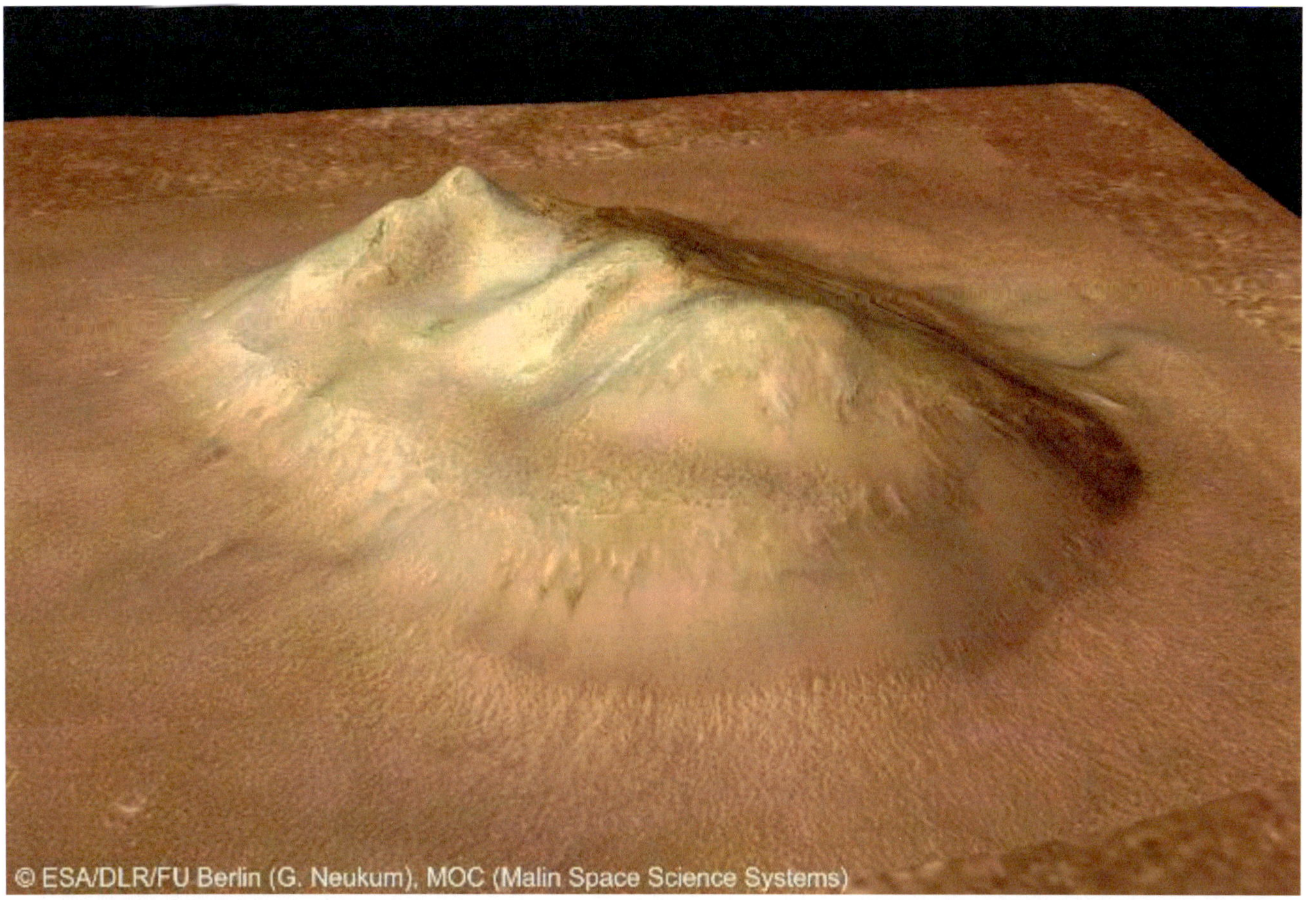
© ESA/DLR/FU Berlin (G. Neukum), MOC (Malin Space Science Systems)

MARS REVEALER

To the untrained eye, someone may have thought all this could not be possible. MSSS hiding color image data. This comparative view indeed shows much more to see than meets the eye. Upon doing some processing enhancement adjustments of the color data perspective view MSSS provided, one can easily bring out the not so easy to see colors within the data. Even knowing about the falsified data distorting the facts of reality and making a distraction, one can see why we were supposed to focus on the bump instead of the color, to not notice or care about its colors.

The color data in what it offers us is astounding, and transforms the Face from what we seen as released to us, into showing diverse colors and signs of materials that are looking just as anomalous as the Face's details itself. Demanding attention and screaming for study, turning this feature that NASA and MSSS claimed to us for decades as being nothing worthy of study as well as claiming the Face as being nothing more than a mere mountain, a mesa, a hill, and also a massif, lava dome into looking like something else altogether. *Something almost beautiful.*

By what is seen in the color data alone, aside again from the fraudulent bump trauma given to the Face, shows that this feature is as looking unlike any mountain or hill or massif that has ever been seen here on Earth, as NASA failed to do in which to compare. No wonder that NASA is as confused as to what to claim publicly what this feature actually is. In order to continue their public position as the Face being a natural feature, they have to go to great lengths to obfuscate its details, it's shaping, as well as obviously, ***its colors***!

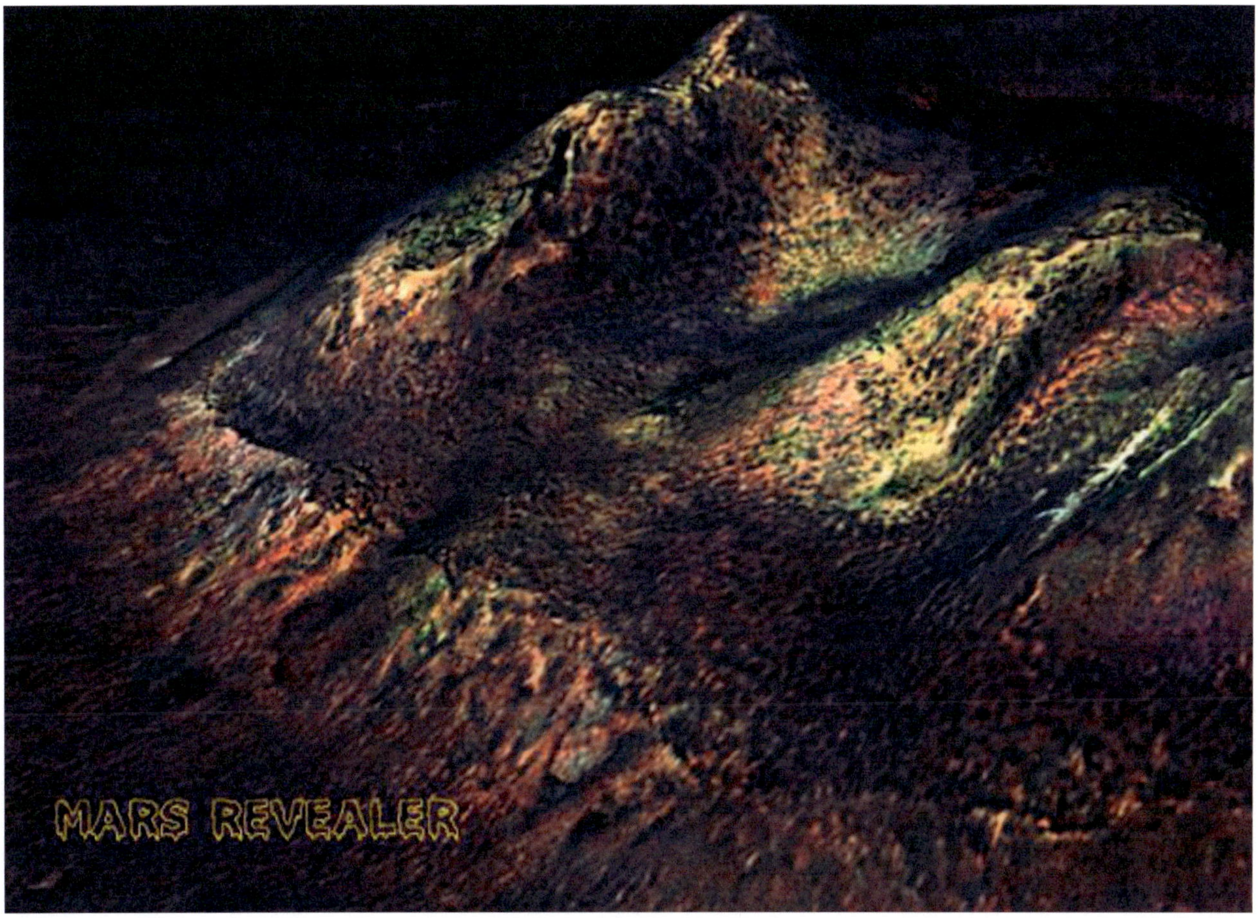

(Closer view of section of the MSSS perspective color view)

As I stated, the colors bring a very interesting new arena of study into the Face. The colors seem to correlate to various Alien Markings of relief features of the Face as well, that appear to be more purposed than coincidental features of natural nature. The colors seems to almost bring LIFE to the Face as well, making it stand out even more as it stares upward into the heavens. To me it only make sense that the colors would also help to reveal the tattoo as it were. Of course that is just my opinion, but my thoughts to me, seem to be accurate.

After all, from what we are told from NASA, the Face is just supposed to be natural rocks and sand. Yet the colors to me anyway, aside from the artistic relief of the Face itself of course, shows me how diverse the colors are and even beautiful and very surprising of what we see, as those colors, just like the Face's artistic relief itself, should not be there! What explains for these colors? The colors do not indicate a natural look. In my opinion, we are not seeing sand and rock, but metals!

As seen in the perspective image provided by MSSS, we can see Greens, Yellows, blues, oranges and purples, reds and pinks, as well as Golds. Certainly nothing like we have ever seen before or have been told about the nature of the Face. I am still in shock that no other independent of mainstream Mars researcher noticed or studied or even spoken about the bizarre nature of the colors. The colors certainly do not indicate support of NASA's natural claims for the Face, as again, these colors do not appear to be natural in its look, but appears to support the Alien Markings techno-signature by its deigning and artistic relief!

In fact so unlike what we have ever been told about or shown of the Face, the anomalous colors in my opinion again, appear to contradict NASA's official claims for a natural reasoning for the Face as being a mountain, a massif, a hill, a mesa or a lava dome, because these things as we know them to be here on Earth are made of rock and dirt with all of them, not showing the kind of beautiful colors that has shiny metallic looks as the Face appears to have and show. No wonder Malin and NASA never released the color version of ***MOC2-283*** *back in 2001.*

It is not a secret that NASA since their public Mars exploration programs began, and images were being given unto the public, that they have always been caught obfuscating the reality of Mars' true colors. This was seen as obviously being true all the more in its Mars lander's missions, especially from 1976 Viking lander missions. For instance just to show you as an example, a Viking Lander image that shows the Martian sky as a red colored hue and the surface color as being the same. But when properly adjusted, we can see that Mars' sky is in reality blue hued in color, and neither is the surface so uniformly colored in a red hue.

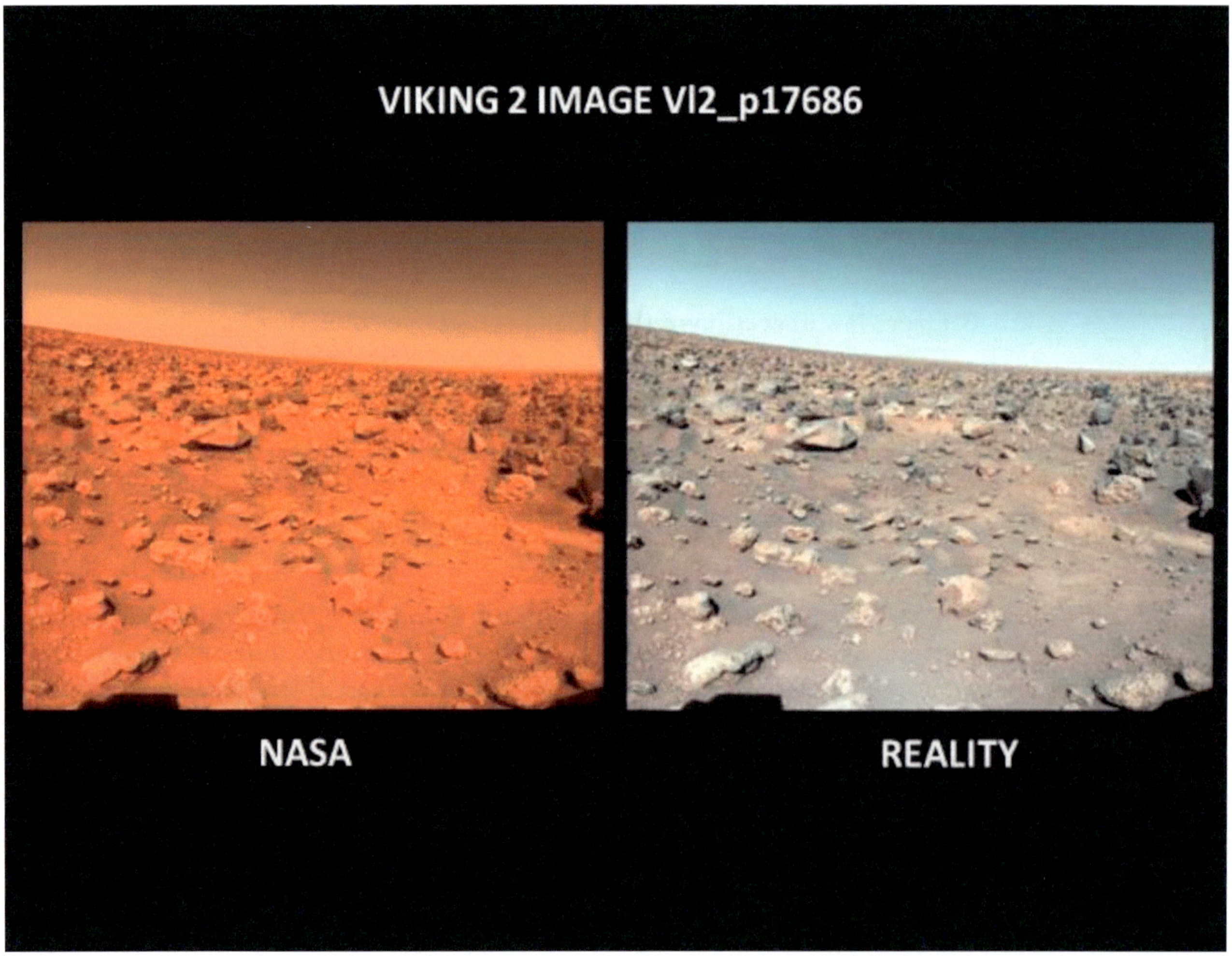

(Viking Lander color image release and correctly adjusted color image.)

So in the minds of Mars researchers, it should not come as a shock or as some far-fetched notion that the colors for what we are presented with in Mars imagery then and ever since, may in fact have much more to them than what is readily seen. To others, this fact may surprise them or shock them as the notion of NASA hiding colors of truth only reveals NASA's true colors to them instead!

We will not be continuing on the ***very important*** subject of Mars' true colors overall, but we will get back to the amazing anomalous colors of the Face, as this side step into this subject was only to make and show you all an important point to keep in mind as we continue in the future to seek and to learn the answers to the true colors of Mars and especially, of the truth of 'The So Called Face of Mars'!

So what we can see from the MSSS color perspective image, it is very intriguing to say the least. Yet we were also provided images from ESA's Mars Express which we also see and also is very intriguing, of which data details show also that contradict the traumatic bump data of the MSSS high resolution perspective image, and also seemingly contradict with the color data as well. Let us look at the image data again with focus of comparative analysis.

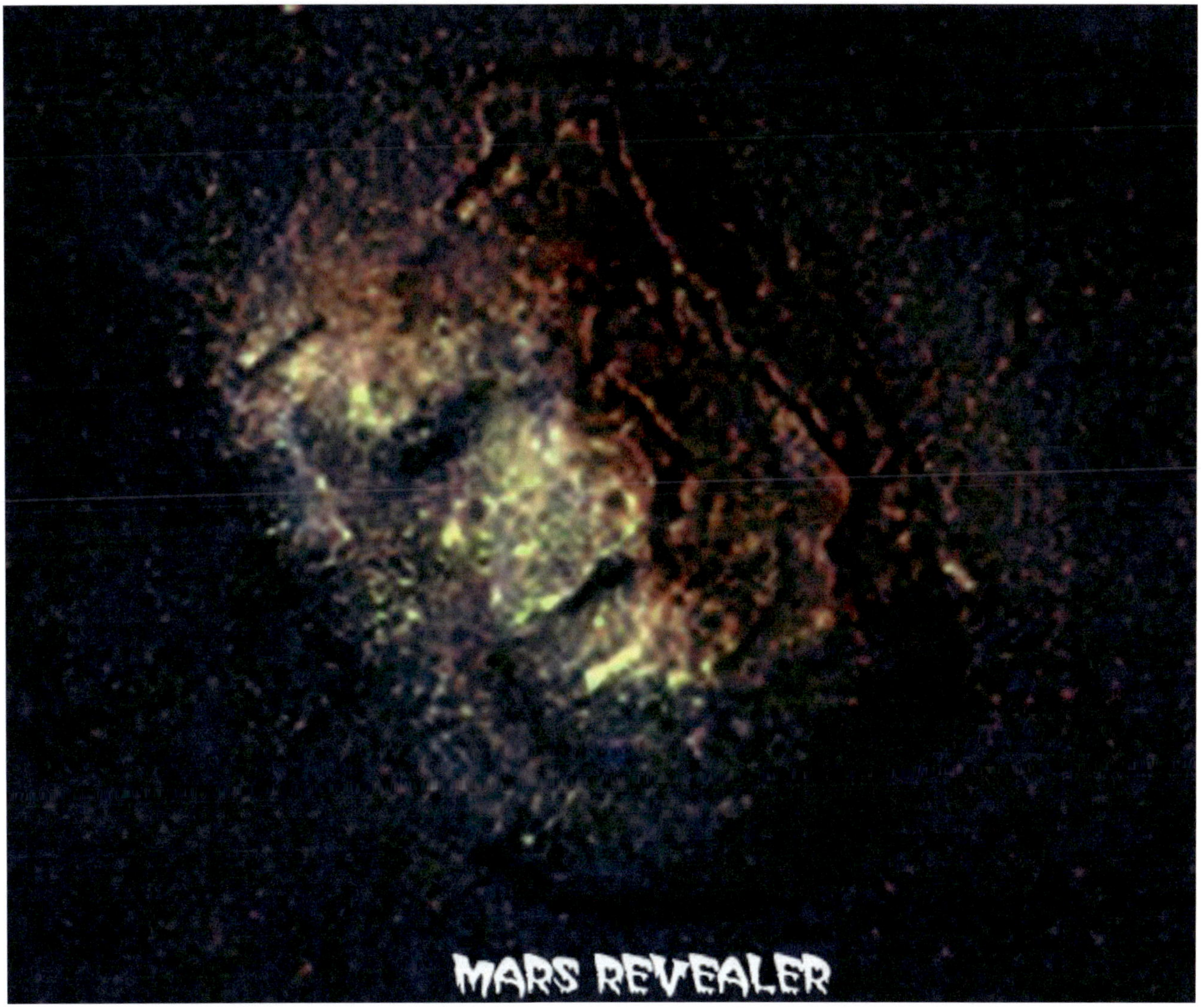

(MARS EXPRESS HRSC NADIR IMAGE #3253 COLOUR VIEW OF THE CYDONIA FACE.)

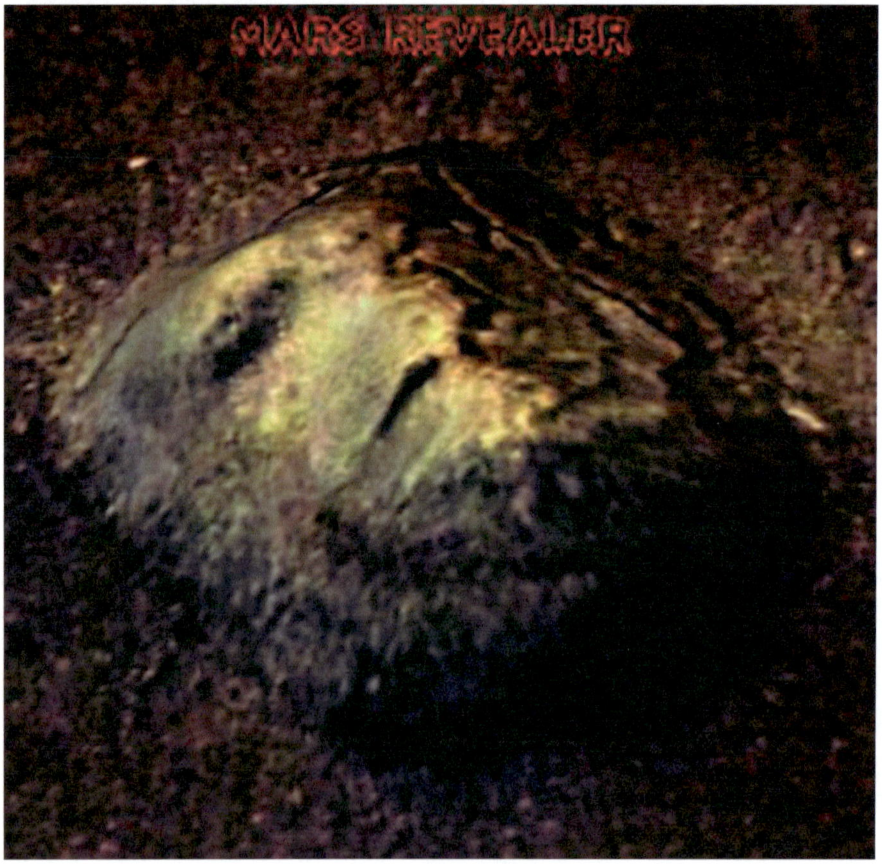
MARS REVEALER

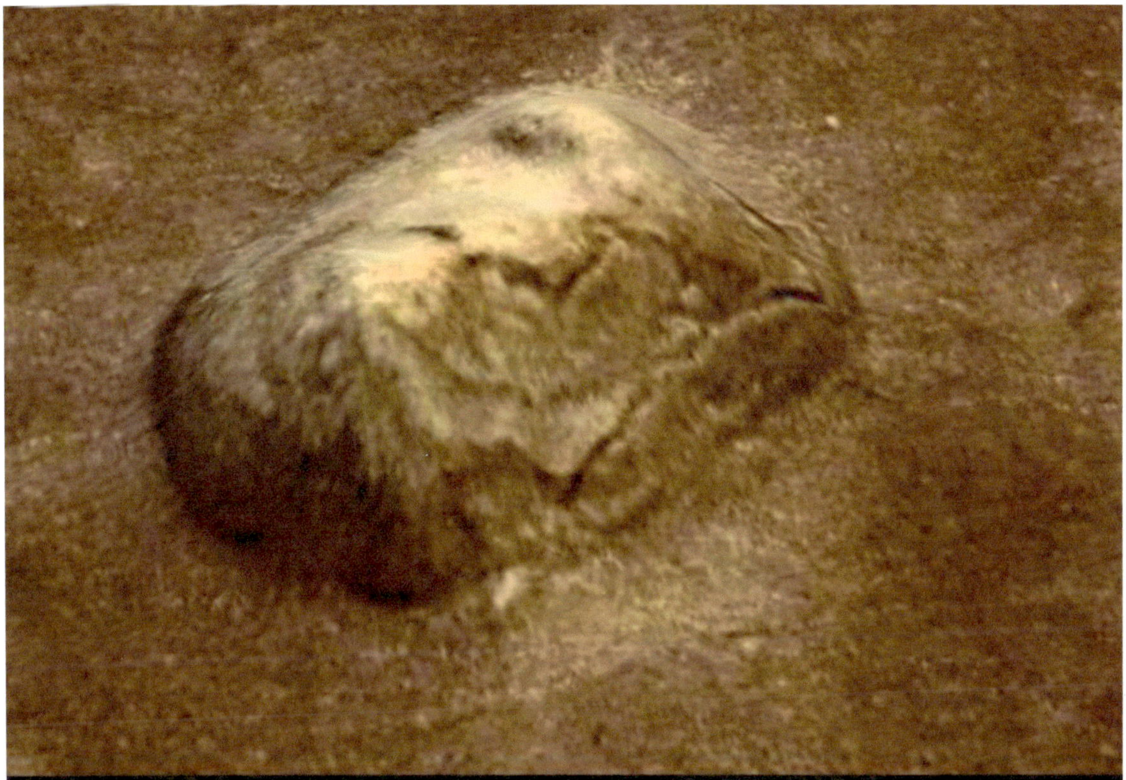

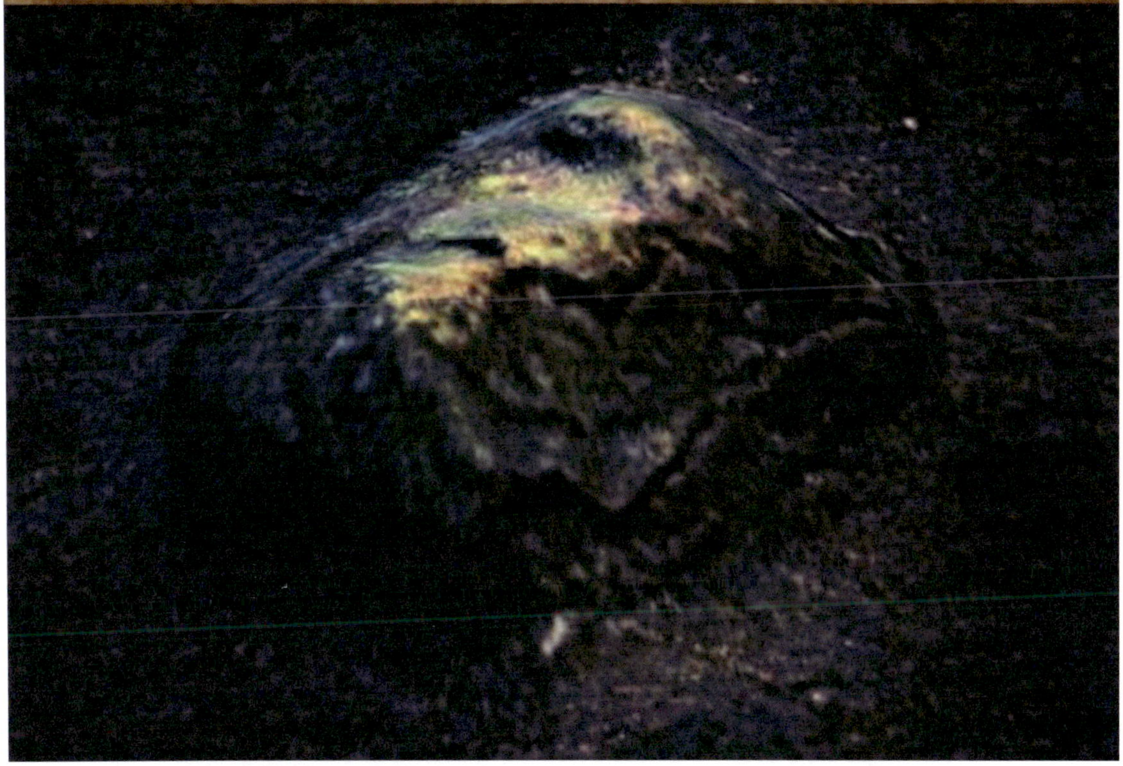

Further processing results in bringing out the colors better observed and within the image data. Even with the changes that further processing exposes of the lower resolution Mars Express color data, it again brings us a more detailed view that transforms what we are given as is, into a more realistic view, which again, does not look natural.

Nor does it look the same as of the color data that was provided to us all by MSSS image ***MOC2-283,*** which was provided for the Mars Express ESA public Hit-Piece release for the Face in 2006. On the *FACE OF IT*, (pun intended), we get more color detail and higher resolution in the MSSS image that brings the Face more to life so to speak, more so than the color data provided by Mars Express HRSC (High Resolution Stereo Camera). Let us compare.

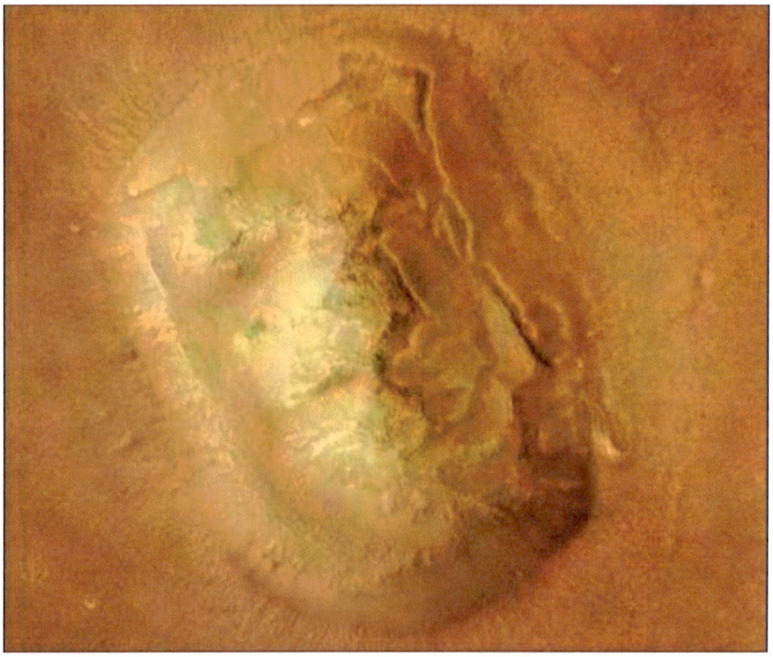

(MSSS MARS GLOBAL SURVEYOR COLOR IMAGE *MOC2-283*)

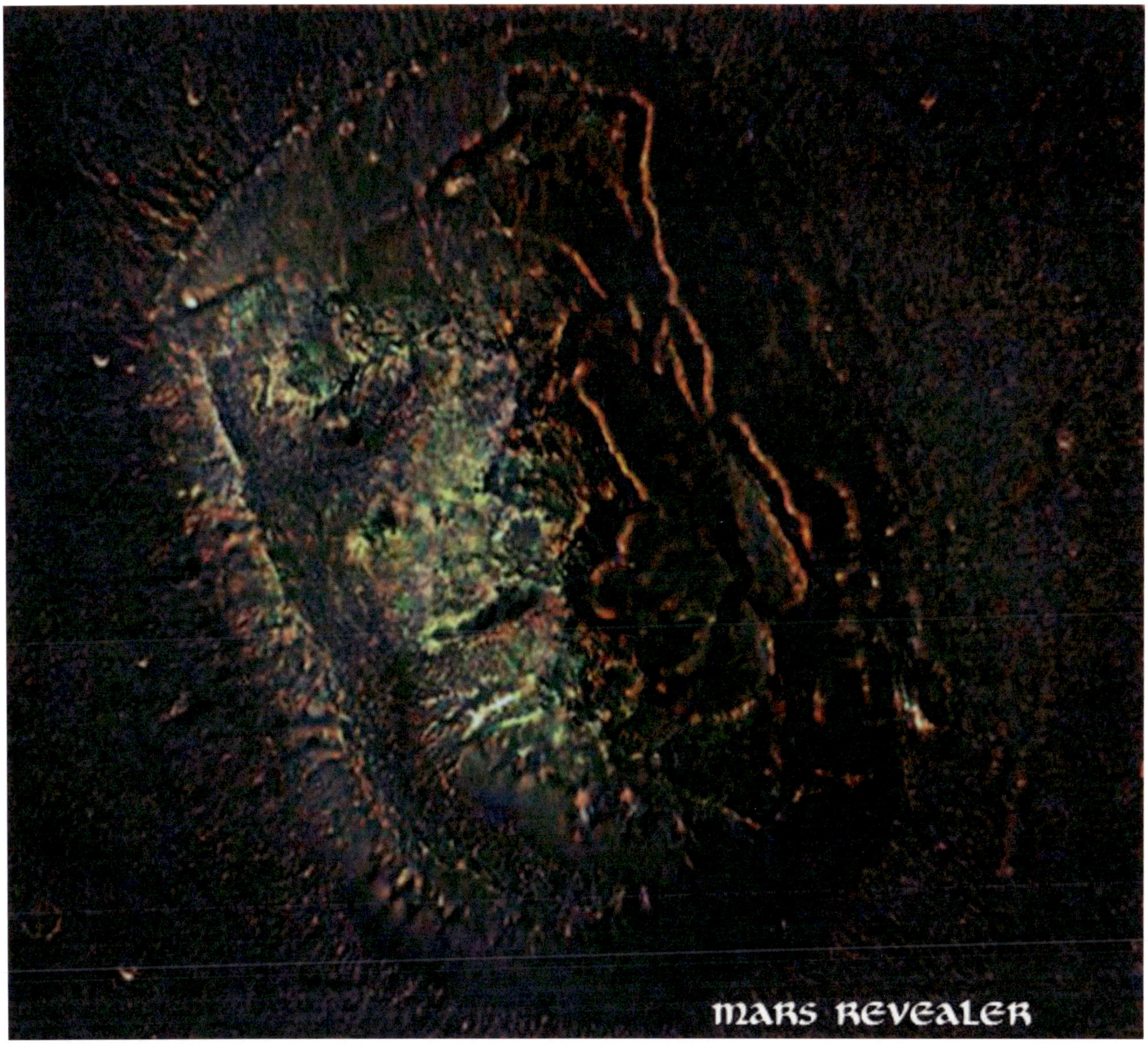

So in comparison between the Mars Express image data, and what is offered from NASA, again, we are faced with a problem of consistency. The vast differences of data facts shown and seen between them both leave us asking the question, "Which space agency image data do we believe? The European Space Agency, or NASA? What data can we trust? One of them, or both of them, has to be incorrect."

Can you ascertain which side put forth the most compelling evidence for revealing what the truth is regards to the shape and the colors of the Face at Cydonia?

CHAPTER 6:

ON THE FACE OF IT

(MARS EXPRESS 2006 Image section of 'The Face' I processed from: H3253_0000_ND3)

After multiple attempts to image Cydonia region from April 2004 until July 2006, were frustrated by altitude and atmospheric dust and haze, the HRSC (High Resolution stereo Camera) on board the Mars Express finally obtained on July 22nd 2006, an image data **showing an unprecedented detailed view of The Face of Cydonia**. The image data was gathered during its orbit of 3253 over the Cydonia region, with a ground resolution of approximately 13.7 meters per pixel.

Dr. Agustin Chicarro, ESA Project Scientist stated "The images of the Cydonia region on Mars are spectacular, they not only provide a completely fresh and detailed view of an area famous to ***'fans of space myths'*** worldwide, but also provide **an impressive close up** over an area of great interest for planetary geologists, and show once more **the high capability of the Mars Express camera**."

It is highly comical that Dr. Agustin also bows to the policy of public suppression as to the reality of the Face, and pays lip service to the policy of downplaying what is seen of certain features of Cydonia as seen in H3253_0000.nd3 to label it all into an insulting category narrative, as being famous to '**fans of space myths**'!

So as I am sure all of you who are reading this book now, according to Dr. Agustin Chicarro, you all are obvious **fans of space myths**! As much as that narrative shows itself to be biased upon this subject, and Dr. Chicarro showing an official position of playing this down as being only just space myths, it is safe to say the reality can be seen beyond mere words of disillusion and policy driven narratives. It is not for which words are to be taken as being factual, but by being what is said needs to solely depend upon the actual image data to come to those words as being valid.

I also like to say that Dr. Chicarro is also wrong as to labelling planetary geologists as having a great interest to this area and its enigmatic techno-signature features, as planetary geologists only share the policy driven narrative that the Face is natural. Dr. Chiccarro is also trying to give position of authority in the public perception unto planetary geologists to determine and to dictate the true nature of the Face of Cydonia.

Yet it is in Dr. Chicarro's words that ring true about what he stated about the High Capability of The Mars Express Camera, as the details seen in the data is what myths are made of, and where truths are hidden. **The Hiigh Capability of the Mars Express camera** that I firmly believe is what Dr. Chicarro was really trying to code to us all, which was hidden in his words of illusion, is information which was beyond what the policy narrative was that is being adhered to, and is hinting to us all to focus on to that **impressive close up image data showing an unprecedented detailed view** of The Face of Cydonia which is to reveal what is '**On The Face Of It**'**!** The truth!

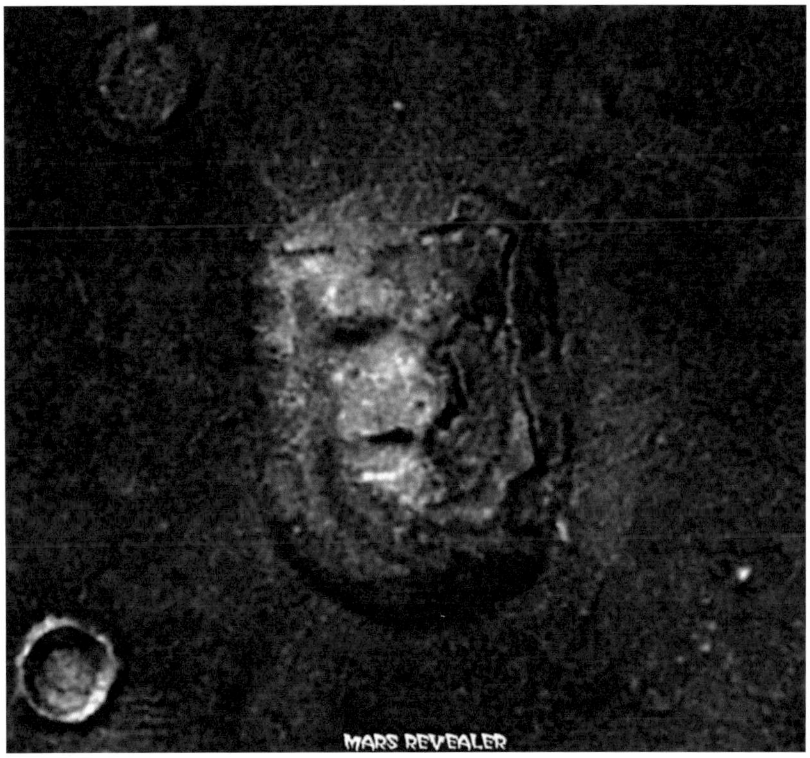

(MARS EXPRESS image H3253_0000_ND3 section of 'The Face' in up-right position.)

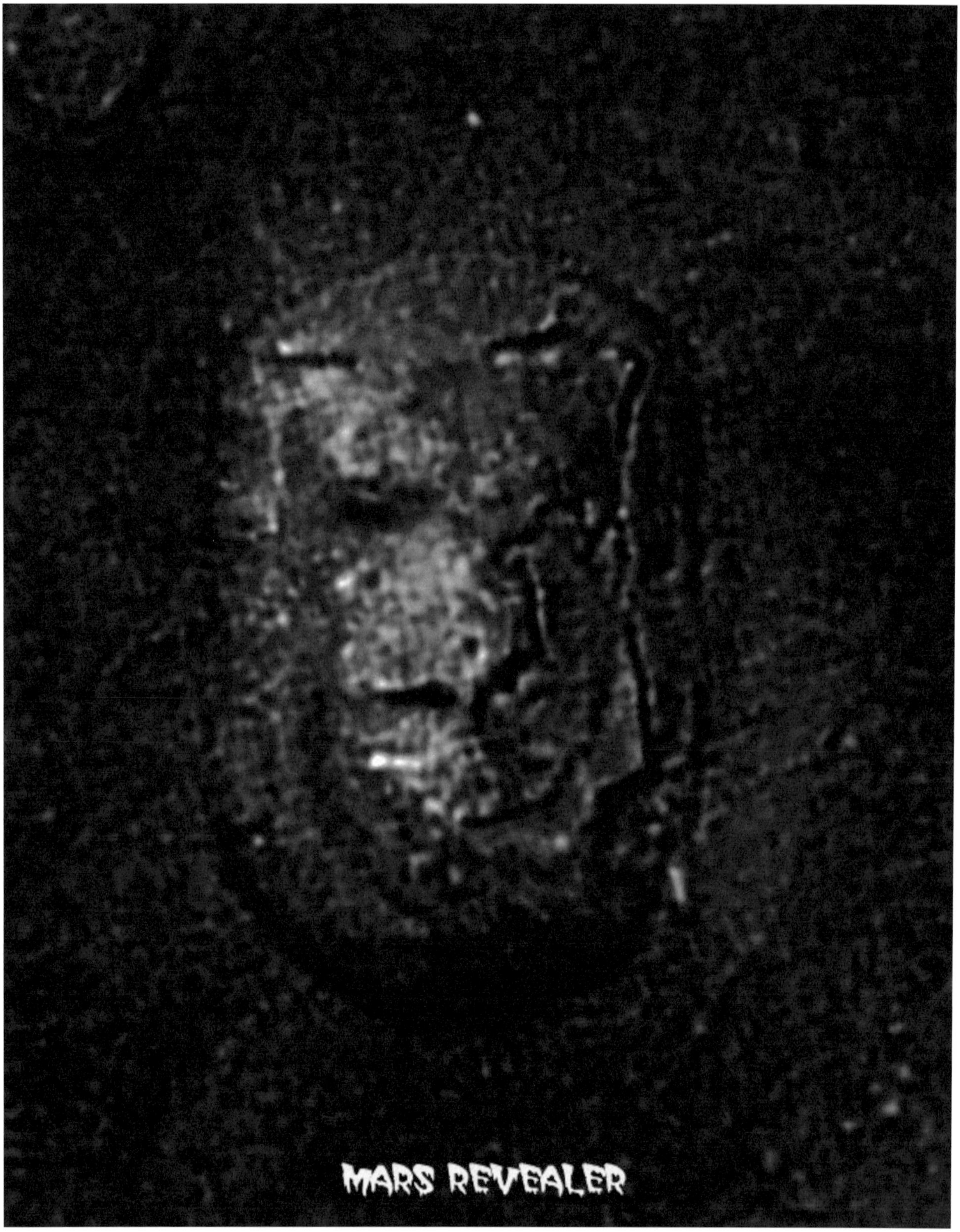

(INDEED AN 'UNPRECEDENTED' DETAILED VIEW OF THE FACE OF CYDONIA: H3253_0000_ND3)

I believe a definition understanding is in order here to grasp what we are being told, and shown, to better understand these **unprecedented** details.

Adjective:
Without previous instance; never before known or experienced; unexampled or unparalleled: *an unprecedented event.*

To me this is so easy to see in words and in image data itself. We are being guided to see this unprecedented event, of never before known or experienced, unexampled or unparalleled **impressive close up image data showing an unprecedented detailed view** of The Face of Cydonia! Through and by **the High Capability of the Mars Express camera**! A play on words? We need to study this further to see if it all is true. We will test this amazing coded statement with what is being focused on itself as the basis of truth, the image data from Mars Express, and compare that amazing unprecedented details and high capability with the only other source to use as a comparative analysis to test Dr. Chicarro's amazing coded words with.

That comparative analysis to test these statements with is of what NASA put forth to public view via the MARS ODYSSEY mission visible imagery. It is nearly equivalent to the resolution of Mars Express. Dr. Chicarro is apparently touting the high capability that is unprecedented in showing details, and poking NASA in its chest daring them to disqualify and nullify Dr. Chicarro's amazing unprecedented statements. That is what I see, and what only counts for this study. No one else has ever said or shown what we will see and study here. This study and this book in and of itself, is **'unprecedented'**.

Before we go into that study in comparing the high capability of the image details from the Mars Express to the 'insinuated' lower quality of NASA's Mars Odyssey Mars visible imagery, stated by Dr. Chicarro's amazing public statements of Cydonia imaged by Mars Express showing in its unprecedented details, we will continue to look at and compare another Mars Express image view taken of the Face at Cydonia.

Between April of 2004 and July of 2006, the HRSC gathered image data from the Cydonia region numerous times. However, high flight altitude, resulting is poor data resolution on the ground (orbits 0262, 2533, 2872), as well as dust and haze in the Martian atmosphere, leading to heavily reduced data quality (orbits 1216, 2872) prevented the acquisition of high-quality Cydonia images.

The only other comparable image to that of the unprecedented details of H3253_0000_ND3, is that of H1216_0000_ND3, image taken on December 29th 2004, of a ground resolution of approximately 12.2 meters per pixel. Though having slightly reduced data resolution, being near the same resolution scale as H3253_0000_ND3, it is still very **'unprecedented'** in the details that are able to still be seen. Most people who looked at the image at that time, and even to this day, did not do their homework in studying the image data, to be able to appreciate it or even understand what is being seen.

As I said before, because '**On The Face Of It**', there are details having various differences of what is able to be seen in H1216, compared to the details seen in H3253, that shows almost a different Face being seen. The people I know who looked at this image back in 2004 when this image was released from ESA, considered the image data was so flawed and degraded and the Face's details and other details of the Cydonia features could not be trusted. Yet that is certainly NOT the case at all, as we will indeed see and compare between the two images!

That being said, there is indeed however some serious data quality issues apart of this H1216 image. It is clearly seen that the data quality is not so much of an issue at all regards to the clarity of the details of the features of the Face and the surrounding features seen in the image, but the obvious data dropouts or glitches, as some people called them, that is seen in the image.

(MARS EXPRESS Image section of 'The Face' and the D&M pyramid I processed from: H1216_0000_ND3)

As can be seen in the image above, you can see the obvious data faults that stand out immediately and destroys the quality data that should be seen there instead. But as can also be seen, is all the rest of the image data that leaves the Face seen far away enough from these data drop outs, and allows us to see and to study the quality data we are given of everything else, including the Face! Which of course, is our point of focus. Despite the data dropouts, this image is very worthwhile in studying with enough clarity and resolution of the image data details to take it seriously.

The Face is admittedly looking much different than what we seen given almost two years later of Mars Express image H3253_0000_ND3, of which people claim is clear image data and acceptable quality as to how some of those people claim and think that they know the Face, and these people seen clearly that the Face looked very different in H1216_0000_ND3. But because it looks so different, it was attributed by some, to as being because of bad image data quality. Again, this is truly NOT the case and is untrue in their wrongful assessment and ignoring that data does not take away the importance of taking the image data seriously, or seeing and accepting that the detail differences between the two images are for a much different reason all together, other than the bad data quality excuse to ignore it, is not even being considered or even discussed really at all.

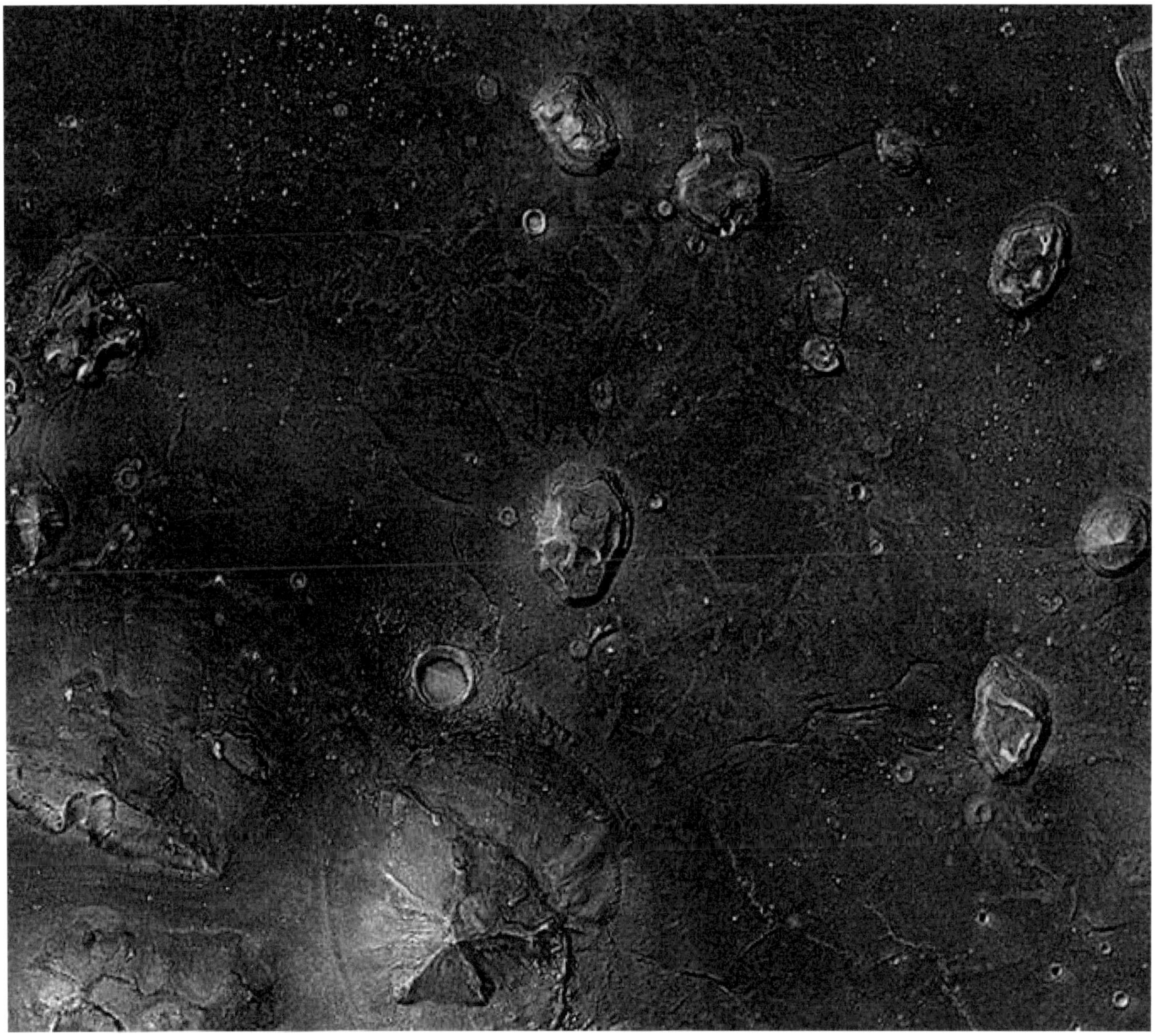

(MARS EXPRESS Image section of 'The Face' and the D&M pyramid I processed from: H3253_0000_ND3)

'On The Face Of It', we must take the details at face value, no pun intended. Then to study these details, and compare it with the details of H3253 and look for consistencies. **The better the image processing done to the images, the more accurate in its scientific analysis can and will be for determining accurate facts about what is seen in the actual image data.** So when at that point, one can verify or rule out the ignorant peoples excuses of bad data quality. I am showing you two examples, both processed properly of the Face area for H1216, and H3253. Let us compare and study them both together. Because the truth is what we determine by and because of the data. Right?

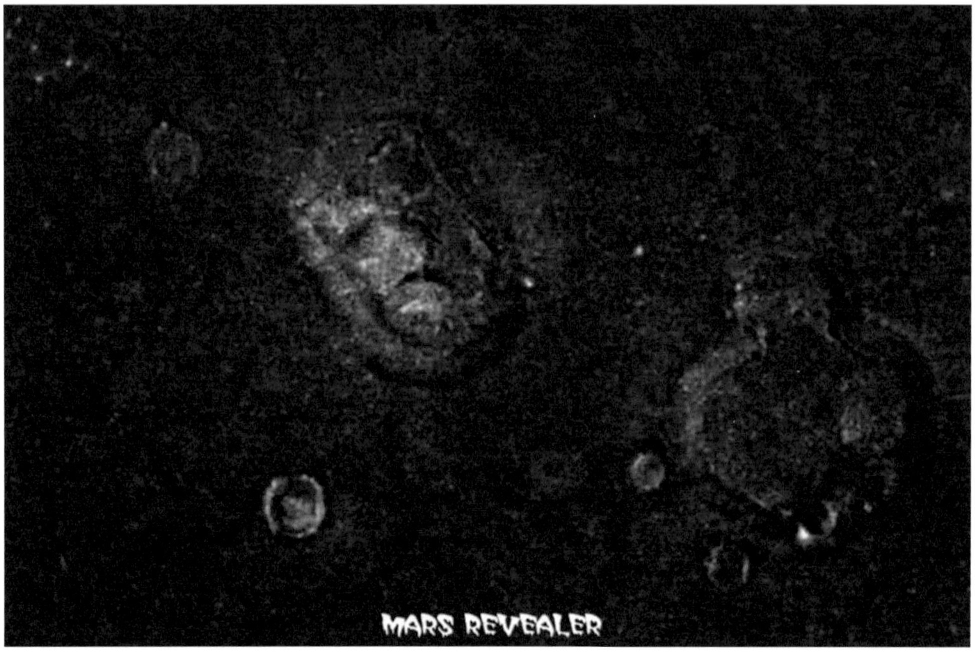

(MARS EXPRESS Image section of 'The Face' processed from: H1216_0000_ND3)

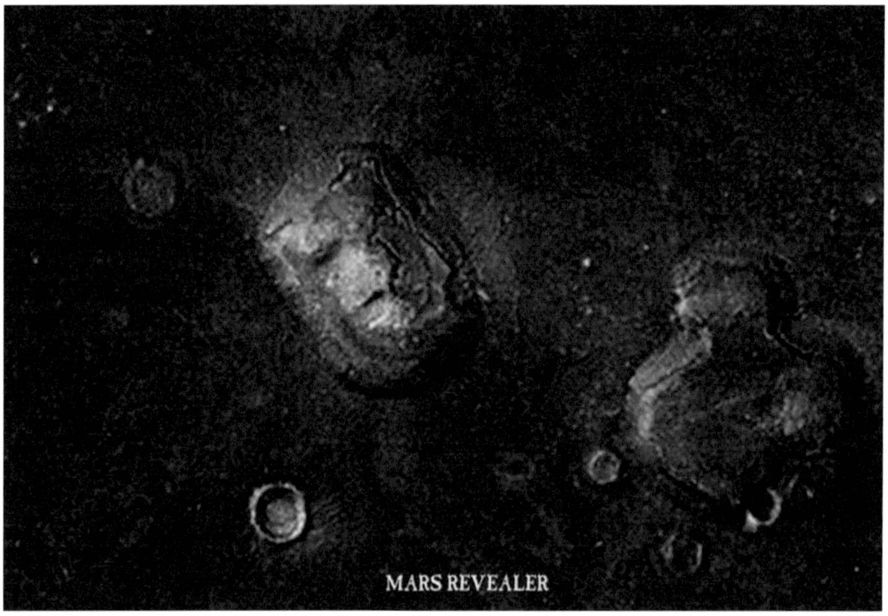

(MARS EXPRESS Image section of 'The Face' processed from: H3253_0000_ND3)

The Mars Express imagery does speak well for the capability of the High Resolution Stereo Camera (HRSC) and especially of the quality of the image data that Dr. Chicarro, was proudly attesting to, and holding his pride of the HRSC up high in the Face of NASA's inconsistent views and inconsistencies of the official positions of the Face of Cydonia over the many years, that shows in unprecedented details, no comparison to any NASA Mars mission imagery of similar resolution. We will get into it a bit later of that comparative study and analysis of the Mars Express image data and NASA's Mars Odyssey image visible data from the THEMIS Camera. The results will indeed show how Dr. Chicarro's words were not misspoken of the unprecedented image details of the Face.

(MARS EXPRESS 2004 image section of 'The Face' in up-right position I processed from: H1216_0000_ND3)

Until then, let us see and study the incredible details and differences between the only two images Mars Express has put publicly put forth, that each having its own unique details are both being seen as unprecedented. The images show us all details which has never been focused upon by NASA imagery before which do not show the amazing details we can clearly see **'On The Face Of It'**. One has to wonder as to why? Again this will have to be looked at and asked again later.

Yet even between the two HRSC images showing the unprecedented views of the Face at Cydonia, they each show vast differences between the two images themselves of the Face that each transforms completely our perspectives and perceptions of how the Face is viewed as what has commonly been seen and accepted and spoken about by NASA mainstream scientists, and media platforms as having only **'Mere Tricks Of Light And Shadow'** to which only gives an impression of there being a false face seen only in certain lighting angles onto what is called only as a natural landform of eroding dirt, sand and rocks. Claiming that it is and has been scientifically proven that this is all it really is. The reality however of their claims is in fact false!

Yet as we see in these HRSC data images themselves, this is obviously inconsistent and therefore untrue of these false claims and statements and official narrative positions that has been laid down upon the public for three decades prior, up and until the releases of this unprecedented imagery that in and of themselves, shows open contradiction to all that we have been told as official positions about the Face, including which it is still being the NASA official positions up and until today as of 2023,and is no longer even being spoken about. Again like it is 'Case Closed'!

However that official narrative was debunked in silence in 2004 and in 2006. No, not by any challenging researchers and scientists or by any self or peer reviewed scientific analysis, but has been debunked in silence by the very HRSC image data itself which are the silent speakers of these facts.

As it is the image data itself that shows us the visual facts and realities taken by highly capable visual instruments from a foreign space agency to allow these images to **scream their contradictions and direct challenges** to NASA's controlled narratives and public official positions. This hints at a power of public control and bias on behalf of NASA and scientific study and findings to which only the public control of NASA policy is enacted and obviously enforced worldwide, and is not simply a national perception and perspective, or which is allowed to be open to any contradicting points of view, but it appears to be an iron curtain covering up and trying to hide these very details from becoming a standalone power to change the world's public narrative and perception by NASA.

When we look at and study the Mars Express HRSC image data of the Face, it is literally derailing official narratives and taking away the deception that the Face at Cydonia is an eroding natural dust heap amid sitting in a desert of sand, dirt, and rocks. The images by Mars Express take away that perception based upon the image data details itself. That is the reason why **NO ONE** has done a serious comparative study and public review of these unprecedented images, as well as the fact most and nearly all of the people studying this subject, do not even know what they are looking at to give an official position, that is why it went mostly ignored.

For the images shows details that have an artistic relief and imprint upon the Face which interacts directly with the light in what it's designing was purposely constructed to utilize and reveal of its techno signature of its sculpted properties and purposes that screams of intelligence, and which is embedded the fingerprint of an intelligent artful creator and its cosmic capable handiwork to be seen upon other worlds. In the following image examples, between the two images of the Face from Mars Express HRSC, the same feature can be seen, but relief details differ between the two images. This is because of lighting angles of the sun, the position of Mars' orbit, and seasonal differences, to accentuate the complex artistic designing of which features seen in one period of time, differ from other features able to be seen. Yet these differences do not ever take away the shape or form of the main structure of which the term 'Trick Of Light And Shadow' applies. However, not to debunk the Face as has been negatively and wrongfully used and done by NASA and controlled mainstream science media to play this down to obfuscate it's reality, but in this 'light', pun intended, the term is used positively and accurately to reinforce the reality of the Face's intelligent techno-signature. The differences are incredible, yet very consistent to main points they both share. We will see each image, and then compare them to a fusion of each, to see greater detail what makes this Face real, and is not because of a lack of data or bad image data. ESA has given us amazing unprecedented evidence of 'Alien Markings'!

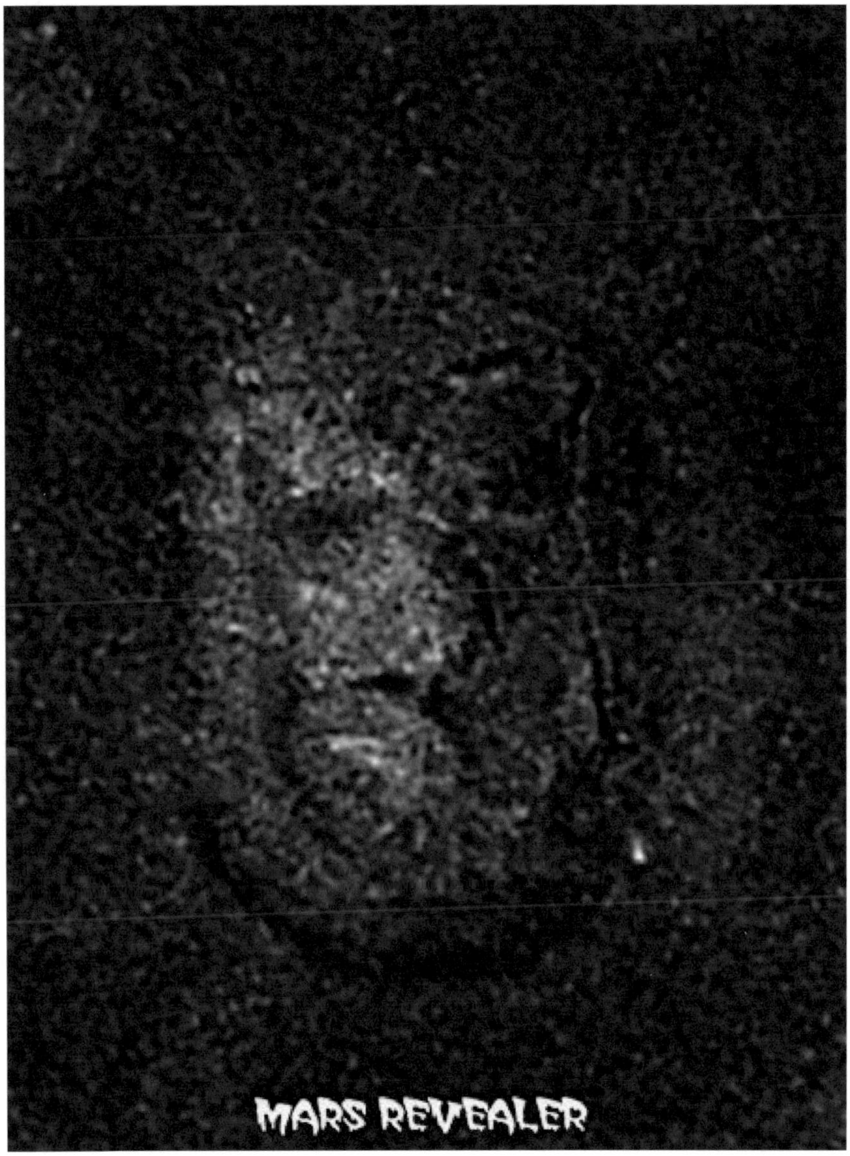

(INDEED AN 'UNPRECEDENTED' DETAILED VIEW OF THE FACE OF CYDONIA: H1216_0000_ND3)

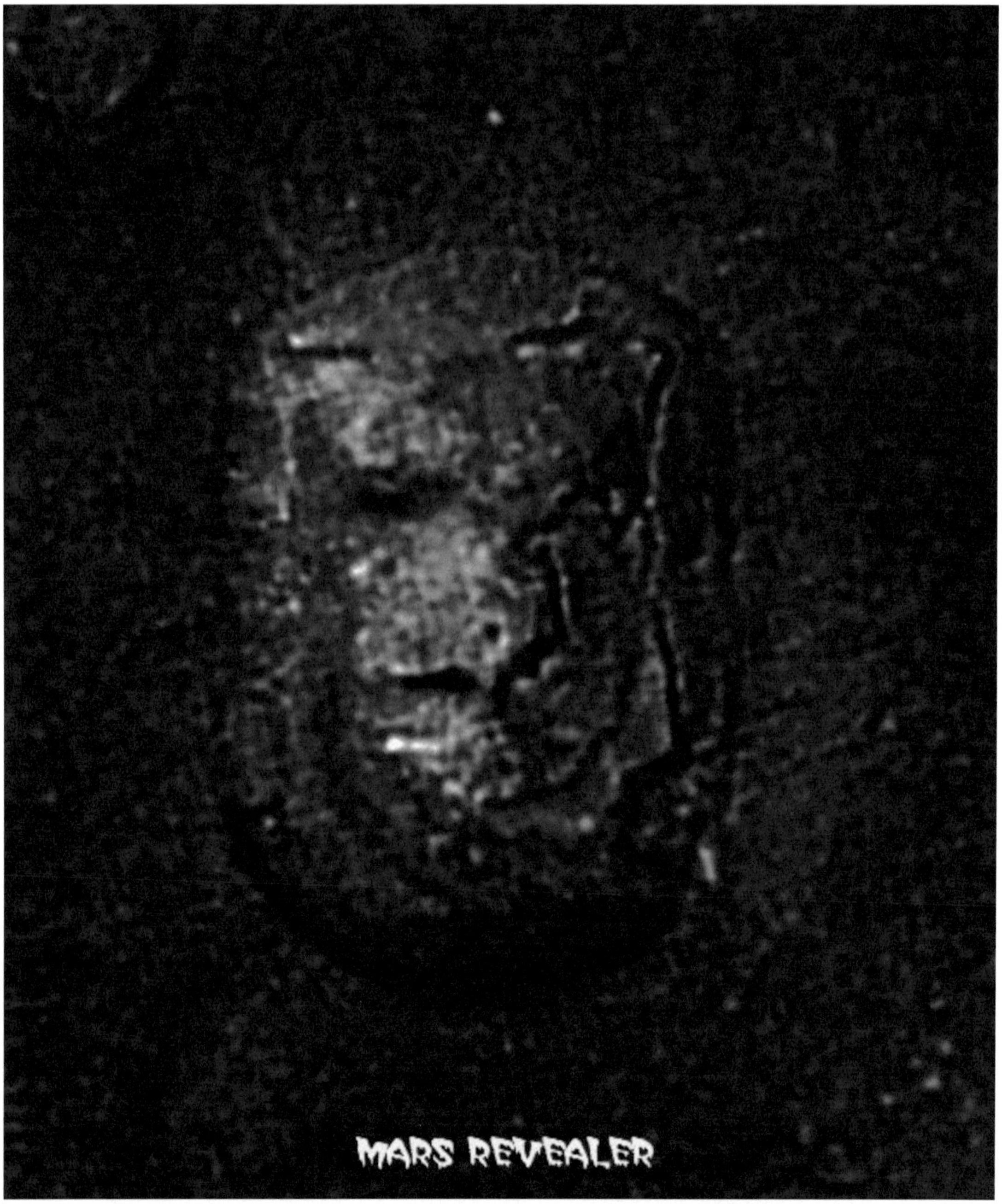

(INDEED AN 'UNPRECEDENTED' DETAILED VIEW OF THE FACE OF CYDONIA: H3253_0000_ND3)

So as can be seen, the Face is still there, although the details show differently, and major structure features can still be seen between them both. Yet at the same time, each show different features standing out separately in each. This allows a great experiment to be conducted, by fusing the two images of the Face together by overlaying one above the other, and balancing the details between them both to allow to bring out a single more solid view of these unique tricks of light and shadow merged into a combined single image view, showing us better evidence the artistically carved features of an intelligent designer, and a genuine techno-signature that is un-mistaking of its reality.

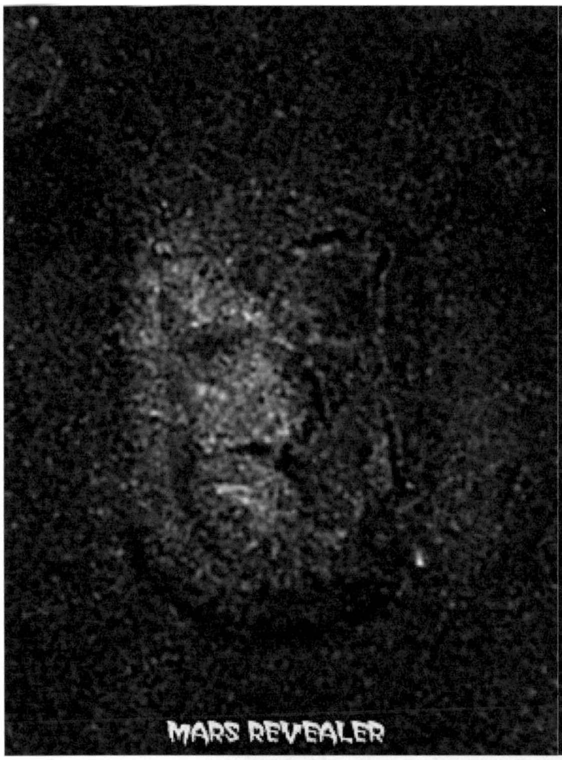

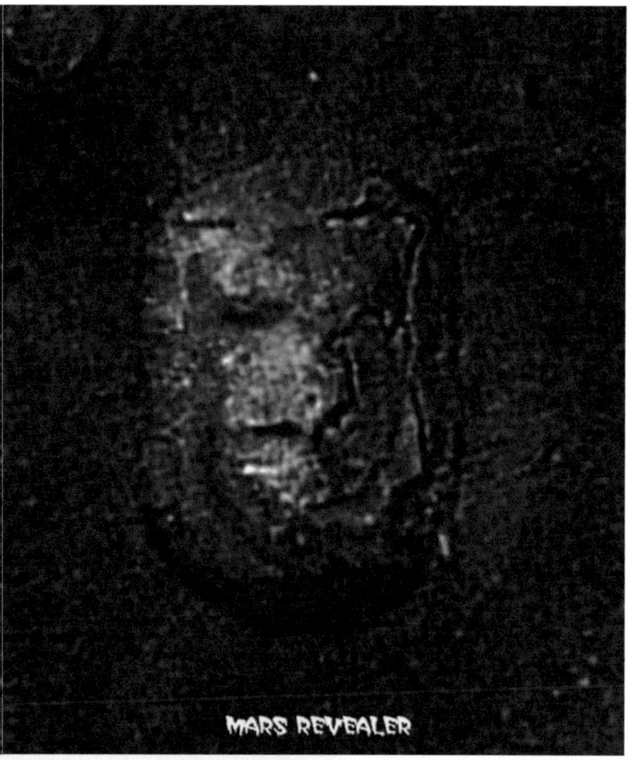

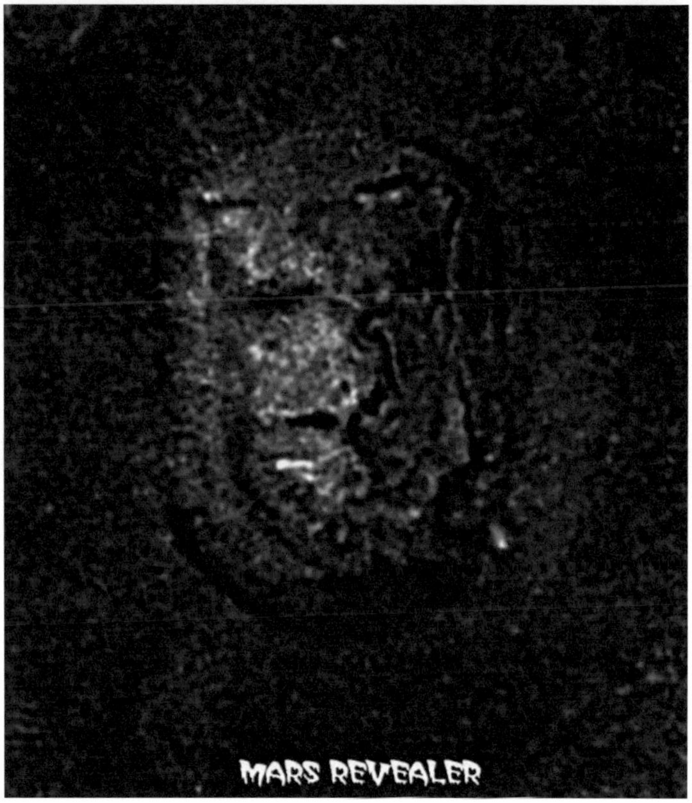

Richard C. Hoagland also stated to me: **Mars Express has taken -- by far -- the MOST accurate Face images to date .... (*September 27th, 2023*)**

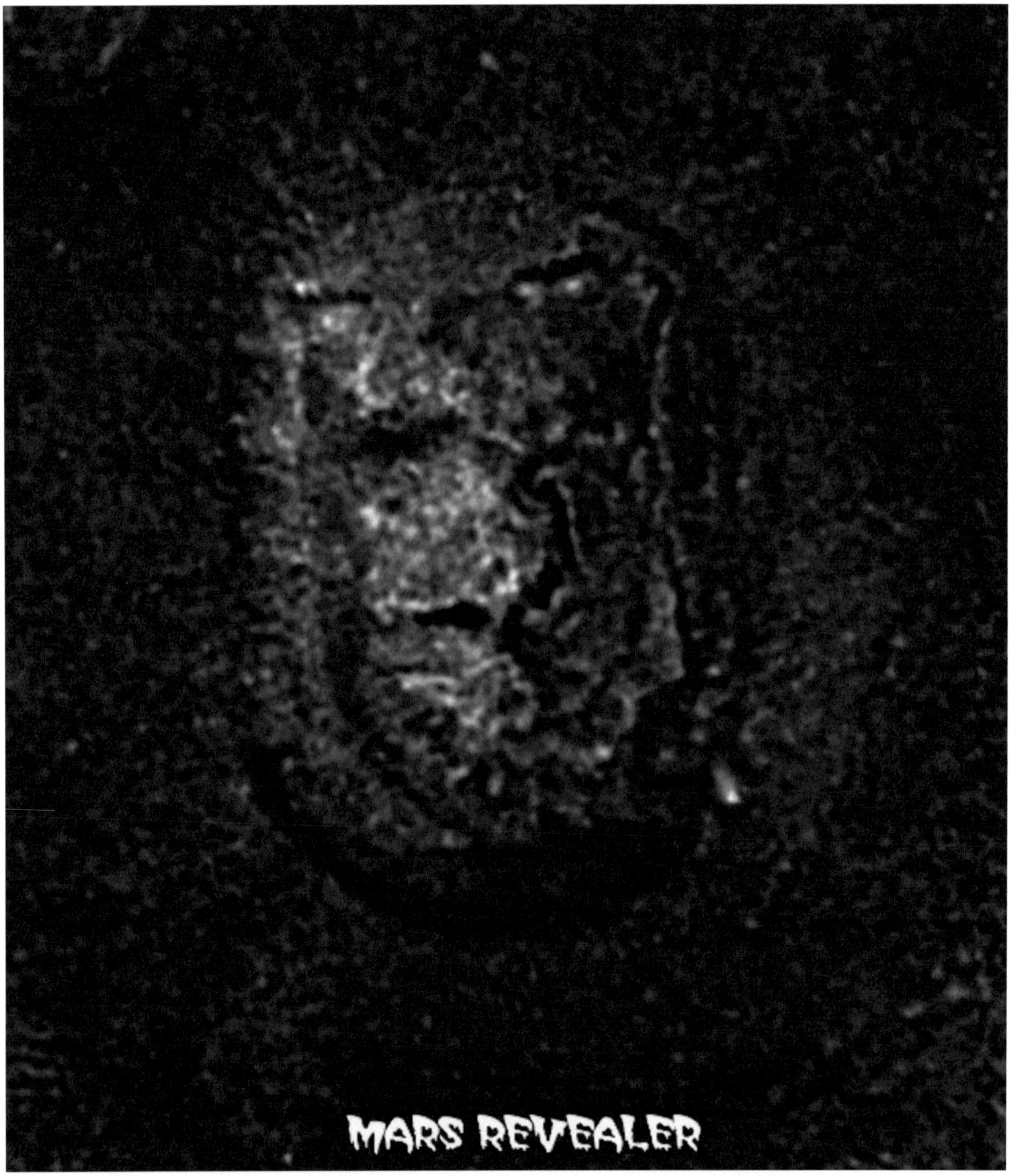

(THE MOST ACCURATE 'UNPRECEDENTED' VIEW OF THE HRSC MARS EXPRESS IMAGES)

Even to this day, no one, but myself actually, has studied the importance of the Mars Express image data in what and how it deserves to be. No one else has done or published these basic comparative analysis results to test and to also prove the features we can see of the Face which deserves far more attention. The image data itself helps us to reveal the truth of its artistic designing. This merging of images allows us to get actually, a third Mars Express image by default! Yet, however, the silence is deafening!

CHAPTER 7:

MARS ODYSSEY: UNPRECEDENTED SILENT DISCOVERIES

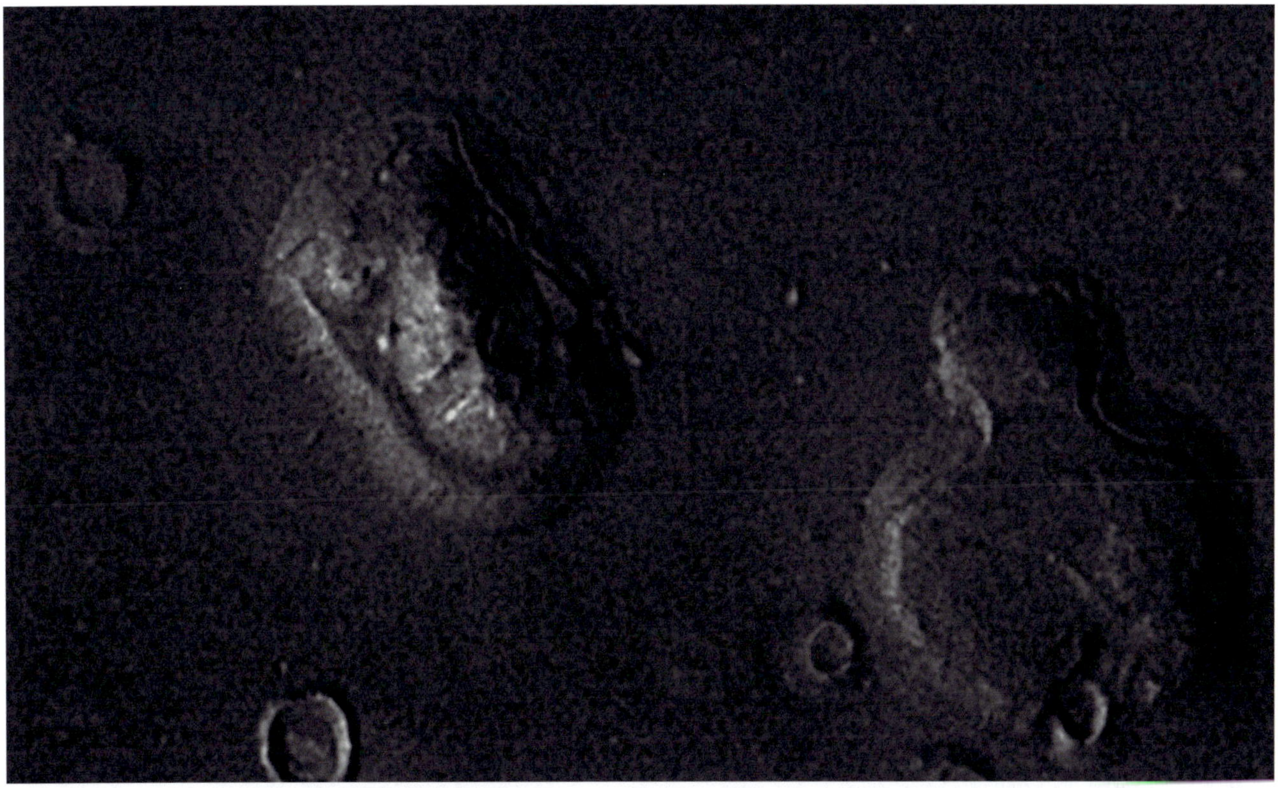

(MARS ODYSSEY IMAGE V01024003 TAKEN ON MARCH 8TH, 2002)

The 2001 Mars Odyssey mission is NASA's longest-lasting spacecraft at Mars. The spacecraft launched on April 7, 2001, and arrived at Mars on October 24, 2001. As of this writing in November 2023, the Mars Odyssey is still in operation orbiting Mars.

The visual images were taken at 59 feet (18 meters) per pixel. This resolution helps fill in the gap between large-scale geological images from the Viking orbiters in the 1970's and the very high-resolution images from the now deceased Mars Global Surveyor. In a comparative image resolution to Mars Odyssey camera, the Mars Express HRSC capability acquired imagery of 10–20 m/pixel.

It makes me really wonder with that comparable image resolution between Mars Express HRSC, and Mars Odyssey Themis camera, why is it that unlike the Mars Express image data of the Face and Cydonia which was put forth and used as the very public NASA and media controlled 2007 hit piece collaboration, Mars Odyssey images are put forth without any slightest interest or even concern, for any NASA official positions and media official positions at all regards to the Face and Cydonia? This to me is a real mystery. Only some independent Mars researchers had some public issue and discussions about Mars Odyssey image data, it only stayed amongst those interested and who follow the research. Never any media publications, or any NASA publications put forth at all of the Face, to add it to the all of the rest of the policy controlled public official positions that was always the norm, in terms of image releases of the Face. It is as if the fact multiple other images were taken of the Face, it was like they not even existed to even have the need or desire to speak about them at all. Let us look at images of the Face from Mars Odyssey and a Mars Express image to compare. Maybe we can see and speculate any reason why? We will compare each of the best Mars Odyssey images of the Face, with that best Mars Express image of H3253_0000. We will start my comparison with Odyssey image V01024003.

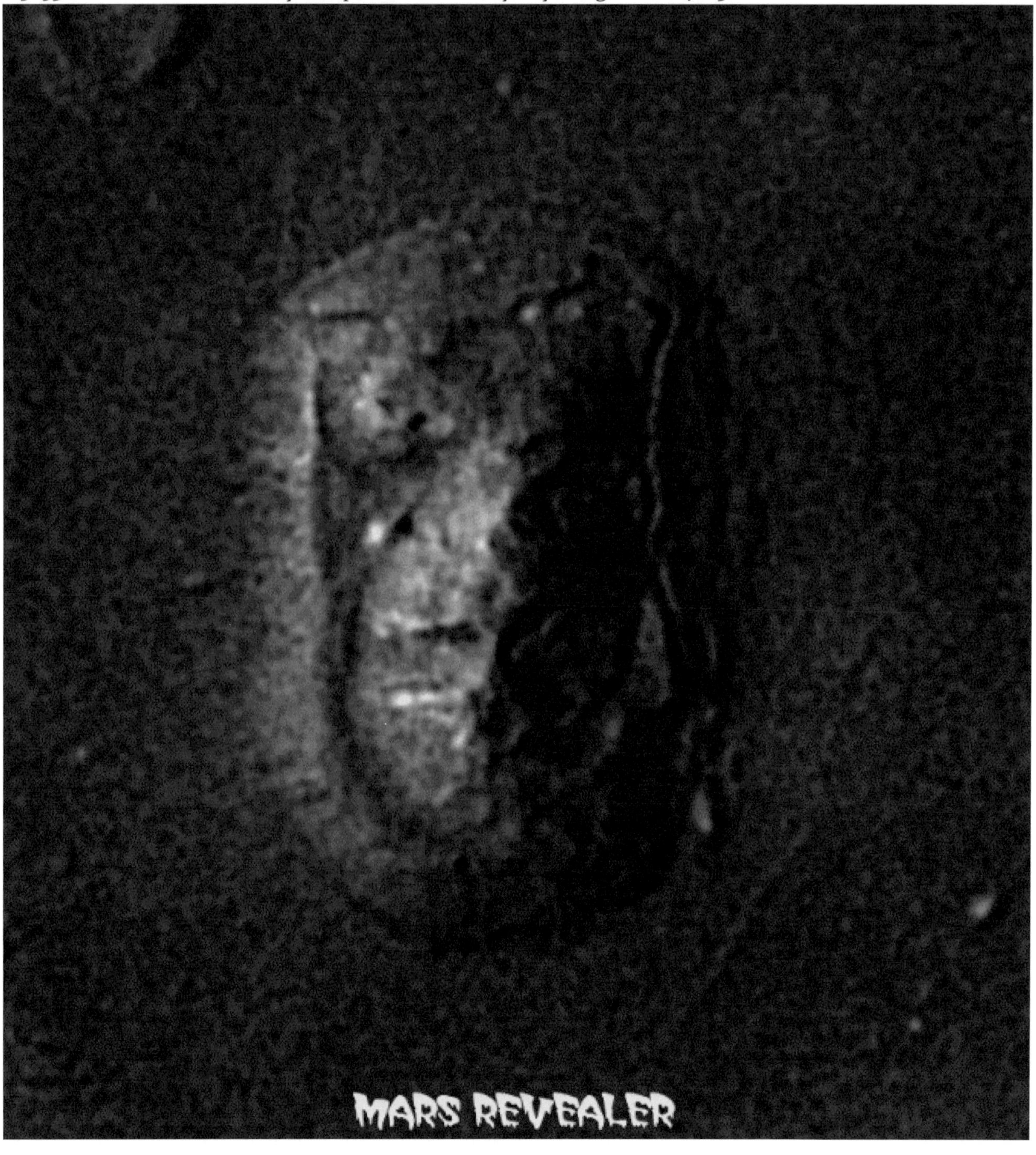

(MARS ODYSSEY IMAGE V01024003 UP-RITE VIEW)

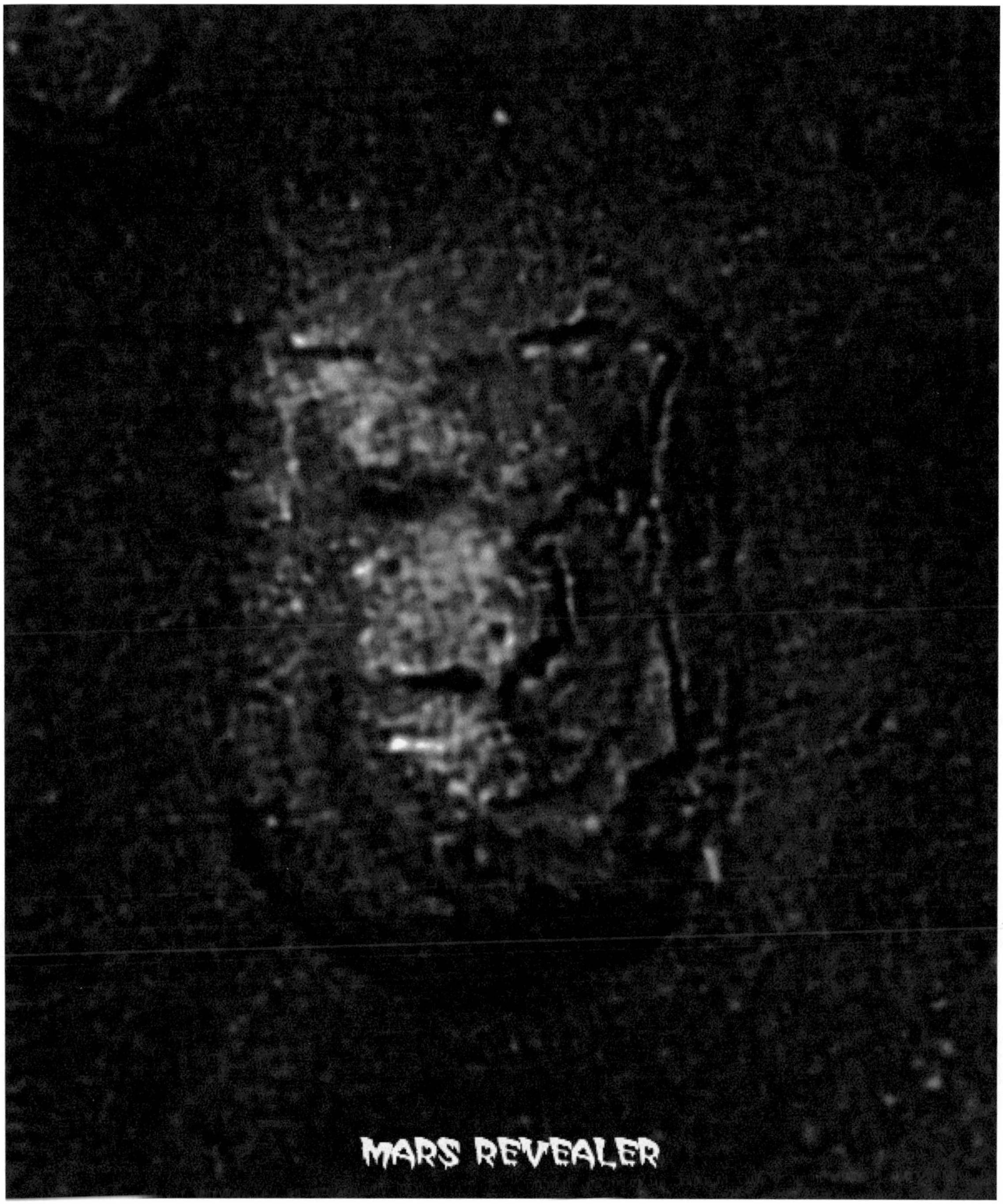

(INDEED AN 'UNPRECEDENTED' MARS EXPRESS VIEW OF THE FACE OF CYDONIA: H3253_0000_ND3)

In comparing between the two images, there is little differences between them, in terms of resolution and quality of image data. Mars Express HRSC is only slightly better resolution than its NASA counterpart, Mars Odyssey. The details of what is seen of the Martian surface features between each differs dramatically, not just because of differing sun lighting angles, but mostly because the images were not taken by the same type and quality of imaging cameras. Mars Express has the higher capability and better capabilities and higher resolution to reveal the details of the Face far more effective to see more easily its alien markings, whereas NASA's Mars Odyssey is apparently not designed to focus (pun intended) on the relief details so much, as is the Mars Express very capable of doing just that.

(AN 'UNPRECEDENTED' DETAILED MERGED IMAGES VIEW OF THE FACE OF CYDONIA: V01024003 AND H3253_0000_ND3)

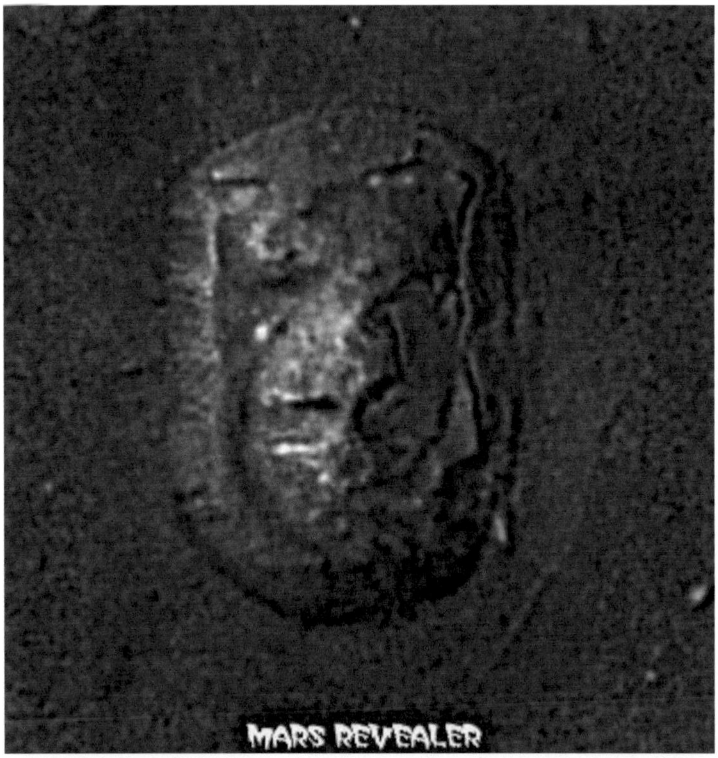

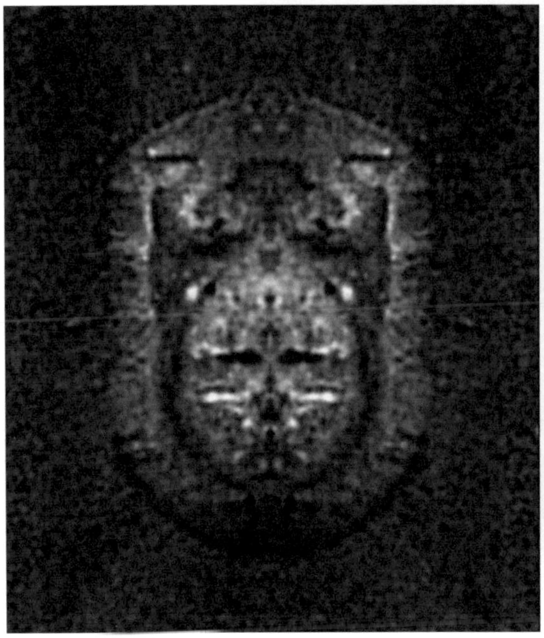

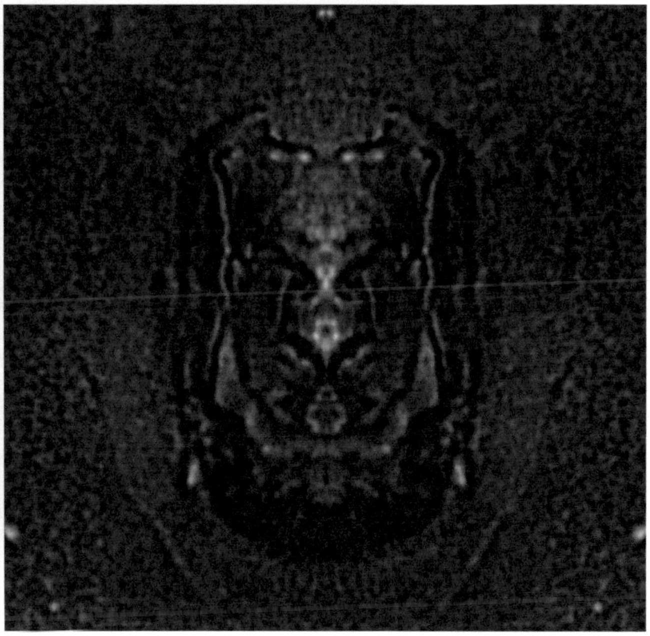

As we continue on in this chapter, you will see each of the three Mars Odyssey best Face image views, and two second best Odyssey image views, showing each with a merged image fusion of Mars Express image H3253. Also each fusion image sample will show views mirrored of their split faced nature, as seen above. As each image varies in details seen of the Face because of lighting conditions and camera ability, showing different details each image inconographic traits of likeness each side represents also will show differences, but all images features being consistent with each other shows us clearly that this Face, is far **'Beyond Mere Tricks Of Light And Shadow'!**

This is not to say that Mars Odyssey imagery is not worthy of scientific study or of being able to be used for comparative analysis. In fact it is very beneficial for comparative analysis. Because of the similar resolution and Nadir angles of the images taken, I was able to fuse the two images together into one image, so we can see a merging of the two images and see what each has to offer by giving us an unprecedented new view of the Face in a single image view. This is very effective and the method is very sound, to get to see a better understanding of what the Face actually shows. This allows us again, to have some unprecedented views, of what we have never seen before of the Face at Cydonia.

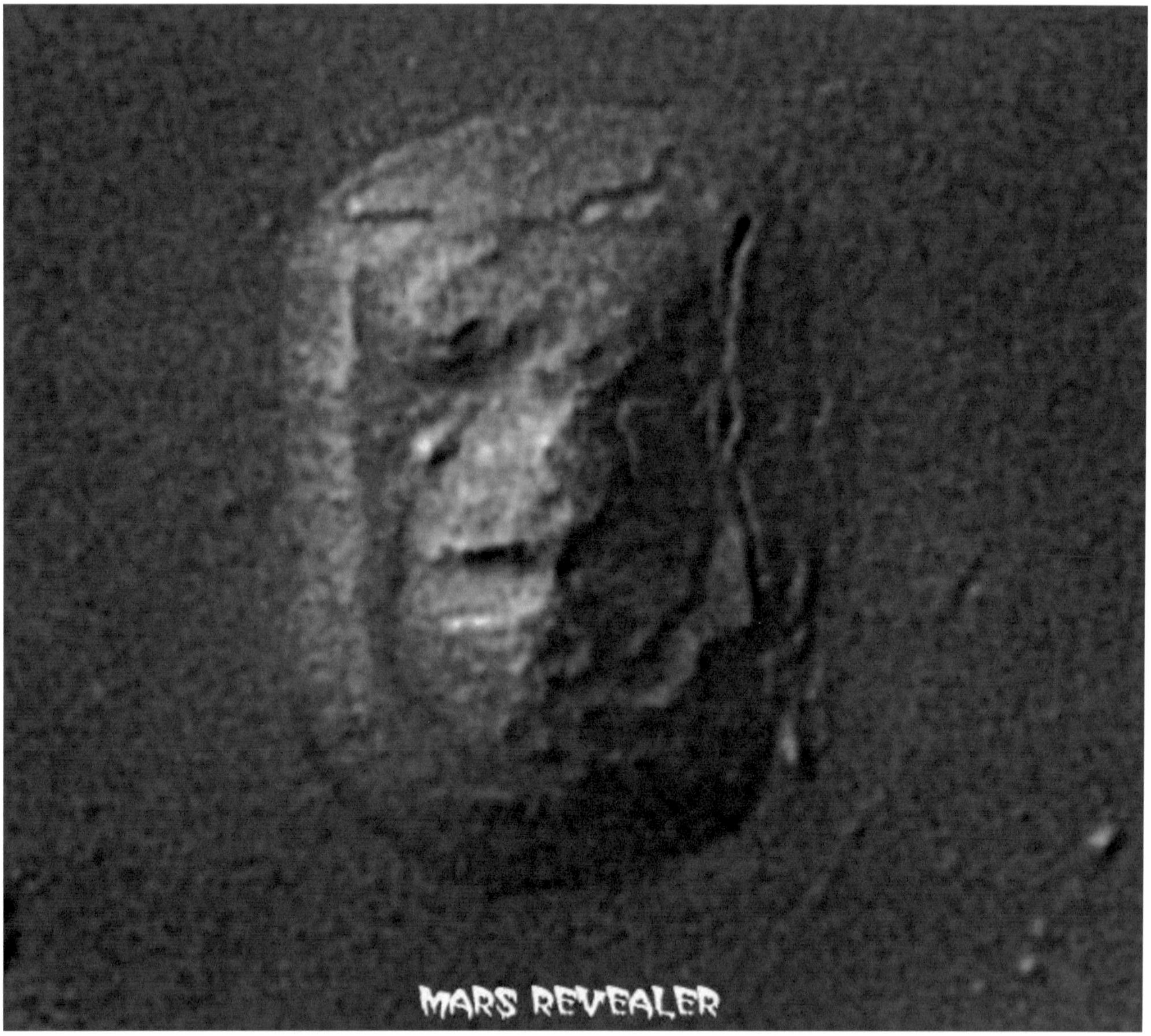

(MARS ODYSSEY IMAGE V10598012 TAKEN ON MAY 5TH, 2004)

With each comparison of Mars Odyssey images of the Face, and from the Mars Express image H3253_0000, we get to see a few different new and unique views, that helps verify what things are seen between them all in each image. So in essence, one cannot say it is false image artifacts, because of unbelief of what is seen the Face actually shows itself as to be, an intelligently designed artistic masterpiece of an extraterrestrial techno-signature. In comparing and fusing the images, data points they each share is seen and matched up to each other, and helps us establish features seen that each has to offer, and create a more real view of what we need to see. This is scientifically consistent and based solely upon the image data itself in what it shows, not what someone wants to believe or not. This is not about belief. This is about the image data, of multiple imaging sources and image sources that confirm each other, and validate itself of its reality.

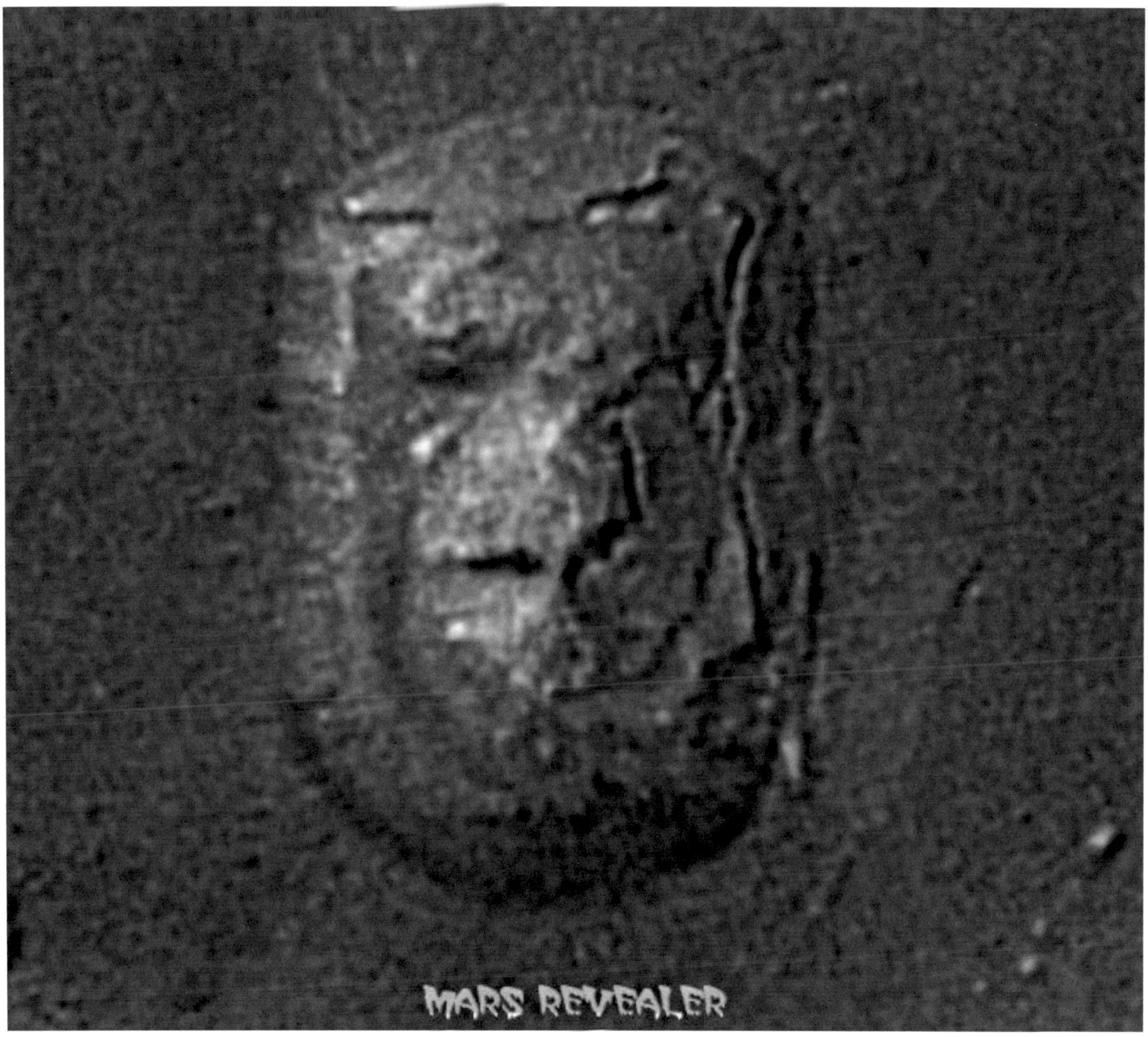

(AN 'UNPRECEDENTED' DETAILED MERGED IMAGES VIEW OF THE FACE OF CYDONIA: V10598012 AND H3253_0000_ND3)

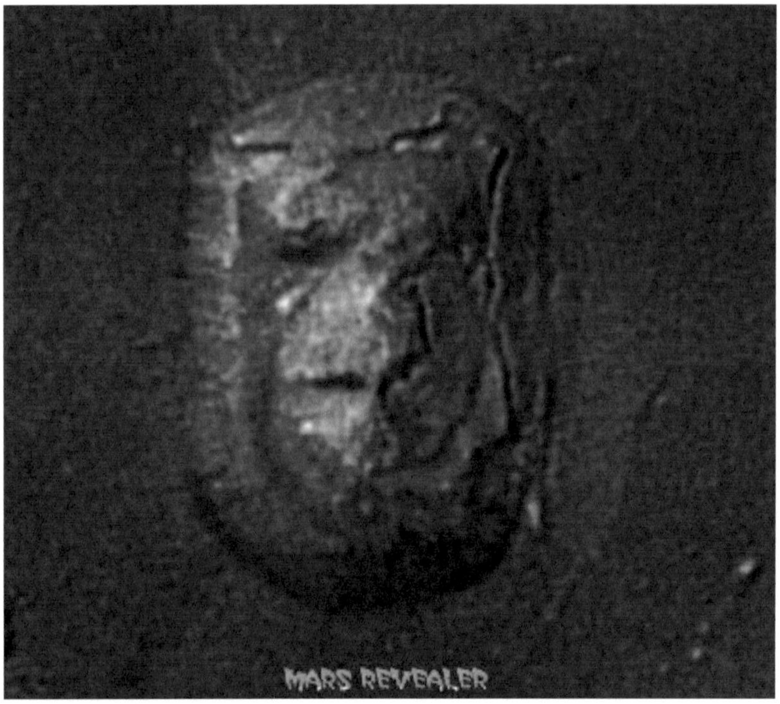

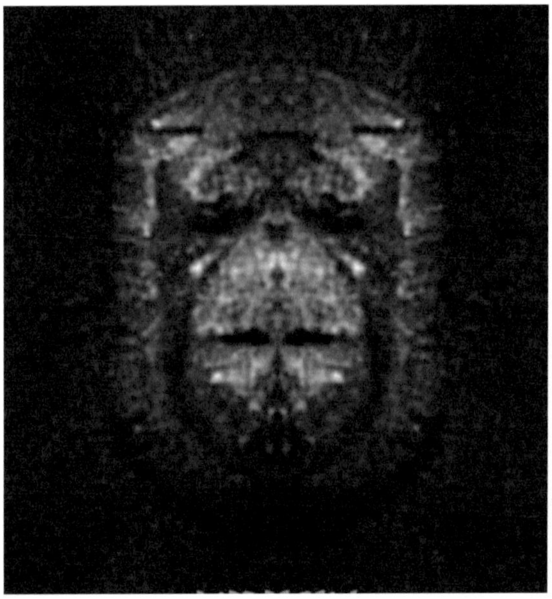

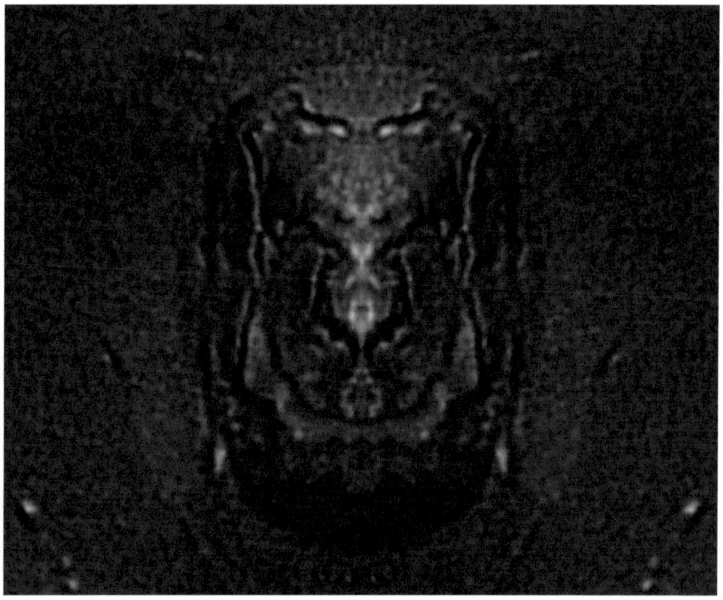

What these compartive studies shows us in these unprecedented views is the viual evidence of an artistically sculpted and molded behemoth of un-natural designing. The Face's releif details is very crucial to try to see in multiple ways and at different times and also again, at varous image resolutions, in order to really try to see and study the Face very deeply in order to understand what it is we are all really looking at. For instance, anyone can only look at one image and say 'yes, it looks like a Face', or someone may claim they not see a face.

That is NOT what this book is about. But this is an unprecedented feature of interest study into using multiple image sources, and using multiple images for purposes if anything but not just to see the Face, but to see and learn of the Face's artistically designed '**Alien Markings**', which again, is it's surface releif features! In my opinion, and analysis, I am attesting that this Face is not existing because merely of coincidence of any natural dirt and rocks.

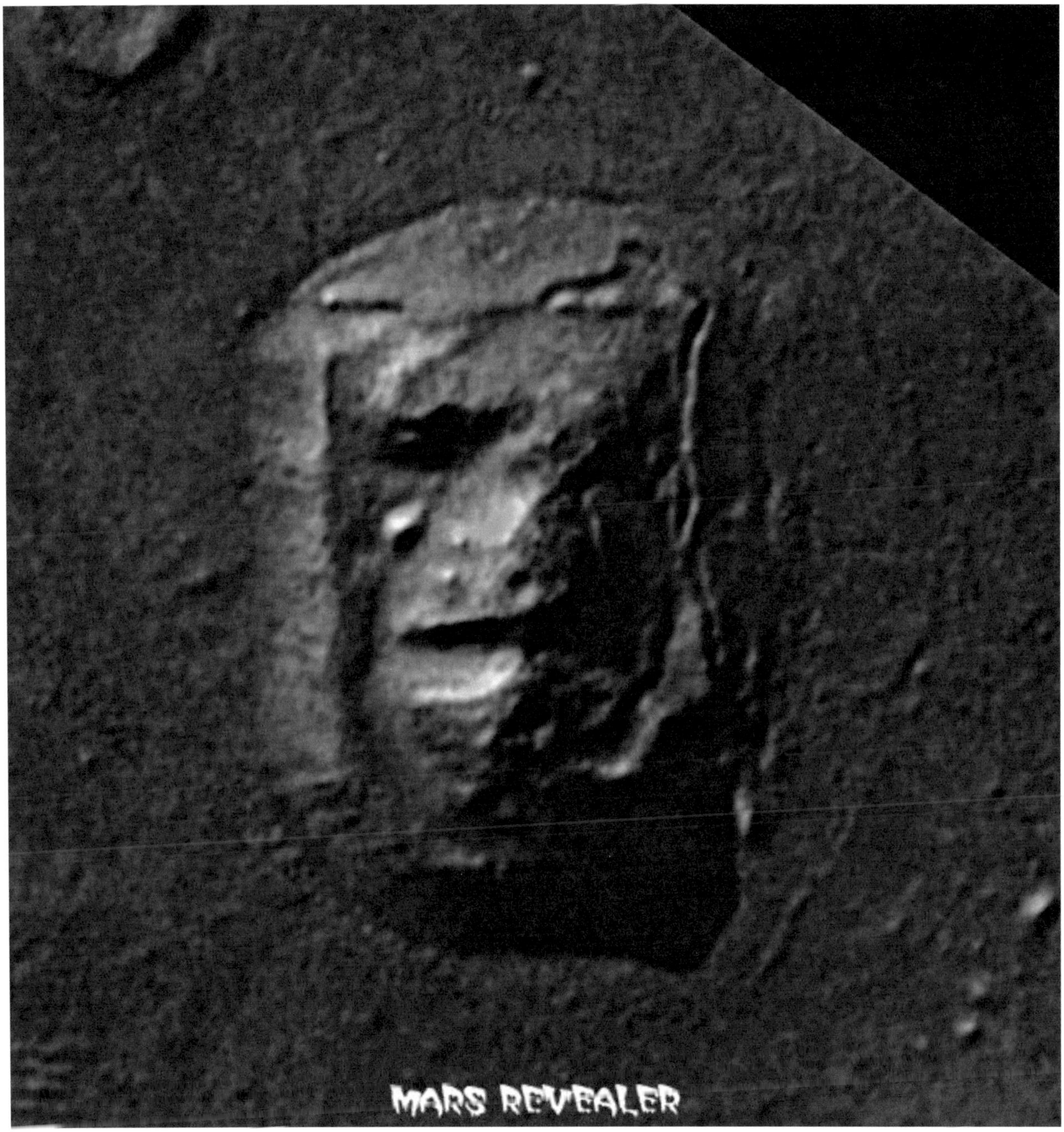

(MARS ODYSSEY IMAGE V54642014 TAKEN ON APRIL 9TH, 2014)

To me, this is my favorite Mars Odyssey image. It is crisp in its details and is starting to cast its famous shadow as seen from the 1st Viking orbiter image that the Face was seen on image number 35A72. As of now also, there has not been a replicated image view of the Face as seen in 35A72 from any Mars mission given to the public. So this particular image itself, is also an unprecedented view. The clarity of the Face even at this resolution is astounding as it brings out details that none of the other Mars Odyssey images show, or barely shows. Again, this is because of various sun angle lighting conditions, which is a good thing, because the more image data we have of the same object of study even at various image resolutions, we get a better analysis judgement overall for the Face when considering all of the image data to be able to consider and enabling us all to see features we may never have been able to notice before. In my opinion, there is no better images comparison to see of these various features than this merging fusion of Mars Odyssey image of **V54642014** and **H3253_0000** from the HRSC from Mars Express.

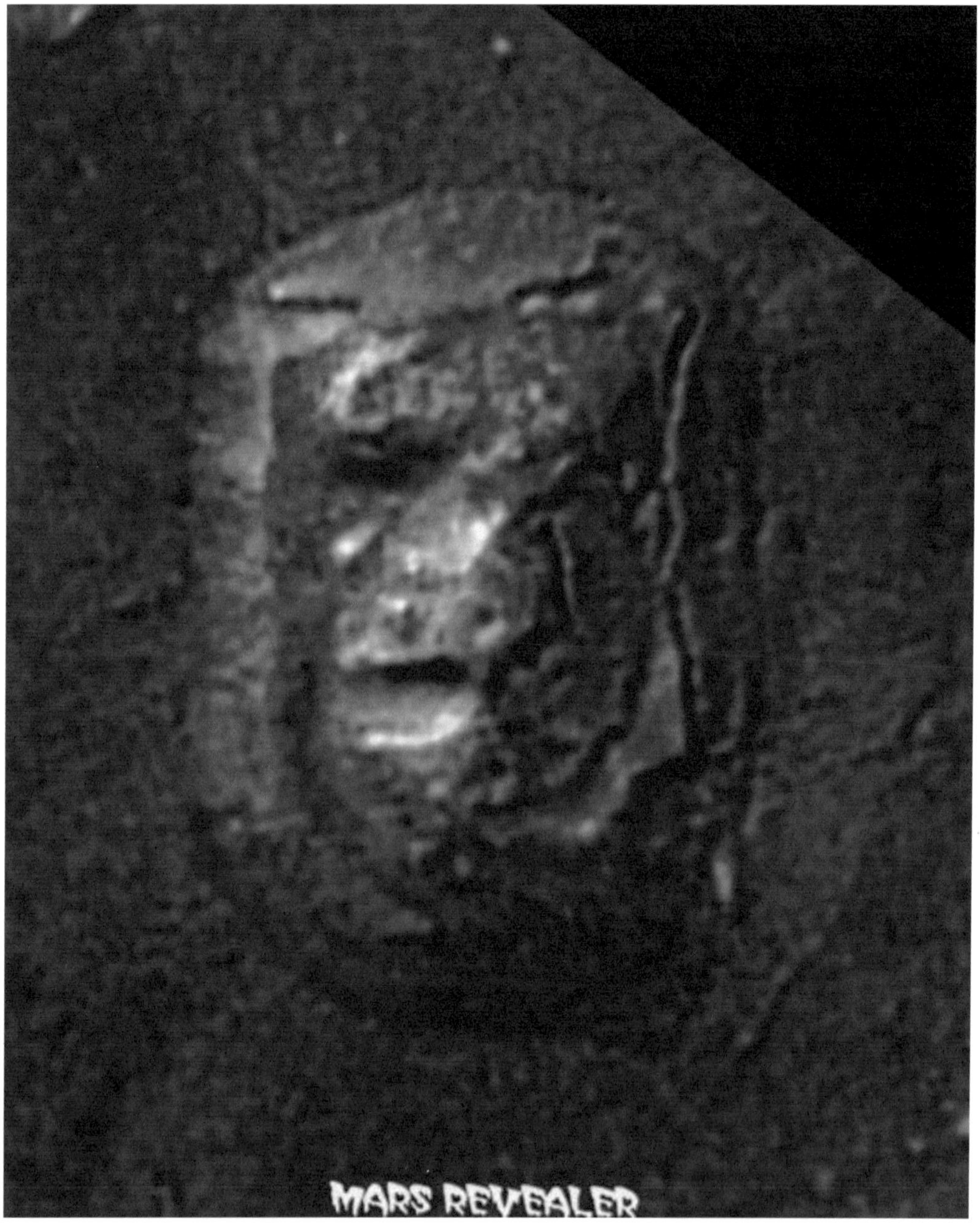

(AN 'UNPRECEDENTED' DETAILED MERGED IMAGES VIEW OF THE FACE OF CYDONIA: V54642014 AND H3253_0000_ND3)

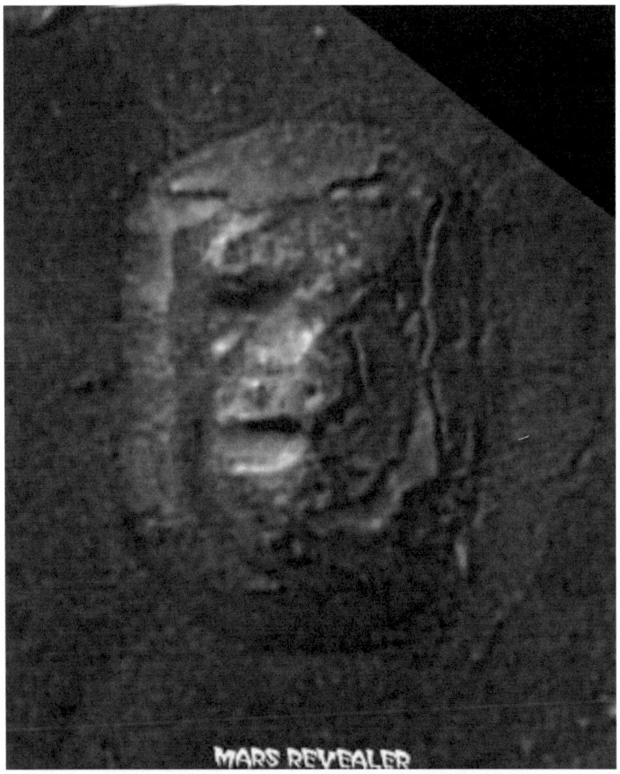

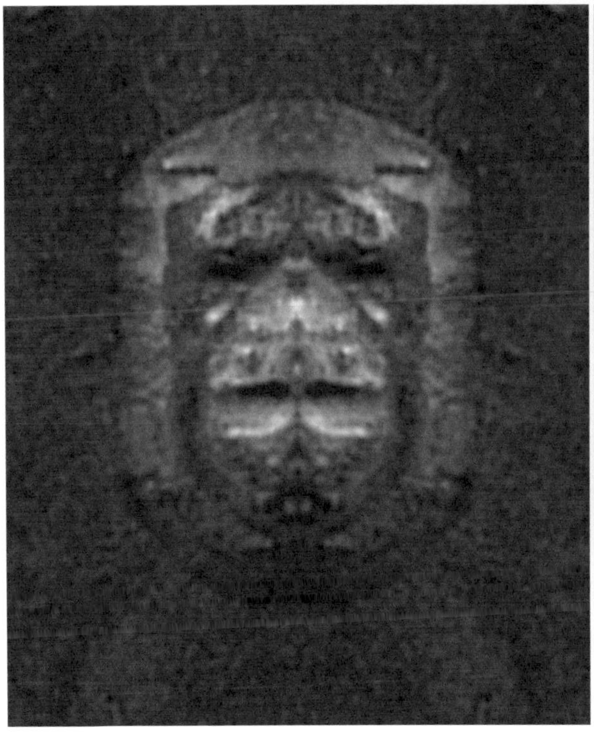

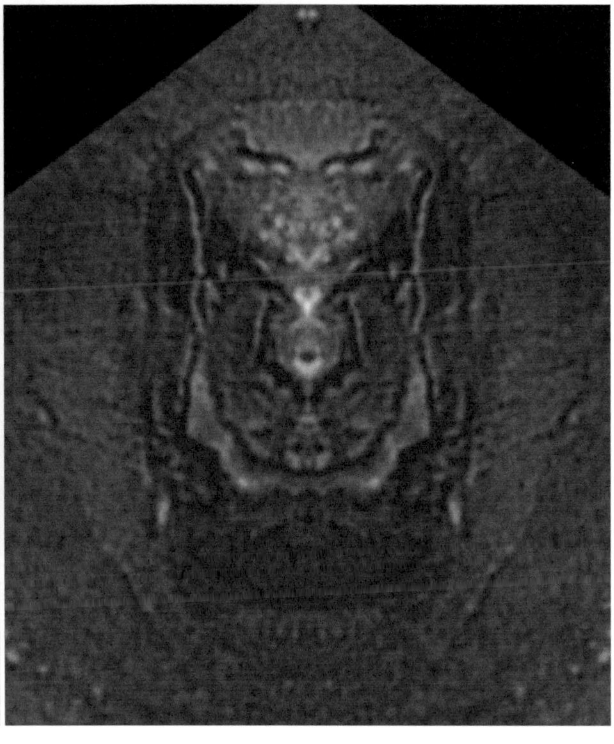

As far as comparisons go of what we seen so far, this image merging of Odyssey image **V54642014** and **H3253_0000** from Mars Express is the most compelling and crucial at least for me in being able to see what I was needing and predicting to see confirmations of. More of that later. But let's look at all three comparisons of the best and my favorites of Mars Odyssey images fused with Mars Express view.

It is a known common scientific model that says scientists usually require 3 sources, that is used to help prove a theory, or to establish something as a fact of science. We have looked at 3 of the best Mars Odyssey images of the Face of Cydonia, and by merging the image data of Mars Odyssey's **V01024003, V10598012, V54642014**, and Mars Express's **H3253_0000** with each of them, **we actually have 6 unprecedented views of the Face that we have never seen before!** Yet this reality is as unprecedented of the 3 Mars Odyssey images were and are, even the majority of independent Mars researchers had not studied or made any big deal for these images, let alone make any comparative analysis between them, and that of the Mars Express image data, to see these amazing **Silent Discoveries!**

To me, without sounding full of myself, I believe this was and still is, because most people still have no idea what they are looking at, or how to see its artistic designing and sculpting of 'Alien Markings'! I have just shown them, and you, an unprecedented analysis with some unprecedented views that they, or you will not find anyplace else from any other known independent Mars researcher and known image analyst studying the Face. That is just also the facts of reality! Which is also why you are here reading and looking at it and hopefully studying this all.

(MARS ODYSSEY IMAGE V12445004 IMAGE TAKEN OCTOBER 4TH, 2004)

The image we are seeing here now, is Mars Odyssey image number **V12445004.** Although being less resolution, this in my opinion is the 4th best Odyssey view of the Face that is also beneficial for comparative analysis due to its sun lighting angle that allows other important features of the Face to be seen that is not so easily seen in the other images. So I decided to do yet another comparative fusion with Mars Express image H3253_0000 to help us to acquire yet another new image view of the Face seen in an unprecedented way. Although being slightly blurred, the details of the Face can be still brought out to notice important differences that needs to be discovered and also appreciated.

(MARS ODYSSEY IMAGE V12445004 IMAGE UP RITE VIEW)

What is unique is that we are using a single Mars Express image on all of the 4 best Mars Odyssey images and with each image, there is consistency of data points matching that show in each image. It is not false image artifacts that is the reason of the things able to be seen on each image of the Face, as it all is collaborating image data with each other that reinforce the data of Mars Odyssey images and all help using the Mars Express image, to get a more accurate detailed view, as to what the Face really looks like. What better way than determining this, than by comparing and studying the image data itself for each one that allows us to have multiple sources and multiple views and multiple sun lighting angles based upon using the scientific method to establish a working model to test and to prove a theory, to become scientific fact! In this chapter alone, we have established and proven the Face is a fact of reality based upon the scientific method and of solely using the image data to yield its comparative test results.

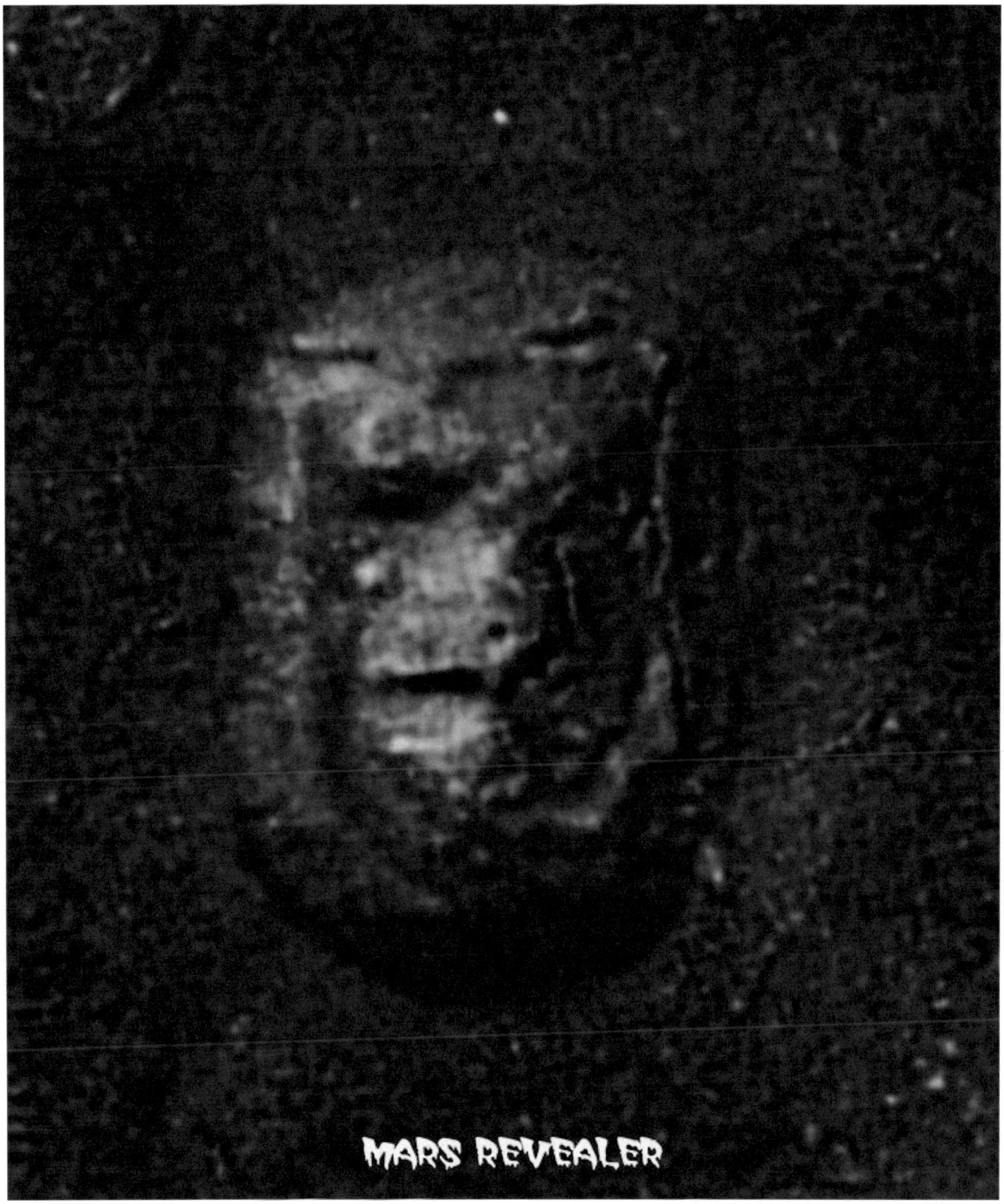

(AN 'UNPRECEDENTED' DETAILED MERGED IMAGES VIEW OF THE FACE OF CYDONIA: V12445004 AND H3253_0000_ND3)

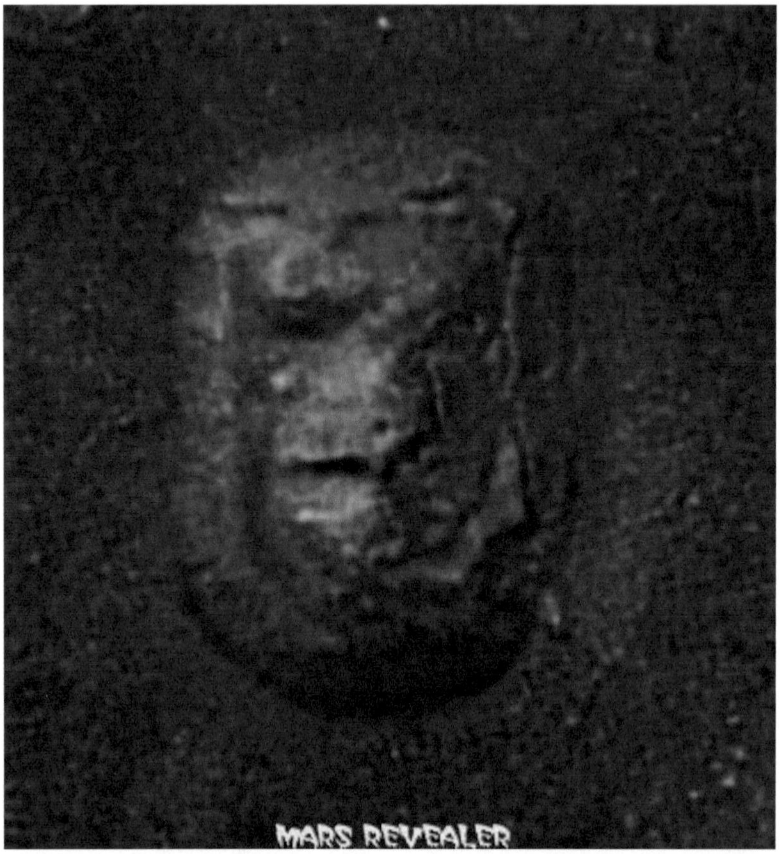
MARS REVEALER

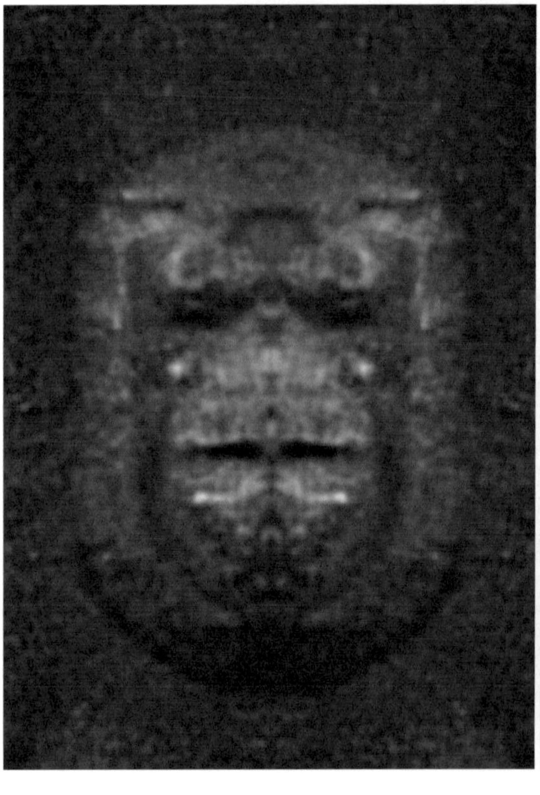

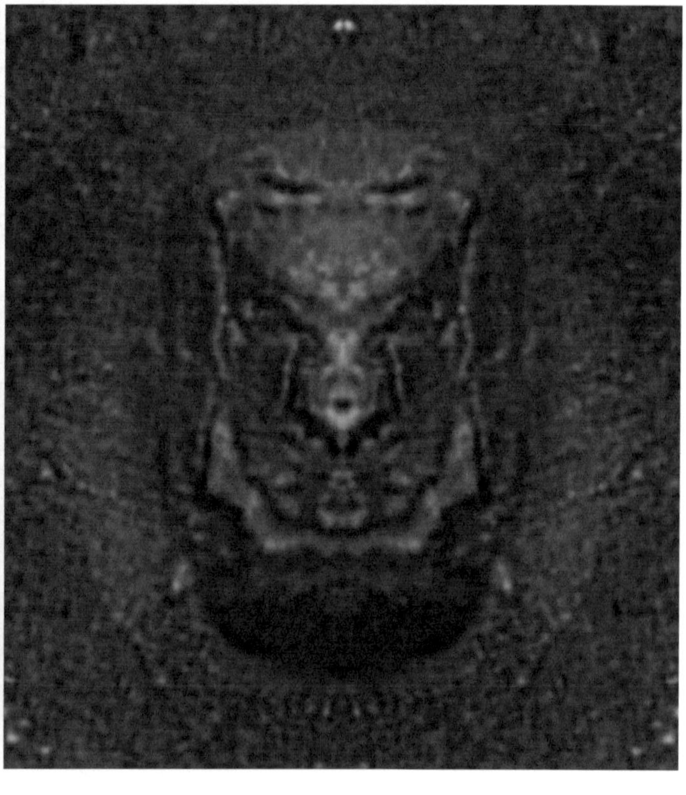

Which brings us to the 5th and final best Mars Odyssey image of the Face, in my opinion, for **unprecedented silent discoveries**! Silent discoveries that are in fact, irrefutable! Let's look at Mars Odyssey image number **V86033017. This image is not only unprecedented, but also absolutely astonishing.** Although the image being taken and shown is near the resolution size of the Viking Orbiter, or just a little bit higher than Viking Orbiters resolution, we are able to see much of the surface relief and artistic detail of the Face. However, not the Face alone is obviously artistic, but even of the surrounding art-scape the Face sits upon as well!

(MARS ODYSSEY IMAGE V86033017 TAKEN ON MAY 7TH, 2021)

There are many detractors of the Face who get so triggered at people who look at the Face and claim that they believe it is a real alien artifact. The triggering the Face causes in certain people is not just laughing at someone, but results the majority of the time to insults, name calling, a form of public shaming to those who asks questions or posts pictures online in various ways and places to try to create some form of discussion with others and potentially speaking about what the images show.

The first thing detractors do in the majority of all places online where any image of the Face is shown or discussed, is by placing up official NASA links and comments about the Face being only a natural geologic feature, that anyone who claims that they believe they see a face, are mistaken and is a subconscious symptom as if it is like a deficiency that humans have to discern reality and therefore are easily fooled. People are labeled and treated as if psychiatrically of an abused diagnosis, that has been created and given a label in which the person who sees a Face, is being tricked into believing what their eyes tell them is something real, but it is not real. This brought forth the diagnosis term "Pareidolia!"

> **pareidolia** /păr″ī-dō′lē-ə/
>
> **noun**
>
> 1. The perception of a recognizable image or meaningful pattern where none exists or is intended, as the perception of a face in the surface features of the moon.
> 2. The tendency to **interpret** a **vague stimulus** as something **known** to the **observer**, such as interpreting marks on Mars as canals, seeing shapes in clouds, or hearing hidden messages in music.
>
> The American Heritage® Dictionary of the English Language, 5th Edition • More at Wordnik

It is interesting to note here that this 'term and definition' happened to come forth around the time in 2001when the 1st full on Face view released by NASA and Malin Space Science Systems (MSSS). In my opinion, it was not a coincidence, and the unwanted and unwarranted psychiatric diagnosis term and definition was really put forth to the public by those trying to control the official public policy and agenda and narrative that has been set in place to keep the Face of Mars in a position that is not up to public debate or questions, that would generate discussions that threaten to contradict and even change NASA's official position of the Face. So yes, I believe this term and definition was created by the NASA official policy makers, and put forth to the public in order to deal with and control the public's imaginations and questions by having things set in place to block and to control the public and contain discussion about the Face. The very definition itself, the very 1st one to be seen after searching for the term on google, show it's deigned bias and purpose to try to cover up the truth that NASA cannot easily hide, or modify enough to fool you all.

Notice in noun number one, the perception they try to portray of its meaning is **NOT** to even address of a Face on Mars being seen, (which is its purpose of being created to begin with), but as instead mentioning that there is a perception of there being a Face seen in the surface features of the Moon! This is hilarious. They not want to even give a person reading the definition, of a hint that there is a Face on Mars to be seen in the surface features of Mars. It is rather comical if you look deep into it, as is also what is said by detractors, to label you or try to shame you, or make you rethink your thoughts, by implying that you are mistaken in your thoughts and opinions, or research, because you are looking too deep into an image or the subject matter, and it is attacked by using the public policy official positions and also using this obtrusive false psychiatric symptom diagnosis narrative forced upon anyone who claims they see the Face at Cydonia as being a real and an intelligently designed artifact or techno-signature. This helps to steer people who may be interested in what is seen or said, and is set in place to act as a blocker for those who may wish to comment or seriously discuss it, but then who is not wanting to be ridiculed or labeled, and not reply, and progress not continues to grow very strong in the public eye or mind set to make any difference or threat to NASA's official policy position.

Also hilarious, is noun number two. Notice that by seeing and interpreting marks on MARS, (now Mars is mentioned specifically), or Martian canals, and seeing shapes in clouds, this mental deficiency symptom effects proper interpretation of human perception because they **KNOW** that you cannot tell the difference between a moving cloud, and a solid ever present surface feature seen **on Mars**. This is comical sarcasm again! This shows its true purpose, to keep the truth from us all as official policy, and purposed by knowing design, of public manipulation to keep their secrets intact! If you are not satisfied of that ***conspiracy theory*** as being a probable fact at this point in reading this book, you will be shown exquisite irrefutable evidence of this fact later on in this book. Even detractors and paid and unpaid operatives who confront interested people about this subject will **NOT** be able to refute it in any understanding way or reason.

There is yet another thing detractors like to use as a means to stop discussion by saying things about that image data itself, as if they know what they are really talking about, and use this as a bullying tactic to stop anyone from using the image data itself as a means of putting up what you see or think or believe is any evidence that the Face of Mars is real. What is that tactic you may be thinking? The detractors love also to claim that the images that are placed up to be seen or discussed is only low resolution, or grainy, and anything interpreted in them by what is able to be seen is just something you are mistaken in seeing or believing again. That you are not qualified by any degrees or acknowledgements behind you to be considered as telling the truth about what you see or think you see in the images. Making excuses even stating that there are images of higher resolution in modern imagery that is available that **'SHOWS IT WAS PROVEN THE FACE IS NATURAL'!** Yet the images you show and discuss about, are the highest resolution, and most modern, and still to this very day, no images proven the Face is natural! Not One!

This is a constitutional threat as well because this official public policy forces compliance and stifles free speech and discussion and beliefs about a subject NASA nor anyone else, has the right to control or to keep from us all of what the truth and facts are of the Face of Cydonia, or anyplace else. They certainly have **NOT** been scientifically consistent regards to the Face of Mars! Especially regards to the ***unprecedented silent discoveries*** of Mars Odyssey!

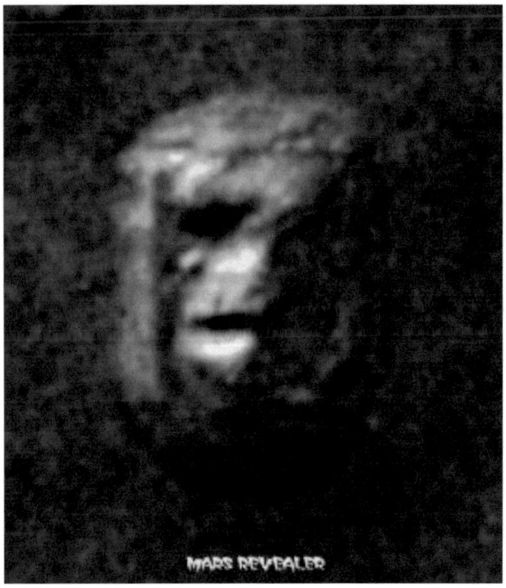

(MARS ODYSSEY IMAGE V86033017 FACE UP-RITE)

(AN 'UNPRECEDENTED' DETAILED MERGED IMAGES VIEW OF THE FACE OF CYDONIA: V86033017 AND H3253_0000_ND3)

As anyone can see, even with and at the various lower resolution images we have from Mars Odyssey, which again, were **NOT** even really looked at or studied or even talked about even by independent Mars researchers, there is so much valuable data to see and compare with each other, as well as the added benefit of the slightly higher resolution non-NASA stereo image data from the Mars Express of ESA, shows they are all consistent with each other in its details of the Face that each image has to offer and connect to what is seen in others. There are **NO TRICKS** about it, there is only multiple sources of image data, and the scientific method of studying the data and doing its comparative analysis between them all.

This is what makes me think and believe the reason why NASA not made any public fuss or discussion of the Mars Odyssey visible images, was and is because NASA knew that the Mars Express camera HRSC images were superior in viewing the Face's ***'ALIEN MARKINGS'***, compared to the limited perceptive reliability of the Mars Odyssey. My accusation is that they knew and know that in doing what we just went over and are seeing here, it shows validity to what the Mars Express imagery data shows, and validity for what Mars Odyssey imagery shows even in its limited capability of seeing detailed relief. It was known already when the Mars Odyssey images were released that there would be too much to see and compare with the Mars Express image data and they would not be able to easily debunk or refute all that image data and its comparative analysis. So NASA and others, stayed silent! Yet as we all can easily see, the image data exists, and is scientifically sound and viable and visually worthy!

I wish to focus on a still somewhat silent discovery from Mars Odyssey that is really incredible. A valuable discovery that shows us all of the Face being seen through an early morning sunlight view. This was Odyssey image number **V03814003.** Although a lower resolution image, it yields forth important information. Only a couple of other independent Mars researchers has looked at this image with any importance given unto it. Even there, no one noticed an important part of this image that is very crucial in understanding what is seen of the Face, and that discovery has to do with helping us to see a glimpse of the true shaping of the Face. ***(We will look deeper into this important discovery in a later chapter).***

However what was exciting for a couple of researchers, and for myself, is the fact that we now/ at that time, finally had an unprecedented image view of the Face showing its usual shaded side of the Face with full early morning sunlight. The usual fully sunlit side of the Face we were all used to seeing it, was now being shaded in darkness. The view definitely draws attention to the Face in a new way not ever seen before. A different view and giving us all a different perspective of the Face to help us to better understand it.

NASA had once stated that after the famous 35A72 image was taken, that they (NASA) had taken another image a few hours later, it had shown the Face was no longer visible. We went over this already, but we know this was a lie by NASA to debunk the Face's existence without using any data, because no such image was ever taken as they described. The truth of this Mars Odyssey image is and of itself, shows that NASA lied to us all way back in their first initial official public position for the Face, as by what is seen in this particular Odyssey image, it shows that the Face structure can never just disappear, but is reinforced by these amazing 'un-natural' tricks of light and shadow!

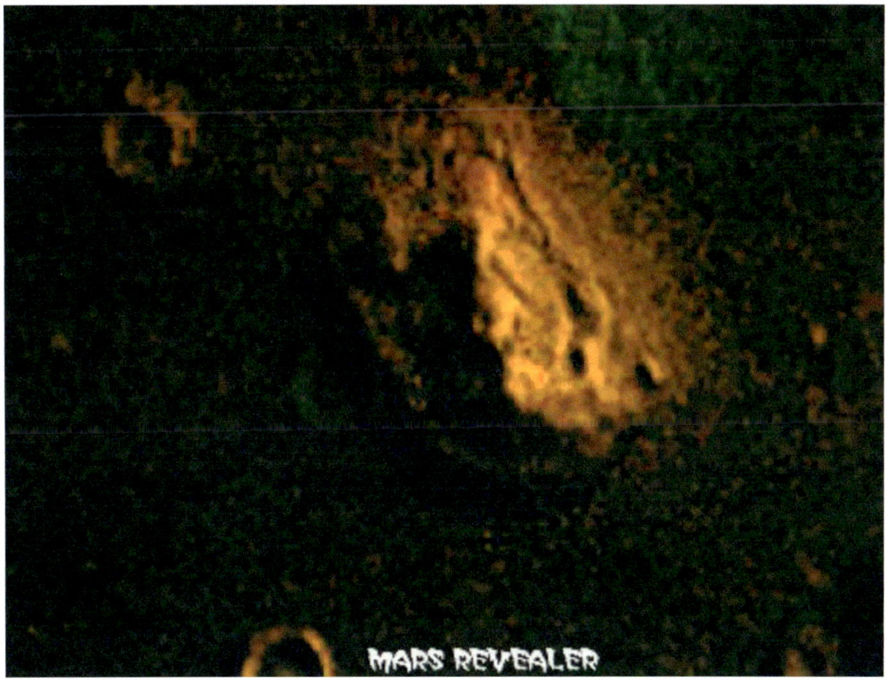

(MARS ODYSSEY IMAGE V03814003 Morning Color)

(MERGED IMAGES VIEW OF THE FACE OF CYDONIA: V54642014 AND H3253_0000_ND3 and V03814003 Morning Color)

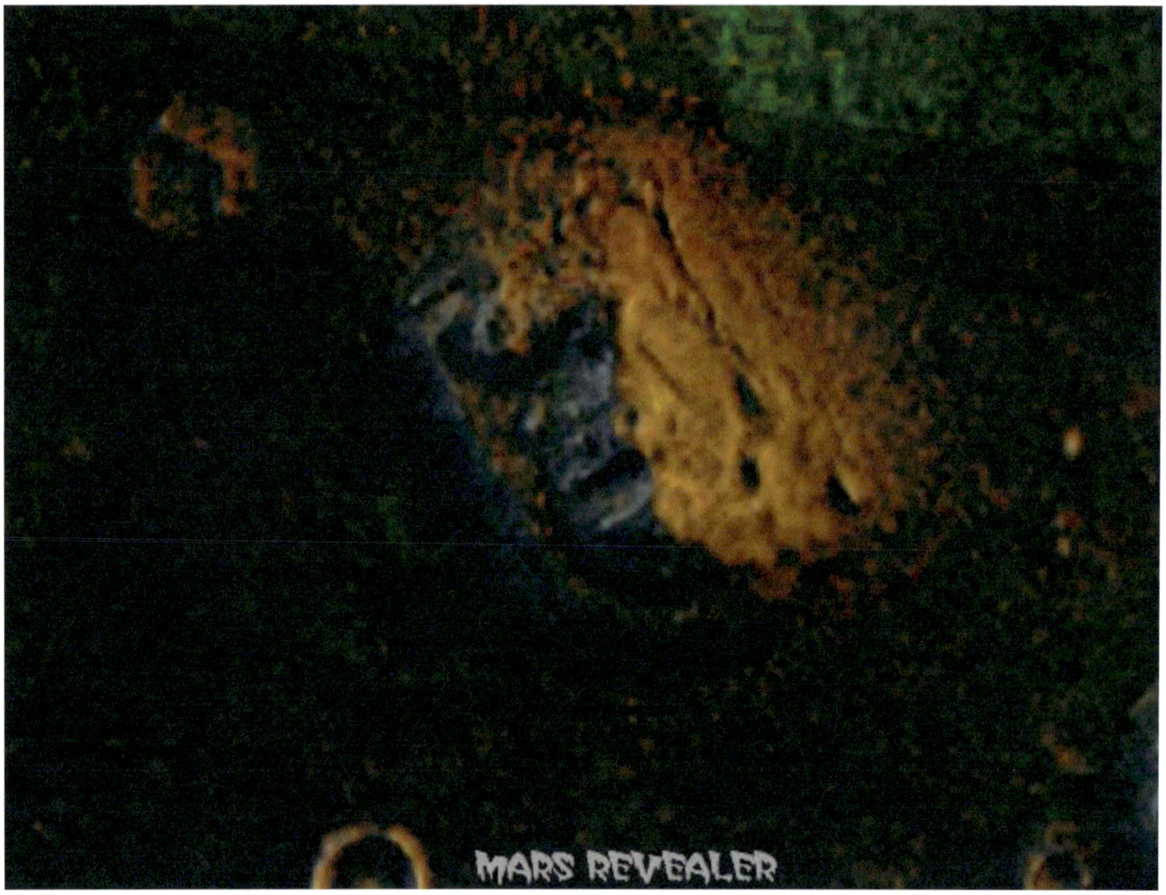

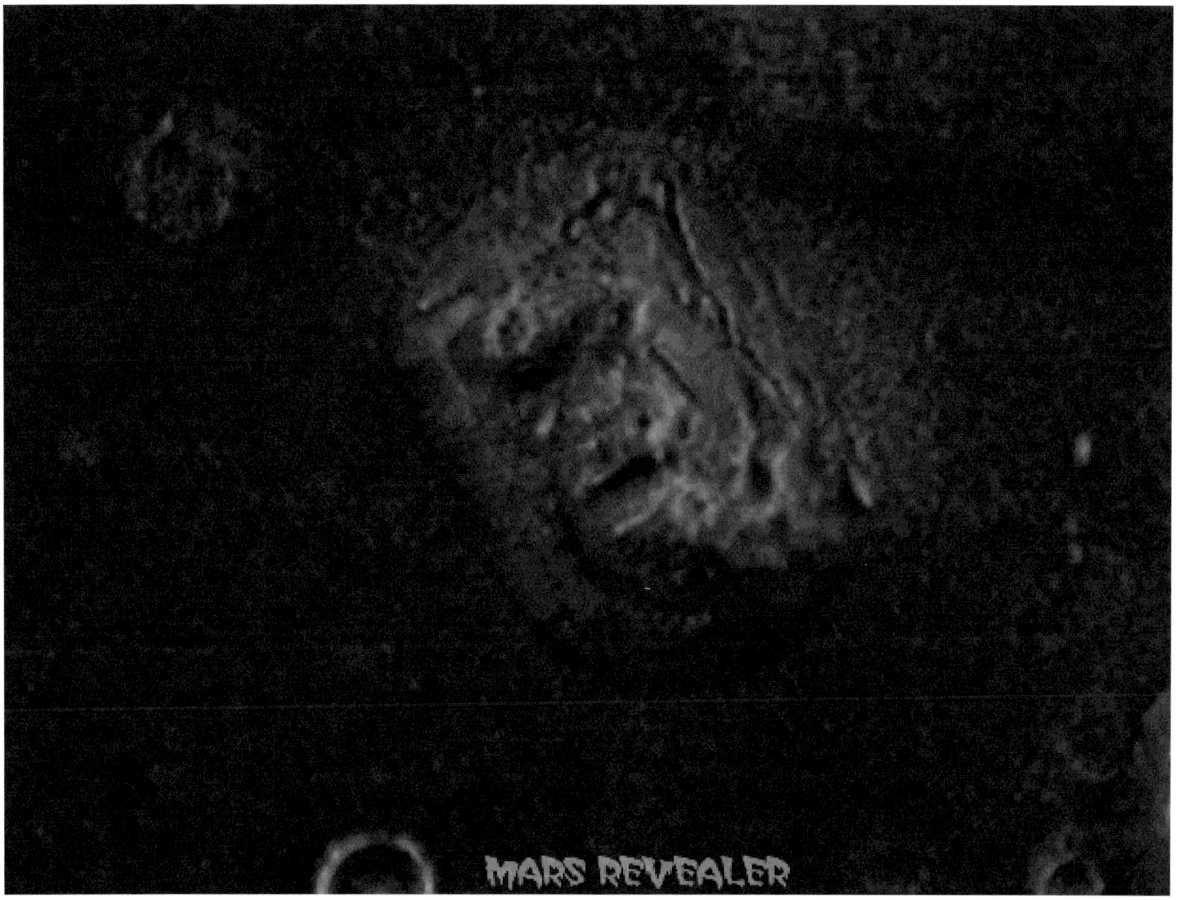

(MERGED IMAGES VIEW OF THE FACE OF CYDONIA: V54642014 AND H3253_0000_ND3 and V03814003 Morning Color black and white)

We also have another incredible unprecedented view from Mars Odyssey image number **V60652003** that shows yet another early morning sun lit view of the Face. I am really surprised NASA allowed these images to be shown at all. Yet more silent discoveries! Now we have yet another example that shows us all immediate contradictions to NASA' official claims especially in terms of what their official position is in regards again, to the true shaping of the Face. As I said earlier, ***we will look deeper into this important silent discovery in a later chapter.***

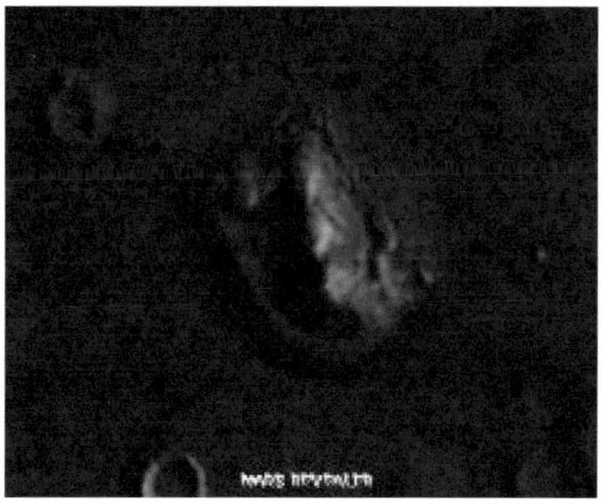

(MARS ODYSSEY EARLY MORNING IMAGE V60652003)

This unprecedented Mars Odyssey view is yet another source and example of seeing the Face in an unprecedented way of seeing it in early morning lighting. As also with **V03814003** Morning Color image, this other image view of early morning lighting again shows us to see a glimpse of the true shaping of the Face. ***(We will look deeper into this important discovery in a later chapter).***

As with Odyssey image **V03814003** Morning Color image, we are going to overlay and make another image fusion of the merged image view of odyssey image **V54642014** and Mars Express image **H3253_0000_ND3** with Odyssey image **V60652003**, to get another yet unprecedented image view of the Face never seen before.

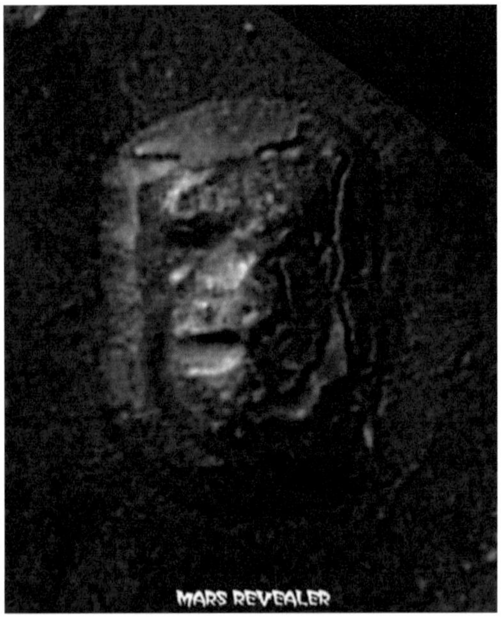

(MERGED IMAGES VIEW OF THE FACE OF CYDONIA: V54642014 AND H3253_0000_ND3)

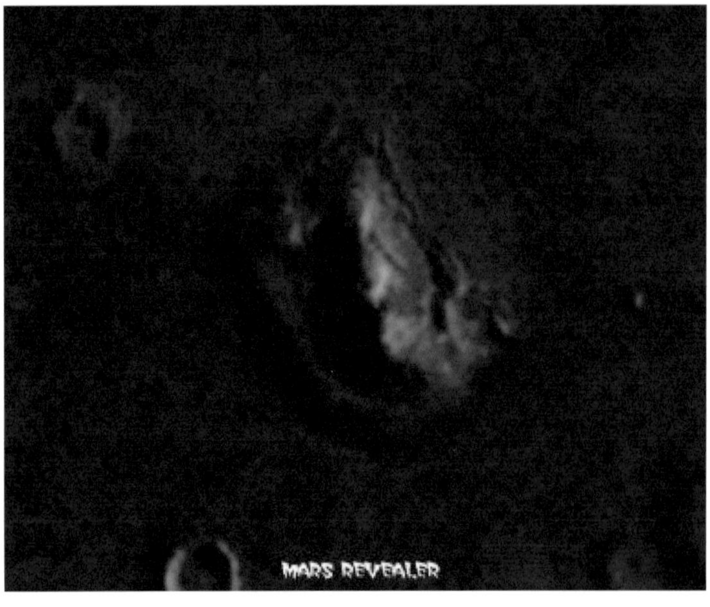

(MARS ODYSSEY EARLY MORNING IMAGE V60652003)

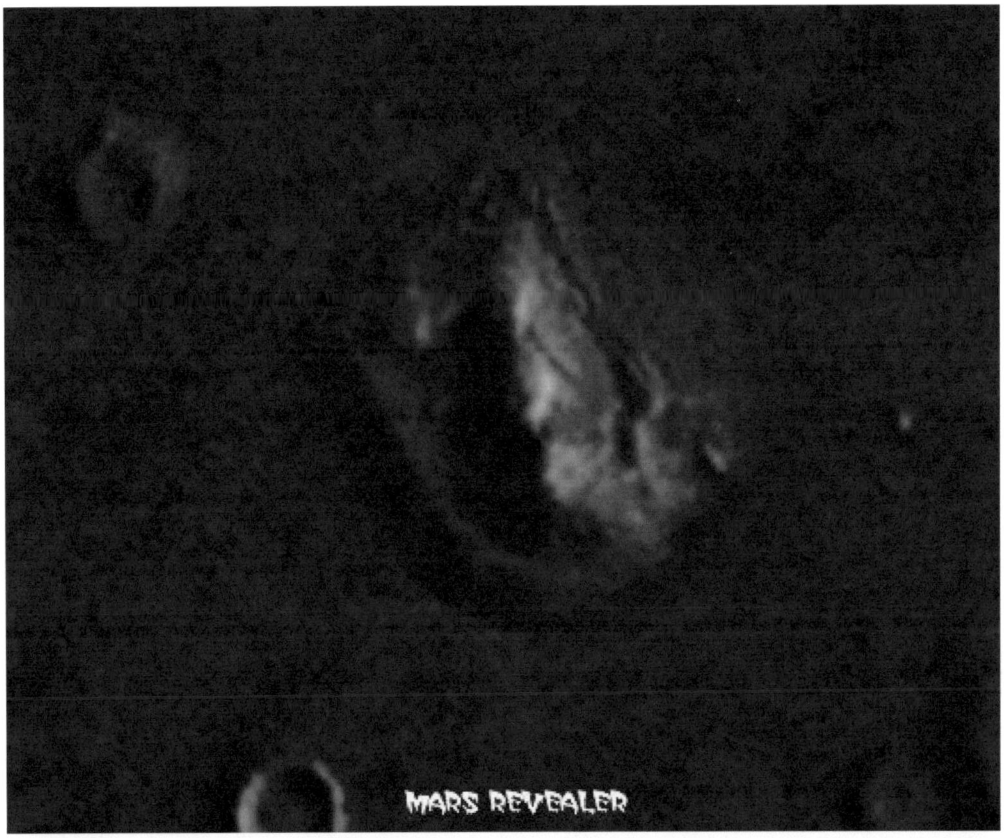
MARS REVEALER

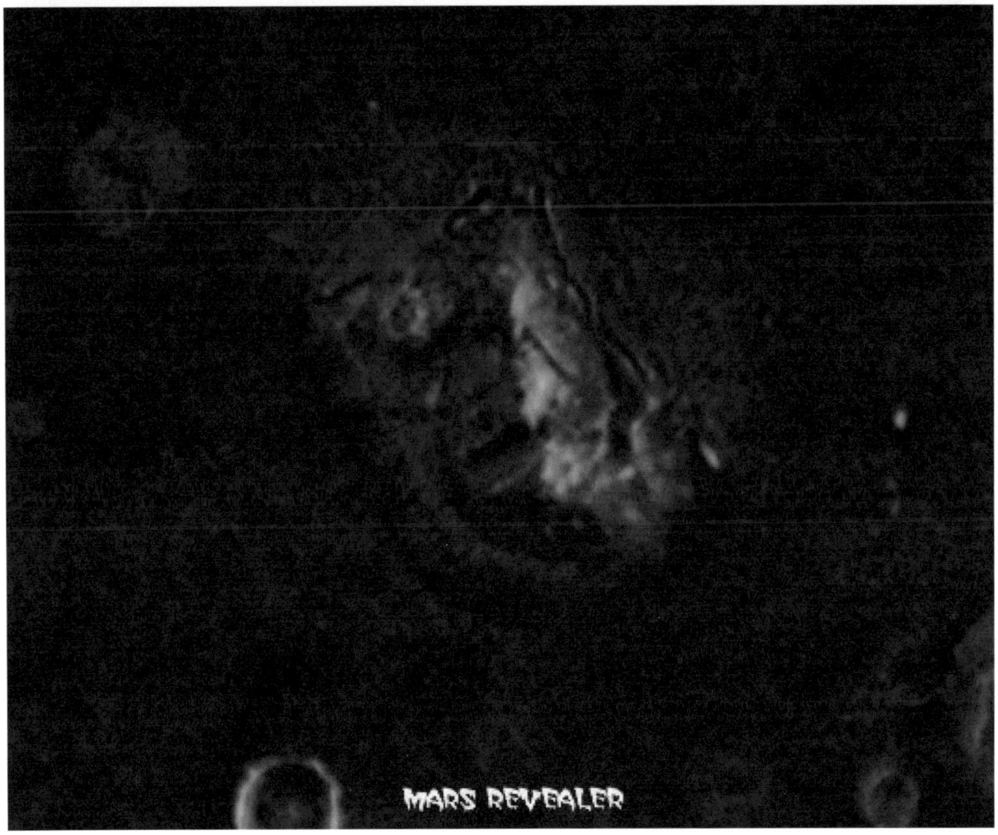
MARS REVEALER

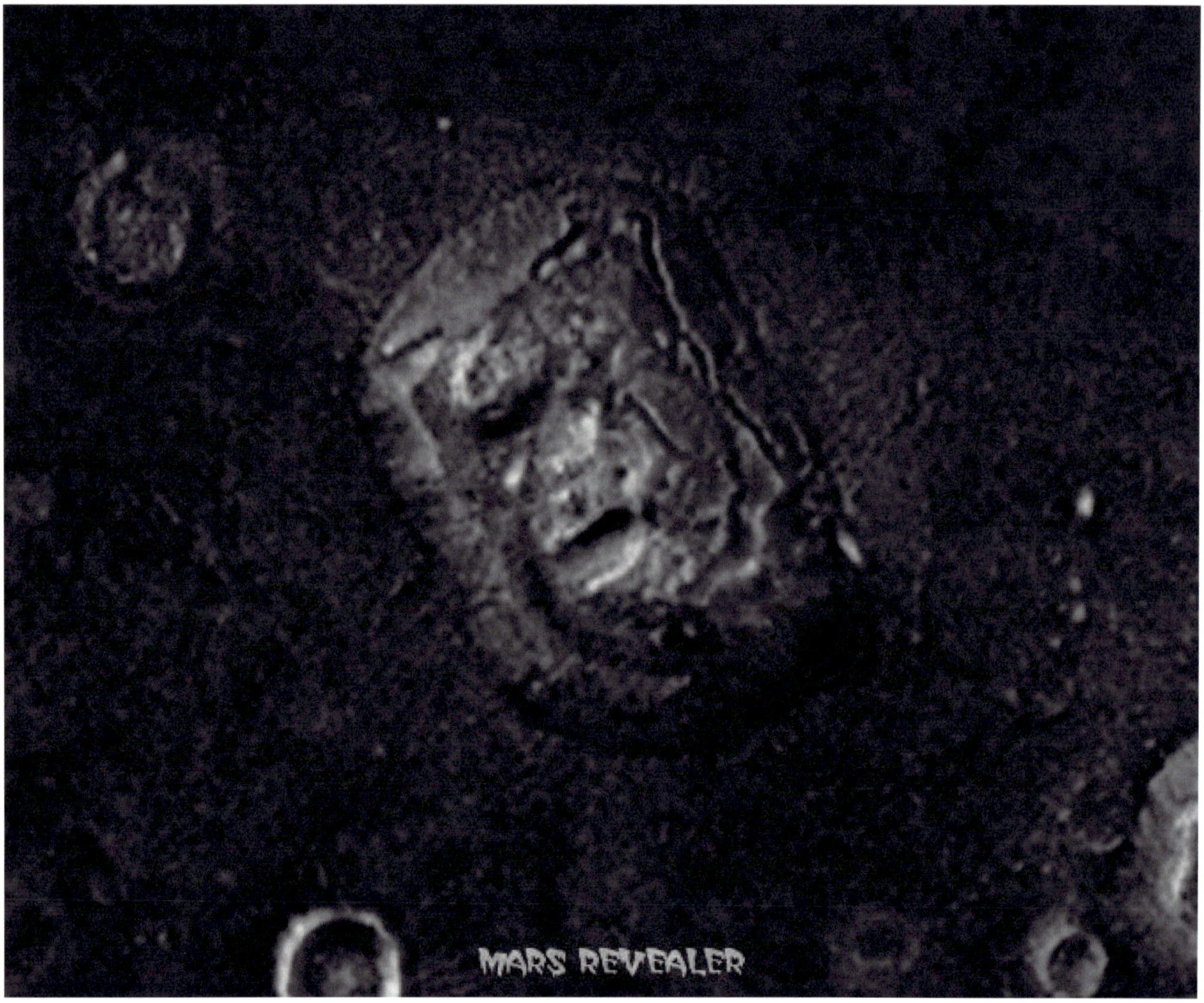

(MERGED IMAGES VIEW OF THE FACE OF CYDONIA: V54642014 AND H3253_0000_ND3 MARS ODYSSEY EARLY MORNING IMAGE V60652003)

It is really important to seriously consider studying the Mars Odyssey image data. So many **Silent Discoveries** and no one else on Earth in the Mars independent researchers community, especially after all these many years, has done a serious look and study and comparative analysis of the Face of Cydonia as you are seeing here. Truly this is an unprecedented study now on record whose time has come, especially when no one else has done it.

After seeing this all, it is easy now to consider why NASA not wanted to focus on Mars Odyssey image data when being Faced with (pun intended), ESA's Mars Express HRSC camera image data during the last official hit piece NASA has released in 2007, which we went over in **Chapter 5: Can This Be True**. I believe NASA not wanted to focus on using Odyssey image data was because they not wanted to give out the idea to people to then focus on the Odyssey image data and compare it with a serious study of the Face with seeing and using the Mars Express image data, and doing the needed serious comparative studies that obviously would go against NASA's natural narrative and label placed onto the Face for the purpose of downplaying the Face and misleading and lying to the world public about the obvious highly intelligent techno-signature the Face really is. They are aware that the similar resolution sizes of Odysseys imager and its limited capabilities helps hide from the public what they not want easily seen, that Mars Express helps to reveal. This unprecedented study here alone, proves NASA's obfuscation.

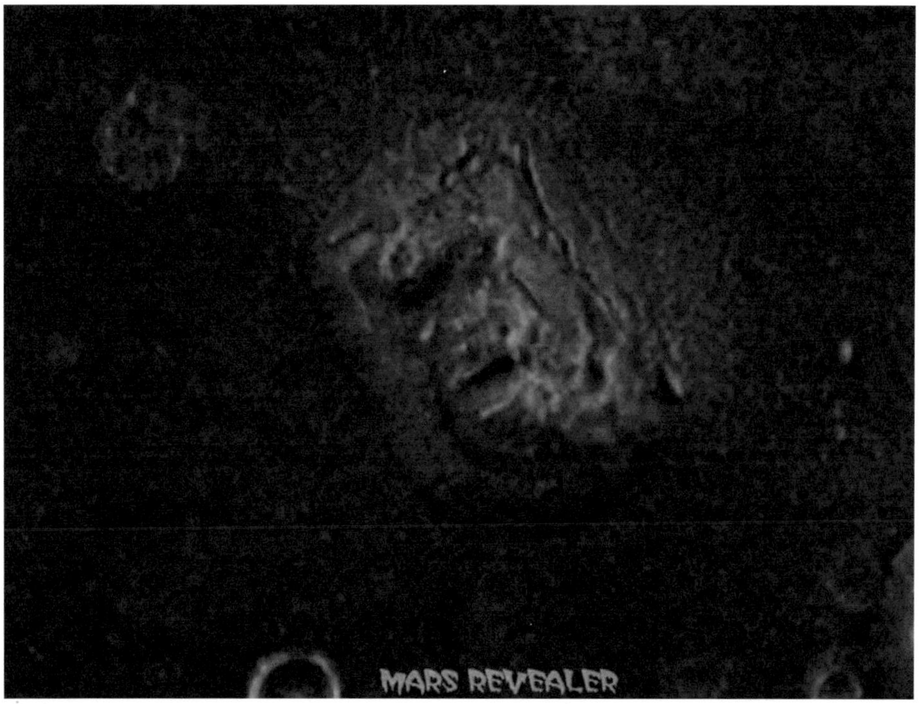

(MERGED IMAGES COMPARISON VIEWS OF THE FACE OF CYDONIA: V54642014 AND H3253_0000_ND3 and V60652003, V03814003 Morning Color black and white)

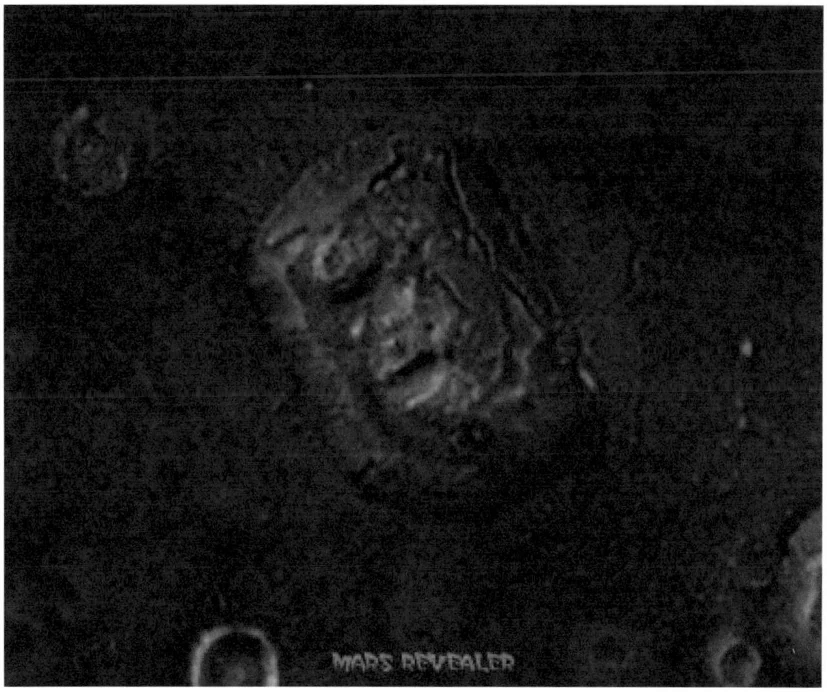

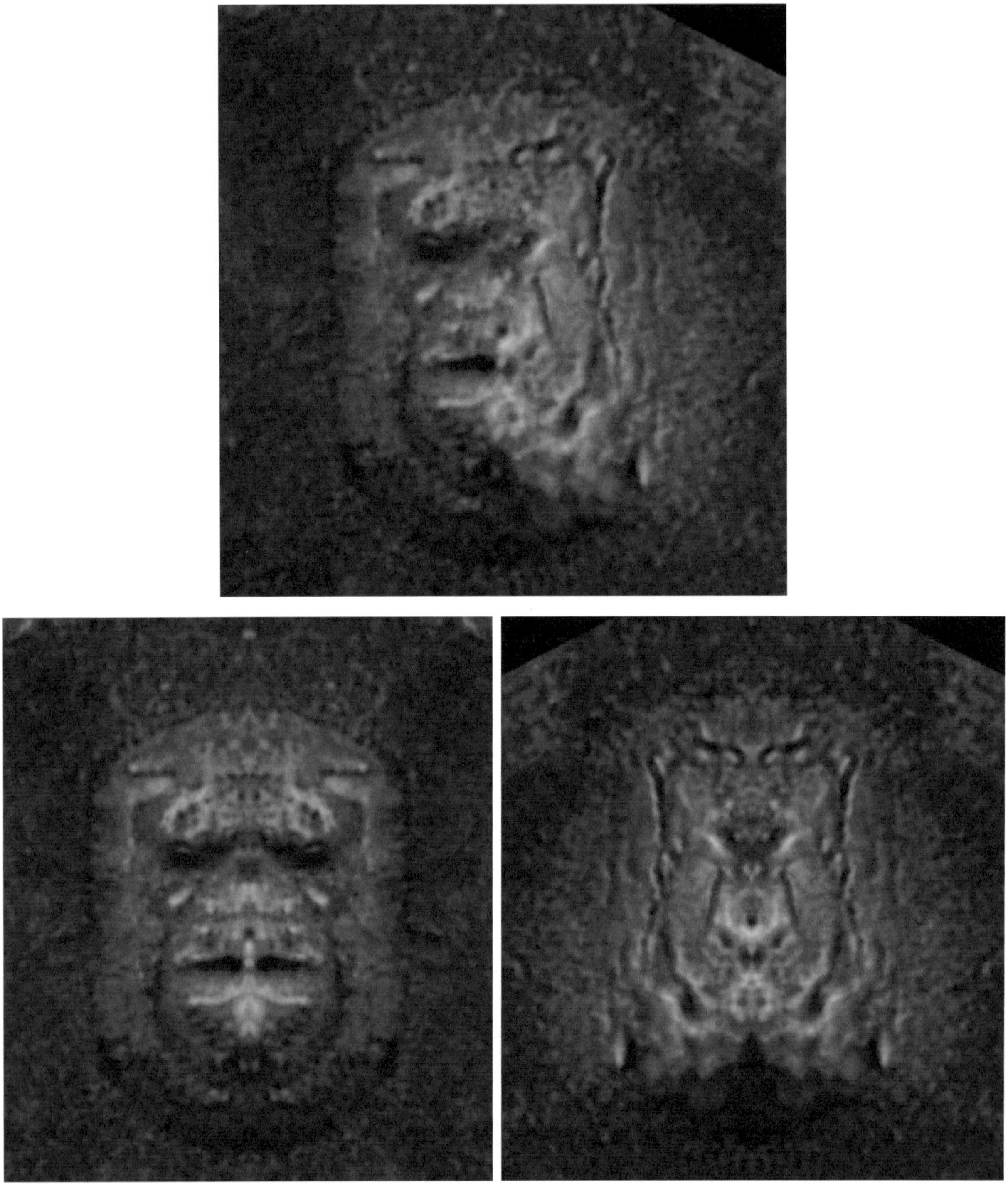

(MERGED IMAGES VIEW OF THE FACE OF CYDONIA: V54642014 AND H3253_0000_ND3 and V03814003 Morning Color black and white)

So many interesting details can be seen in various sun angles and lighting conditions which again, Mars Odyssey offers great insight upon what we are really supposed to be seeing here on and as the Face.

The Mars Express image data used for the comparative analysis is also adding the missing and needed touches for the Odyssey images.

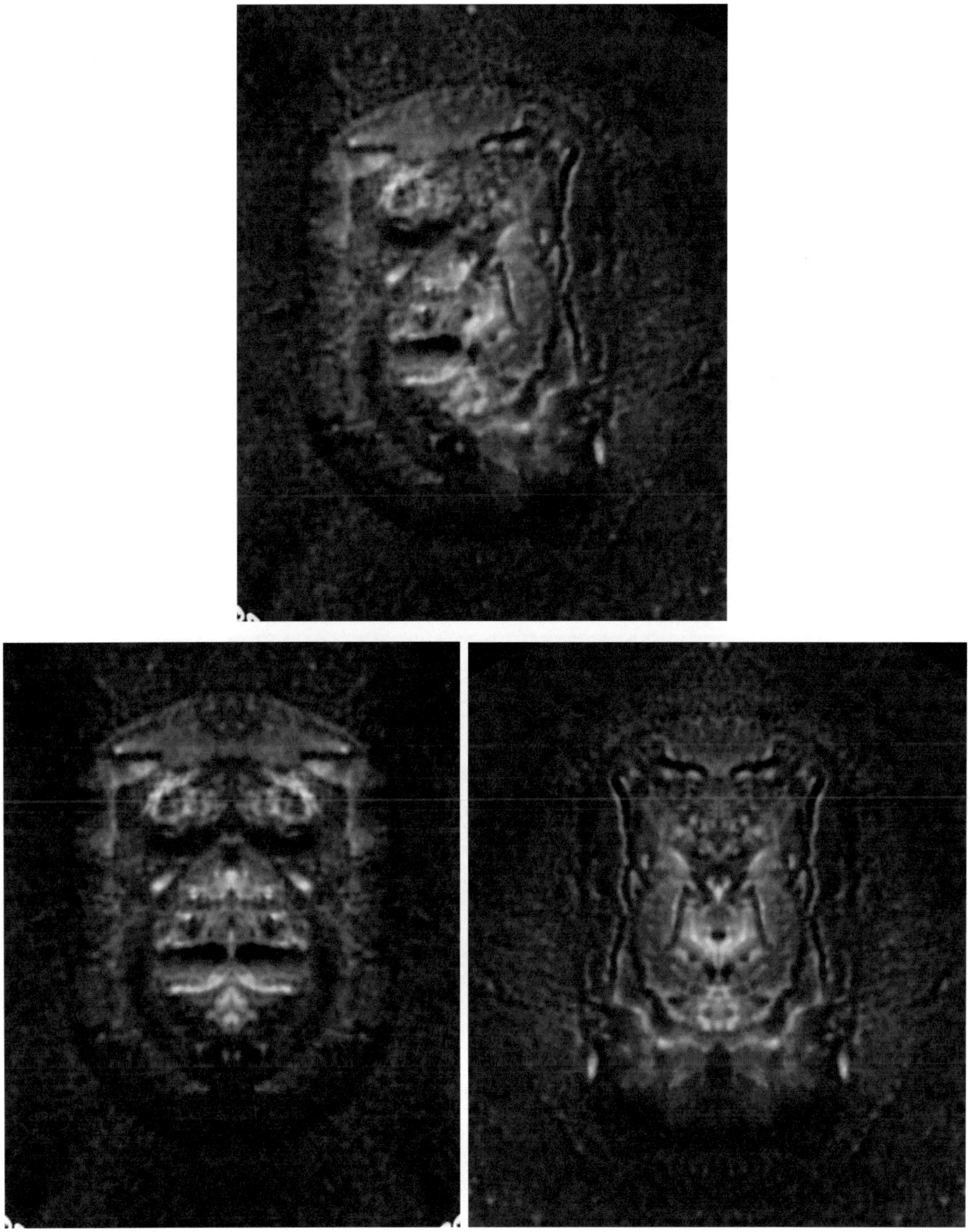

(MERGED IMAGES VIEW OF THE FACE OF CYDONIA: V54642014 AND H3253_0000_ND3 & ODYSSEY EARLY MORNING IMAGE V60652003)

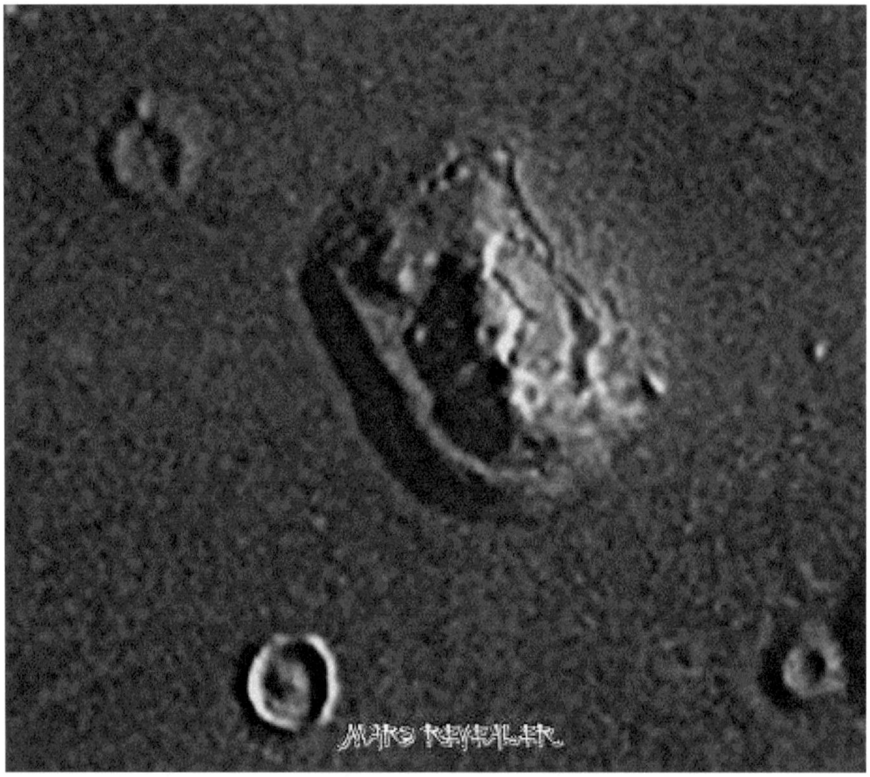

(MARS ODYSSEY EARLY MORNING IMAGE V96459003)

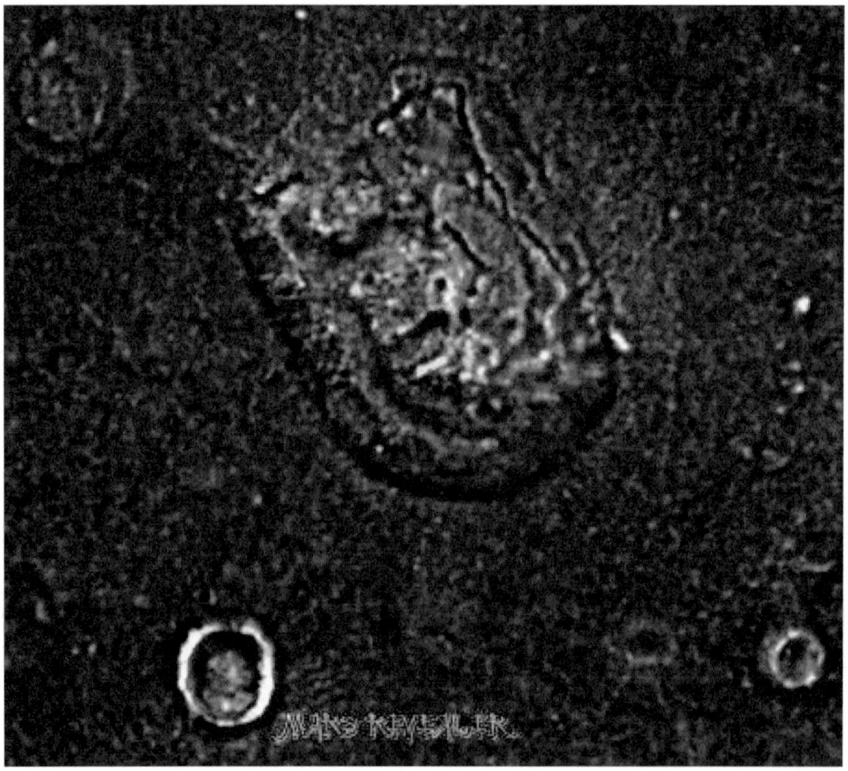

(MARS ODYSSEY EARLY MORNING IMAGE V96459003, H3253_0000 MERGED)

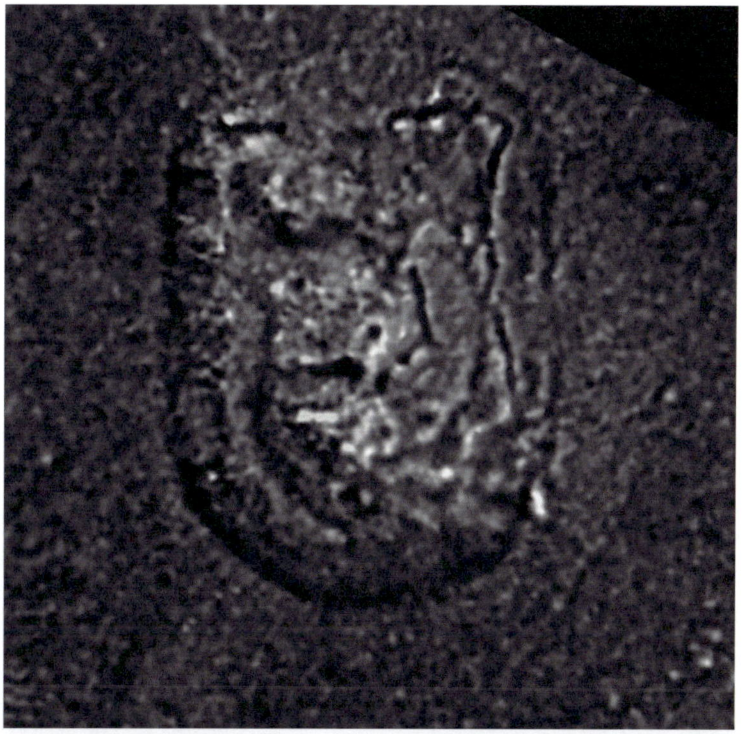

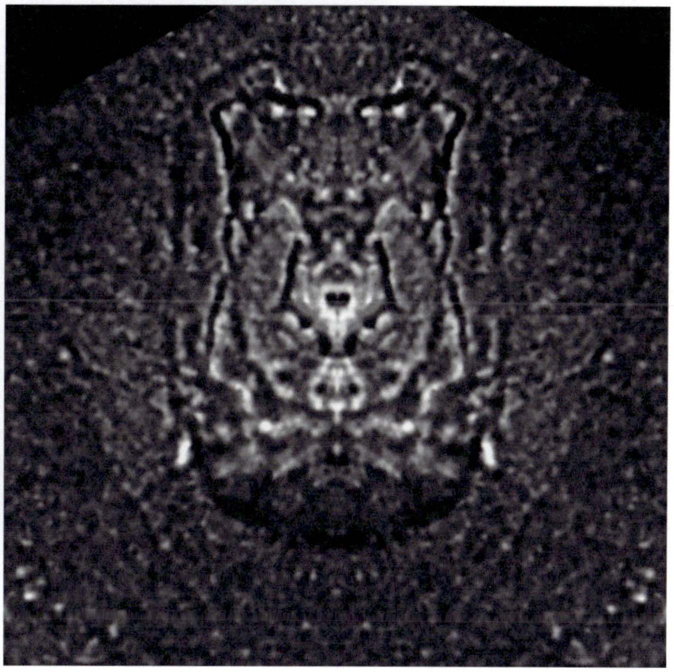

Indeed such amazing **Silent Discoveries** for an unprecedented analyis and study which is finally brought forth inviting public scruitiny and challenge to refute what is seen, or to validate. I pressented my case. **Someone had to!**

CHAPTER 8:

MARS GLOBAL SURVEYOR VIEWS OF THE FACE

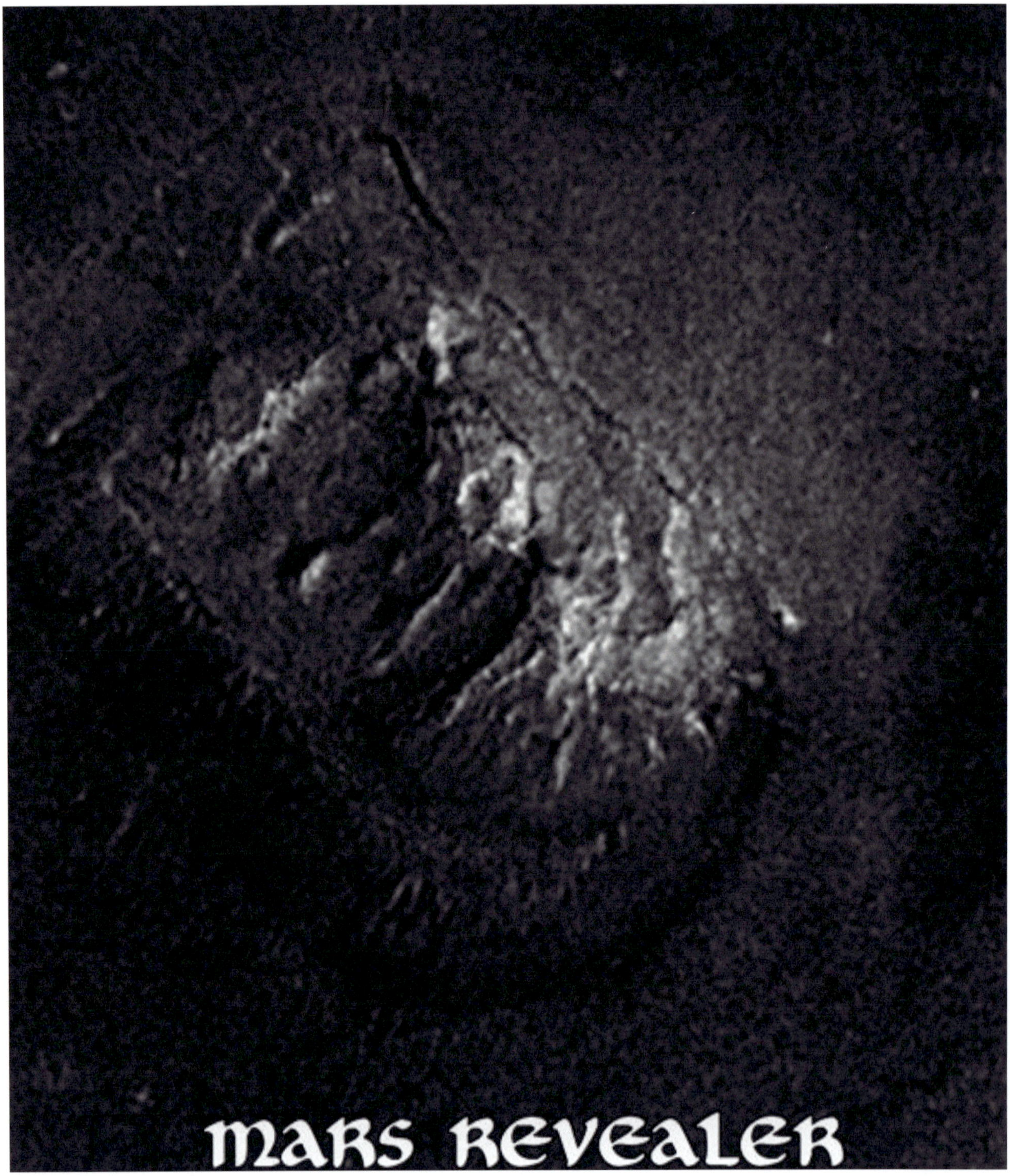

(Processed & Ortho-rectified image of the Face from MGS image SP1-22003 released April 6, 1998)

We are at a great advantage to be able to see all of these many views of the Face from Mars orbital missions we have so far seen in this book. Especially when considering the first Viking Orbiter image views of the Face taken in 1976, and then having to wait 22 years until we received another image view of the Face in April 1998, from the high resolution MOC (Mars Obiter Camera). After Mars Global Surveyor had taken many images of Mars in a well-placed orbital mission, taking images for a couple years, then we had the Mars Odyssey arrive and image Mars in much lower resolution than the Mars Global Surveyor, and also ESA's (European Space Agency) Mars Express also arriving and taking images. It is quite astonishing and amazing really how much advantage these missions and various qualities and quantity of image data gives us of the Face.

After 22 years, of having nothing and to anxiously having to wait for better images of the Face, we now get 3 missions and imagers and receive multiple images, all at almost the same time with each mission working in conjunction with each other. At this point in time of 2024, we still have 2 of the 3 original Mars orbiting missions of that time of the early 2000's still working and orbiting Mars. The Mars Global Surveyor, sadly is not any longer still with us.

So it is now time for us to focus on all of the Mars Global Surveyor images taken of the Face, since we have seen all of the publicly released images of the Face that is worth seeing in this book from Mars Express and Mars Odyssey. Mars Global Surveyor had given us all 10 images showing the Face from 1998 to 2006.

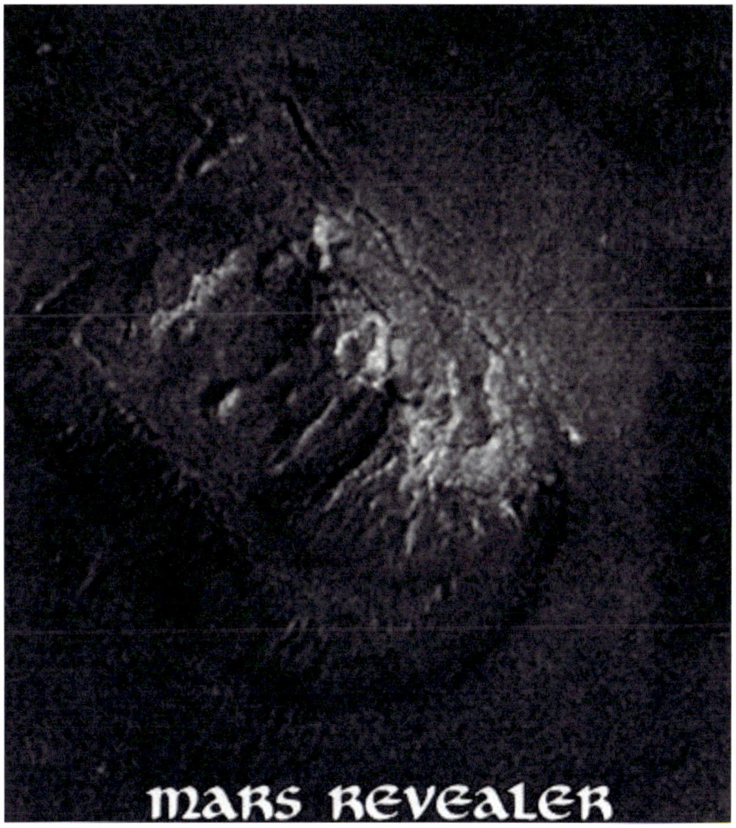

(MOC Image: SP122003 taken April 5th, 1998)

https://viewer.mars.asu.edu/viewer/moc/SP122003#P=SP122003&T=2

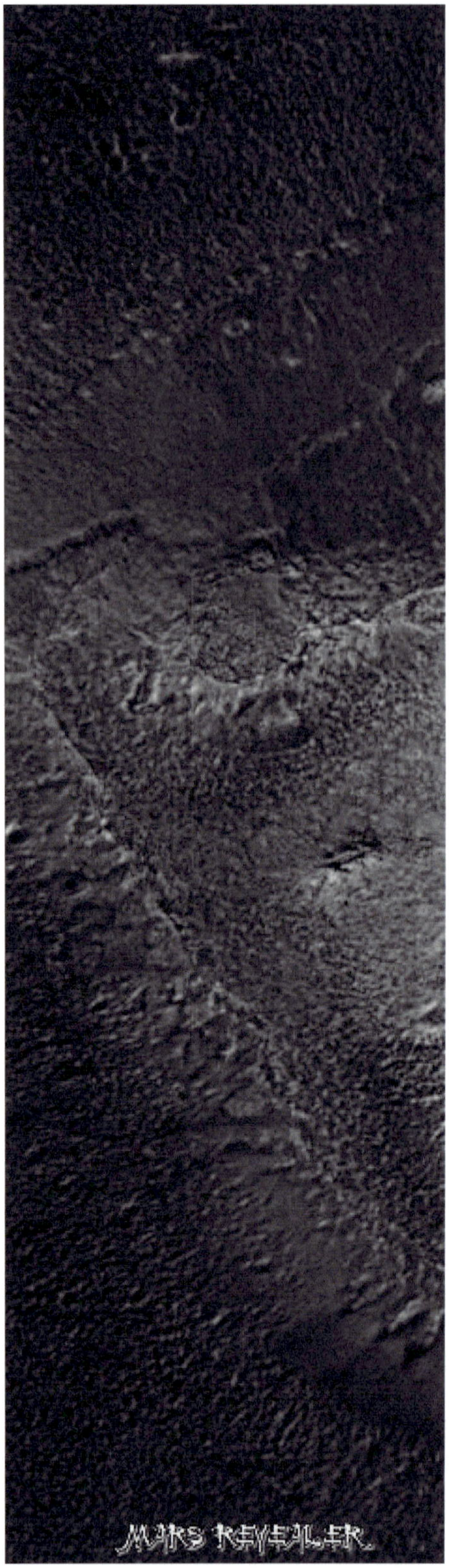

(MOC Image: M1600184 taken June 3rd, 2000)

https://viewer.mars.asu.edu/viewer/moc/M1600184#P=M1600184&T=2

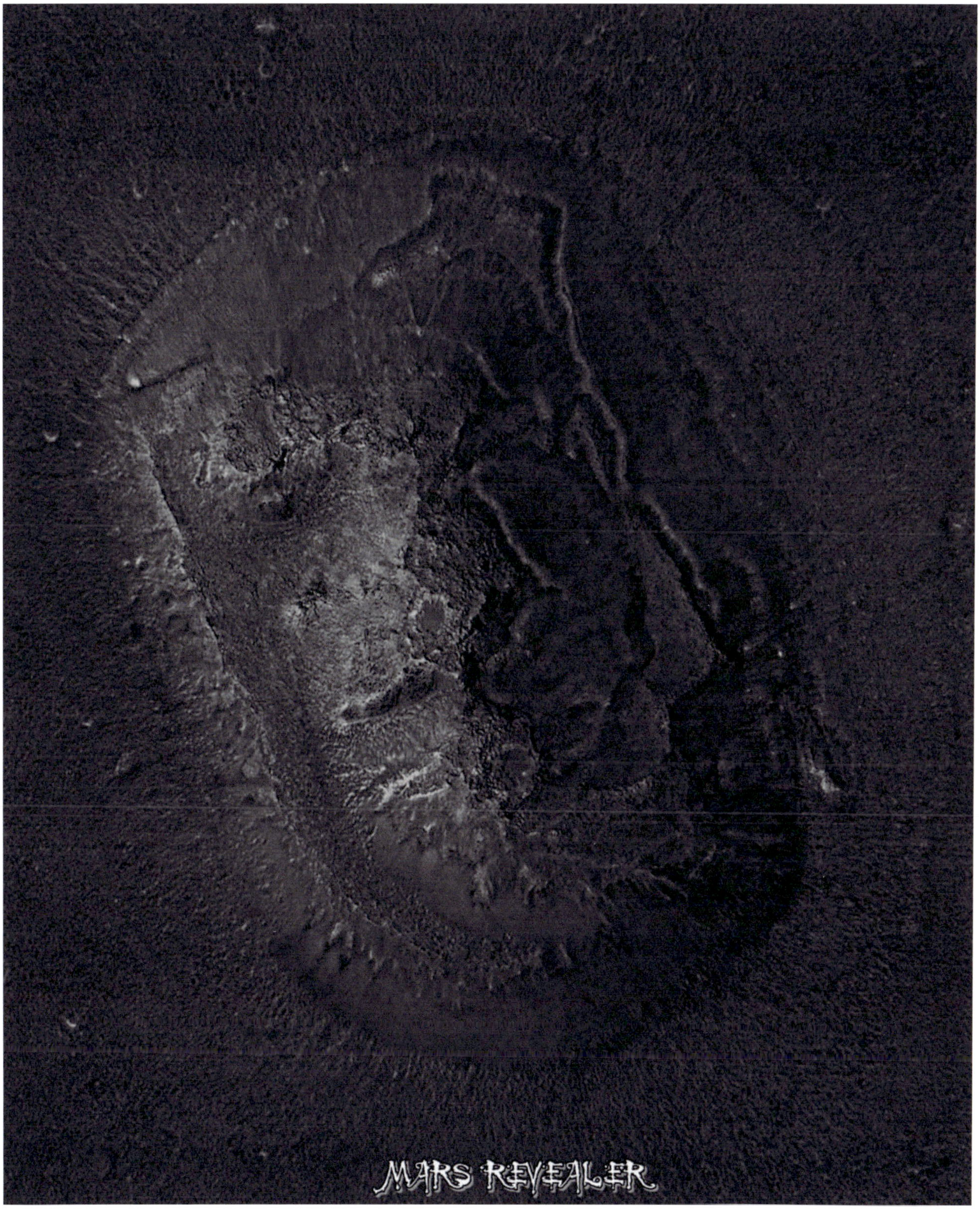

(MOC Image: E0300824 taken April 8th, 2001)

https://viewer.mars.asu.edu/viewer/moc/E0300824#P=E0300824&T=2

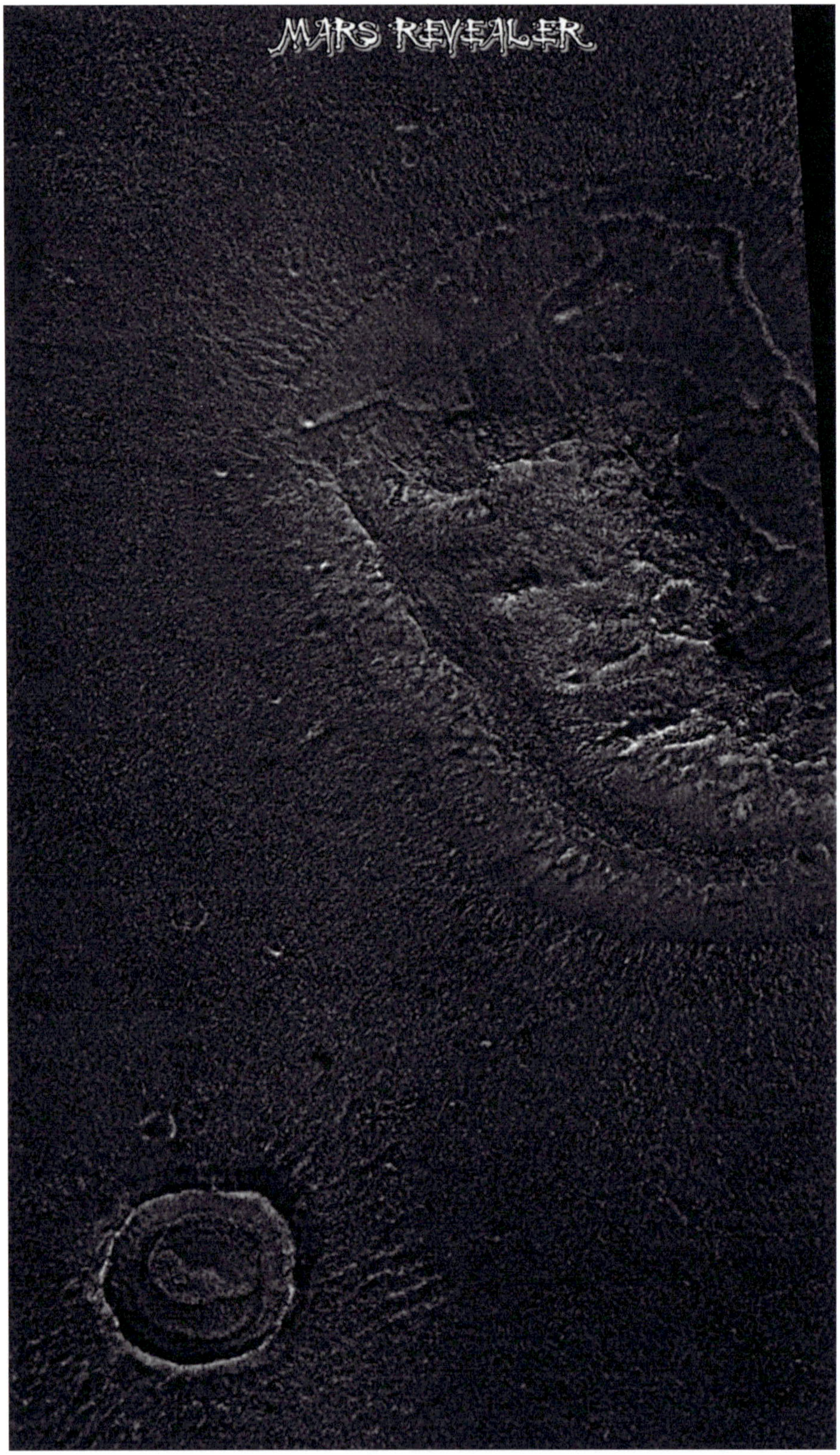

(MOC Image: E1003730 **taken November 27th 2001)**

https://viewer.mars.asu.edu/viewer/moc/E1003730#P=E1003730&T=2

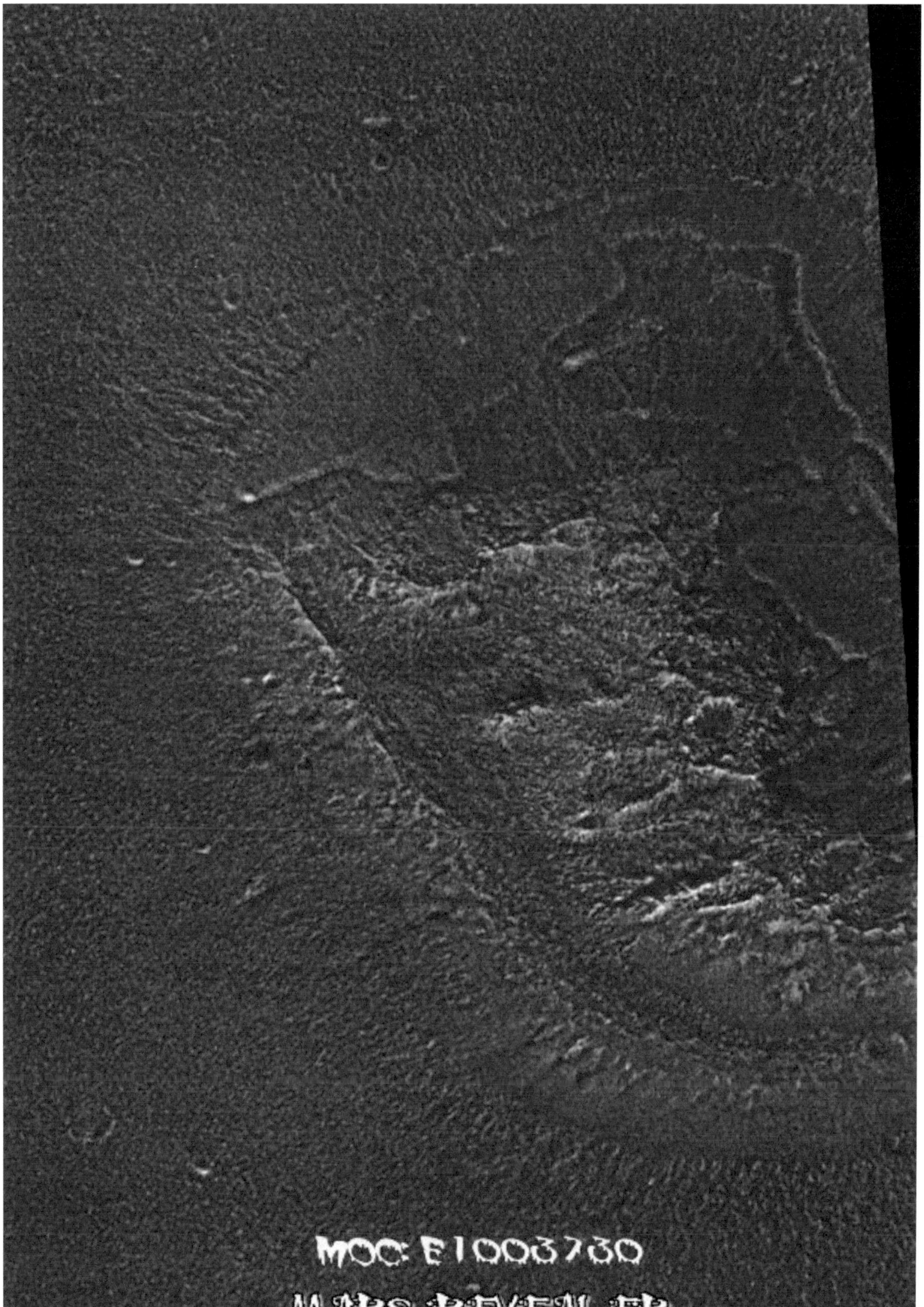
MOC: E1003730
MARS REVEALER

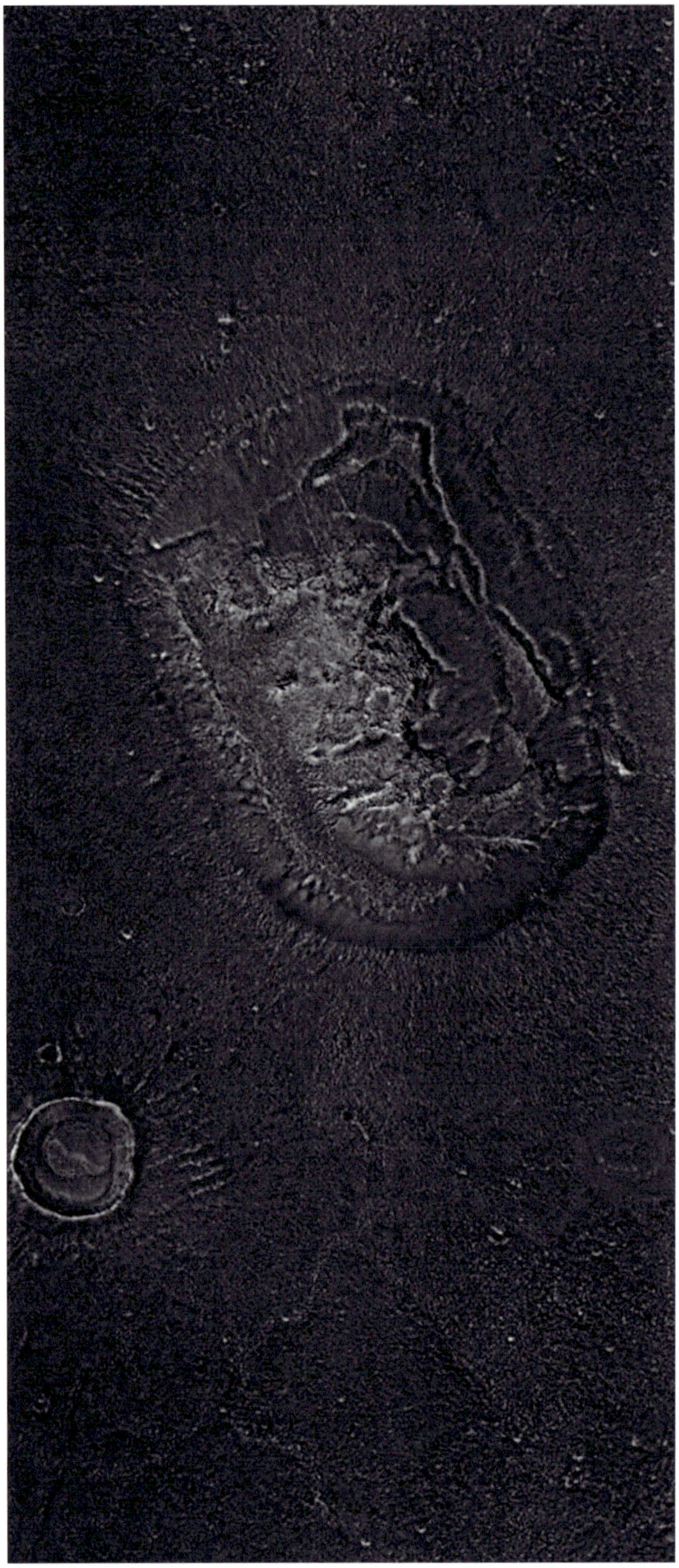

(MOC Image: E1701041 taken June 16th 2002)

https://viewer.mars.asu.edu/viewer/moc/E1701041#P=E1701041&T=2

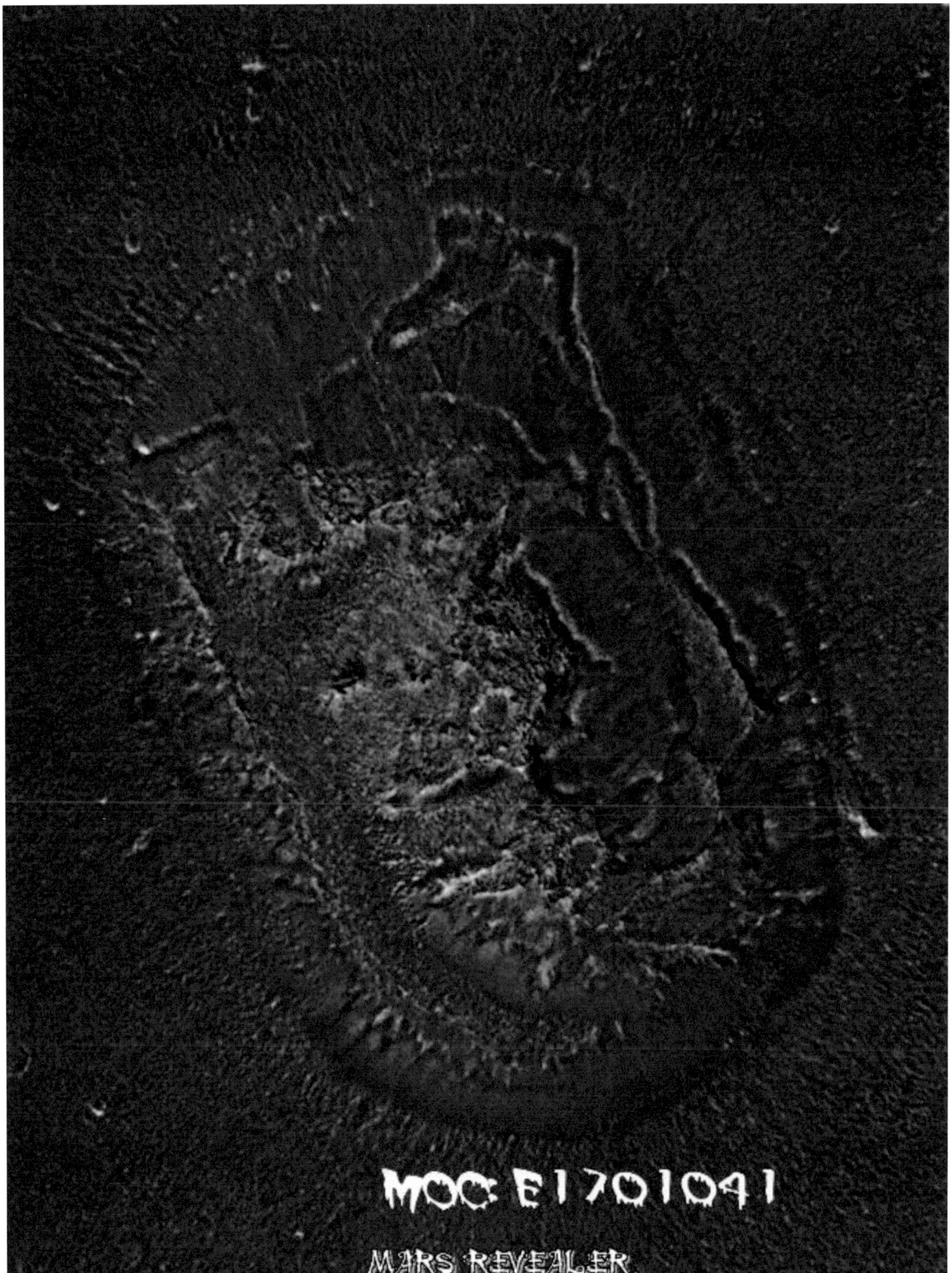
MOC: E1701041
MARS REVEALER

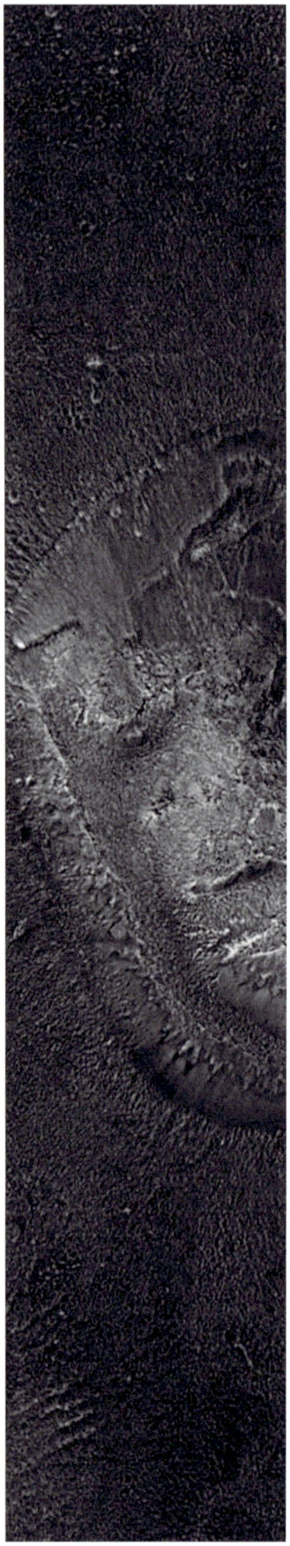

(MOC Image: E2001532 taken September 27th, 2002)

https://viewer.mars.asu.edu/viewer/moc/E2001532#P=E2001532&T=2

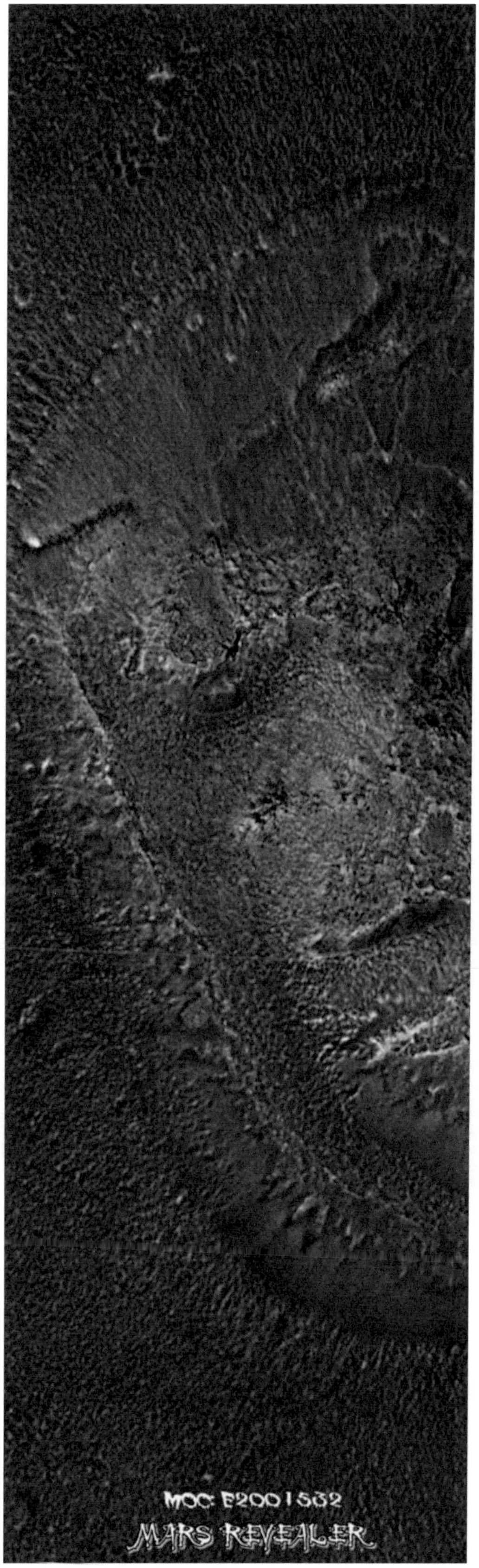
MOC E2001562
MARS REVEALER

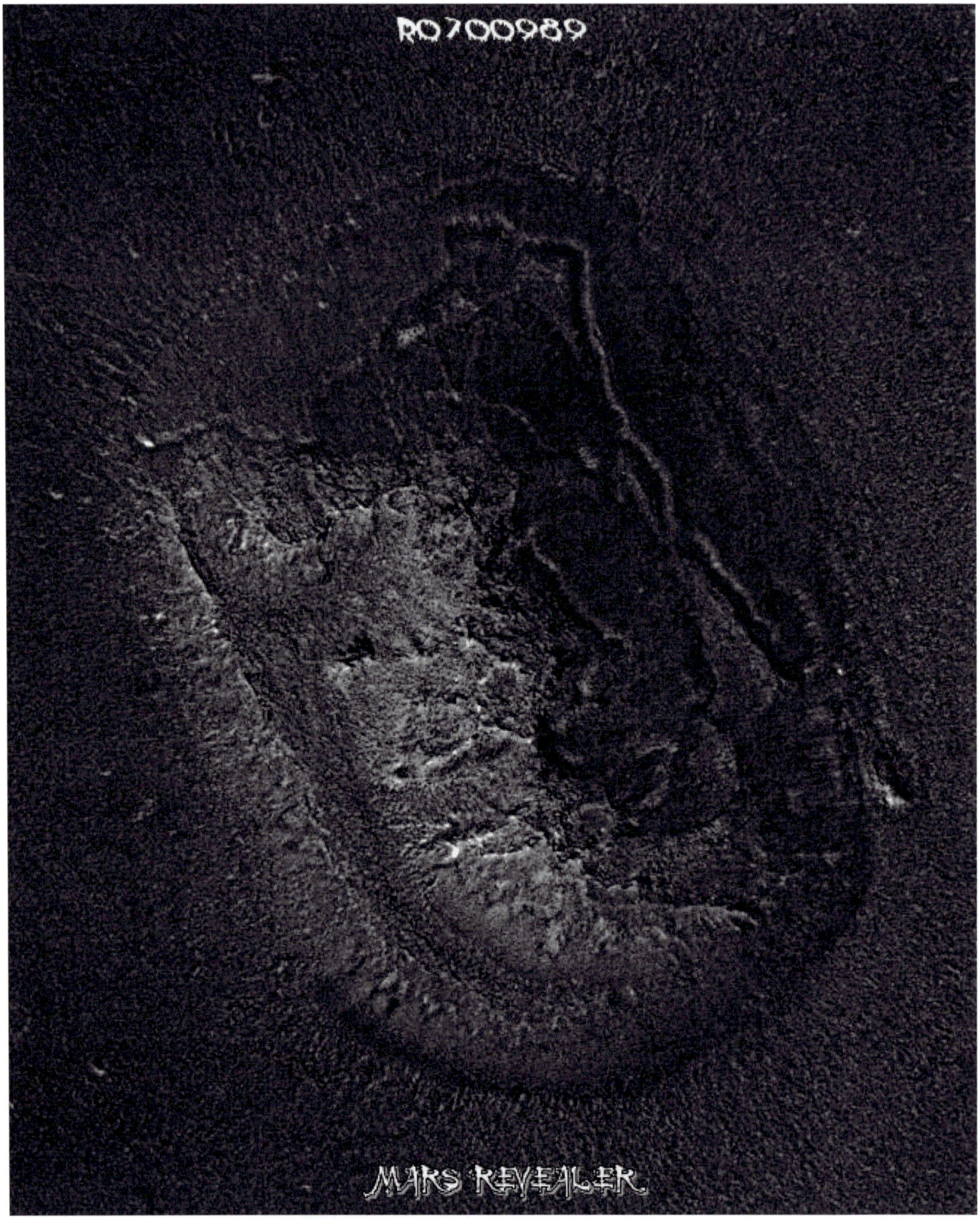

(MOC Image: R0700989 taken July 13th 2003)

https://viewer.mars.asu.edu/viewer/moc/R0700989#P=R0700989&T=2

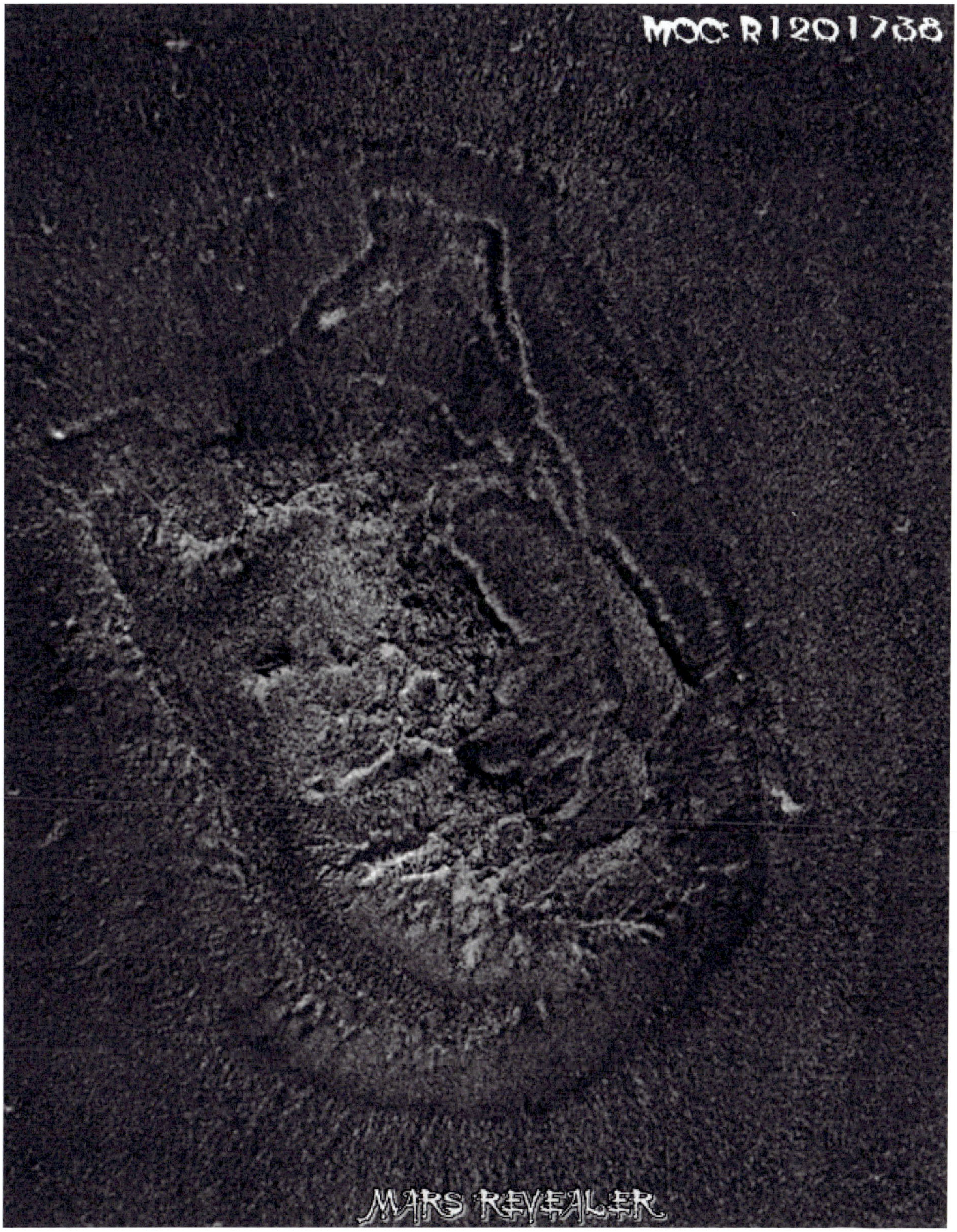

(MOC Image: R1201738 taken December 16th, 2003)

https://viewer.mars.asu.edu/viewer/moc/R1201738#P=R1201738&T=2

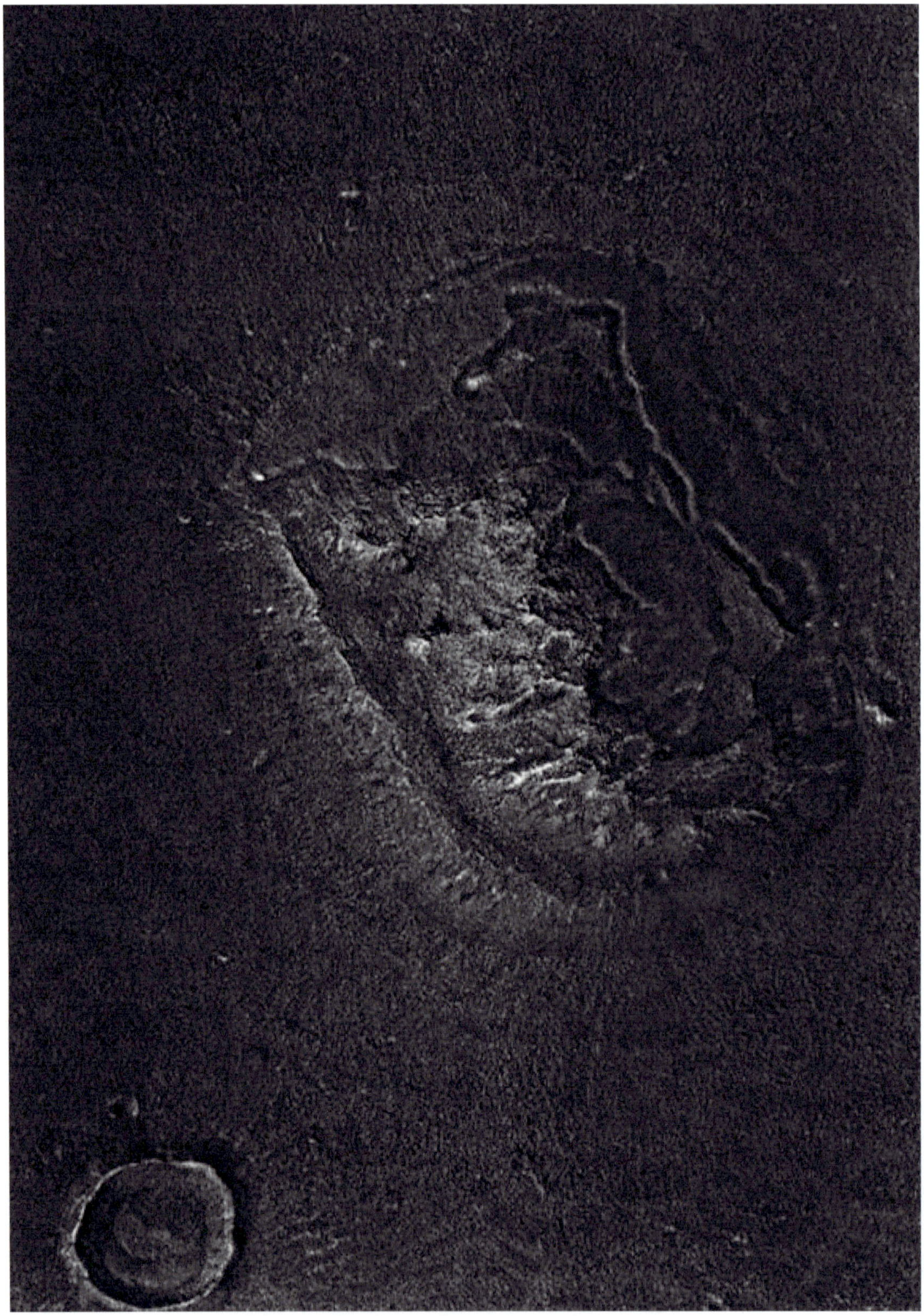

(MOC Image: S0601061 taken May 11th, 2005)

https://viewer.mars.asu.edu/viewer/moc/S0601061#P=S0601061&T=2

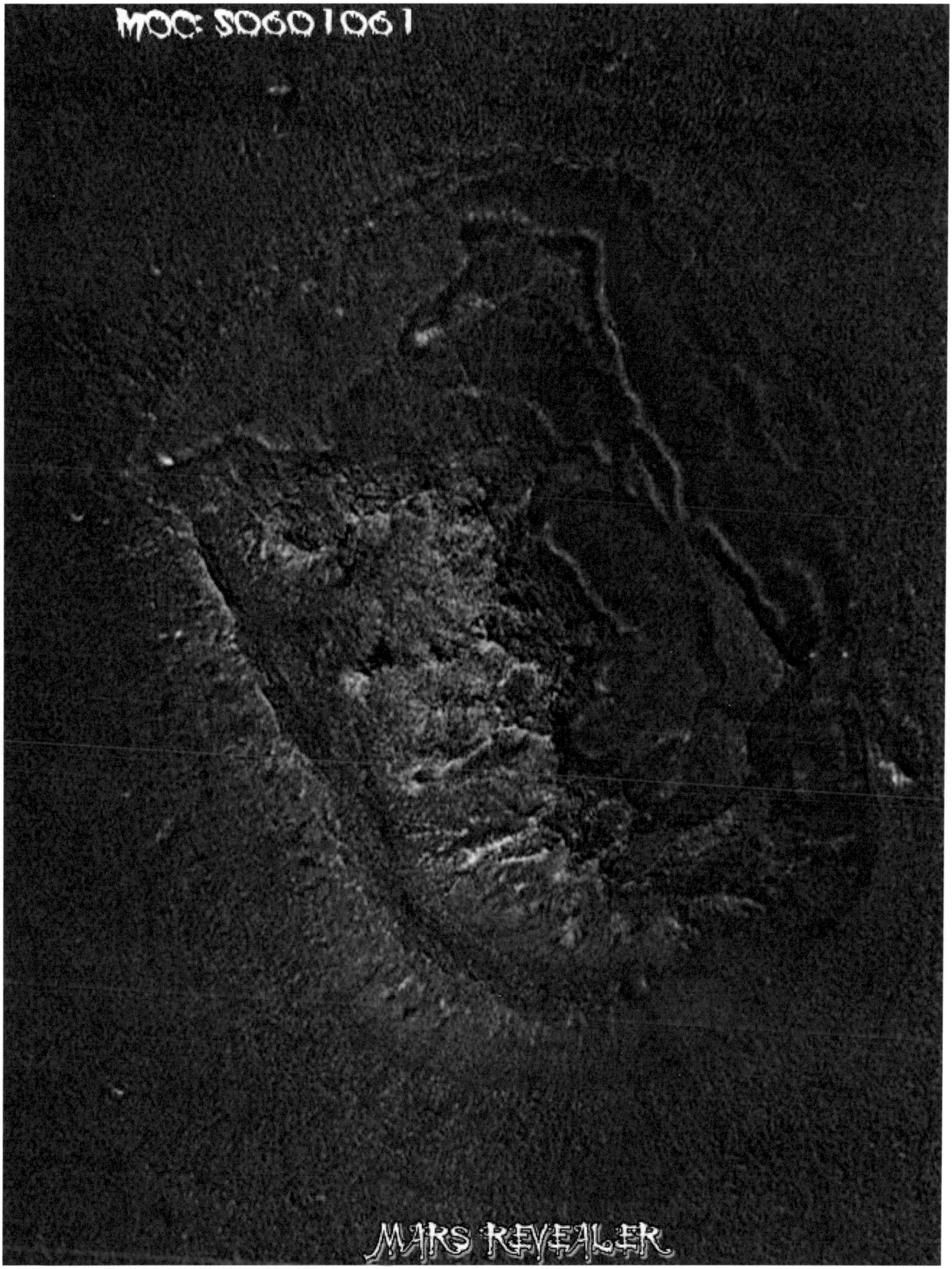
MOC: S0601061
MARS REVEALER

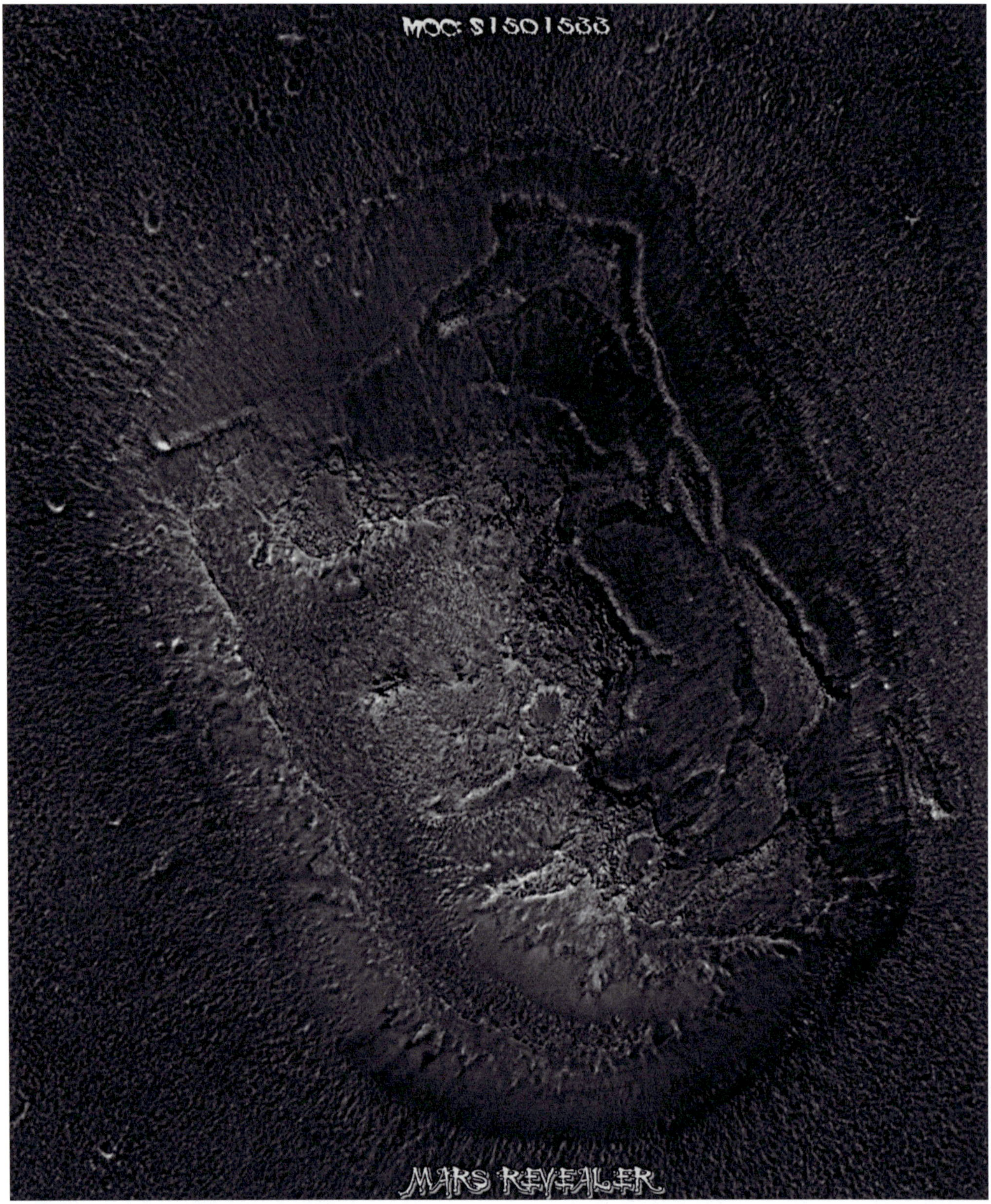

(MOC Image: S1501533 taken February 15th, 2006)

https://viewer.mars.asu.edu/viewer/moc/S1501533#P=S1501533&T=2

CHAPTER 9:

MODERN RECONNAISSANCE VIEWS...

OR...

CONTINUED MISLEADING & OBFUSCATION?

It is an amazing thing to be able to look at Mars' surface and especially of Cydonia and its features of intelligent techno-signatures in imagery from multiple sources. As we have seen so far from the NASA Viking Orbiters, from Mars Odyssey, Mars Global Surveyor, and from ESA's Mars Express, we have met the scientific model of having at least three sources to see, study, and compare data to show us and tell us that in at least regard to the Face at Cydonia, the Face is now an **established scientific fact** of its existence.

Although NASA and the narrative policy controlled mainstream will ignore that science that shows that the Face as being a scientific fact, and instead continue to mislead us by telling us that is in fact not a scientific fact. Outright lying, misleading of information and false narratives of official positions and manipulation of image data is not science, and is in fact obfuscations. These types of obfuscations we have seen by NASA and come to have learned of up until this point. Remember that according to NASA, the Face is a case that is officially closed since 2007, because officially, **THE FACE IS ONLY A NATURAL OBJECT!** NASA forcefully controlled the narrative of the Face and NASA purposely closed the door officially to the public to speak about the Face or having to answer for any questions in terms of its artificiality of techno-signature of intelligent design.

No matter what you see or can show using various sources scientific image data to prove otherwise, including NASA's image data, according to NASA, the case is closed. You can actually hear them saying, **"We are NASA, and only our official narratives are science and we own that science and we are that science. We give you what we want you to see and believe is science and natural! Never mind what you think you see or think you can prove! You are wrong! Nothing to see here."**

There appears to be a double standard between science, and truth, and between data and facts, and the ignoring of scientific data and facts where and when it comes to the public at large who think and dare to ask questions to NASA, and question the science. Yet we are being provided and led to new sources of image data from a powerful mission to Mars, of data gathering **to show us?**,... Again of the scientifically established Face at Cydonia, to give to us yet much more image data of the Face to continue with our continued comparative analysis that has been established here at least in this unprecedented book looking at this all. More image data being brought forth to us all by another modern Mars mission called, **The Mars Reconnaissance Orbiter!**

All of the images we seen of the image data so far of the Face in this book, all stem from many years collectively. So this book again, is a book whose time has come with this unprecedented collective study, providing us all with yet more image data to add to all we seen here, now from an unprecedented Mars mission.

The Mars Reconnaissance Orbiter launched from Cape Canaveral Air Force station on August 12th, 2005. After reaching Mars, it began Mars orbit insertion March 10th, 2006. The orbiter carries a new High Resolution Imaging Science Experiment camera, known as Hi-RISE.

The big and powerful HIRISE camera takes pictures that cover vast areas of Martian terrain while being able to see features as small as a kitchen table. This new camera has the capability of photographing the surface of Mars in unprecedented detail, allowing researchers and the public to view objects that enable them to distinguish objects on Mars and to study the morphology (surface structure) in a much more comprehensive manner than ever before. From an altitude that varies from 200 to 400 kilometers (about 125 to 250 miles) above Mars, Hi-RISE acquires surface images containing individual, basketball-size (30 to 60 centimeters, or 1 to 2 feet wide) pixel elements, allowing surface features 4 to 8 feet across to be resolved. These new, high-resolution images are providing unprecedented views of layered materials, gullies, channels, and other **science targets**, such as the **Face of Cydonia!** Well, at least,... **IT SHOULD!**

Along with the Hi-Rise imaging camera, we also have the Context Camera, called CTX, which is another imager that provides a big-picture, background view of the terrain around smaller targets that are studied by other cameras on the Mars Reconnaissance Orbiter, such as Hi-Rise. The CTX (Context Camera) **'makes observations simultaneously'** with high-resolution images collected by HiRISE. (*This is an important fact we will address later.*)

Its images are taken in black and white (grayscale); a broad band of visible light from 500 to 800 nanometers in wavelength. Its image size is 5064 pixels-wide x varying length (due to a 256 MB DRAM buffer, so that it can acquire pictures that have down-track lengths greater than 99 miles (160 kilometers). The CTX has a resolution of 20 feet (6 meters) per pixel scale from an altitude of 186 miles (300 km) over a swath 18.6 miles (30 km) wide.

The Hi-RISE camera began its mapping orbit around Mars on September 29, 2006. It was not until the following year, on April 5th, 2007 that the Hi-Rise and CTX imagers had taken a high resolution portrait of the Face of Cydonia. The public and scientific community would receive its first high-resolution portrait of the Face of Cydonia. On April 11, 2007 that finally happened. We waited to be officially given the portrait of the Face 6 days after the imagers taken the images. The Arizona State University team finally posted the first MRO HiRISE image of the Cydonia Face on their public web site.

The MRO HiRISE image PSP_003234_2210 was aptly titled **'Popular Landform in Cydonia Region'.** The image was taken in the early afternoon, almost directly overhead, with an amazing resolution of 25 cm per pixel. Unlike previously released image files like those of the Mars Global Surveyor, Mars Odyssey, Mars Express, which offer images at limited resolutions, the full Hi-RISE portrait is so large, there is no way to fully show its portrait in this unprecedented book, but we will show examples showing how much closer and greater clarity we are actually getting.

Something to notice and to point out is that it was on the 21st day of September of 2006, that the special and final NASA narrative 'HIT PIECE' press release was given out. This was 7 months before this unprecedented portrait of the Face of Cydonia was imaged and released to the public

The new Hi-Rise image of the Face was now being released with a new narrative name change by official position from calling it THE FACE ON MARS, or THE SO-CALLED FACE ON MARS for decades previously since 1976 up until the final Hit-Piece official position of September 2006, (we went over this all in chapter 5: Can This Be True?), the official policy of NASA having to answer the public with continued attention and controlled news hit-pieces addressing the Face lying to the public over and over, was now officially over. NASA wants you to think and see and believe and accept that the Face is considered now as being CASE CLOSED, ONLY A NATURAL OBJECT! Implicating somehow that it was somehow proven by them that the Face is natural. When nothing and no one of any kind of status or position or any space organization has shown any proof of that at all.

Now when 7 months later after that narrative ending hit-piece, we now get the highest resolution image of the Face ever taken, and it shows of the now new NASA official narrative policy which is to simply ignore the issue with bias to the public, with now the Face being released to us titled as **'Popular Landform in Cydonia Region'.**

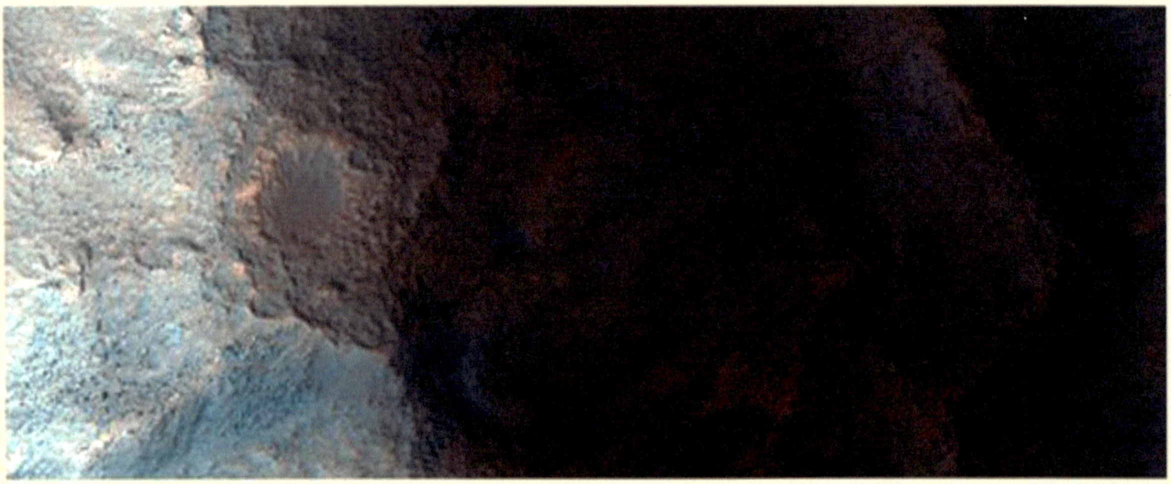

NASA/JPL-Caltech/UArizona

Popular Landform in Cydonia Region

PSP_003234_2210 Science Theme: Landscape Evolution

HiRISE captured this image (2007) of an eroded mesa made famous by its similarity to a human face in a Viking Orbiter image with much lower spatial resolution and a different lighting geometry.

Compared to the original Viking image, HiRISE shows incredible detail, even

WALLPAPER
800
1024
1152
1280
1440
1600

This is very comical and hilarious and so revealing to how now NASA's new public narrative will NOT even mention the Popular Landform In Cydonia Region as The Face on Mars. Even though that is what the popular landform in Cydonia region is publicly and famously known for, and even WAS officially named and known for again by NASA itself, since 1976! Nothing else makes any sense to this reasoning accept what I have laid out and hopefully that you all have come to see in this book, that simply,... **NASA LIED!**

NASA will not ever again address the issue of the Face officially, or deal with the controversy about it ever again! **CASE CLOSED, ONLY A NATURAL OBJECT!**

We shall see how many people will keep believing that BIG LIE!

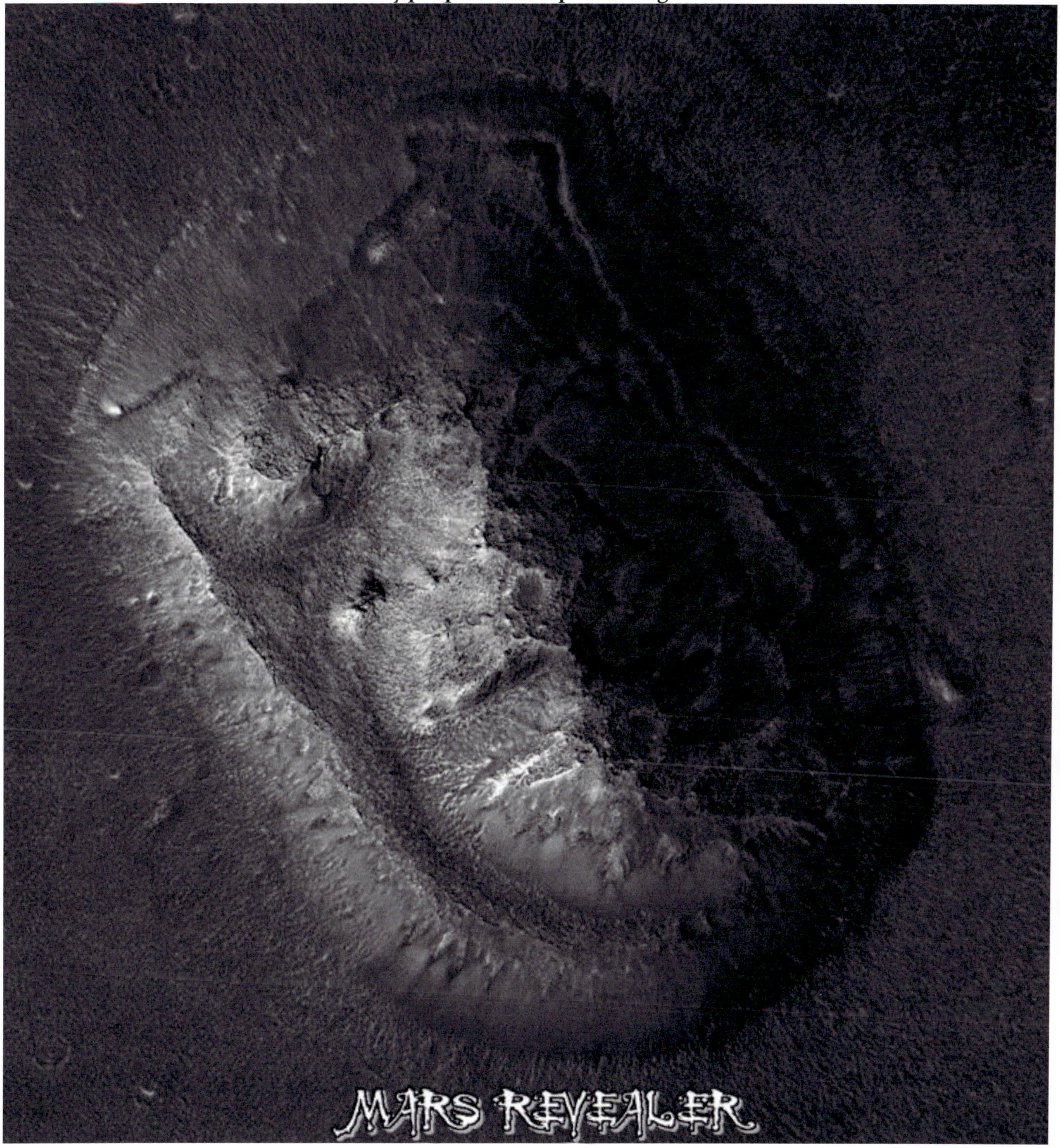

(MRO HI-RISE IMAGE PSP_003234_2210 IMAGE TAKEN APRIL 5TH 2007)

IMAGE ARCHIVE LINK: https://www.uahirise.org/PSP_003234_2210

(MRO HI-RISE IMAGE PSP_003234_2210_MRGB TAKEN APRIL 5TH 2007)

The April 11th 2007 image release was unprecedented in clarity of surface details, and in terms of the size of the image and resolution, that shows its awesome artistic detail everywhere you look on the Face giving us all so much to see and study closely. We also were able to get a partially covered RGB color view for the Face as well. This color data allows us to add another dimension to the Face to see its un-natural nature, as well as to help bring out the tattoo as it were that literally completely covers the Face. A more real view to that dimensional shift of seeing the Face in what we were and still are needing to see of its artistic un-natural relief features, as we all have been used to just seeing and studying the Face in black and white, and that also in my opinion, is another form of obfuscation by default used by NASA to keep us all from seeing the Face in color and in high resolution. All with the official purpose and manipulative intent to hide the Face's Alien Markings from the public sight and public knowledge!

NASA knew they had to end their official public discussion and narrative of control and denial policy and move it to a case closed position for the Face of Mars, as after denying and lying about it for 30 years up until 2006. NASA has not been able to successfully control the natural explanation narrative for the Face or keep explaining their position over and over having to come forth to discuss about the Face and only always using misinformation and lies and manipulated image data that was not so very convincing to begin with. Especially in light of the Mars Reconnaissance Orbiter in the images for the Face it will provide that NASA knew would only keep them being forced to have to publicly deal with the Face and still, not being able to really prove their case the Face is natural. In essence they are tired of losing Face!

They had to officially kill the 'Face of Mars' in the eyes of the public. But not to me and some others like myself, who study the Face and see what is really there to see! NASA not wants to fight against their own superior image data to only make them look more incapable of being able to tell us what the Face really is. Their continued public policy of control and denial has been used up of their official position discussions over the decades **without** showing proof that the Face is natural, and they ran away from it never wanting to address the issues about the Face anymore to the public.

According to NASA, the discussion was now over, laid to rest in the eyes of the public and just claiming without merit or data, that it was proven to be natural, positioning that the Face was successfully debunked! Saying something over and over again does not mean that what is being said over and over again, is the truth! They want to keep '**THE BIG LIE**' going and in a controlled suppression of information and deliberate withholding of image data, or obfuscations of surface details in image data that only contradicts and derails their official lies about the Face. These official control policy actions come from all past Mars missions, and especially in modern Mars missions.

Which begs us to look into this a bit deeper.

Why is this powerful Mars mission, called the **Mars Reconnaissance Orbiter**? Reconnaissance is a military term. Meaning to go out and conduct scouting of 'foreign' terrain. Now in the official press release, it is bragging about MRO, and all of the instruments it is carrying, and the camera capabilities, and touting to be able to be able to pull in the most data gathered and be able to send the most data returned to Earth than all of the other Mars missions combined.

Yet what was interesting is that in the MRO official press release on page 16, of 48 pages, says that **the Hi-Rise Camera is the most powerful camera ever built for use at a 'foreign planet'!**

WHAT?...

'A FOREIGN PLANET'?

Science Instruments

Mars Reconnaissance Orbiter will carry six science instruments. Two additional investigations will use the spacecraft itself as an instrument.

❑ The **High Resolution Imaging Science Experiment** will photograph selected places on Mars with **the most powerful telescopic camera ever built for use at a foreign planet.** It will reveal features as small as a kitchen table in images covering swaths of Mars' surface 6 kilometers (3.7 miles) wide. Combining images taken through filters admitting three different portions of the spectrum will produce color images in the central portion of the field of view. Paired images of top-priority target areas taken from slightly different angles during different orbits will yield three-dimensional views revealing differences in height as small as 25 centimeters (10 inches).

(MRO ARRIVAL PRESS KIT OF MARCH 2006 LINK BELOW)

https://www.nasa.gov/wp-content/uploads/2015/03/143619main_mro-arrival.pdf?emrc=0bbe04

I think it is here at this point, that we need to reflect on the importance of the wording in the official press information, in order to better understand what this implicates of what we are trying to be told. So I believe we need to see the definitions of **reconnaissance**, and **foreign:**

reconnaissance

or **re·con·nois·sance** [ri-**kon**-*uh*-suhns, -*zuhns*]

noun

1. the act of reconnoitering.
2. *Military.* a search made for useful military information in the field, especially by examining the ground.
3. *Surveying, Civil Engineering.* a general examination or survey of a region, usually followed by a detailed survey.
4. *Geology.* an examination or survey of the general geological characteristics of a region.

BRITISH DICTIONARY DEFINITIONS FOR RECONNAISSANCE

reconnaissance **reconnoissance** / (rɪˈkɒnɪsəns) /

noun

1. the act of reconnoitring
2. the process of obtaining information about the position, activities, resources, etc, of an enemy or potential enemy

foreign [fawr-in, for-]

adjective

1. of, relating to, or derived from another country or nation; not native: *foreign cars.*
2. of or relating to contact or dealings with other countries; connected with foreign affairs.

foreign / (ˈfɒrɪn) /

adjective

1. of, involving, located in, or coming from another country, area, people, etc: *a foreign resident*
2. dealing or concerned with another country, area, people, etc: *a foreign office*
3. not pertinent or related: *a matter foreign to the discussion*
4. not familiar; strange
5. in an abnormal place or position: *foreign matter; foreign bodies*
6. law outside the jurisdiction of a particular state; alien

A reconnaissance mission, to a **foreign** planet! The language is quite clear and in your face, pun intended. (Where's those illegal aliens?) So who is NASA trying to tell us all, really owns Mars? If it is not the beings who used to be there, it is of an alien nation? A **foreign** planet, its right in the official language in the official document.

It is interesting how exactly matching the objectives for Mars **Reconnaissance** Orbiter are in the official wording that helps us able to see and understand what this mission is in reality, and is set out to achieve. Let us study the definitions for each and putting the bigger picture together to grasp the information we are being clued into seeing.

1: Military: A search made for useful military information in the field, especially by examining the ground.

2: Surveying, Civil Engineering. A general examination or survey of a region, usually followed by a detailed survey.

3: Geology: An examination or survey of the general geological characteristics of a region. (And the archeological!)

4: The process of obtaining information about the position, activities, resources, etc., of an enemy or potential enemy. (Looking for positions, activities, and the resources of an enemy, on Mars?)

5: adjective
Of, relating to, or derived from another country or nation; not native: foreign cars. (Or in this case, alien foreign vehicles not native of Mars?)

6: Of or relating to contact or dealings with other countries; connected with foreign affairs. (Already known alien government/s that has been contacted, and having unmentioned dealings connected of foreign affairs? Implicating!)
7: Of, involving, located in, or coming from another country, area, people, etc: a foreign resident. (Known alien peoples from another planet, or place that did not originate or was not indigenous on Mars, in this instance?)

8: Dealing or concerned with another country, area, people, etc: a foreign office. (Dealing with an alien government, or leadership from someplace else, or from and of another planet, other than Mars, that is concerning military and governments on Earth?)

9: Not pertinent or related: a matter foreign to the discussion. (An alien government or leadership of alien peoples, which is not related to the indigenous Martians? Or also possibly including not related to humans? A matter foreign to the discussion? Of what? Of the official public scientific discussion and narrative policy perhaps?)

10: Not familiar; strange. (An alien foreign not indigenous Government and leadership of peoples who are NOT familiar, and are strange, to indigenous Martians? Or not familiar and strange with humans of Earth? Which is dealing with them, because of contacting them or their foreign offices? Where? Here on Earth? Or, on Mars? Strange indeed!)

11: In an abnormal place or position: foreign matter; foreign bodies. (The strange aliens peoples government or leadership not related to humans, is in an abnormal place or position of foreign matter? An alternate universe maybe? Located in a foreign place or position, on foreign planetary bodies in an abnormal place or position? A foreign matter, possibly meaning a WAR of the Worlds? Or from a foreign matter of contact with the strange and not familiar alien people whose government is on a foreign planetary body, in an abnormal place or position, by the human military and governmental leaderships of Earth?)

12: Law outside the jurisdiction of a particular state; alien. (A foreign matter, possibly hinting of A WAR?,... Of foreign bodies, in an abnormal place or position, that is outside the jurisdiction of a particular state? Can this mean a secret war that is being waged outside the jurisdiction, or laws of a particular state, or rather, a particular Government? Outside the jurisdiction and law of that particular state? Maybe the Deep State is running this matter foreign to humans of the discussion for public knowledge for this mission of so called science, outside of the jurisdiction and law of the particular state of government? All this to consider alone, can be a subject for an entire book in and of itself.)

After seeing all 12 of these definition clues, of **reconnaissance** and **foreign**, to act as again, like a code being given to provide us hidden information where we can glimpse the truth, it raises the question here, that if all that I, and many others are seeing here on Mars, in all the imagery of Mars from all missions, and from the MRO imagery are in fact happen to be NOT ALL RUINS we are looking at, but is showing us all that somebody, is still home? Again the official language itself allows us all to contemplate this potential possibility.

However the image data we see does not tell or show me that anybody is still home. We do not see lights, or technology or vehicles moving the inhabitants around here and there on the surface, or in the skies above the surface. I have never seen before of Mars having been expressed in any place, as a foreign planet!

This all tells us that somebody thinks that **A**, there is intelligent stuff there. **B**, it is owned by somebody. **C**, somebody is still there, and **D**, they are perhaps an enemy? Yet thinking in terms as if they are perhaps an enemy, than it is interesting to note that at least publicly, that nothing has happened in over 50 years of probes from Earth invading Mars orbit, its skies and surface. Yet at the same time, we are apparently seeing mile after mile of archeological ruins in vast land surfaces that looks devastated by past cataclysmic events, or of a single event. Mars appears in images to us all as a planetary ruin, of a foreign planet. Of which a non-human strange unfamiliar people are governed from an abnormal place and who are from another planetary body or dimension, are residing there? In the least, very implicating of how we come to see this information, and certainly interesting to at least think about.

Obfuscations come in various forms and not just come from the outright lying and deliberate hiding of the truth also. Yet there are other ways for NASA to continue also obfuscating the truth without NASA having to say anything at all anymore officially about the Face. That obfuscation is quite literal even in the very images that are publicly released on the NASA image archive pages and may seem so innocent in the eyes of the public who have no clue or anything else which to compare. Especially to the people who come to these archive pages having no idea of what controversies exist seen within the images that are there for all to see, or rather for all to NOT-SEE?

Even for the people who know of the controversies, such as the Face, most people have no real knowledge about it, or any real knowledge of the details of the Face as we have shown and seen here, because all they may have seen is what NASA officially releases to the public on their image archive pages which is not fully processed, or incorrectly processed on purpose in order to obfuscate details. Most people also have no real idea that what is seen in the public release images are showing the truth or not. Most people accept it as truth mainly because it is NASA which provides us with image data, as well as the lies and misleading about controversial features such as the Face of Cydonia.

Most people would only ask why would NASA lie, or hide the surface image details of Mars, or manipulate surface image data in order to lie? It all just sounds like a crazy conspiracy theory. Maybe, just maybe you are also such a person also who asks that question, and think that those who make claims against NASA is just a crazy conspiracy theorist? If you still think that and feel this way after reading this book and seeing what has been presented so far, allow me please to show you in YOUR FACE, that this is exactly what is and has been going on and you can see clear and present evidence and outright proof that this is in fact the case as told to you and being presented by this crazy conspiracy theorist. Or in terms of today's terms, this crazy NASA FACT CHECKER, to show you all that this is a NASA conspiracy fact!

It is true NASA is known for being forthcoming with image data from its image instruments sent abroad on various probes to other worlds etc. That being said, that does not mean that you are getting the best quality or true views of surface images in terms of what the surface features really has to offer us and show in the images taken and being released to us. Especially from images of Mars, and especially of the Face of Cydonia.

Without getting into all technical specifics, the images are taken by digital imagers. As the images are taken by the imagers, they are captured and stored in RAW image formats that are compressed and sent back to Earth and needing to be opened, and processed by image software that can open and modify the particular RAW image formats to calibrate and process them into a series of black and white, and color products, including browse images and JPEG2000-compressed images and tools to make it easy for everyone to explore these enormous images. Quality viewable images seeing all the surface features, and then make them in most common standard downloadable image formats such as JP2, PNG, TIFF, or JPEG, just to mention some of the most used formats in downloading images by viewers.

So the processed images are processed, named, and stored and linked for viewing and downloadable from NASA public image archive pages on the internet. So when someone browses through these image archive pages and wants to see the images without having to download the images to see them, it is not a problem. People can choose to download various images at their leisure and choice and also be able to download the actual RAW image data. You will NOT be able to open or view them in most cases because of the RAW image formats that use certain programs not used by majority of the people. It also is not an easy task if not impossible, for anyone unfamiliar with working with RAW image formats to even open them let alone knowing how to process them. So the majority of people only can see and use and show, what is officially released as being seen in NASA's so called processed images that are supposed to be scientifically valued and accepted as showing us the truth.

It seems NASA is continuing at using their own **Tricks of Light and Shadow**, to obfuscate surface details on Mars, especially of the Face, and continue their policy of misleading and denial by literally outright 'covering up' what is really there in the data the people are NOT getting to see. Especially regards to the Face at Cydonia. Yet the improper processing effects many and most other surface features seen in the images as well, not just the Face.

Before we continue on to show you the truth of my claims, I would like to bring up an ironic point regard to this mission that is laughable at this point, considering all we seen and learned so far. I hope you appreciate the comical irony about this as believe it or not?,... The Mars Reconnaissance Orbiter mission to Mars has been given to **THE PEOPLE!**

Called **"The People's Camera"**, HiRISE will revolutionize what we know about Mars. **The general public is encouraged to join scientists and participate in this exploration by assisting in HiRISE targeting and data analysis.** At the beginning of 2010 we released Hi-Wish (**http://www.uahirise.org/hiwish/**), opening HiRISE targeting to anyone in the world with Internet access, and already hundreds of public suggestions have been acquired. HiRISE has proven very popular with the public and science community.

This is incredible irony and a cover up operation is in place and in play. It seems what all we were guided to see as hidden information from the definitions of reconnaissance and foreign, seem to be more accurate and fulfilling. This MRO mission is in fact a military operation as we were clued in to see and learned, that is being used by the military **in an abnormal place, or position, (a public Mars mission labeled and positioned as the people's camera?), that is Law outside the jurisdiction of a particular state? (Meaning possibly military law?)**

It appears to be that the MRO, is being used as a dual purpose mission. Being the people's camera on the Face of it, pun intended, acting as a public purpose of science and exploration with NASA outreach included with public participation, yet **its real intended purpose, is conducting its military operations** doing the things we were guided to see by NASA's own official statements. I thought I would point that irony out. You cannot make this up!

However, there is added irony and this is the part which is comical about the people's camera. **The general public is encouraged to join scientists and participate in this exploration by assisting in HiRISE targeting and data analysis.** This is so in your face contradicting to the reality. Though being used as a military prime objective, the MRO is officially the people's camera. **WE THE PEOPLE... OWN IT! WE THE PEOPLE... PAID FOR IT!**

That being said, the data quality of the images are supposed to be of top priority and of top standard of quality for **OUR** Mars imaging mission, and including the releases of **ALL** imagery seen as high quality images and correctly processed images revealing **ALL** of the surface in crisp detail in the best capability. I have also '**assisted**' exploration by selecting some areas for image target requests, to which I have had a few Hi-Rise images taken of my request selections. I also do data analysis on the RAW imagery and process them myself, so I can see in the RAW data what the public cannot see at all in the public releases, which is the details of the surface features itself!

I have sent '**data analysis**' questions to the Hi-Rise team showing them my processed versions of their RAW image data for areas of interest, and I ask them why are they not using the people's money to give the people what we deserve to see from the people's camera, again which is the details of the surface features being able to be seen clearly, or even at all. I ask them to please take care and process the imagery so there are no obfuscations of surface details. I would show them the examples of images they released to be seen by the public, of the Face of Cydonia, and I would how them my processed version of the same image, and I never get a reply back. **WHY?** It is not like as if it is not a valid question to ask them.

It is the MRO CTX imagery that is subject of this issue, though the same problem is there for Mars Odyssey image releases also, that is not being given to us all as it should really be seen. Let you be the judge of my concerns.

I am going to reveal over a dozen MRO CTX examples of my claims and use them **ALL** as evidence and proof, that NASA is deliberately forcing the continuation of the policy of deliberately misinforming the public by deliberate obfuscation without having to officially lie to the public anymore in their official statements. These images are also all of the MRO CTX images taken of the Face that is known and released. As one can see in the image examples just seen, you can see the way NASA officially releases its MRO CTX (context) image showing (or barely showing rather), the Face of Cydonia in the top portion of the image. It is highly misprocessed and over contrasted to wash out much of the surface features relief that obviously can be seen, if correctly processed. **Why?**

You cannot legitimize this very serious data quality issue, when and as compared to my processed image from the RAW image data. You all also cannot be very impressed with how **WE THE PEOPLE** are receiving our high quality image products. Go online and check for yourself.

CTX Image archive link:
https://viewer.mars.asu.edu/planetview/inst/ctx/B01_010143_2216_XN_41N009W#P=B01_010143_2216_XN_41N009W&T=2

We will see each of the CTX Face images of all that has been taken so far and compare all of these correctly processed examples to that of the NASA public releases.

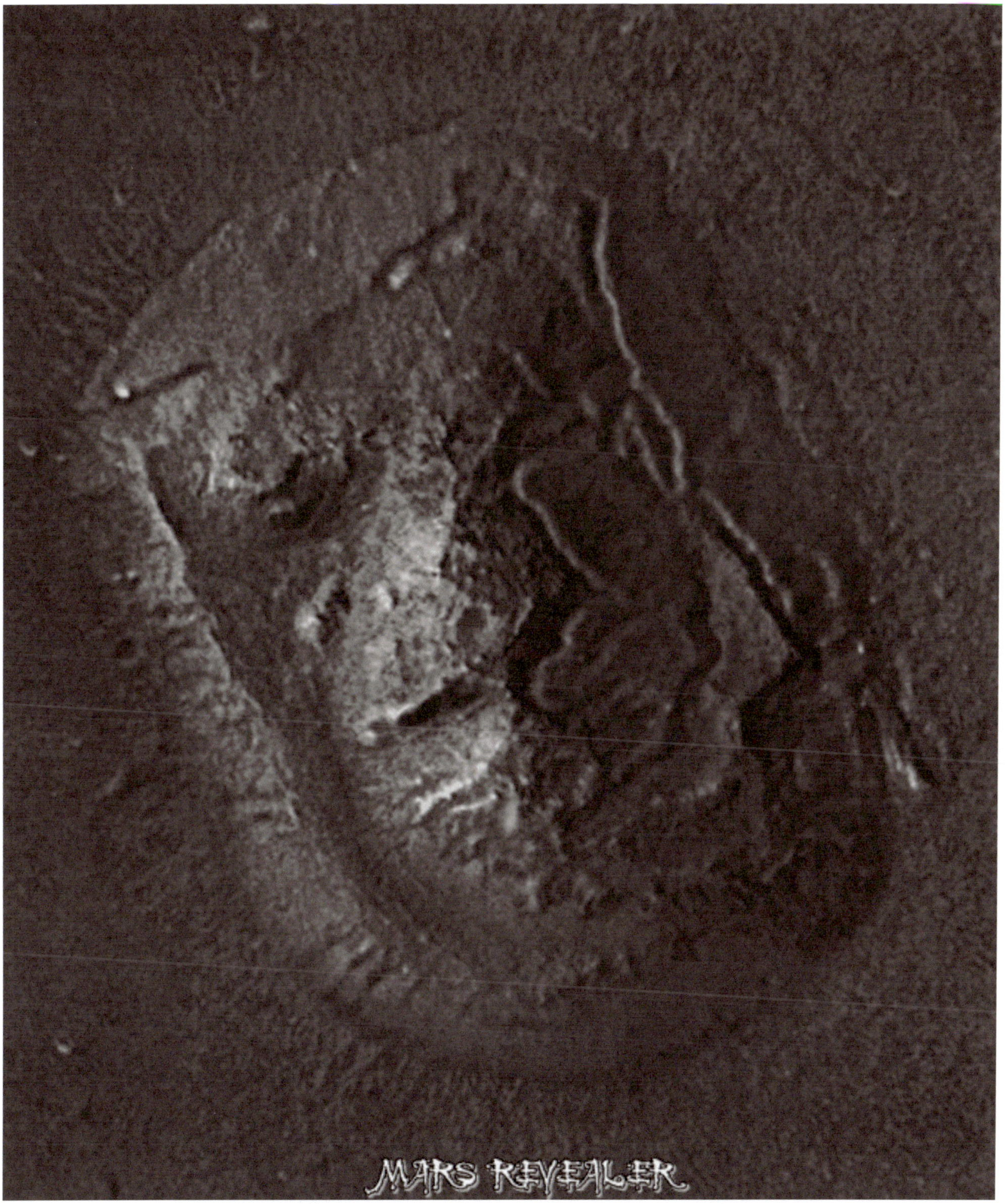

(MY PROCESSED RAW CTX IMAGE: B01_010143_2216 **IMAGE TAKEN: SEPTEMBER 24TH, 2008)**

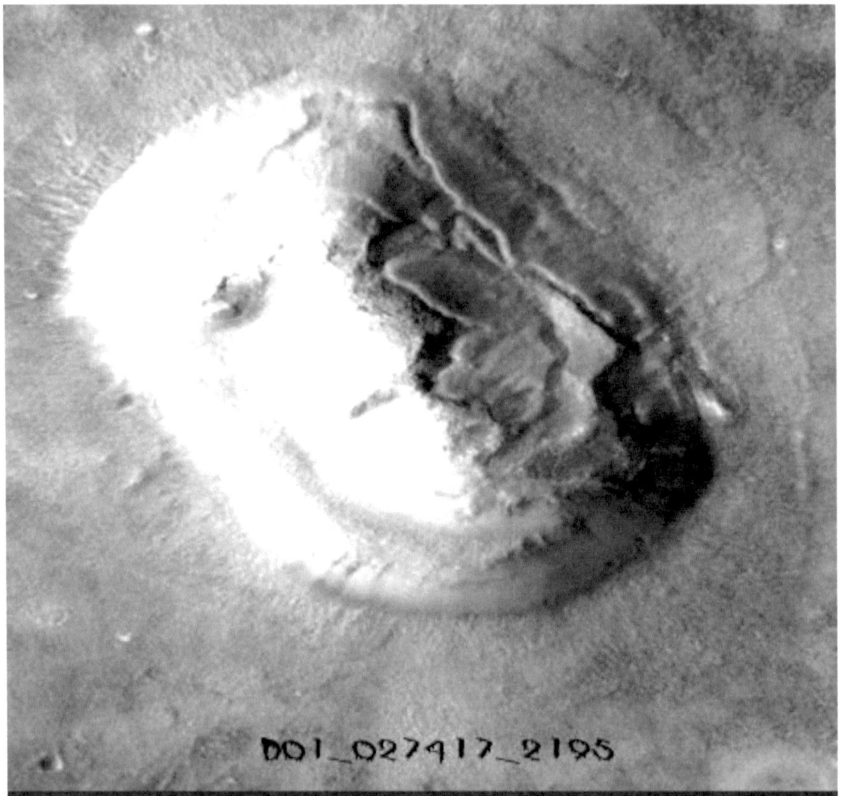
D01_027417_2195

MARS REVEALER

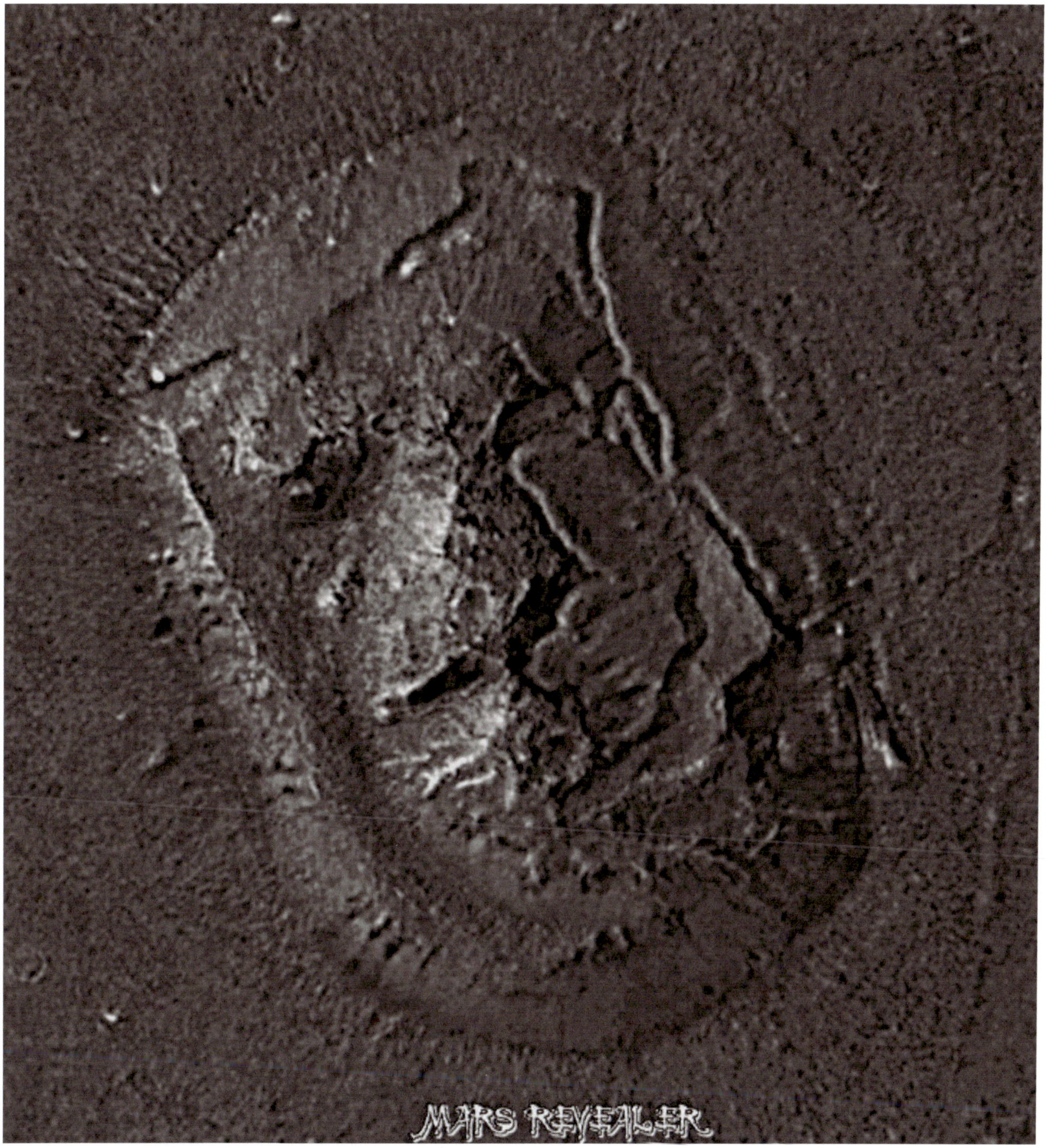

(MY PROCESSED RAW CTX IMAGE: D01_027417_2195 IMAGE TAKEN: JUNE 1st, 2012)

(MRO CTX IMAGE LINK:)

https://viewer.mars.asu.edu/planetview/inst/ctx/D01_027417_2195_XN_39N009W#P=D01_027417_2195_XN_39N009W&T=2

VERY IMPORTANT IMAGE WILL BE DISCUSSED AND SEEN LATER

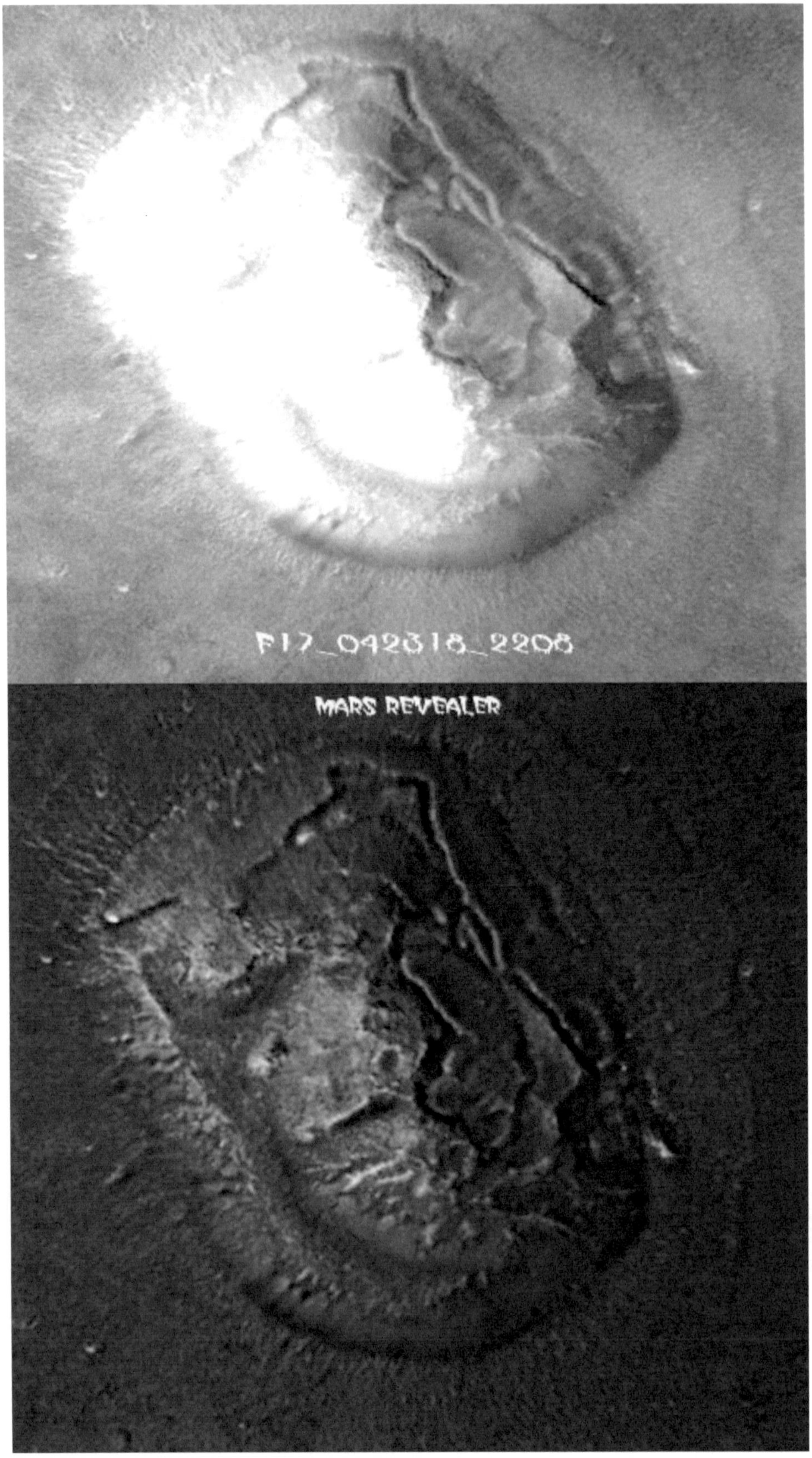
F17_042318_2208
MARS REVEALER

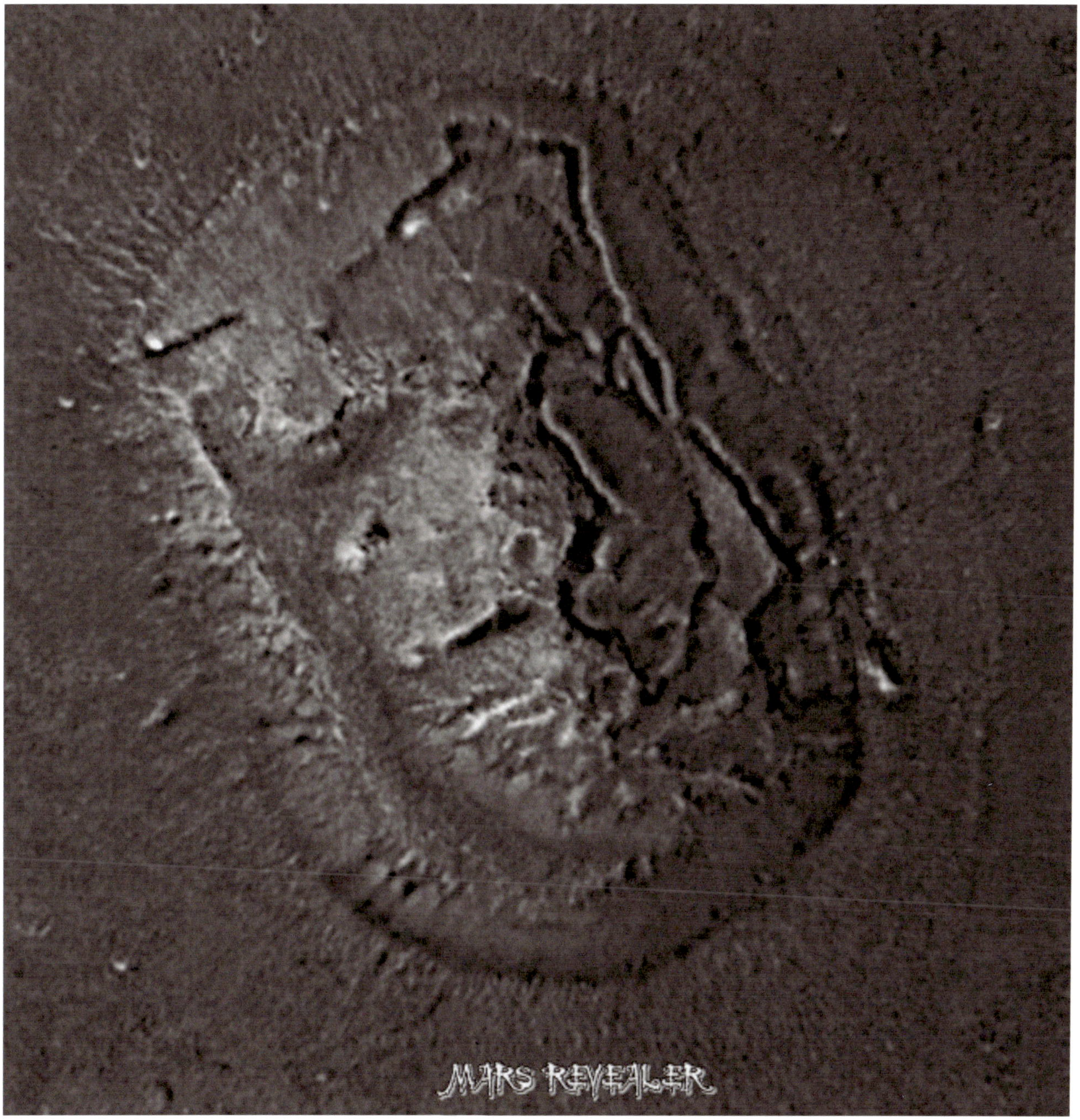

(MY PROCESSED RAW CTX IMAGE: F17_042318_2208 IMAGE TAKEN: AUGUT 6TH, 2015)

(MRO CTX IMAGE ARCHIVE LINK)
https://viewer.mars.asu.edu/planetview/inst/ctx/F17_042318_2208_XN_40N009W#P=F17_042318_2208_XN_40N009W&T=2

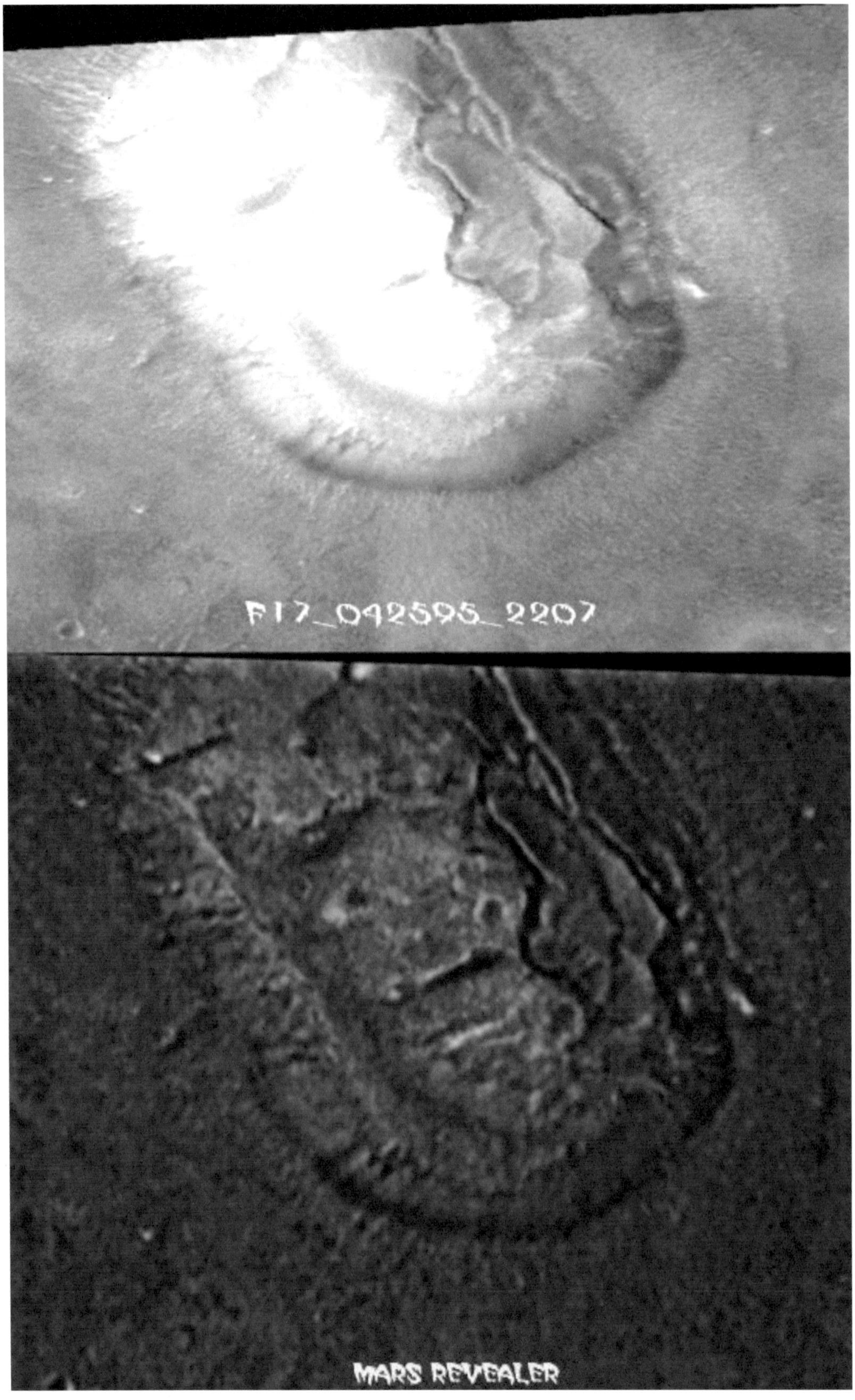
F17_042595_2207
MARS REVEALER

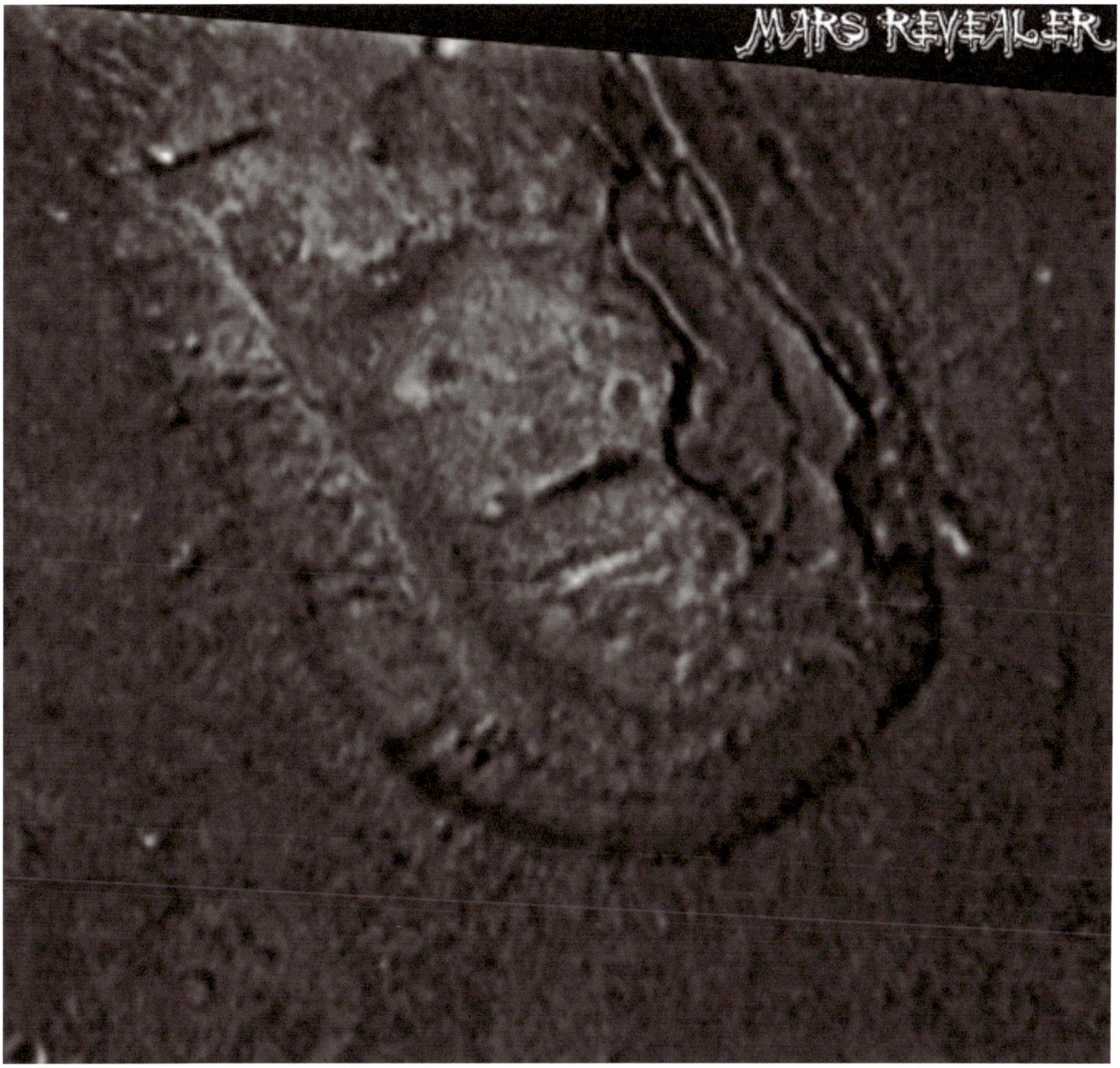

(MY PROCESSED RAW CTX IMAGE: F17_042595_2207 IMAGE TAKEN: AUGUST 28TH, 2015)

(MRO CTX IMAGE ARCHIVE LINK)
https://viewer.mars.asu.edu/planetview/inst/ctx/F17_042595_2207_XN_40N009W#P=F17_042595_2207_XN_40N009W&T=2

As you can see in all of these MRO CTX image comparisons and examples, you can see that the Face has been imaged at various angles. A couple are images on NADIR (meaning directly above), but most are imaged OFF NADIR (meaning not directly above). There are many reasons to image the Face this way. I believe it is to somehow figure out the true accurate shape of the Face structure. I also believe it is to see the artistic Alien Markings that cover the Face as its surface relief. This makes sense as NASA knows the un-natural tricks of light and shadow can reveal relief features at different lighting angles and lighting conditions. We will reveal the importance of this truth later, **especially regards to the true accurate shaping of the Face.**

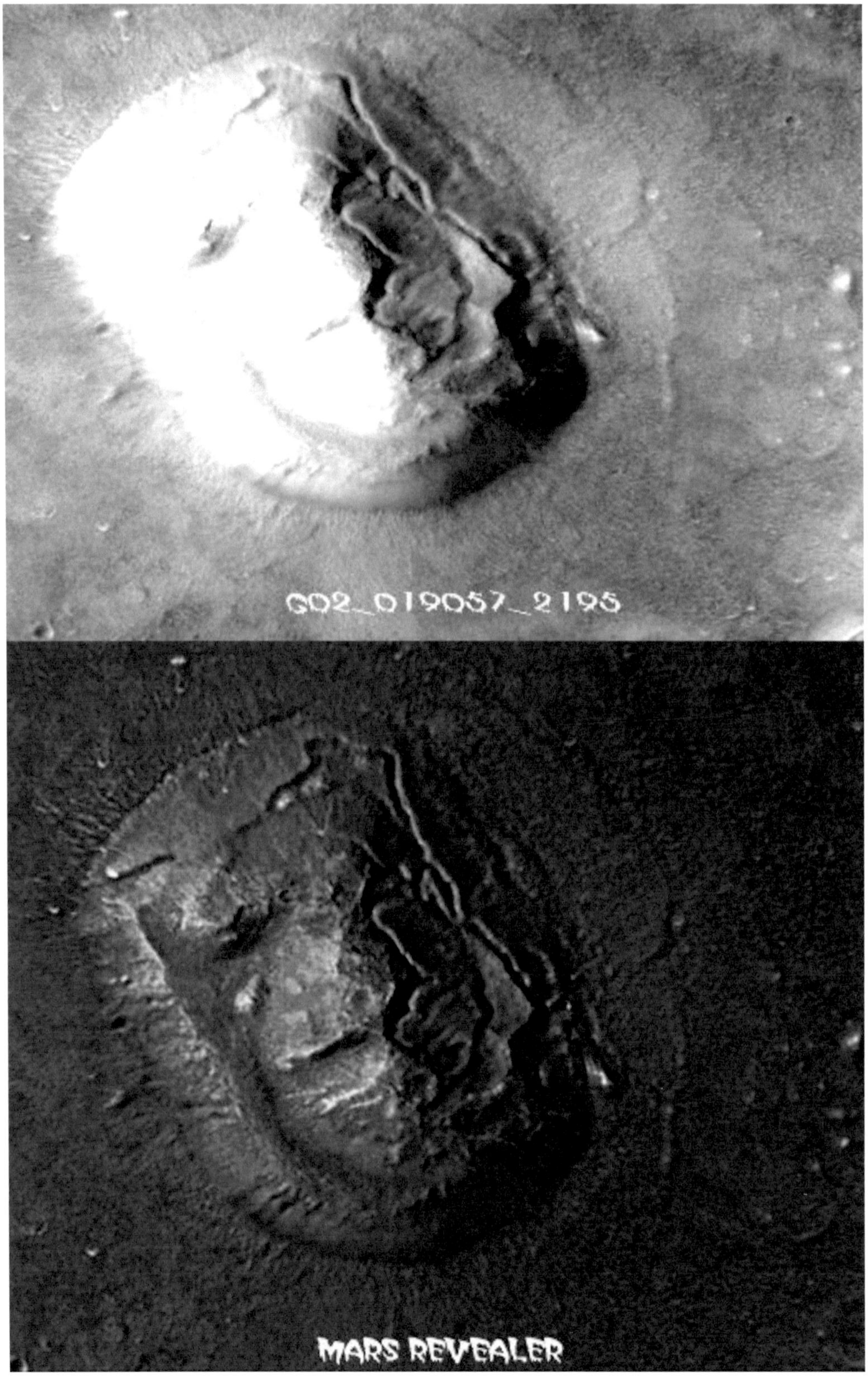

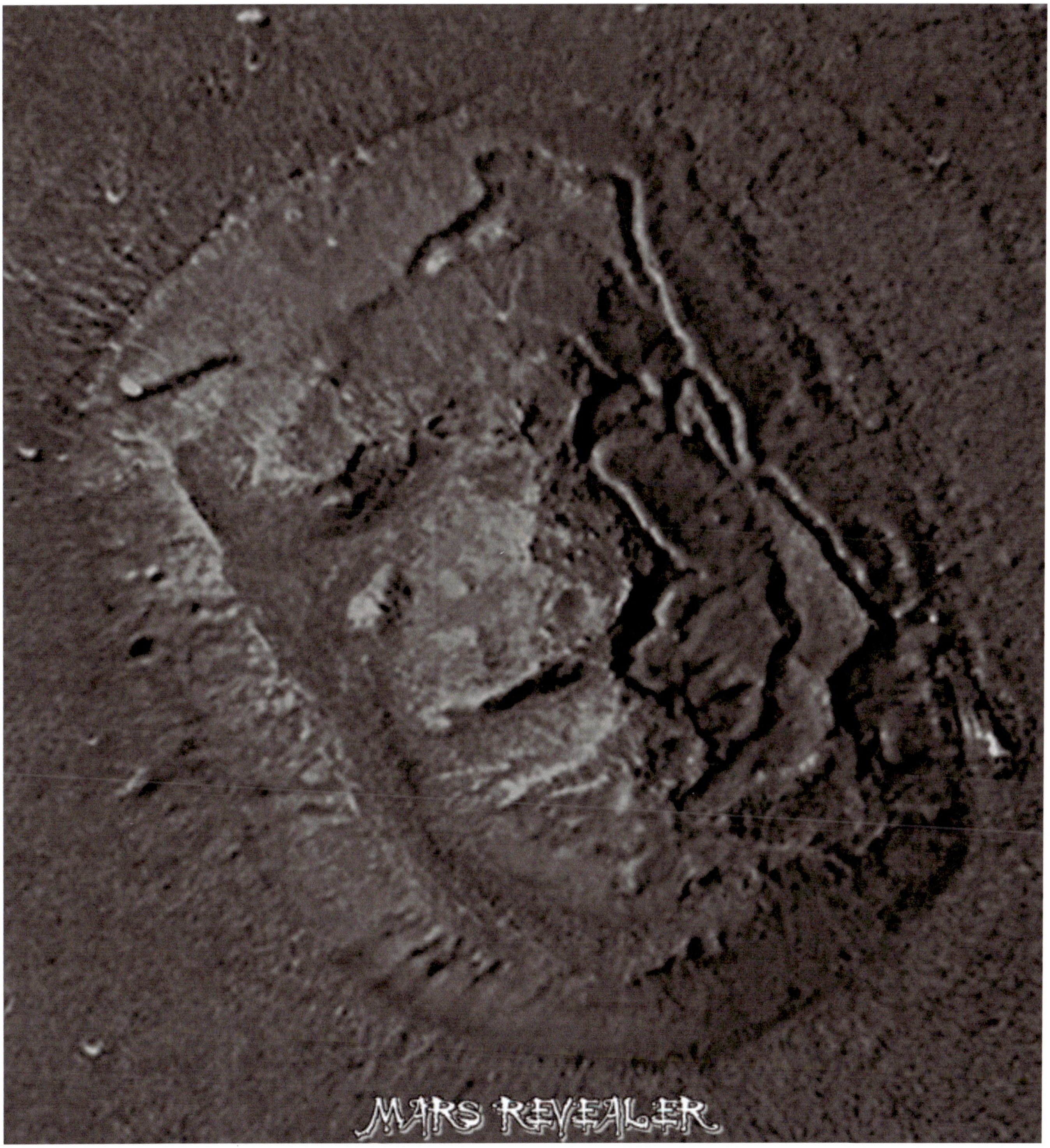

(MY PROCESSED RAW CTX IMAGE: G02_019057_2195 IMAGE TAKEN: AUGUST 20TH, 2010)

(MRO CTX IMAGE ARCHIVE LINK)

https://viewer.mars.asu.edu/planetview/inst/ctx/G02_019057_2195_XI_39N009W#P=G02_019057_2195_XI_39N009W&T=2

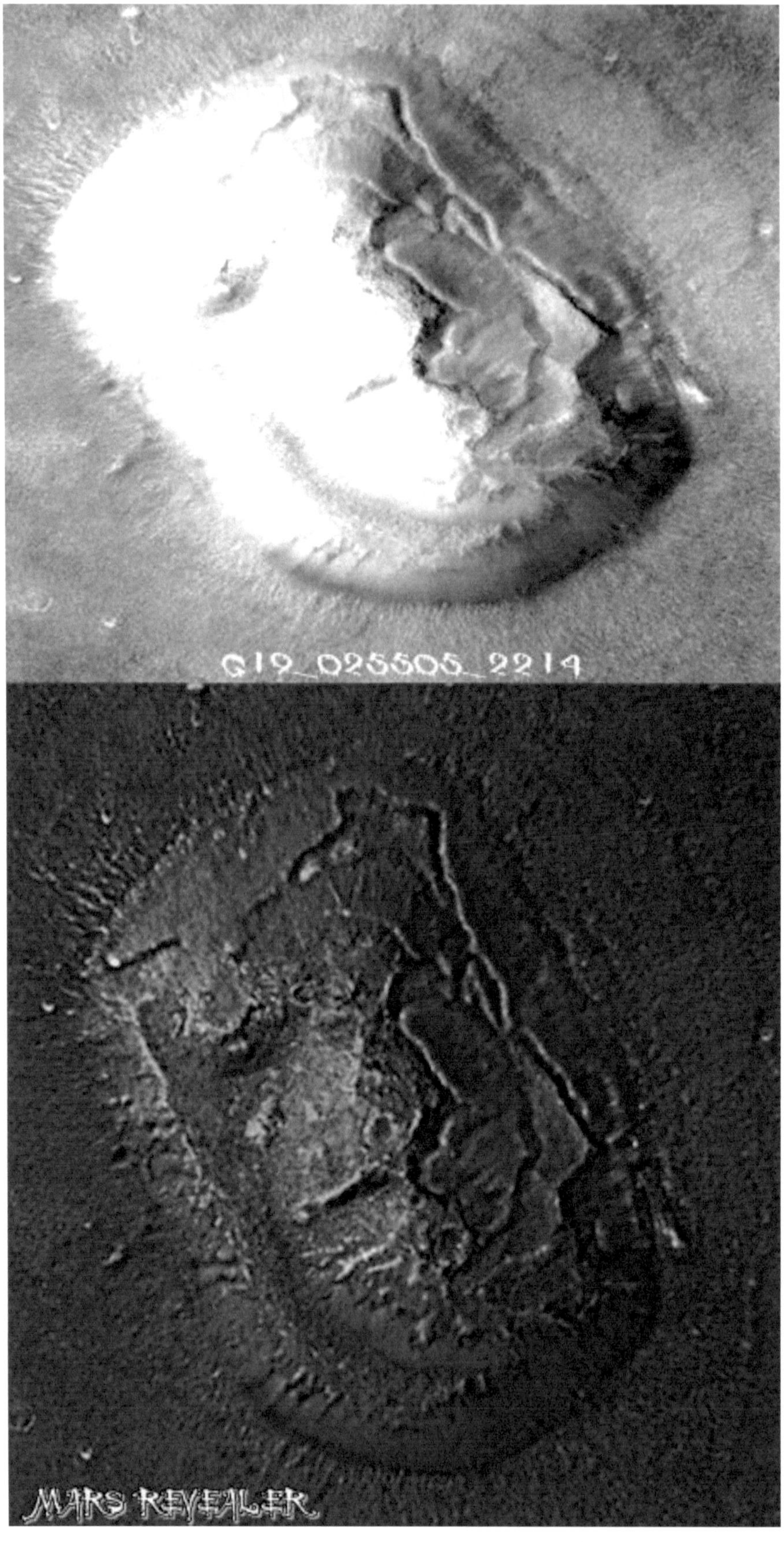
G19_025505_2214
MARS REVEALER

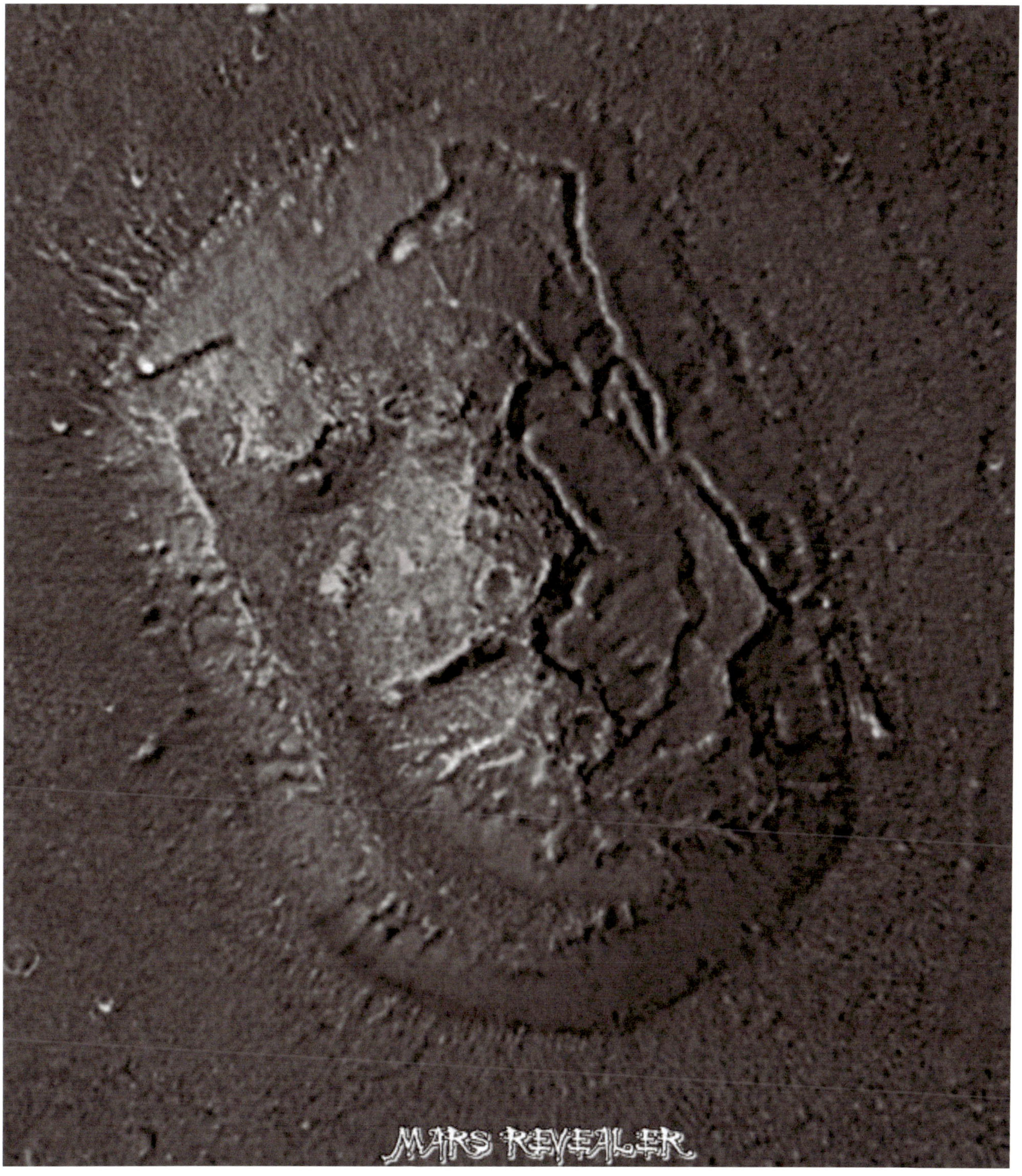

(MY PROCESSED RAW CTX IMAGE: G19_025505_2214 IMAGE TAKEN: JANUARY 4TH, 2012)

(MRO CTX IMAGE ARCHIVE LINK)

https://viewer.mars.asu.edu/planetview/inst/ctx/G19_025505_2214_XI_41N009W#P=G19_025505_2214_XI_41N009W&T=2

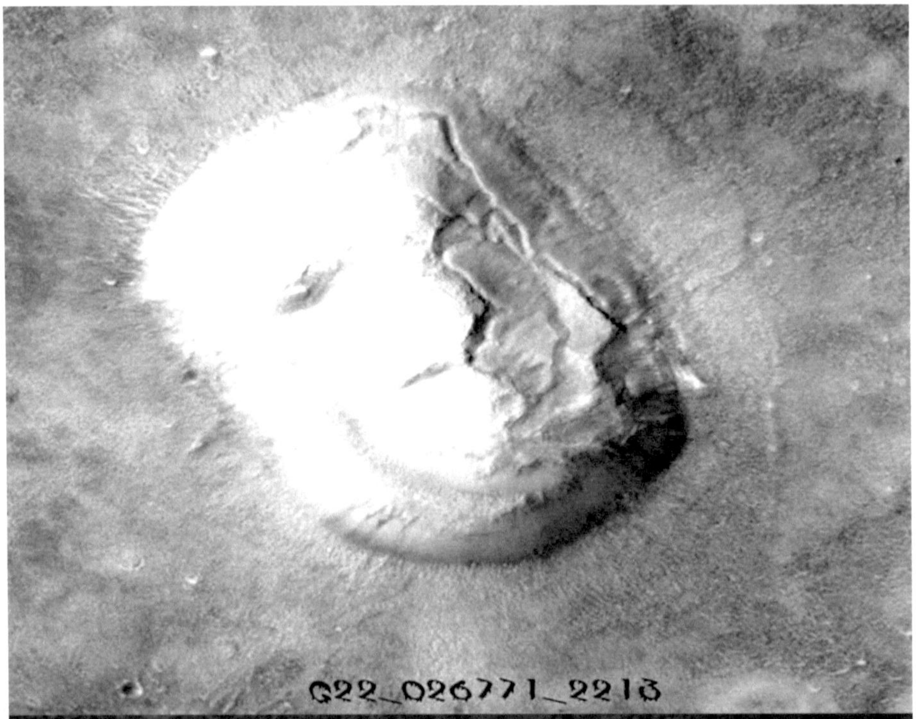
G22_026771_2213

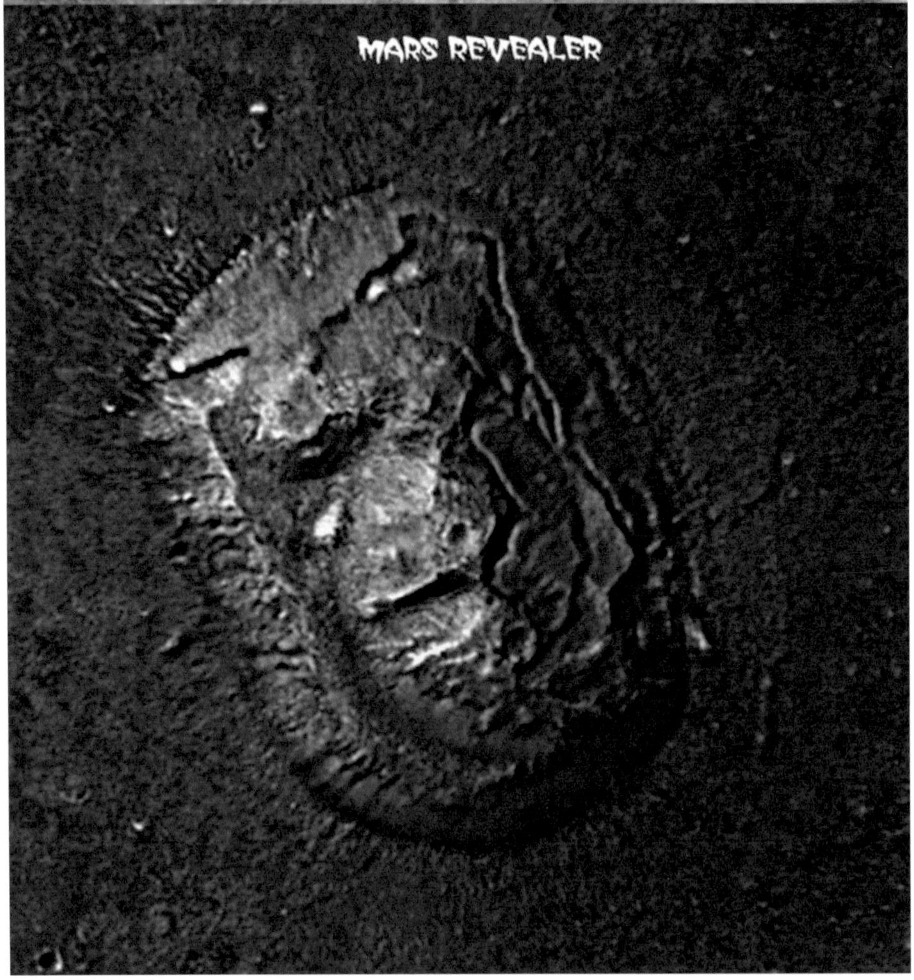
MARS REVEALER

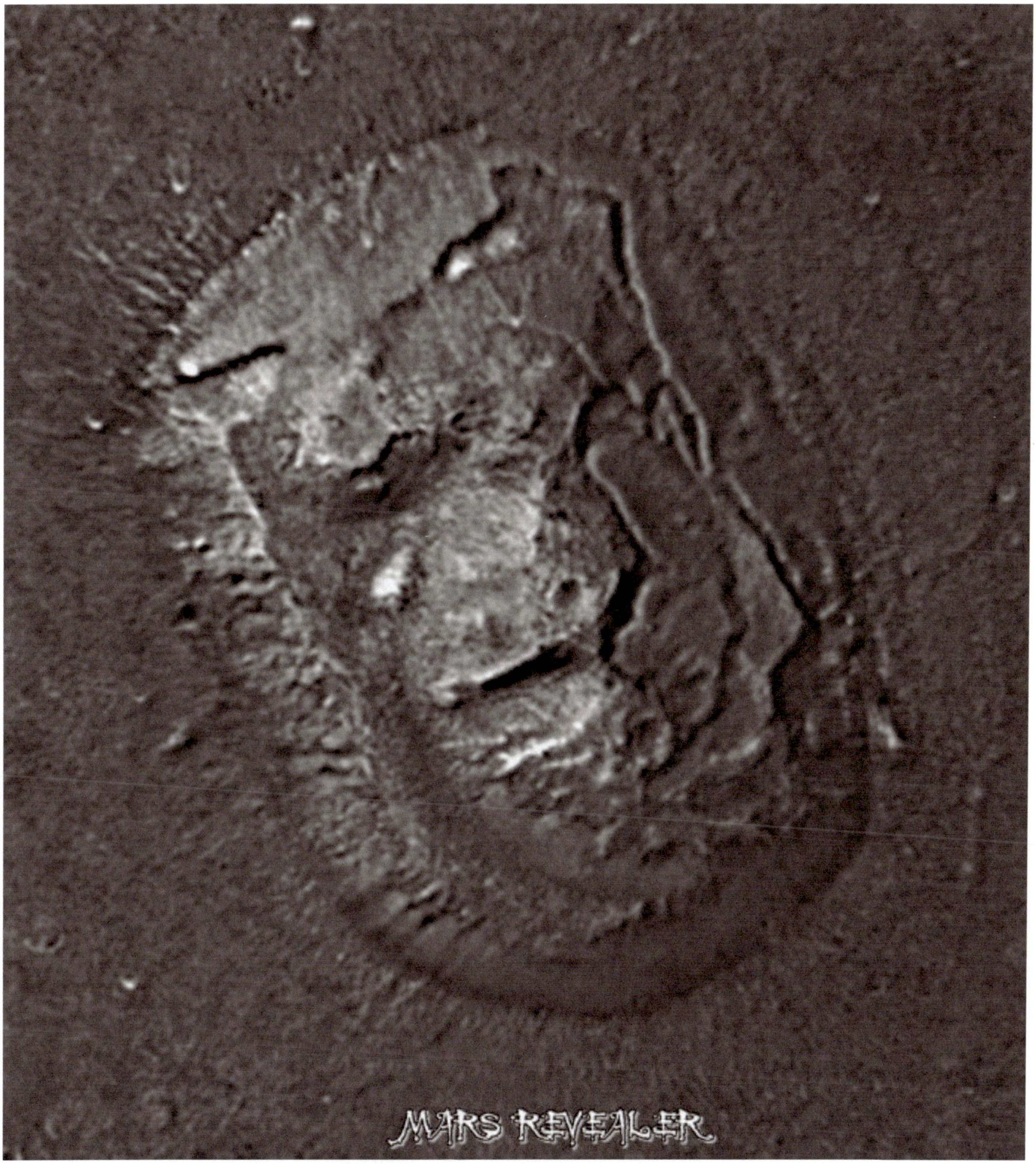

(MY PROCESSED RAW CTX IMAGE: G22_026771_2213 IMAGE TAKEN: APRIL 12TH, 2012)

(MRO CTX IMAGE ARCHIVE LINK)

https://viewer.mars.asu.edu/planetview/inst/ctx/G22_026771_2213_XI_41N009W#P=G22_026771_2213_XI_41N009W&T=2

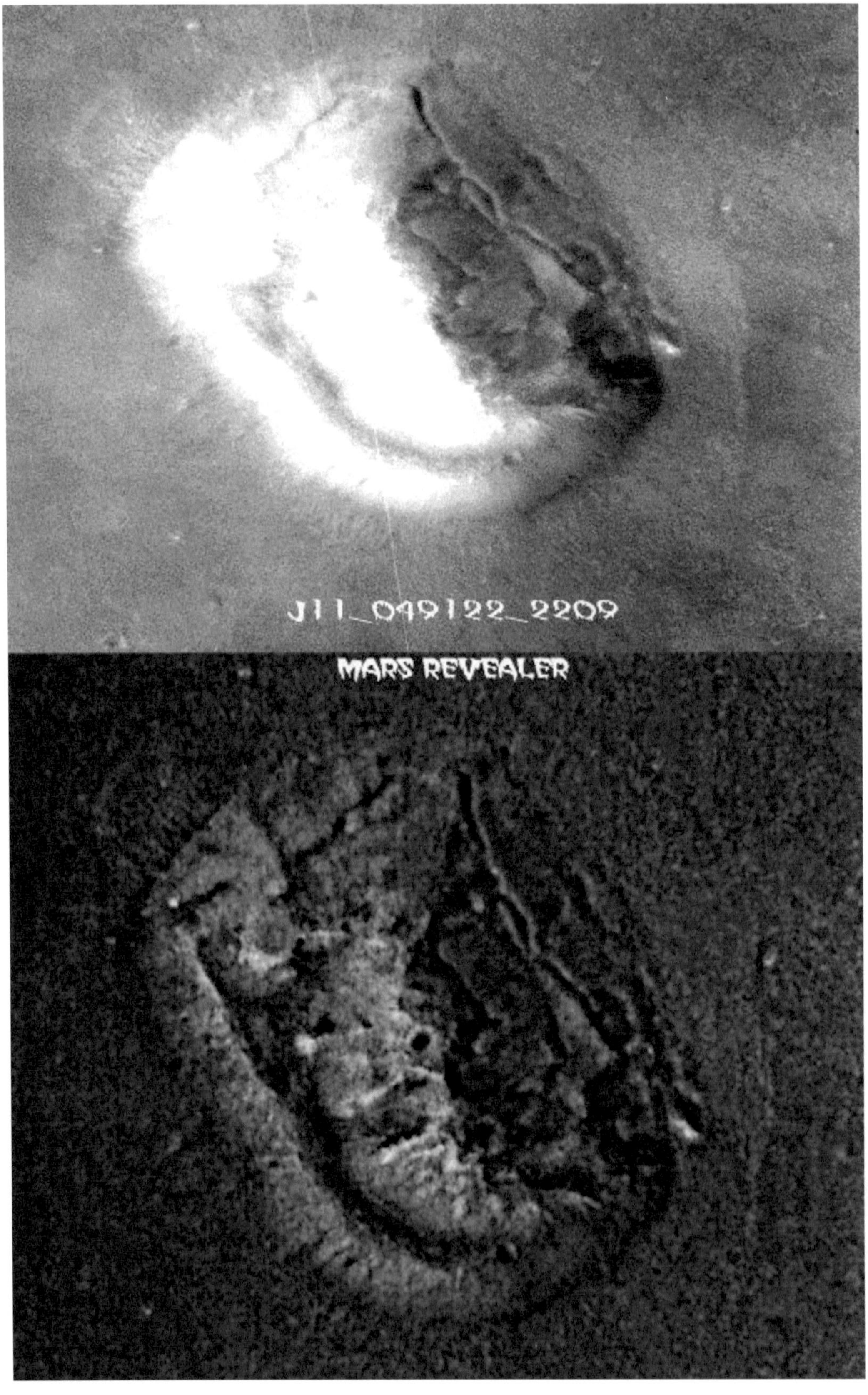
J11_049122_2209
MARS REVEALER

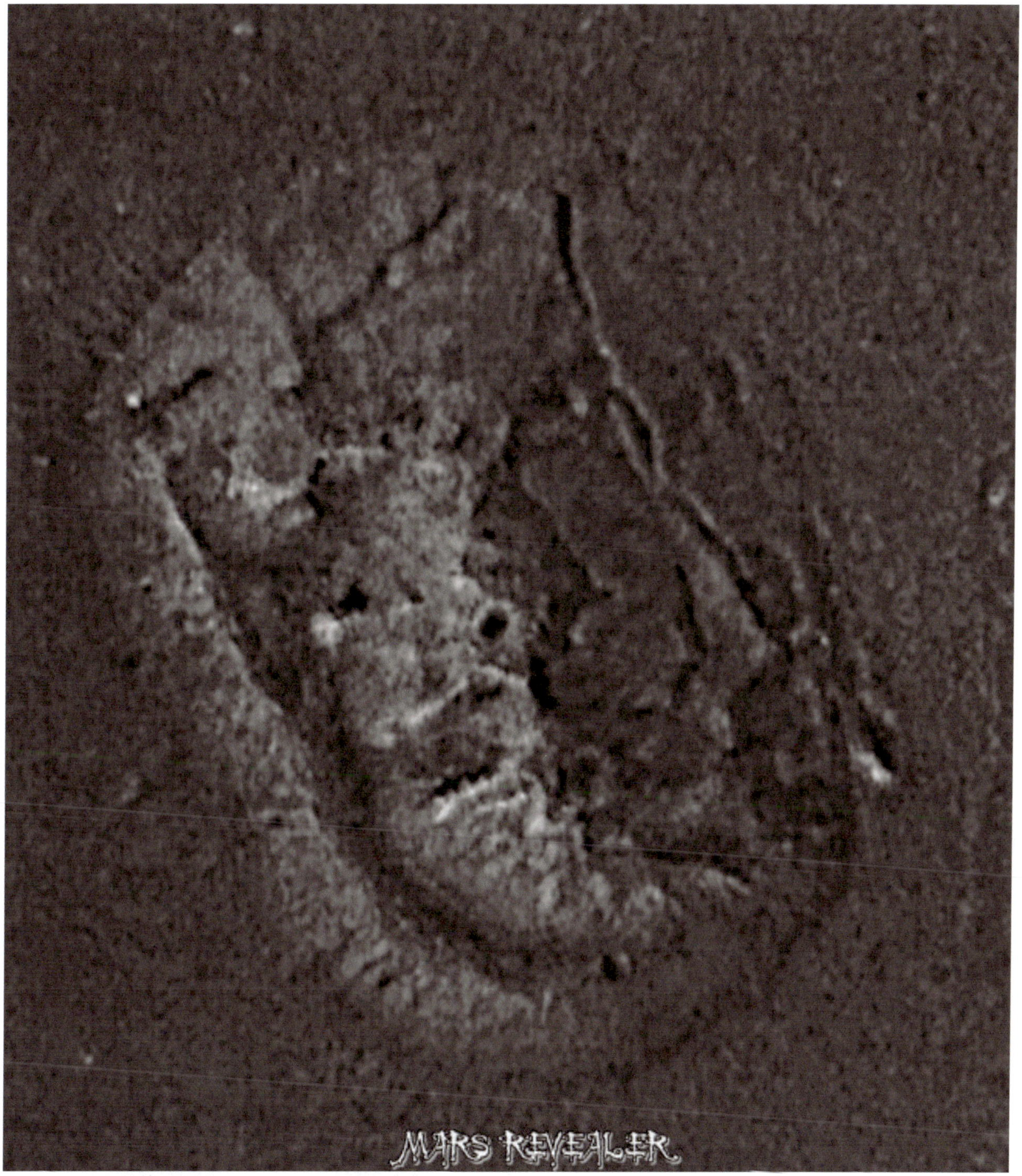

(MY PROCESSED RAW CTX IMAGE: J11_049122_2209 IMAGE TAKEN: JANUARY 18TH, 2017)

(MRO CTX IMAGE ARCHIVE LINK)

https://viewer.mars.asu.edu/planetview/inst/ctx/J11_049122_2209_XN_40N009W#P=J11_049122_2209_XN_40N009W&T=2

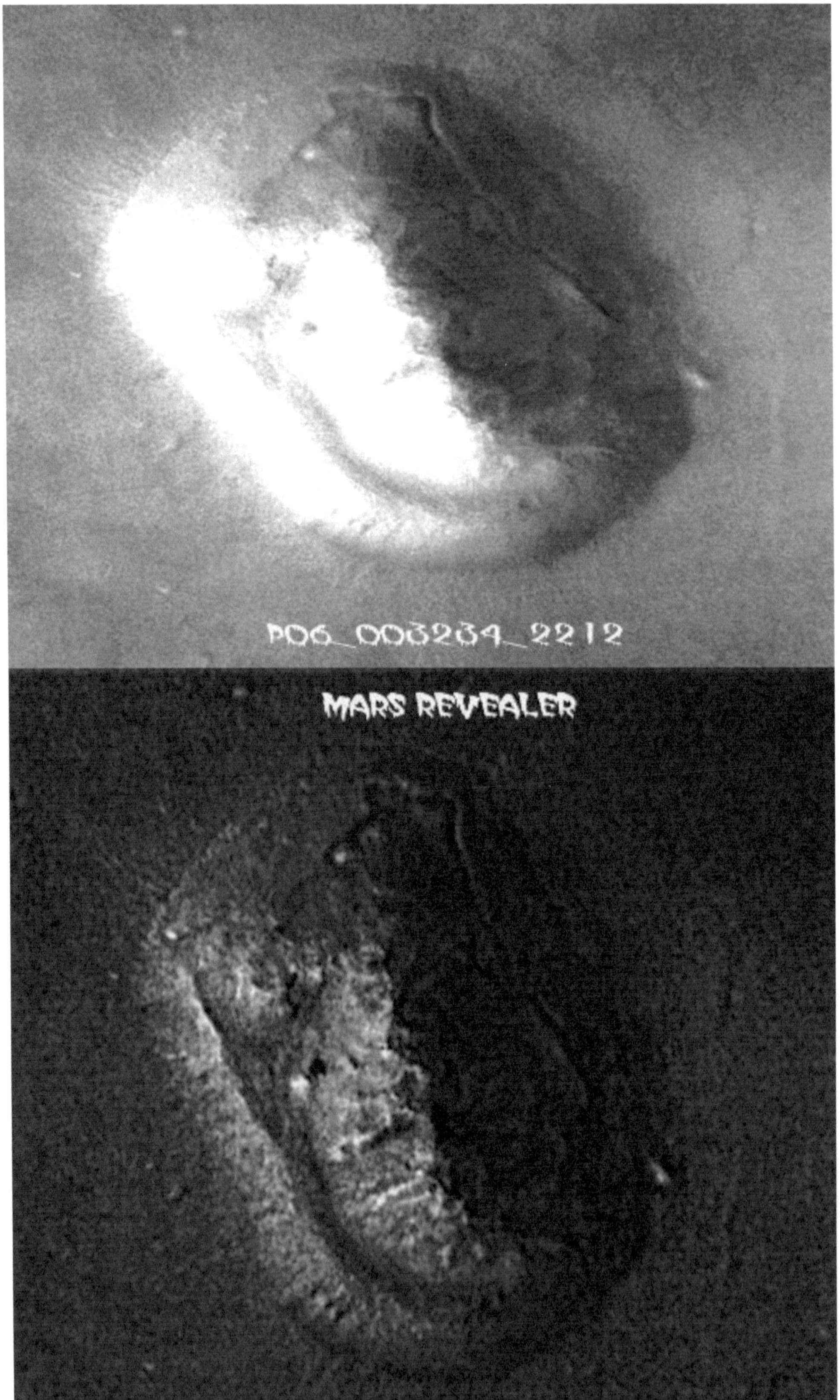
P06_003234_2212
MARS REVEALER

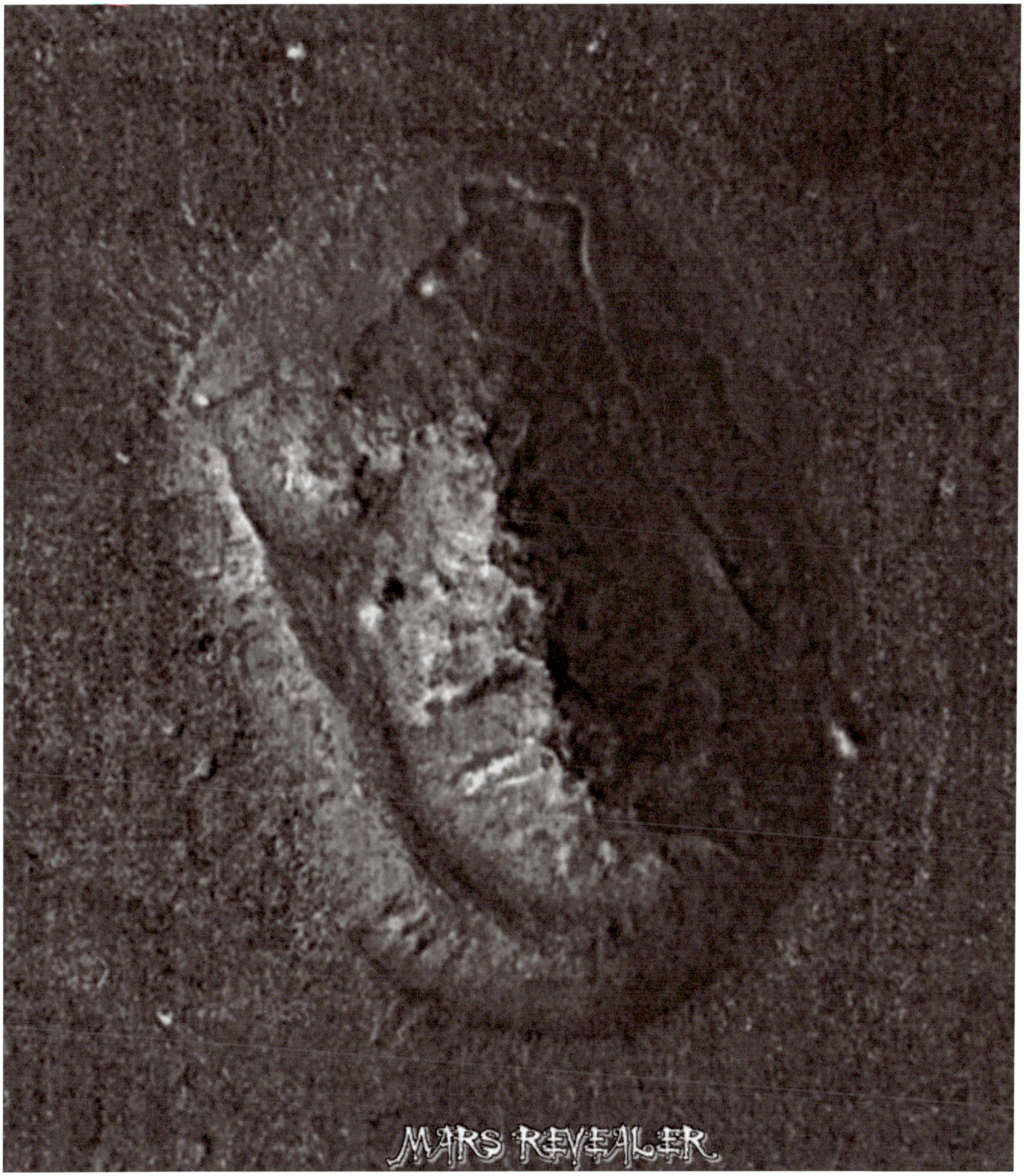

(MY PROCESSED RAW CTX IMAGE: P06_003234_2212 IMAGE TAKEN: APRIL 5TH, 2007)

(MRO CTX IMAGE ARCHIVE LINK)

https://viewer.mars.asu.edu/planetview/inst/ctx/P06_003234_2212_XN_41N009W#P=P06_003234_2212_XN_41N009W&T=2

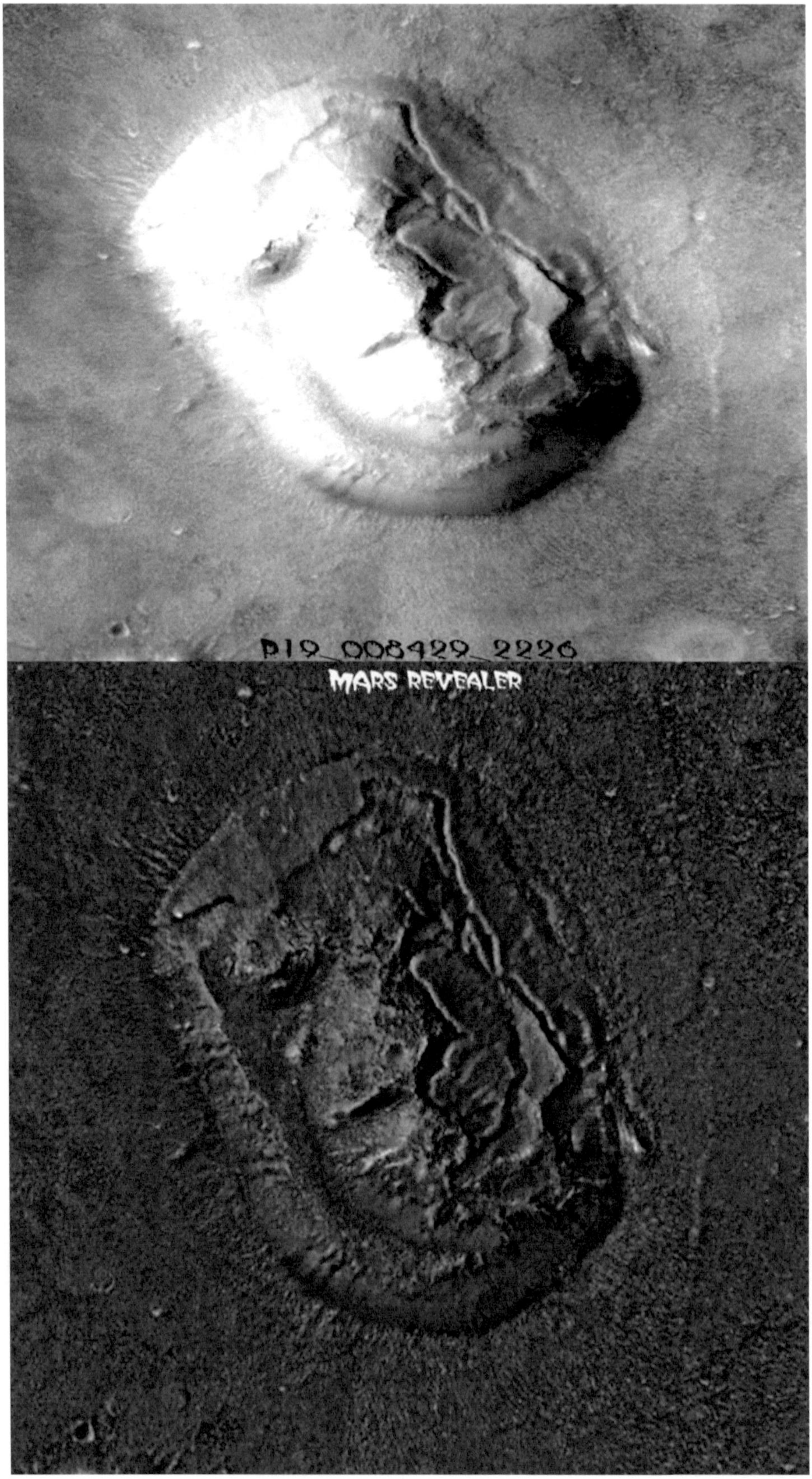
P19_008429_2226
MARS REVEALER

(MY PROCESSED RAW CTX IMAGE: P19_008429_2226 IMAGE TAKEN: MAY 14TH, 2008)

(MRO CTX IMAGE ARCHIVE LINK)

https://viewer.mars.asu.edu/planetview/inst/ctx/P19_008429_2226_XN_42N009W#P=P19_008429_2226_XN_42N009W&T=2

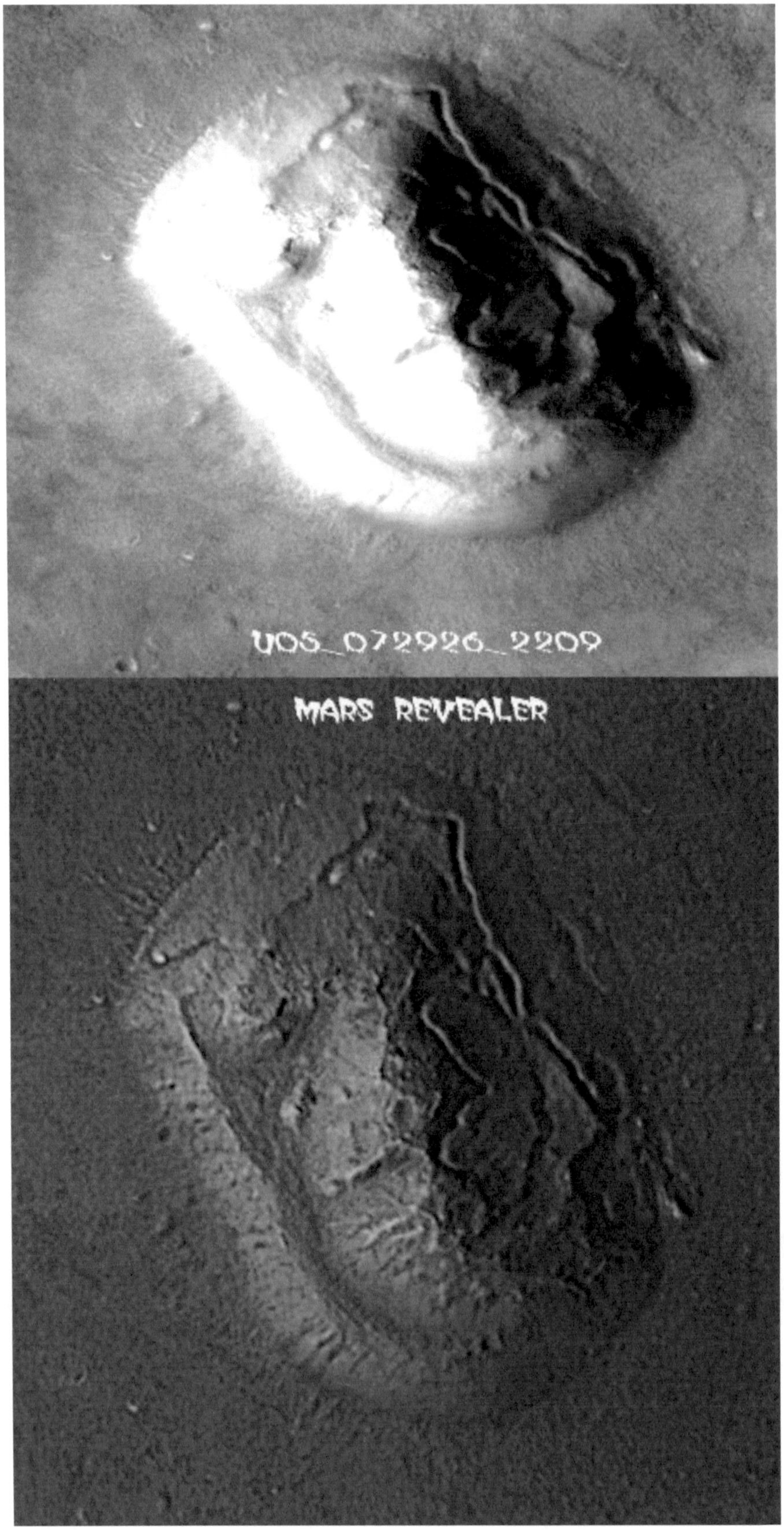
U05_072926_2209
MARS REVEALER

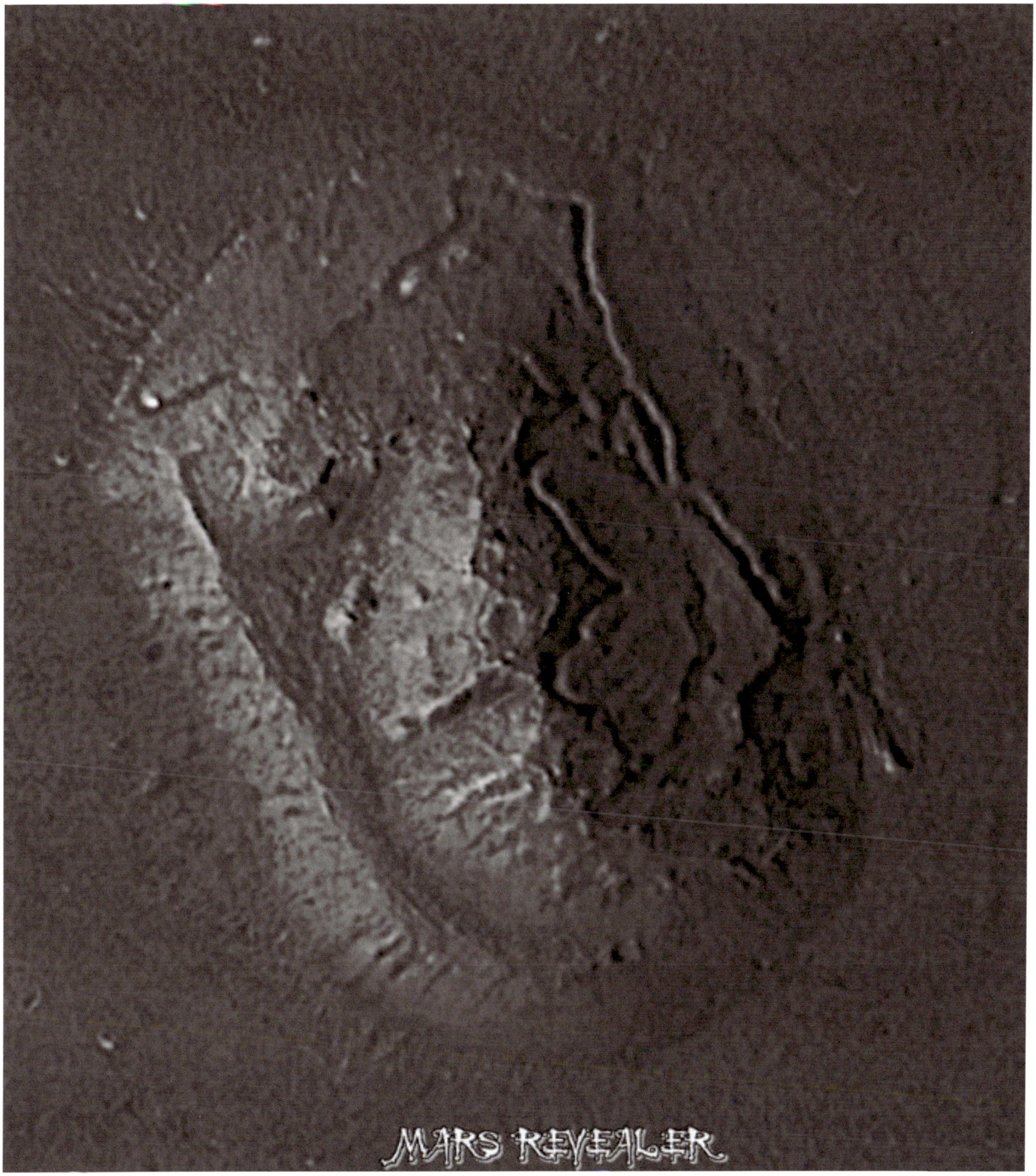

(MY PROCESSED RAW CTX IMAGE: U05_072926_2209 IMAGE TAKEN: FEBRUARY 15TH, 2022)

(MRO CTX IMAGE ARCHIVE LINK)

https://viewer.mars.asu.edu/planetview/inst/ctx/U05_072926_2209_XN_40N009W#P=U05_072926_2209_XN_40N009W&T=2

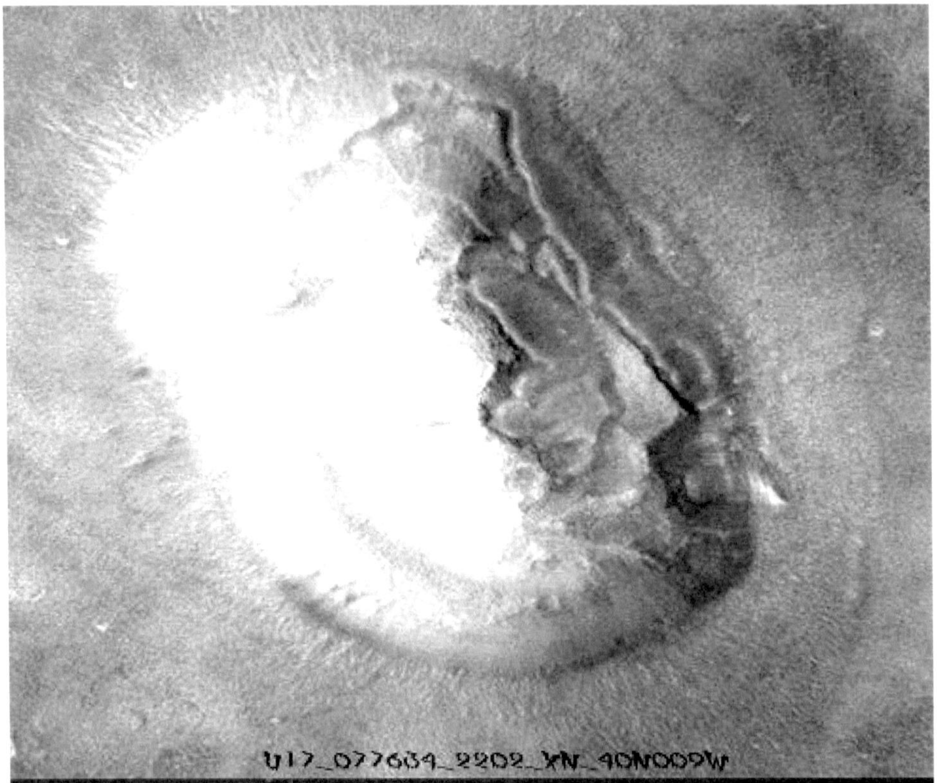
U17_077634_2202_XN_40N009W

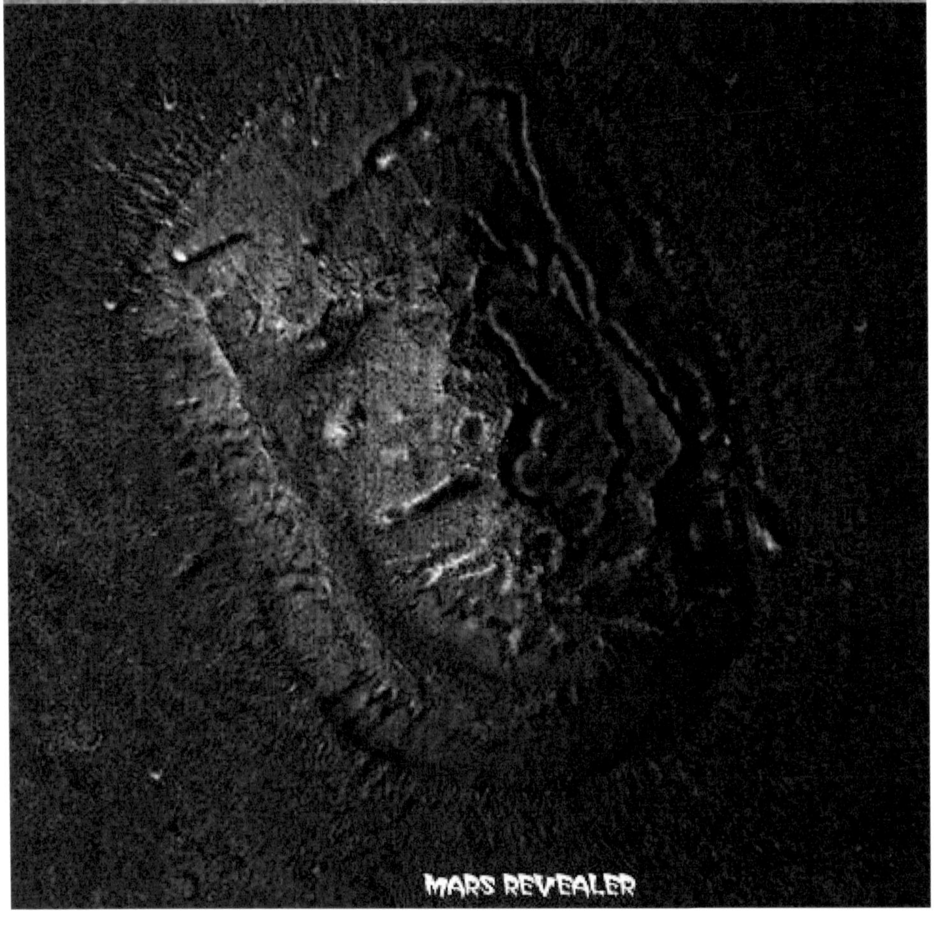
MARS REVEALER

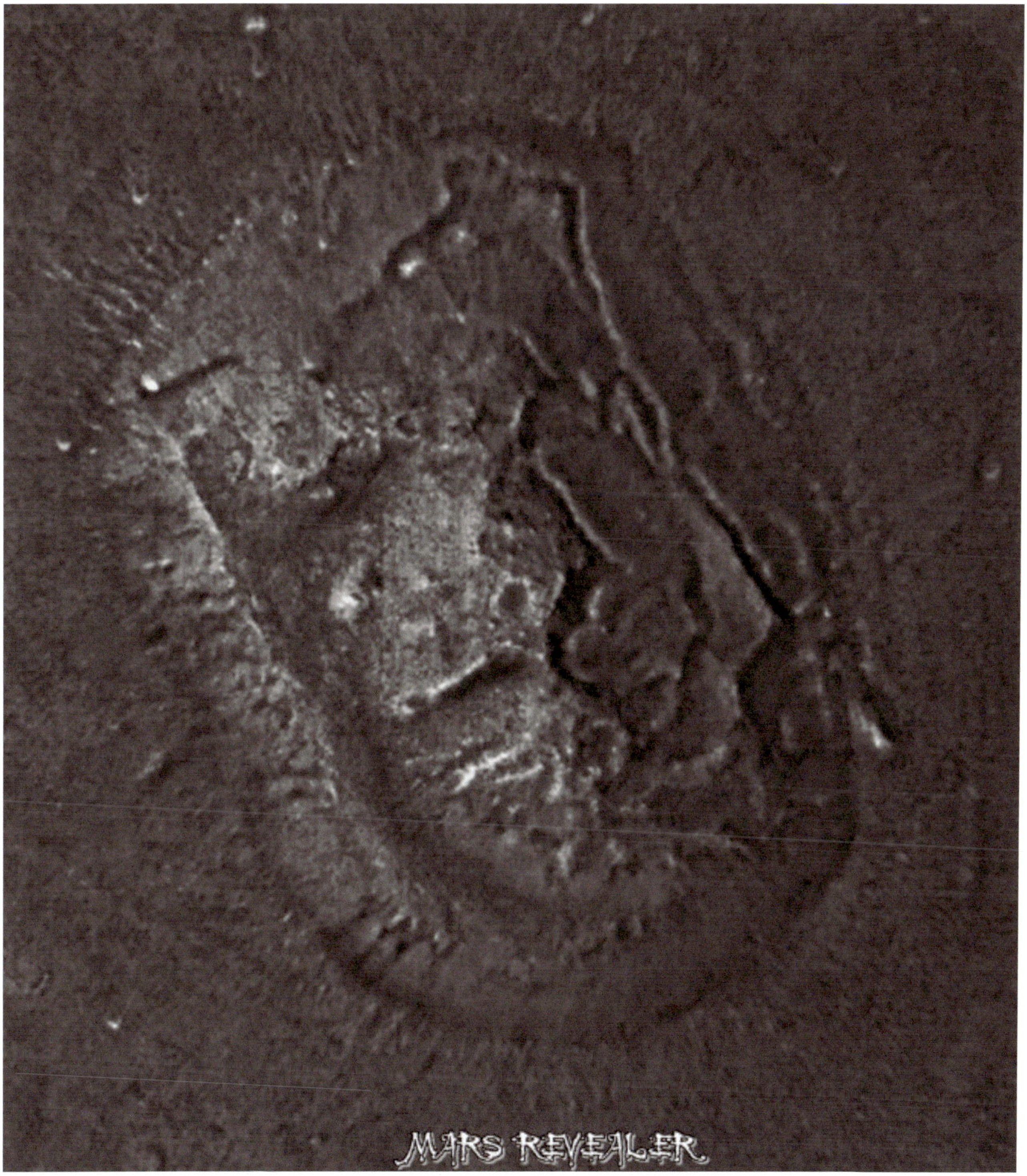

(MY PROCESSED RAW CTX IMAGE: U17_077634_2202 **IMAGE TAKEN: FEBRUARY 17TH, 2023)**

(MRO CTX IMAGE ARCHIVE LINK)

https://viewer.mars.asu.edu/planetview/inst/ctx/U17_077634_2202_XN_40N009W#P=U17_077634_2202_XN_40N009W&T=2

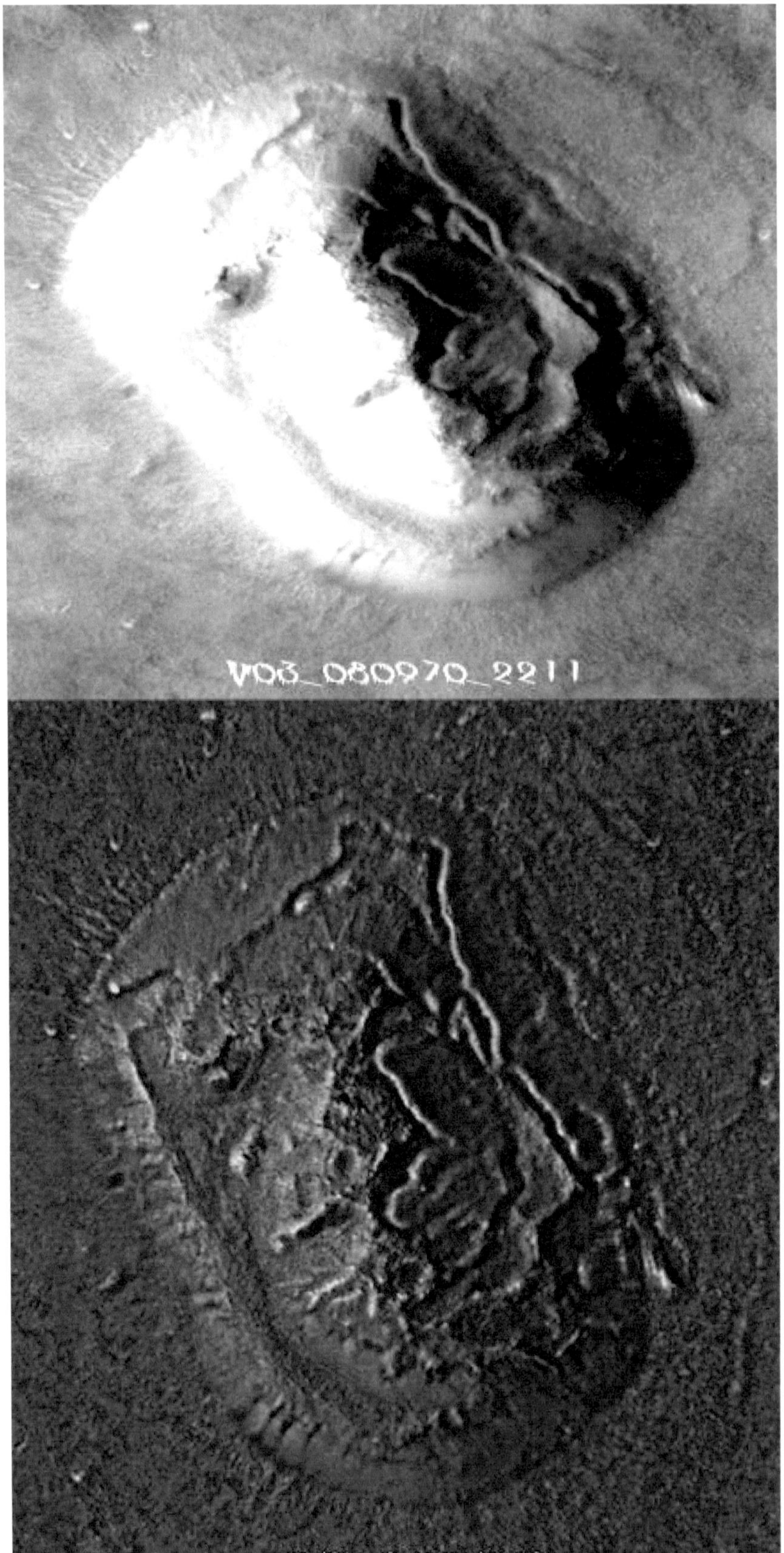
V03_080970_2211
MARS REVEALER

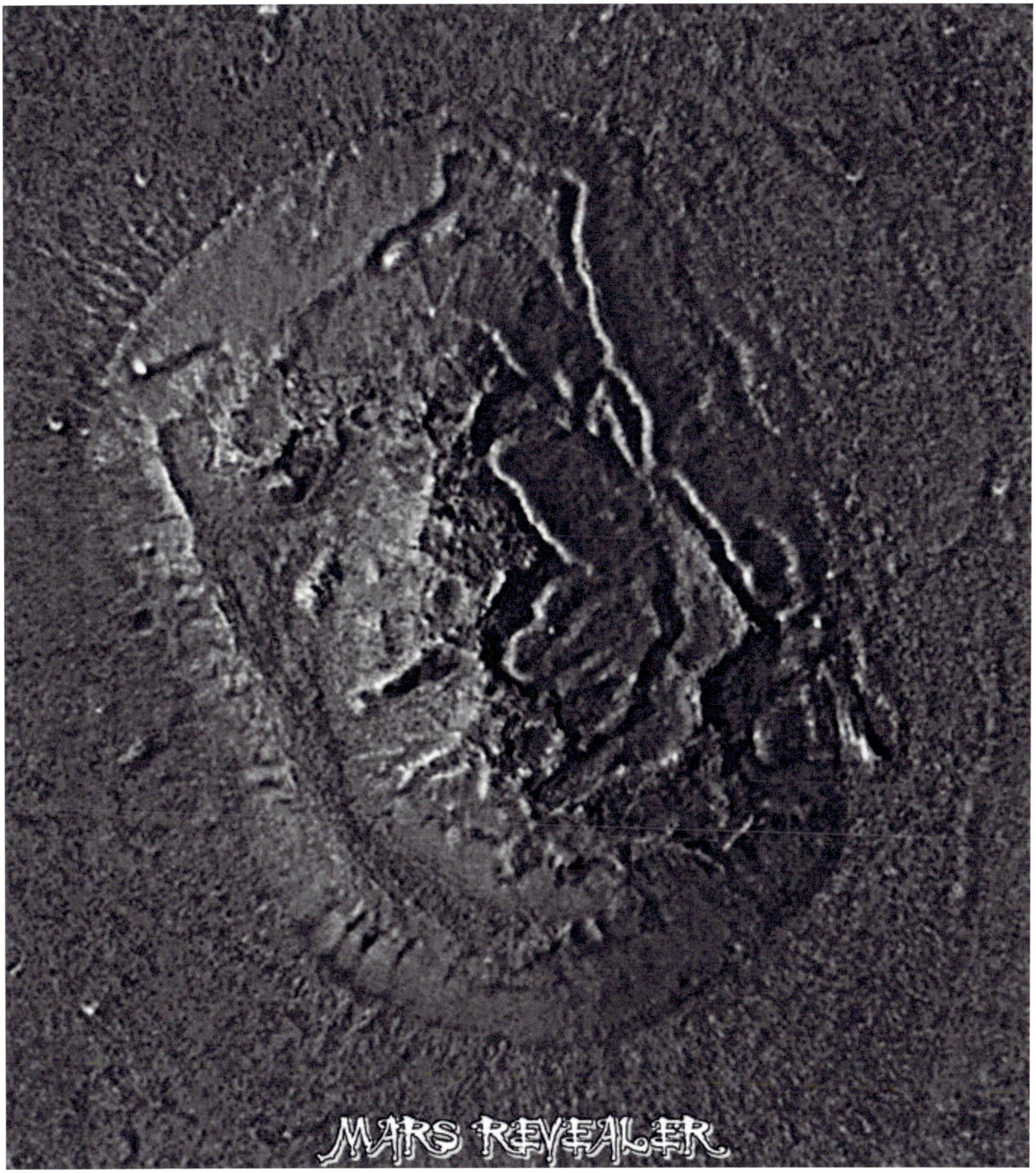

(MY PROCESSED RAW CTX IMAGE: V03_080970_2211 IMAGE TAKEN: NOVEMBER 4TH, 2023)

(MRO CTX IMAGE ARCHIVE LINK)

https://viewer.mars.asu.edu/viewer/ctx/V03_080970_2211_XN_41N009W#P=V03_080970_2211_XN_41N009W&T=2

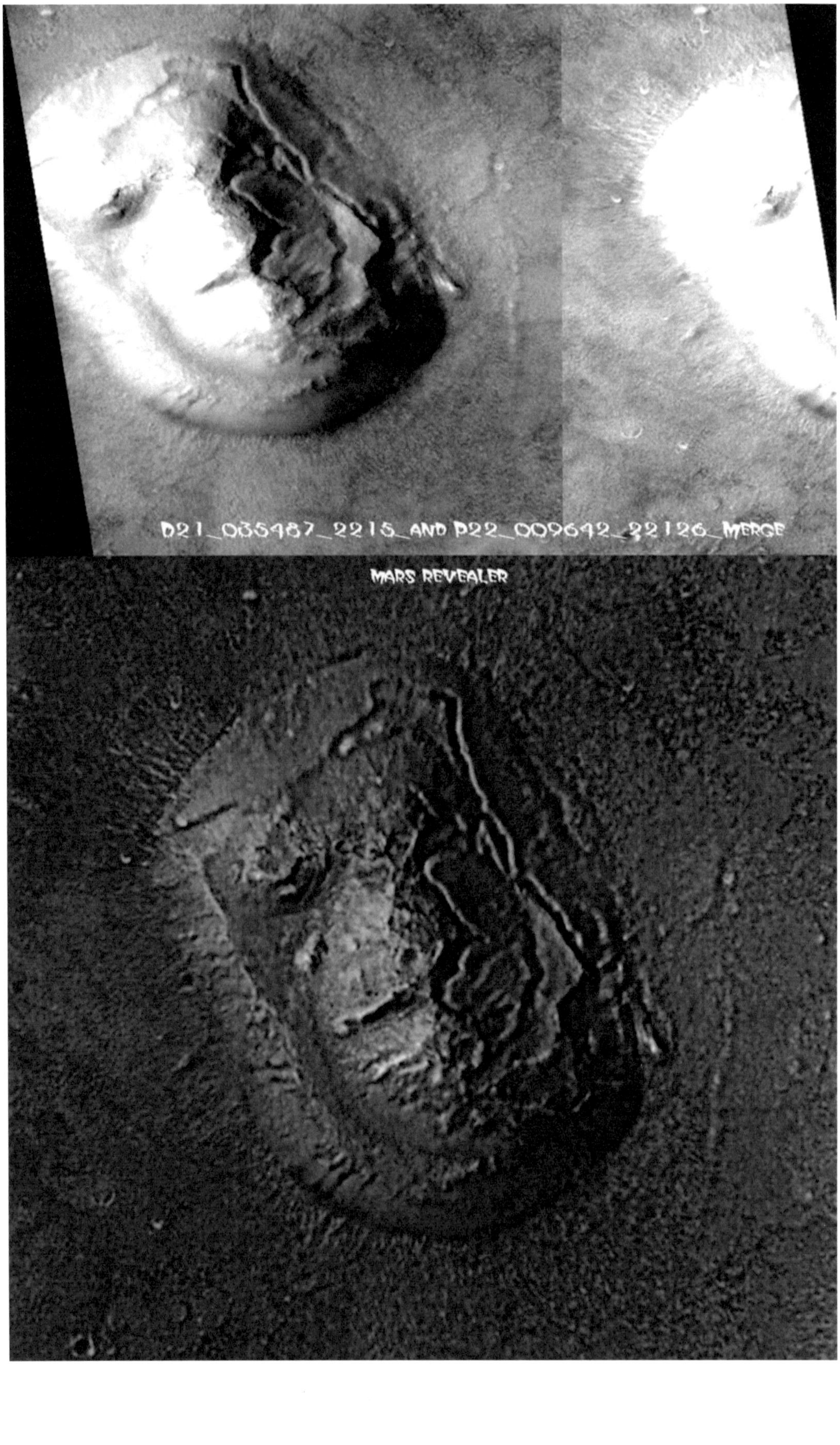
D21_035487_2215_AND P22_009642_22126_MERGE
MARS REVEALER

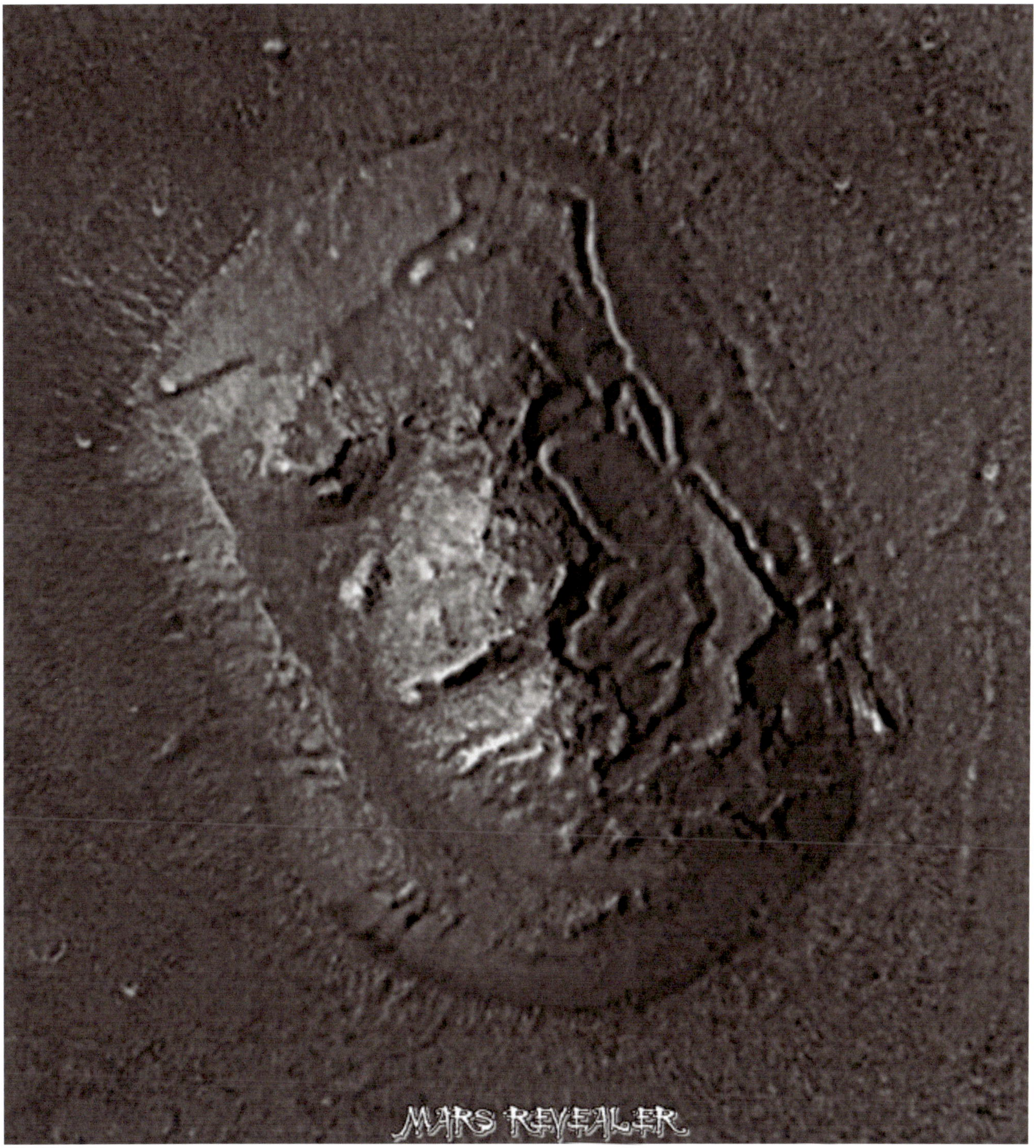

(MRO CTX IMAGE ARCHIVE LINKS)

https://viewer.mars.asu.edu/planetview/inst/ctx/D21_035487_2215_XN_41N009W#P=D21_035487_2215_XN_41N009W&T=2 (image taken: FEBRUARY 20TH 2014)

https://viewer.mars.asu.edu/planetview/inst/ctx/P22_009642_2216_XN_41N009W#P=P22_009642_2216_XN_41N009W&T=2 (image taken: AUGUST 16TH 2008)

In all of this that is shown and implicated of NASA's conspiracy of misleading and obfuscation regards to the Face of Cydonia we have seen here so far, as well as us all '**considering**' the hidden coded aspects of information being brought through to us by NASA's own official wording to help us see a hidden reality opposite of their public official narrative and plausible deniability, it becomes possibly clear that maybe it is NOT NASA which is forcing this policy of misleading and obfuscation, but possibly the military acting outside of the Law and Government of the state who are wanting to do this?

It is safe and easy to speculate about the possibility that the dual purposed Mars reconnaissance Orbiter mission with one side of the mission which is pushed as the mainstream public narrative that it is for science and exploration and studying Mars purposes for the people, and the other side is which NASA's official wording has given us the clues to see and understand that it is mainly a priority military mission acting outside of the law and government of the state detailing doing the things what all we learned of foreign and reconnaissance. We obviously know and clearly see that NASA presents misrepresentations and misinformation, lies, propaganda, and obvious obfuscations to deliberately mislead the people as we can see by the evidences we all have just looked at.

So how am I connecting these images experiences with the other information? Let's consider this more closely. Again, NASA's own official wording on the surface, allows us to consider this bizarre speculative possibility of information to see under the surface and grasp it, providing us with obvious secret coded information, which obviously would be denied by NASA and we would be called crazy for considering such a thing. Yet the official language is the code key that allows us all access to information we would not have even considered at all, if it was not for their official language to begin with. Is here also the old art of saying it without saying it? Speaking in code.

If the military which is in control of the MRO mission, and running it outside of law and government of state, it is they who may be forcing NASA to abide by what they want the public to officially see and officially know? I am only speaking in defense of NASA's position here, and considering that they provided us with secret information as it were, coding to us what is really going on inside their official statements, because they cannot officially tell us about it themselves officially and publicly. I come to this train of thought only because of how they present these obfuscated jokes of high quality images to the public in order to hide the details and surface data truths that they hold within that data, and which are able to indeed be seen, by the also correctly processed publicly released RAW image data!

Think of this, if this is actually as we were led to see not just now of the MRO mission, but throughout the entire history of the many years of official coded statements by NASA we looked at and went over in this book, of the **so called the Face On Mars** narrative that we had to **unmask**, this makes sense also to consider because this shows and is telling us that it is the same military acting outside of Law and government of the state, that is and was also behind and in control of this driving force behind all the past policy of what the public was to see and know and told by NASA and media as the mouthpiece of plausible deniability? It is important to consider.

The fact is that we are being obfuscated of details on the surface of Mars in terms of public viewing and data narrative, of what the public is getting to see as official image data that is officially released. Yet NASA provides us all with unseen though publicly released RAW image data where you can literally process it yourself, and come out with much better results. This is just baffling to me and should be to you all also in trying to follow this book so far.

It is a great point. WHY? Why all the things done over these many years, to hide and lie about this all, yet under the surface of their official statements of narrative, we get bits of truth and facts, and information that is nuts to consider as really happening, yet comes with a believable story line, or consistent in the least, that seems to always point to someone else of power above themselves as preventing us from seeing the truth of the image data and also from NASA themselves of what the Face really is and shows? It is a long winded question, but still a great one!

One must wonder, because we are given unseen information in the official words, and given unseen data online that clearly shows the truth once downloaded and processed by anyone who knows how, but the results is not ever publicly acknowledged by NASA of the truths seen in the RAW image data when brought out to be shown to the public, but yet they publicly released and make available of that unseen RAW image data by NASA for what? **To be obviously seen? By someone?**

Yes, I would agree all of this seems like such great efforts by NASA to reveal truths and unseen or hidden information and image data, without saying it or admitting it? This tells us that NASA possibly, wants to tell us these truths, but we are being guided to see whose fault it really is? Telling and showing us that they are not the ones to blame for **this conspiracy against WE THE PEOPLE**? Yet if so, why would this military control working outside Law and government of state in real control of MRO then allow NASA to release the publicly unseen RAW image data where anyone in the public can download and open and process and reveal what is really there? That is a bigger question to ponder I admit. **One not to be answered now yet I feel.**

So I am NOT doing this work that I do as an independent Mars researcher, or am writing this book only because of my bias of NASA. I believe you all can see well enough that NASA lied, misinformed the public and obviously obfuscated image data to literally cover up information and surface details of Mars. I think that is a fair and safe and obvious assessment at this point. I hope you agree. Plus considering these potential crazy facts of coded information provided by NASA, in their own official wording and narrative over the years, where this shows a possibility of redemption. Yet again, they will never admit to the crazy things I am saying and what we are considering here.

I admit, it is the only good defense they can hopefully hope to have, because I cannot think of anything else or see any other benevolent possibility here on NASA's behalf, on playing their part in deliberately doing this all to us outside of the Law of government of state, like the constitution I am mainly thinking of here in this regard! Especially considering when we are dealing now with **THE PEOPLES CAMERA!**

So no one can rightfully say I am just an anti-NASA nut. As you see, I offered up a possible defense, as crazy and bizarre as it is, defending them based upon their own data of their official words that enabled me to think of and construct this defense for them. Other than this, who can debunk my claims or what I am showing you all of the same official narrative data, and especially regards the image data, again, that they released themselves over the years and in modern times dealing with directly with the Face on Mars? Certainly these anomalies would be considered as being **Beyond Mere Tricks of Light and Shadow**, **of NASA!**

Speaking of anomalies, considering all of the MRO CTX imagery we looked at, not only the features of the Face and its Alien Markings are the anomalies in those images, but the very images themselves have some anomalous attributes to them that to me, not make much sense, or which seems logical. I am speaking of the image capture dates that each image was stated on each of the image release archive pages for them, provides the image capture date and time for that particular image data. I provided those dates for you all also for each image shown.

In my mind, here is the anomaly, their sequence numbers, or file numbers. Notice they all CTX images start with a Letter. For instance **B01**, then **D01**, then **F17**, and so on and so on. Go back to the images again to look for those image numbers and see related dates for each. So what is the anomaly? The anomaly is that the dates do **NOT** correspond or sync up with the letter for the image file. The lower letter starts the sequence the images were filed since the mission began its science and obvious 'reconnaissance' phase and captured the images. Some dates of these images show later dates than should be for the image file letter it was stored in. I will not get into all the technical details, as all you all have to do is go back to all the CTX Face image pages in this chapter, and look for yourself. It is self-explanatory. Plus it is nice for you to do some fast self-research and test if my words are true or not. Maybe there is a great reason for these anomalies? Or is this just more **misleading and obfuscation?**

CHAPTER 10:

HI-RISES FOR THE FACE OF MARS

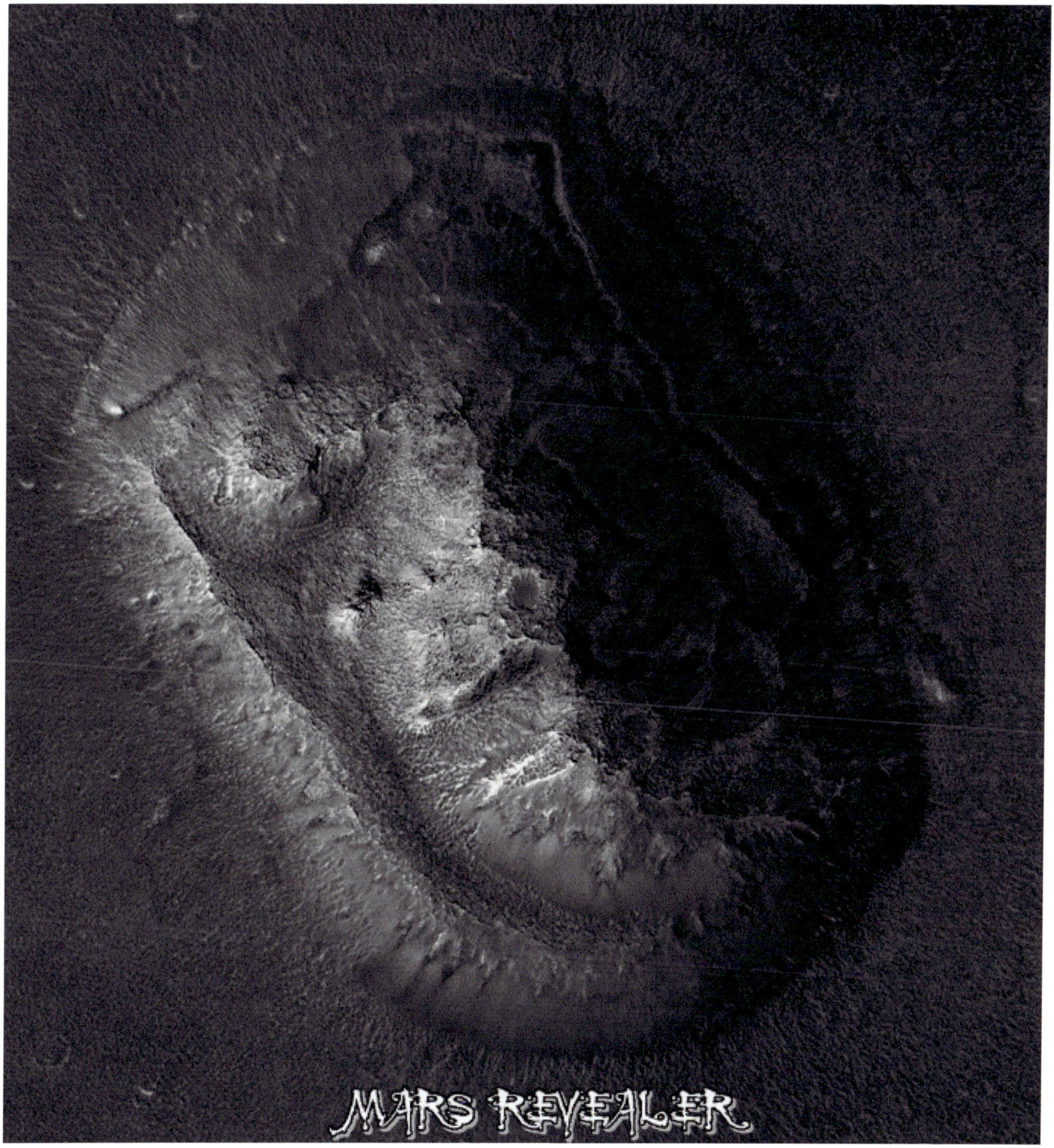

(MRO HI-RISE IMAGE PSP_003234_2210 IMAGE TAKEN APRIL 5TH 2007)

IMAGE ARCHIVE PAGE LINK: https://www.uahirise.org/PSP_003234_2210

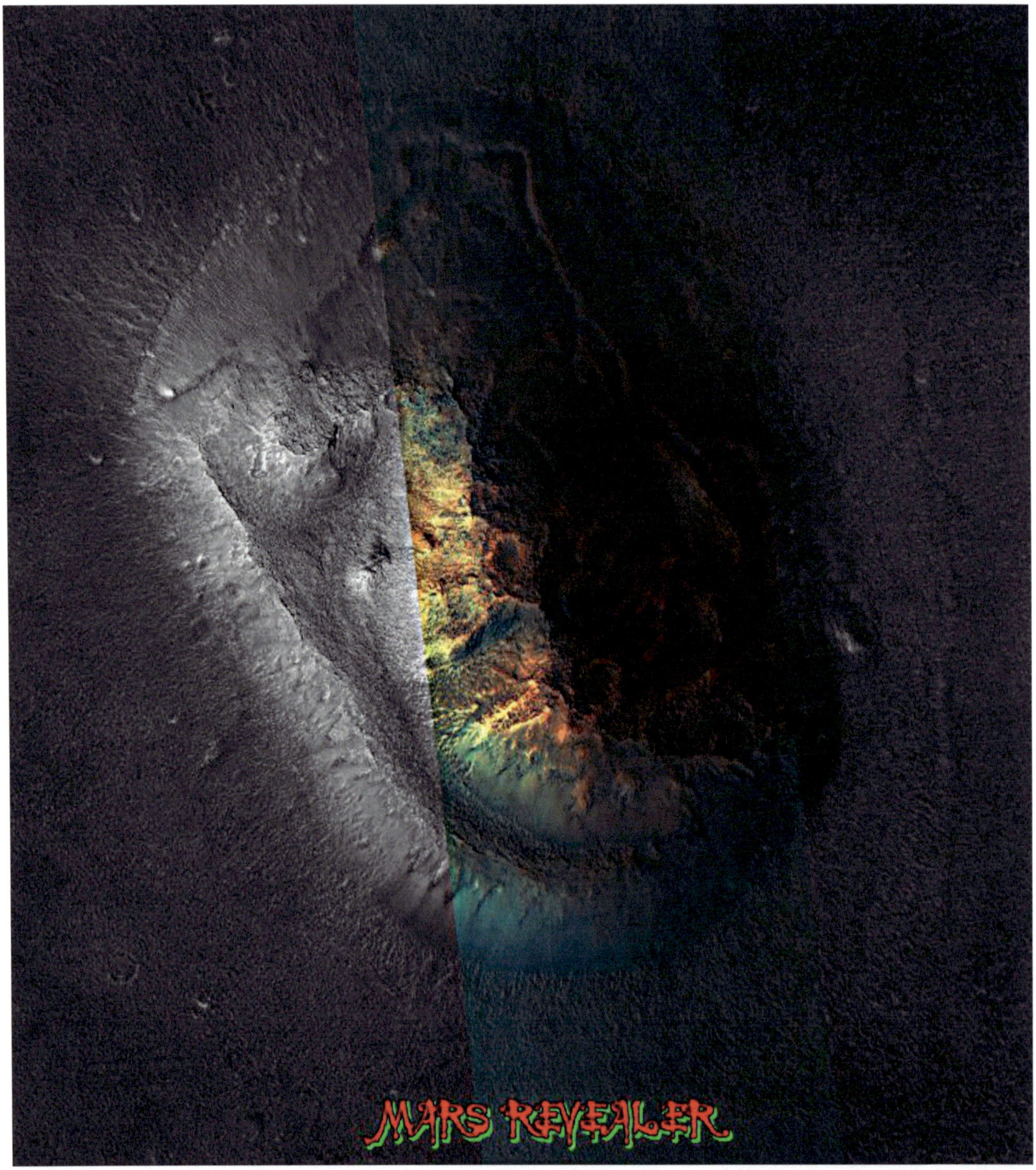

(MRO HI-RISE IMAGE PSP_003234_2210_MRGB IMAGE TAKEN APRIL 5TH 2007)

IMAGE ARCHIVE PAGE LINK:

https://hirise-pds.lpl.arizona.edu/PDS/EXTRAS/RDR/PSP/ORB_003200_003299/PSP_003234_2210/PSP_003234_2210_MRGB.abrowse.jpg

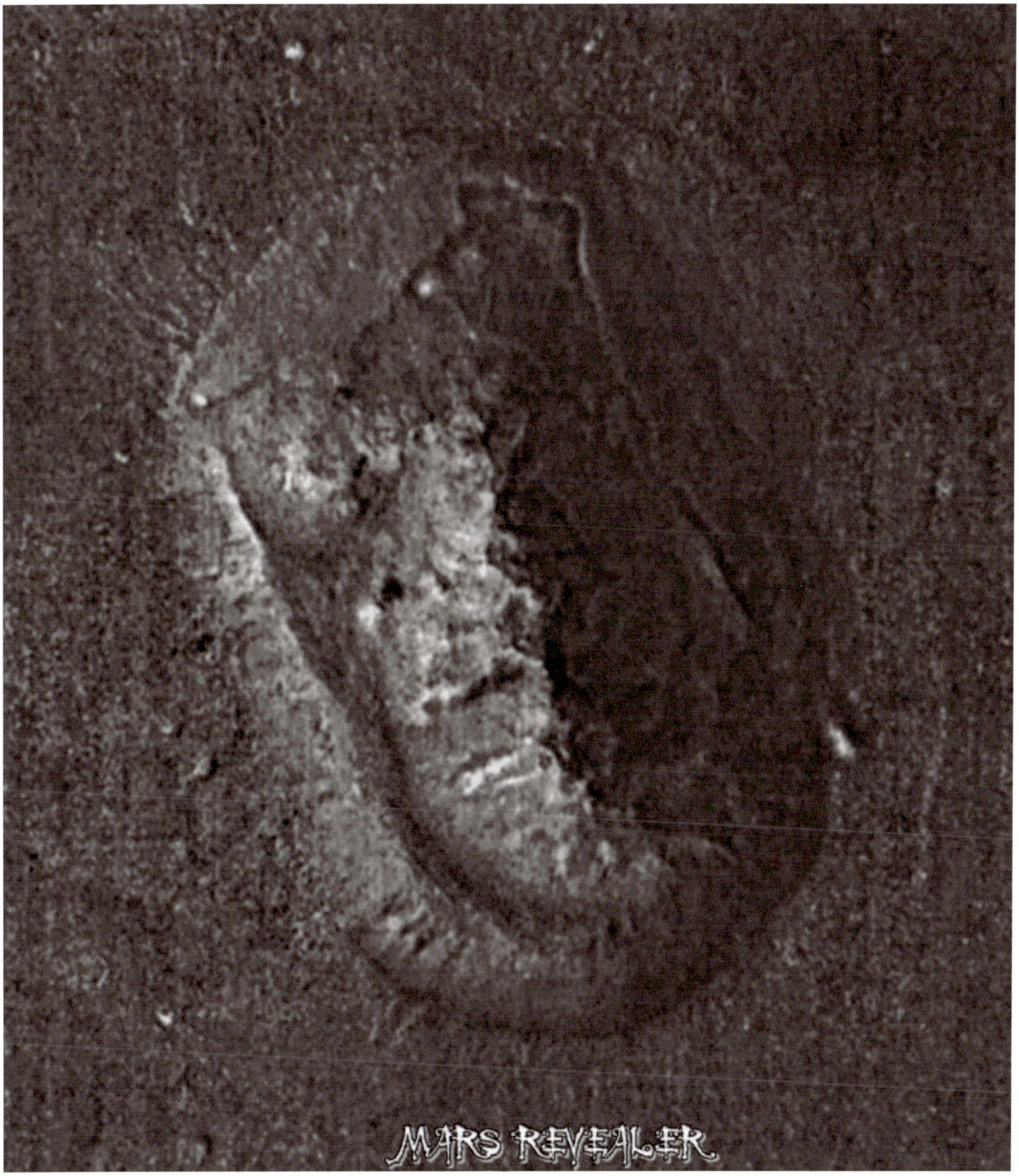

(MY PROCESSED RAW CTX IMAGE: P06_003234_2212 IMAGE TAKEN: APRIL 5TH, 2007)

(MRO CTX IMAGE ARCHIVE LINK)

https://viewer.mars.asu.edu/planetview/inst/ctx/P06_003234_2212_XN_41N009W#P=P06_003234_2212_XN_41N009W&T=2

To add further insult unto all of our injury, after that amazing first MRO Hi-Rise image of the **'Popular Landform in Cydonia Region'** from 2007, not only have we been forced by official narrative and policy to no longer call the Face of Cydonia **'The Face'**, we have also been very neglected in terms of continued Hi-Rise observations and study of **THE FACE!** One must wonder and ask WHY?

Why would only one single MRO Hi-Rise image of the 'Popular Landform in the Cydonia Region' be the only Hi-Rise image taken of **'The Face'** for 16 very long years of having plenty of ability and opportunity as well as enough time, to have given us at least a dozen, and in the least half of a dozen of Hi-Rise images of the **'Popular Landform'**. **But none ever came.** Even though JPL has had many public target requests for the for the Hi-Rise camera to image **'The Face'**, now called **'Popular Landform in Cydonia Region'**, through the public Hi-Wish public target request program. Again, because after all, remember, this is **THE PEOPLES CAMERA!**

Here is a view of what examples of target requests shows on the public website. The yellow outlines and yellow place markers are examples of the requests made to request to target and image the Face. This only represents a few of the many that are actually requested. I am showing this as the only example that they provide. But one can see the entire Popular Landform is covered.

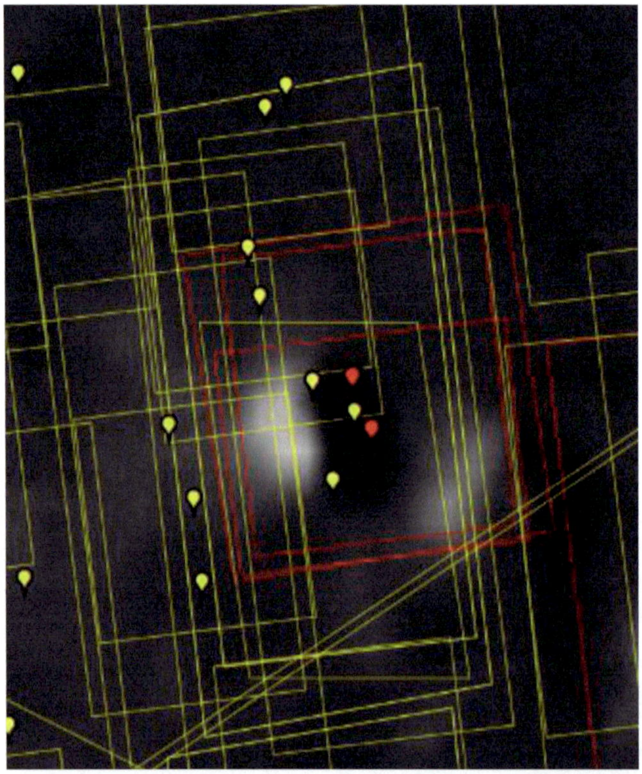

I feel like it is also a slap onto all our faces. Again, I believe this has been done on purpose to have NASA continue to mislead us all into thinking that there is actually nothing much or worthy to really see there. Their thinking is that by them making the impression, that if they ignore it, that we will also. **They are sadly mistaken.**

You can also notice that there are red place markers also being seen. These red markers represents Hi-Rise images taken of **'The Face'**, **'Popular Landform in Cydonia Region'**. Which now brings us all up to what I want to show next, which is new information which was released as this book was being written. We had received **two** new Hi-Rise images of the Face, **after 16 long years**, released to the public. Here is a closer look at the red placement markers indicating the images taken of '**The Face**'. Which brings us to now having actually three Hi-Rise images of The Face, instead of only one.

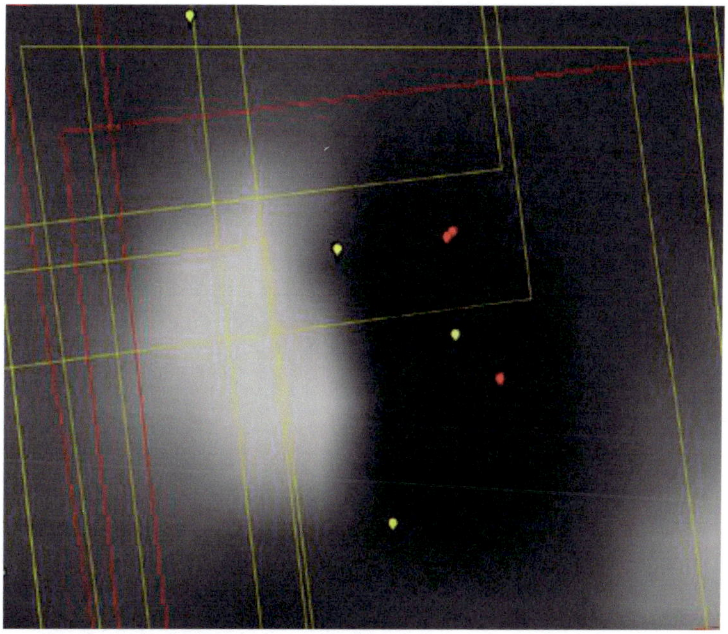

It was indeed a surprise, one that works to my advantage in fact, as while this book is being written, we were given 2 new images, two months in a row, of the Face in November of 2023, and also in December of 2023. Again, given to us all **16 long years** after the first Hi-Rise image was released to us in 2007. Despite how much to their chagrin, and also to ours, these two new Hi-Rise image releases could not have come at a better time, which was just in time for the thanksgiving and the Christmas holiday season. So to me, it was very symbolic, and I was certainly thankful for their releases, and I also felt like it was indeed a Christmas present, to **ME!**

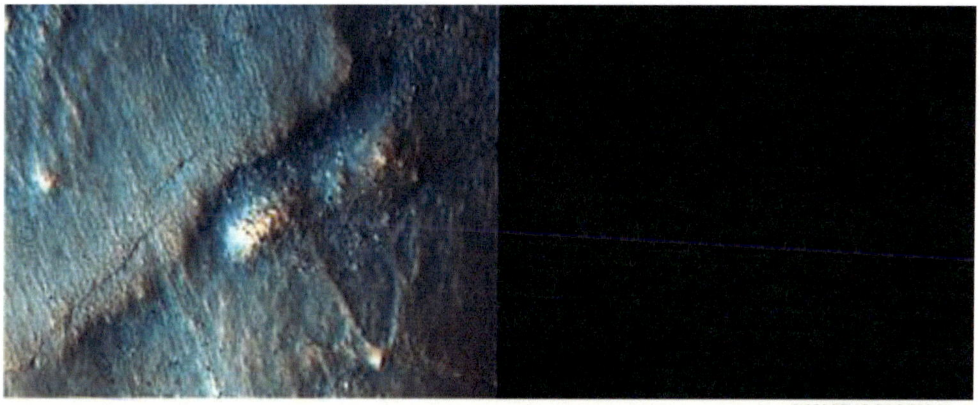

NASA/JPL-Caltech/UArizona

Terrain Sample in Cydonia Region

ESP_080535_2210 Science Theme

This is a stereo pair with ESP_080970_2210.

The first of these Hi-Rise images releases was ESP-080535_2210, image taken in October 2023 and released in November 2023. This time however, the **'Popular Landform in Cydonia'**, was now called **'TERRAIN SAMPLE'!**

(MRO HI-RISE IMAGE ESP_080535_2210 IMAGE TAKEN OCTOBER 1ST 2023)

IMAGE ARCHIVE PAGE LINK: https://www.uahirise.org/ESP_080535_2210

(MRO HI-RISE IMAGE ESP_080535_2210_RGB IMAGE TAKEN OCTOBER 1ST 2023)

IMAGE ARCHIVE PAGE LINK:

https://hirise-pds.lpl.arizona.edu/PDS/EXTRAS/RDR/ESP/ORB_080500_080599/ESP_080535_2210/ESP_080535_2210_MRGB.abrowse.jpg

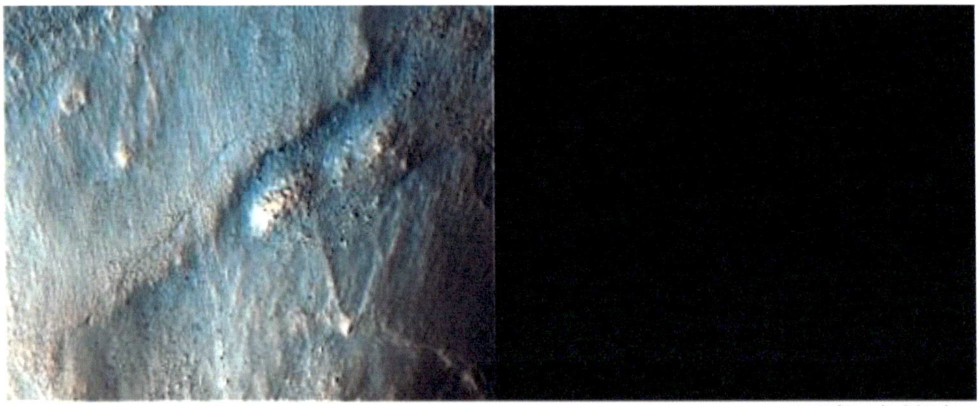

NASA/JPL-Caltech/UArizona

Terrain Sample in Cydonia Region

ESP_080970_2210 Science Theme Landscape Evolution

This is a stereo pair with ESP_080535_2210.

The second of these recent Hi-Rise images releases was ESP-080970_2210, image taken in November 2023 and released in December 2023. This time however, the **'Popular Landform in Cydonia'**, was also being now called **'TERRAIN SAMPLE'!** It is incredibly elusive and obfuscating and also very insulting to us all of how **'The Face'**, is demoted of its actual factual reality and manipulated by fear and hate and disdain to a **'Popular Landform in Cydonia'**, and now demoted even further to take away any attention to it at all, to be now termed **'Terrain Sample in Cydonia Region'**, by those who are supposed to be and are supposedly our so called leaders of science and deliverers of the truthful knowledge from discovery; NASA!

To me it is easy to be annoyed or angry at NASA about this deliberate misleading and deliberate attacks of facts and reality, because of my work and vested interest, and position in this research field. Yet when we consider what we have learned to see and understand in this book, we now know who and what is the true culprit responsible for these attacks against **'The Face'** and to us all as well, and which are the ones who control NASA and its budgeting, which is the government of THE UNITED STATES OF AMERICA! It is obviously the same who are also directing the deliberate avoidance over the many years of using the Hi-Rise camera aboard the Mars Reconnaissance Orbiter to image **'The Face'** at every opportunity. Especially even more so for it being such an admitted **'Popular Landform in Cydonia Region'**.

Its obvious true nature of never disappearing features only become stronger over time, and even over multiple resolutions of the images taken of that Popular Landform, as we have seen in this book that reinforces the existence of this amazing intelligent techno-signature staring up into the heavens. Therefore, it is easy to see and conceive that **'The Face'**, is obviously being imaged and studied at every opportunity by the other side of this mission, the **reconnaissance side** of this mission, its **military side** of this mission!

'THEY', the powers that be, want it obscured from the public in official capacity due to a controlled policy of government and military control over how the public sees and hears of this **'Popular Landform in Cydonia Region'**. The question is....WHY? The answer is obvious as well, because **'THEY'**, are seeing an obvious extraterrestrial design and artifact of intelligent creation and obvious techno-signature that demands their attentions at the moment more so than the public's attention and knowledge. This also tells me, that there is much more we are not being told and shown of '**The Face**' that famous '**Popular Landform in Cydonia**', now degraded down to it as just being a '**Terrain Sample in Cydonia Region**'. It is obvious there is a reason for the US government to keep this truth hidden.

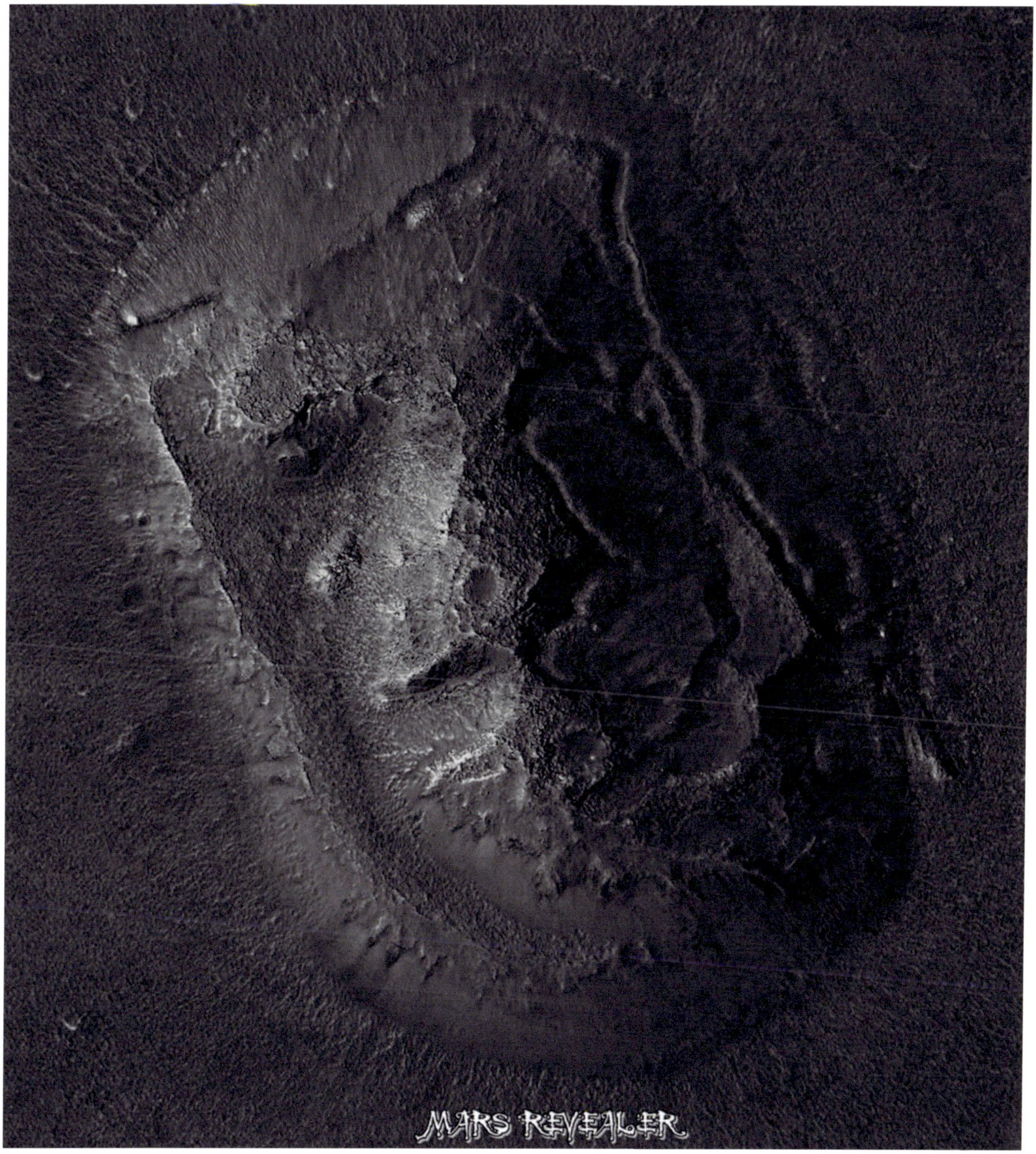

(MRO HI-RISE IMAGE ESP_080970_2210 IMAGE TAKEN NOVEMBER 4th 2023)

IMAGE ARCHIVE PAGE LINK: https://www.uahirise.org/ESP_080970_2210

(MRO HI-RISE IMAGE ESP_080970_2210_RGB IMAGE TAKEN NOVEMBER 4th 2023)

IMAGE ARCHIVE PAGE LINK:

https://hirise-pds.lpl.arizona.edu/PDS/EXTRAS/RDR/ESP/ORB_080900_080999/ESP_080970_2210/ESP_080970_2210_MRGB.abrowse.jpg

CONTEXT IMAGE FOR HI-RISE IMAGE ESP_080970_2210

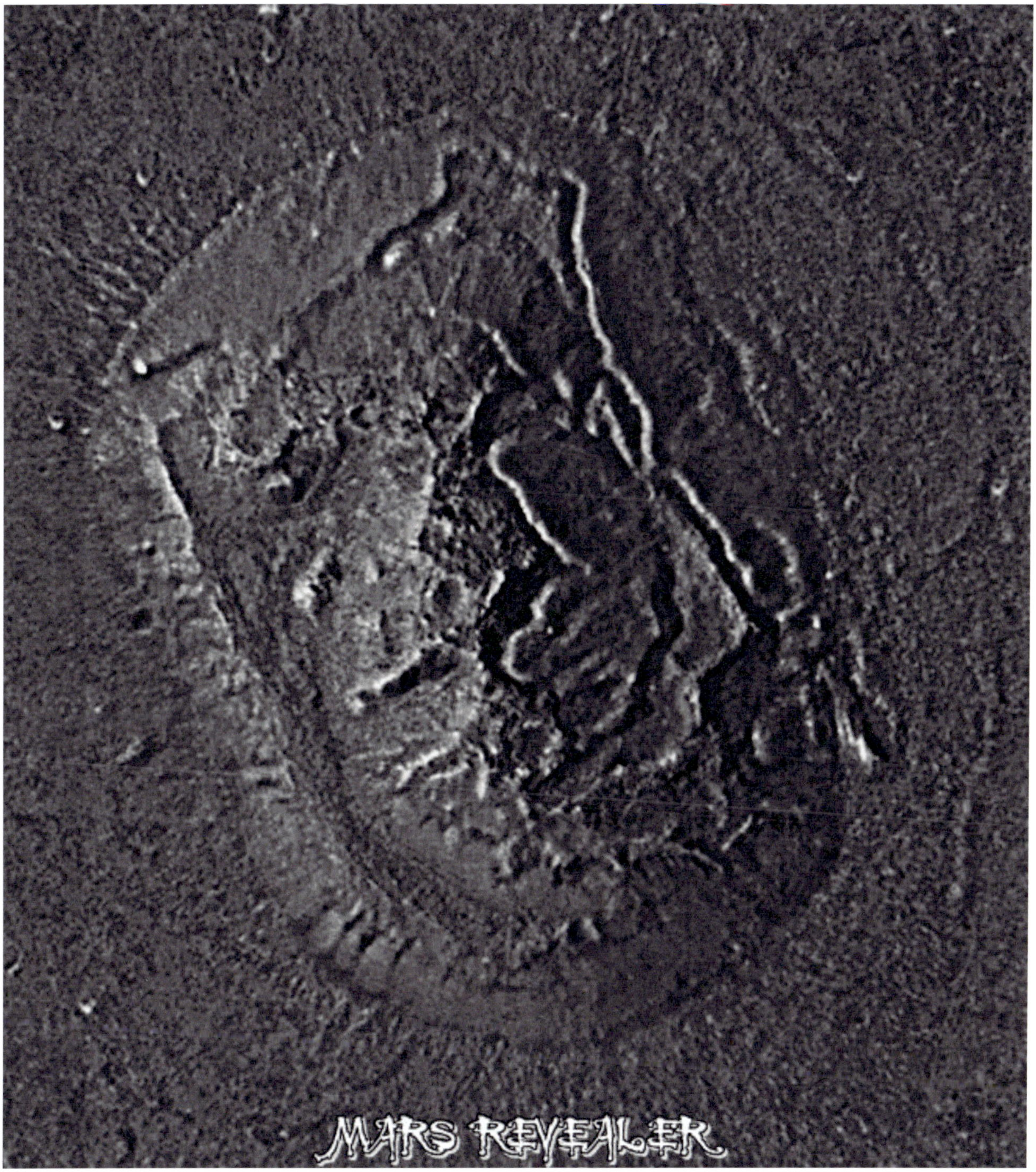

(MY PROCESSED RAW CTX IMAGE: V03_080970_2211 **IMAGE TAKEN: NOVEMBER 4TH, 2023)**

(MRO CTX IMAGE ARCHIVE LINK)

https://viewer.mars.asu.edu/viewer/ctx/V03_080970_2211_XN_41N009W#P=V03_080970_2211_XN_41N009W&T=2

One may notice that I have also included the Mars Reconnaissance Orbiter CTX (context) images for these Hi-Rise images: **PSP_003234_2210**, which is CTX image number **P06_003234_2212**, and also for Hi-Rise image **ESP_080970_2210,** which is **CTX** image number **V03_080970_2211**. However, that is only two CTX (context) images being referenced out of three Hi-Rise images ***ever given*** to us all from NASA/JPL. **Why?**

It seems once again, that NASA is obviously being obfuscating and is currently deliberately withholding CTX (context) image data, and preventing we the people from getting access or public release ***(as of now September 5th, 2024)*** to one of the two new Hi-Rise images of the Face released to us last year in 2023. The first image released to the public in November of 2023, which was **ESP_080535_2210**, and the second Hi Rise image was released to the public in December of 2023, which is **ESP_080970_2210**.

To understand why I am saying this about NASA again, is not just because I simply wish to point out controversy or conspiracy, but because of the bizarre actual facts connected to these very images themselves. It is NASA/ASU policy that when a Hi-Rise image is taken and released as special releases, I have read (someplace) that after the Hi-Rise image is publicly released, it will take six months until the corresponding CTX (context) image for that particular Hi-Rise image would then be publicly released.

So to describe this accurately to easily understand, we received after 16 long years the second Hi-Rise image for the Face ever taken and released publicly in November of 2023, (again after 16 long years), which was again, image number **ESP_080535_2210.** Now six months from that Hi-Rise image release, would bring us to May of 2024 for its corresponding CTX (context) image to be publicly released and then being available to view and download its image data according to the image public release policy.

There was **NOT** a CTX (context) image release for **ESP_080535_2210** when May of 2024 came! To me, as I have come to see and try to understand all the skullduggery of NASA's actions over the many years regards to their handling and mishandling images and statements for the Face, I was not really surprised! However, I felt anger, and hints of frustration as having to keep asking myself why this would keep happening, even with a CTX (context) image **where transparency is to be assured by the policy which is in place**, specifically stating that this is 'THE PEOPLE'S CAMERA', and therefore by default, it also is 'THE PEOPLE'S IMAGES'!

I literally would keep checking each and every weekday of May 2024, in the public NASA image archives on the internet to see if it would be added for the world to access it, so I can download it and process its raw image data myself. I wanted and I still want to view this historic correspondence image to the Hi-Rise image for the Face for reference, and for comparative analysis study of what each has to show. So for the entire month of May 2024, nothing ever came of it to the public image archive.

Now considering the second Hi-Rise image release of the Face given to us one month later, **ESP_080970_2210,** in December 2023, we have a different situation that not causes any stress or anger or ideas of skullduggery and lack of transparency. That is because in June of 2024, exactly six months later after the public release for that Hi-Rise image release, we **DID** receive that corresponding CTX (context) image, **V03_080970_2211**, at the beginning of June 2024. So I was very happy and excited to see that transparency of that image data was given and adhered to, according to their own policy.

Yet I still am very puzzled over this, because why did we receive this CTX (context) image **V03_080970_2211** exactly on time, and not any CTX (context) image at all for the other Hi-Rise image released on November 2023? The thoughts run across my mind constantly and I cannot think of a single logical or legitimate reason why we would not have received the image by now. We have a CTX (context) image **P06_003234_2212** for the Face for the first Hi-Rise image for the Face released in 2007, yet we have never received the all so important CTX (context image ***number unknown***), for the historic second Hi-Rise image of the Face **ESP_080535_2210** after 16 long years?!

I apologize for sounding and being redundant over mentioning about the important historic nature of the second Hi-Rise Face release of **ESP_080535_2210**. This is, in my mind a milestone for such an important subject and what this subject represents and considering all of the many years of the knowledge of the existence of this most extraordinary unique and obvious intelligent techno-signature ever discovered off of this Earth, which has been also the most extraordinary cover up and policy of plausible deniability and denial of science and deliberate public suppression and scientific persecution of a reality which is the most sought for by humanity of answering the ages old question known to humankind of at least human recorded history.

The question is far beyond the question of if there is life on Mars, to a much higher important standard of pondering, which is, **Are we alone in the universe**? It is a criminal travesty to keep this subject swept under the rug only to be walked over and not realizing what is hidden beneath of what is being trampled upon. Yet there are those who do know what is really there and want to keep this knowledge and reality hidden. We have went over various aspects of NASA's obfuscations and demeaning of the seriousness that this subject and this feature in Cydonia really is. Over the many years people have cried conspiracy theories about NASA's public handling of the Face at Cydonia. Many of which as we seen some of in this book, are sad but true. Just because claims of conspiracy theories exists, does not mean that there is not a conspiracy in place about this.

Yet once again, even in this modern day of transparency and public interest and participation involved supposedly with regards to the Mars Reconnaissance Orbiter, especially being equipped with the so called 'Peoples Camera' and the 'Peoples Images', it appears once again, there is a new conspiracy theory established regards to images being deliberately withheld from the people. It is sad that I myself, am making that accusation myself and am needing to include this new conspiracy theory here now in this book. I did not expect this situation, nor wanted it to be occurring but I must in all fairness include it here now, so you all can know about it, and possibly agree with me, that this is fitting of me including it here now. Even if you will not agree with me about this conspiracy theory, you should at least see my reasoning as to why I am and would be very concerned about it, to include it here now.

In attempts to try to understand and figure out why we do not have a CTX (context) image for Hi-Rise image **ESP_080535_2210,** I decided to take this serious matter into my own hands, since it is my job and expectation to do so, and people asking me to get answers about this, as their interest also in the missing CTX (context) image has them also wondering and also asking why we did not receive the image data, as it was being hoped for and expected to be released to us all. That never happened.

What I decided to do, was to go directly to the source who control the mission and image releases to seek answers, the jet Propulsion Laboratory via the media contacts. So on July 8th 2024, I called the media video and image correspondent named Rexana V. Vizza, phone number 818-393-1931, where I left a message for her briefly stating why I was calling and left details about what I was seeking, and provided my contact information. The next day on July 9th 2024, I receive a phone call from someone named Andrew Good, who turns out to be the Media Relations Specialist for Mars, phone number 818-393-2433. It became obvious that Rexana Vizza passed my message off to Andrew Good to reach out to me. He asked me briefly about what I was needing information about and told me to email him as much details as I can about what I was seeking. I thought it best to just put that email here for all to see. Email request sent July 10th 2024:

Hello Andrew. It was good speaking with you today. Sorry belated e-mail reply.

OK... You asked me to provide information for what I am needing. I am wanting to know the 'status' of the CTX image for Hi-Rise 1st stereo pair image for ESP_080535_2210, titled in the special releases as '**Terrain Sample in Cydonia region'.** This Hi-Rise image was released in November 2023. The CTX image for that has **NOT** yet been released and is far over the 6 month wait for it to be released. We are at 8 months now. We should have had the CTX image for ESP_080535_2210 in May 2024. **WHY HAS IT NOT YET BEEN RELEASED?**

I am reaching out for myself & for others who requested me as the host of THE MARTIAN REVELATION show, to help get those answers, as well as the CTX image immediately released to us all, since it is the People's Camera, and the people want their images.

In December 2023, we also had Hi-Rise image ESP_080970_2210 released to us. Also titled '**Terrain Sample in Cydonia region'.** We **DID** receive the CTX image for that in June 2024, at the 6 month norm wait. So it is very perplexing and confusing as to why the CTX image was not already released for ESP_080535_2210 ahead of the CTX image for Hi-Rise image ESP_080970_2210.

Here is the CTX (context) image archive link: https://viewer.mars.asu.edu/viewer/ctx/V03_080970_2211_XN_41N009W#P=V03_080970_2211_XN_41N009W&T=2

So we are now at 8 months, and still NO CTX IMAGE ARCHIVE RELEASE FOR ESP_080535_2210. **WHY**? And **WHEN WILL WE RECEIVE THAT CONTEXT IMAGE RELEASE?**

Please reply back to me with the important needed information, and please provide an answer as to WHEN we shall receive the CTX image for ESP_080535_2210.

Thank you Andrew for your call & time & help & care in resolving this issue.

Respectfully;

Gary Leggiere

Host of THE MARTIAN REVELATION show

In the following couple of days, I received no reply back from Andrew Good. I then e-mailed him again on July 12th 2024, and asked him if he received his requested e-mail of information I sent to him, and I asked him to please reply back acknowledging having received my e-mail that again, he requested me to send to him. No reply ever came. I waited for about another two weeks before I tried reaching out to him again, by calling Andrew Good (phone number 818-393-2433), in which I left a phone message for him asking him to please call me back and acknowledge my e-mail he requested me to send him and also my phone message. No reply e-mail or return phone call ever came. This is perplexing that I received no reply at all and I started to feel annoyed as I was starting to think I and the issue he was supposedly going to help me with, was being blown off.

I then decided to call Rexana V. Vizza once again, (phone number 818-393-1931), and I left a phone message again with her. I had let her know all that transpired with Andrew Good, and me not receiving any e-mail reply or return phone call about this from him. Her phone message recording states that she will return the call. No call ever came. I figured maybe she would call me back, or at least have Andrew Good reply back to my e-mail or call me. Since it was she, who I first called about this and she passed my message off to Andrew Good to begin with. No reply or phone call from either person ever came.

Now I was feeling certain I and this important issue was being blown off. I was thinking that they must have looked into it, and seen that the image numbers that I had provided to them for what I was seeking of the CTX (context) image to be released to the public, had to deal with showing the Face of Cydonia. Therefore, for some reason or other, I and the important valid issue I was bringing forth to both of their attentions, was being deliberately ignored by them both. **WHY?** They are the ones who are literally listed for any media and the public for whom to contact about such issues and inquiry of Mars image data, so I was following the correct protocol of inquiry.

After another week of waiting, now in the second week of August 2024, I decided to try to call Andrew Good once again, and leave yet another phone message, as I was feeling frustrated and a little angry of still not having received no e-mail reply or phone call reply. This should not be so difficult or take so much time. However, this time, Andrew Good picked up the phone. I feel I must tell you all the entire conversation of this bizarre phone call, because the details of this phone call show that something is incredibly wrong here and is lacking professionalism and lacking of transparency regards to public releases and availability of the Mars Reconnaissance Orbiter CTX (context) camera imagery.

I said: "Hello Andrew, this is Gary Leggiere, who sent you the information in an e-mail that you requested weeks ago for me to send to you, of which you told me on the phone when we first spoken that you will see what you can find out and get back to me soon. It has been over one month since we have spoken and I e-mailed you, and you still have not gotten back to me by e-mail acknowledging having received my e-mail of all the information you had requested of me, nor have you called me back from my messages I left to your phone. I find it a miracle you even picked up your phone now! "

Andrew Good: **"Hello Gary I am still trying to help you get the information and image data you requested".**

I said: "Ahhhh I get it, it's all those pesky tricks of light and shadow! I understand." (*My sarcasm shown my frustration in my voice*). "But before you go on sir, I just wish to say that growing up, my parents taught me at an early age that it is expected and understood that its polite to reply back to phone messages left on the phone by someone, and applies also to replying to e-mails in a timely manner, especially when asked to send information requested by someone, and told that a reply would be coming back soon. I just wish to let you know this, in case you may have forgotten that life lesson. No offence sir, but I left you phone messages as well as e-mails and you never are replying back, knowing full well that for over one month I was waiting for you to get back to me, as you claimed you would so called soon."

(With no reply addressing my frustrated comments and sarcasm, and not even apologizing for not getting back to me as he said he would, he instead asked me a few fast questions).

Andrew Good: **"Gary, what exactly are you wanting? What are you looking for? What is your interest in the image data that you are wanting?"**

I said: "Holy Shit Andrew?! Are you kidding me? Did you not receive the e-mail I sent to you immediately upon your request for me to do so, so you can help me by giving you all the information I can about what it was I was seeking, with as much details as possible, you requested me to provide even? Did you receive my e-mail sir?"

Andrew Good: "Yes I received your e-mail."

I said: "Okay great, nice to at least finally know that now from you. I was not sure if you received it at all. Okay that being the case sir, why then are you asking me now what exactly I am wanting and looking for? I provided you with everything that you requested and needed from me to easily be able to find things out, to reply back to me with in an appropriate manner and an appropriate time frame. And you still never replied to my e-mails or my phone calls. **WHY?** "

Andrew Good: "Let me ask you this way, what is it an image of? What is it that interests you regarding the image you are wanting?"

I said: "Again Andrew, are you kidding me or what? I am trying to understand what you are asking me. You have the information I provided to you, so you can easily know by looking at it all what is showing in the image. (In my e-mail he requested me to send, that I already shown, I did **NOT** mention that the Face at Cydonia was the subject feature of the image I was seeking).

I said: "The title for the image even tells you what it is, ***'Terrain Sample In Cydonia Region'***, so I am confused as to what exactly you are trying to ask me? I need the CTX (context) image for that Hi-Rise image of *ESP_080535_2210*. It not matters what my interest is regards to what is seen in the image to you Andrew! Why would you even ask me that? That not even makes any sense at all! That is **NOT** a requirement to get images or information about image data that should have already been put forth in the public image archives already, and you also never asked me to provide you with that information to be sent in the information needed in the e-mail you requested me to send you. So why are you asking me this now? That's a really weird question Andrew? Just answer me as I deserve answers as I been waiting over a month because of your lack of replying back to me as promised, ok? So where is the image? WHEN WILL IT BE RELEASED TO THE PUBLIC ARCHIVES?

"That was and is all I want and need to know! I am here on the phone with you now, calling you, you picking up the phone, and me needing to ask you of your progress in getting 'We The People' **OUR** image data that we are seeking and waiting to be released for us all. Why has not the image been released by now as we are now at 9 months since the Hi-Rise image *ESP_080535_2210 was released, and by policy, again as I stated to you in information requested in my e-mail to you, it is now 3 months overdue for release. So please Andrew, when will the image be released?"*

Andrew Good: ***"Well Gary there are lots of possible reasons why the image has not been released yet".***

I said: "OK, then what are those many possible reasons why Andrew? I really want to know those possible reasons, since that tells us all that you not obviously really know why? So please tell us, as we all would appreciate knowing it!"

Andrew Good: "Well for instance, there are literally thousands of images to be put forth and it is hard to get everything ....."

(I immediately cut him off from what he was saying and going to say).

I said: "What are you trying to say now Andrew? That there are so many CTX images there, that there is not enough time to release an image that was taken in October 2023, for corresponding Hi-Rise image ESP_080535_2210, which was released at the beginning of November 2023? The CTX images has a 6 month waiting period for public release according to the policy of transparency to the public. Are you telling me the CTX (context) image for Hi-Rise image ESP_080535_2210 is needing beyond that time of the policy of transparency, that even now at 9 months, we still not have that image released to us yet? Is that what you are telling me Andrew?

"But before you answer me sir, please know also that for some reason, in November 2023, another Hi-Rise image ESP_080970_2210, which is also showing of the same features seen in ESP_080535_2210, also being titled '***Terrain Sample In Cydonia Region***' , being one of two images for use of creating an anaglyph image, had been released to us all at the beginning of December 2023. And then exactly 6 months later, we DID receive the corresponding CTX (context) image **V03_080970_2211** at the beginning of June 2024. "So how was it not enough time to release the corresponding CTX (context) image released one month earlier in November of November 2023, to which we should have seen by being released and available to view and download in May of 2024? Yet, there is still no image release for it at 9 months? So no offence Andrew, but what you are trying to tell me sounds like you are trying to give me a **BULLSHIT** excuse!"(He not replies back to me answering about that logic, and instead goes on to make yet another bullshit excuse).

Andrew Good: "You need to understand how this works. First, the scientists have to look at it and determine what is being seen and....."
(I cut him off once again, as I seen what he was going to say. I could not believe what I was hearing).

I said: "Again Andrew, are you kidding me? Now you're telling me another **BULLSHIT** excuse? Now are you really trying and going to tell me that scientists, meaning more than one, are needed to look at the image first and determine what is seen? For what? For whom? Who cares anyway? That also does **NOT** apply to the policy of transparency of image data to be released to 'We the People'! There is no language of policy rules for using scientist's statements to be made unto the images to make any claim about what is seen in the image data what so ever before, or upon, or even after the image is released! So again, I am not understanding yet what possible excuse there is for the image to not yet be released to us all! To me it just sounds like you are only making excuses and not wanting to tell me when the image will be released and why it has not been released."

Andrew Good: "Why do you really want this image? What is it are you wanting to see? What is in the picture of that makes it so important to you?"

I said: "Why are you messing with me man? You obviously know by now what it is a picture of and what I am wanting to see in it. It just seems I feel that you are not wanting me to see it? Or anyone else! **Why?**

It also not matters what I want to see or why I want to see it! That is not a requirement for me and all else to have to answer in order to receive the image data being asked for. Let me remind you Andrew, this image comes from a mission where it is called 'THE PEOPLE'S CAMERA', and 'THE PEOPLE'S IMAGES'! We not have to answer questions as we paid for it! It is ours, which is why the policy of transparency is set in place, in what the process is in how and when we are to receive our image data! It is that simple really Andrew! Why are we even having this conversation?"

Andrew Good: "I just want to know what you want to see in the image, and why you are so interested in it."
(I was furious at this point and not held back my frustration, nor was I so polite anymore).

I said: "You know damned well what is in the fucking image of what I am wanting to see Andrew. You know damned well it is of the Face of Cydonia that was changed its famous name to 'Popular Landform in the Cydonia Region', and then renamed again to just 'Terrain Sample in Cydonia Region'! I am not going to address and discuss with you what my research findings are, or share my thoughts about it, as I not have to answer to you or anyone else about that. Who cares what I want to see in this image? It can be a big hole in the ground you guys love to take images of over and over, for all you care, it has no bearing at all of why we are here now speaking about releasing the CTX (context) image and when we can expect it to be available on the public archives web site! So can you please tell me when that image be available for public download please?

Andrew Good: "I am trying to help you, but you will not tell me anything about why you are so interested in it."

I said: "If you were trying to help me sir, you would have replied to my e-mail you requested I send you with all of the details that you asked me for and needed, in order to find out the information to help me get an answer to begin with. The information I given you, I am sure it was real easy for you to find out what it is being seen in the image. If you were really trying to help me, you would have replied long ago to my e-mail and my phone messages. Let me just tell you now Andrew, that our call is being recorded, and I intend to play our conversation here on my show THE MARTIAN REVELATION!"

Andrew Good: "You are recording our conversation and intending to play it on your show?"

(His tone of voice sounded nervous and seemed to not like what I had said to him.)

I said: "Yes Andrew I am going to play this recording on my show, so I can show all my listeners to hear my efforts on their behalf to try to find out why we have not received the image and find out from you of when it will finally be released! We have the right to have access to that image data as it is ours and NOT YOURS, and I nor anyone else has to explain to you or anyone else as anything at all about it, as it is all of our images to be provided to us all without reasons or questions given! I intend to show everyone that your words and actions about this is obfuscating and avoiding helping us all, and failing to provide any answers and most importantly failing to provide us all with the important image data that we are supposed to be here talking about! You have given me multiple bullshit excuses and asking me questions instead beyond what you originally asked me for to provide to you in order for you to help me, and now you are avoiding answering me of these simple questions I have. Even if you not wanted to answer anything about why it was not released then fine, but you are not answering the main question that you need to answer which is the when, the date that the image will be released to us all?! Instead of appearing as if you are really trying to help, you are appearing and sounding as acting complicit in suppressing this image data and also appearing culpable in obfuscating that image data to prevent us from being able to see and have access to that image data! You obviously know the image I am seeking, and you obviously know what I want to see in it, as you are acknowledging the image itself now for all to hear!"

(It was at this point that Andrew Good, had hung up the phone.)

I then called my friend Francis Knize asking him about what I could do to force an answer from Andrew, or force the image to be released to us all. Francis Knize was someone I can come to ask about this issue because in 2001, he had made a case to the Data Quality Act on behalf of Richard C. Hoagland regarding a Mars Odyssey image controversy at that time. I told Francis I believe I have another image issue for the Data Quality Act! Long story short, Francis asked me for Andrew Good's phone number and e-mail address. I provided both to him. He told me he will reply back to me once he gets a reply or speaks to Andrew.

One hour later, Francis called me back, and told me that they had spoken and that Andrew told him that the CTX (context) image being sought and questioned about would be released in one month. Which would bring us to September 2024. No image was ever released as promised it would be, as said to Francis by Andrew Good. Around September 8th 2024, I decided to call and leave phone messages again to Andrew Good, and Rexana Vizza, asking them why no one was replying to me by phone or e-mail about when the image will be released and asking to please return my call. No reply came.

Instead on September 10th 2024, I received an e-mail from **beautifulmars@uahirise.org.** I was surprised, as it was also my birthday, yet I was to discover it would not be a birthday present being given to me. The e-mail states as follows:

HiRISE Observation: ESP_080535_2210 was acquired on October 1, 2023 and released to the PDS on October 31, 2023. CTX did not coordinate on this observation, and there is no corresponding CTX observation for this HiRISE observation. CTX did make one observation earlier on the same orbit as the HiRISE observation, V02_080535_1908_XN_10N007W, but this image does not overlap with the HiRISE image at all. You can find the CTX data volume on the PDS imaging node website and verify that CTX did not acquire a corresponding image on the corresponding orbit.

HiRISE Observation: ESP_080970_2210 was acquired on November 4, 2023 and released to the PDS on November 30, 2023. CTX did coordinate with this observation, and the corresponding CTX ID is V03_080970_2211_XN_41N009W. The CTX observation was released to the PDS on June 1st of this year and the raw IMG format file can be downloaded from the PDS imaging node website.

This is all the data that HiRISE and CTX acquired at this location for the dates and times that you are asking about. No other imagery was acquired.

HiRISE Media

(I replied back to the email the next morning September 11th 2024, stating as follows):

WOWWWWW! So you're telling me the MRO, NOT ACQUIRED A CONTEXT TARGETING IMAGE FOR THE HIRISE IMAGE ESP_080535_2210 AT ALL? THAT NOT MAKES ANY SENSE AT ALL!

After 16 fucking long years of only having ONE Hi-Rise image for the FACE taken and given to the public in 2007, there is no context image for acquiring another Hi-Rise image, the 2nd for THE 'FACE' of Cydonia, especially after so long a time?! We were all under the impression the context imaging camera is utilized in targeting Hi-Rise imagery? Why not for this opportunity?

Why there wasn't any information about this at all provided to me in all this time by Andrew Good, or anything provided anyplace online about this, especially when I was trying to deal with this issue for 2 months with Andrew Good? Why didn't he just immediately tell me about this?

Something is wrong here. Sorry and no offense, but I am not buying that reasoning, and the e-mail excuse you end me lets you off the hook for the trouble caused to the public in getting any answers? I have Andrew Good on recording now Mind you, making MULTIPLE B.S. EXCUSES for why we may not have received it publicly yet. NONE OF THOSE REASONIGS WERE BELIEVABLE OR ACCEPTABLE! He even stated to my associate on the phone over one month ago (in August 2024), that in one month from then (in September 2024, times up now), the image would be released to the public!

So if no image was acquired by the CTX imager for the Hi-Rise image ESP_080535_2210, why did Andrew say it will be released to us all in one month? (Again times up now). That also is an admittance by Andrew Good that such an image exists! Was he lying? Or are you? No offence, but this question must be asked!

I think and feel I am going to have to sue for further information and transparency about this now.

What is your name, so I can attribute validity in who you are on behalf of Hi-Rise media, with this unbelievable excuse, and not providing me with your official name? All of this situation is being placed as we speak in a book I am currently writing now, and once again, it looks like an obfuscation tactic to avoid providing image data and transparency, utilizing deception, on part by Andrew Good, whose name will be placed in the book, and also the recording to be aired on my show THE MARTIAN REVELATION! As I said to Mr. Good... He looks culpable, and not helping. So someone.... Is lying to us all here. Which is it?

(No reply was ever received as my email had been blocked apparently and my e-mail reply was returned to me. I tried 3 different times to resend it but kept getting the same result.)

Mail Delivery Subsystem <mailer-daemon@googlemail.com> Wed, Sep 11, 9:41 AM
to me

Message blocked

Your message to **beautifulmars@uahirise.org** has been blocked.
See technical details below for more information.

The response from the remote server was:

```
550 5.7.1 Access denied
```

So as you can all see, we have yet another conspiracy theory to add to the already long list of conspiracy theories over the four and a half decades long conspiracy set in place as official policy to mislead, lie, obfuscate and hide the facts and truth of the Face of Cydonia from the American people, who paid for the truth, not lies! We may never get to see the CTX (context) image, as we are obviously now told that it not exists.

We heard this reasoning also given before regards to the Face! I did not really want to place this information here in this book taking up pages, but it is a rather important note for historical and research purposes, to show also that once again, NASA's policy and U.S. government control of the public transparency for the Face at Cydonia, is Beyond Mere Tricks of Light and Shadow!

CHAPTER 11:

THE TRUE SHAPE OF THE FACE OF CYDONIA: NOT MERE TRICKS OF LIGHT AND SHADOW!

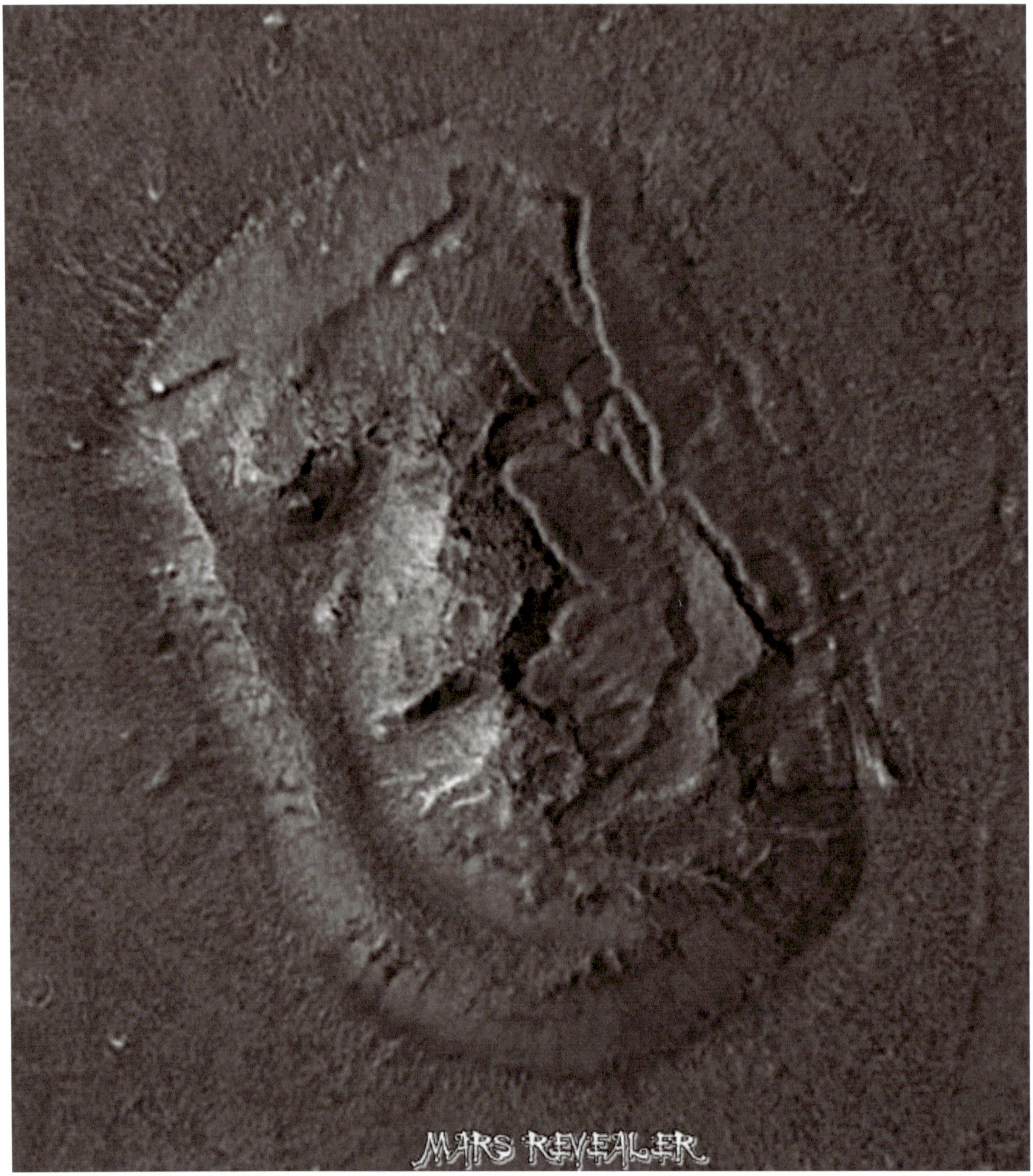

(MY PROCESSED RAW CTX IMAGE: D01_027417_2195 **IMAGE TAKEN: JUNE 1st, 2012)**

Even after seeing all of these images presented within this book, we are left with imagining to ourselves of what the Face of Cydonia actually looks like beyond a two-dimensional view. Many image products showing the Face, at varying resolutions, only help us to conceive a speculation in our mind of what the true shaping of the Face may look like. It is hard to get a good idea or conception of its shaping reality as would be seen in a 3-D image data view, from only 2-D image data views.

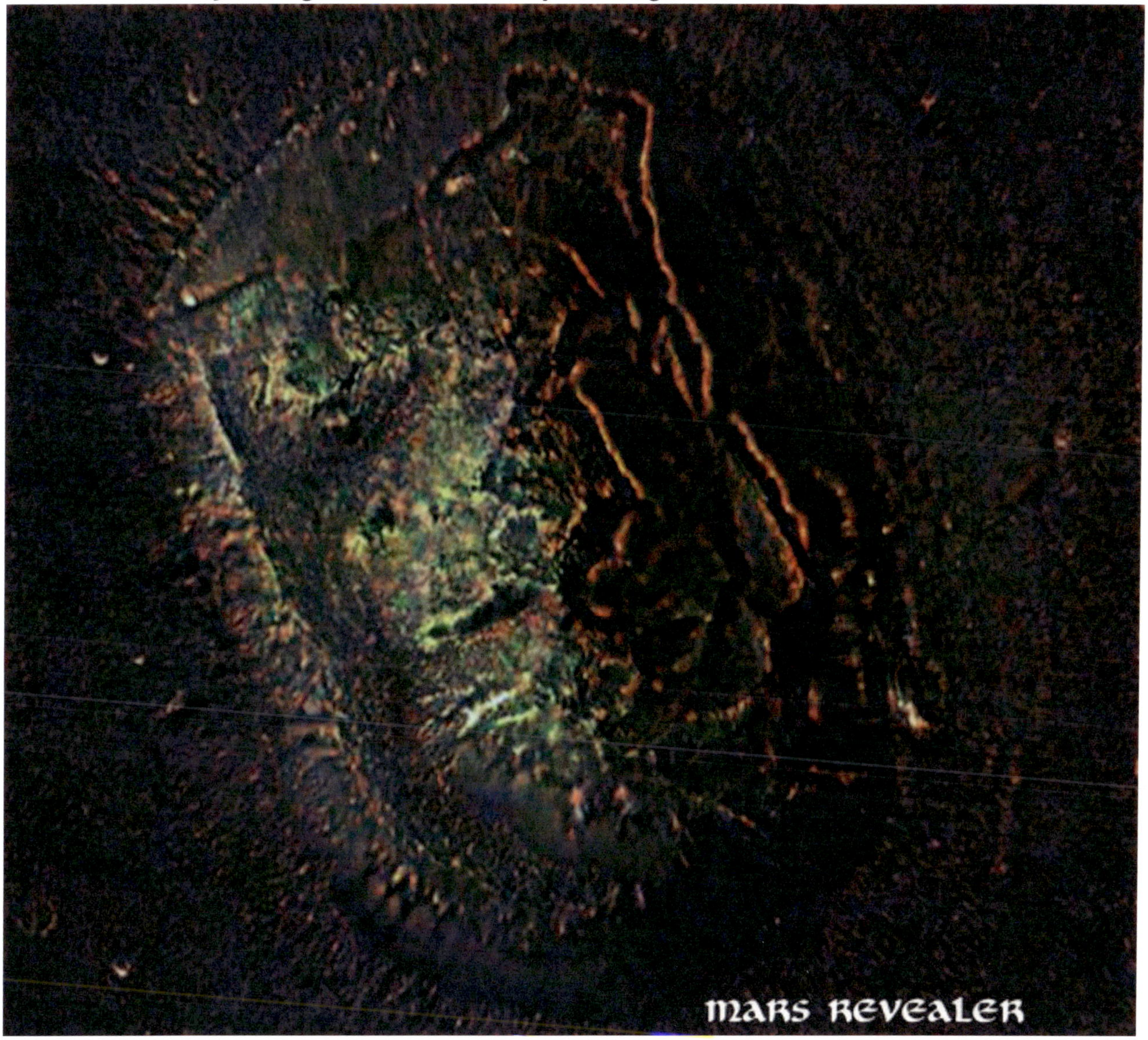

I would like to stress that for starters, the Face and surrounding area features are **NOT** flat and smooth, but have heights and also surface relief that also have highs and lows and dips and spaces as well as various surface relief strata directions etc. NASA wants us all to think and believe that what they put forth to the public as official scientific fact that the Face is smooth, and smoothly eroded on places of it, with some highs and lows and looking **almost** natural.

This is a continuation from Chapter 5: CAN THIS BE TRUE? We saw the so called 3-D model view they portray as the true shape of the Face, based on the **unreleased** color version of **E03-00824.**

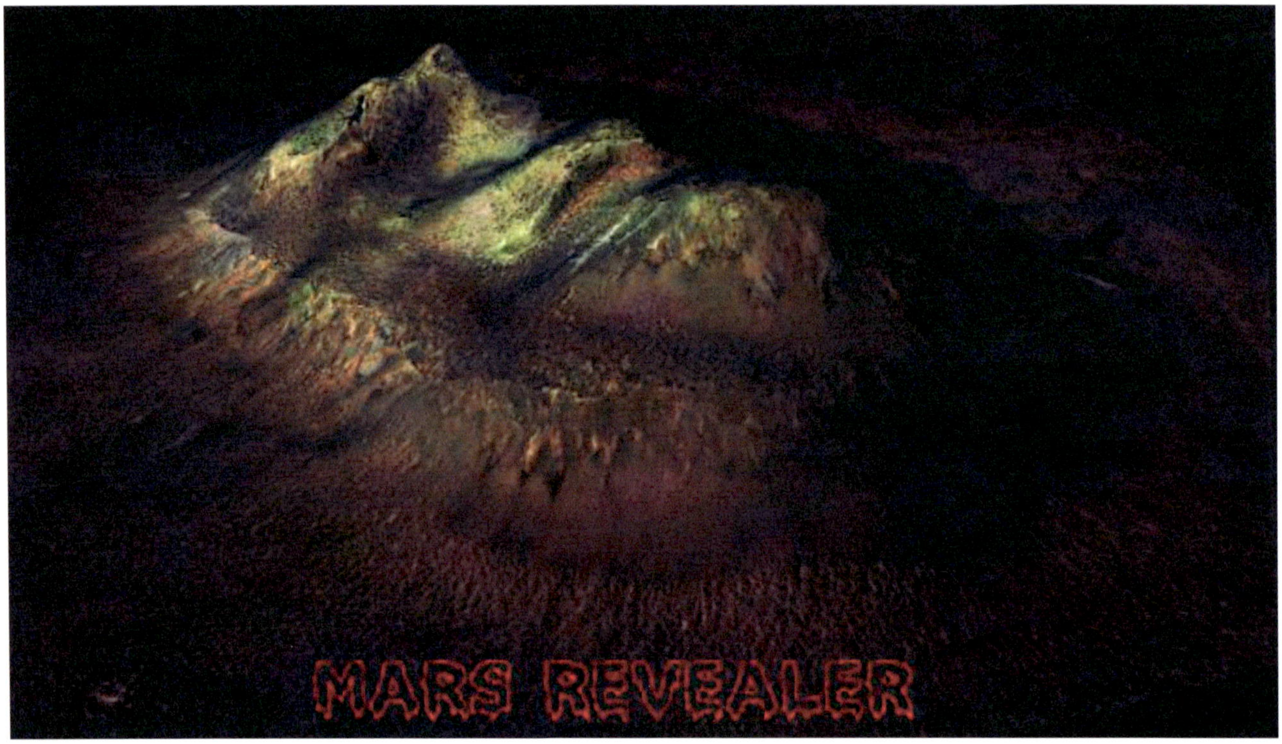

In my opinion, it lacks accuracy, and it not matches up very well to its artistic designing details and most importantly, its surface relief details as I see it in the 2-D imagery. Not to what I believe the Face is truly really showing and must be shaped as. There is no way to match them up, let alone be able to get an accurate or even a realistic 3-D image view from them based upon the useless unrealistic and false manipulated data of MOLA! It is not realistic nor scientifically accurate, but what we need is actual evidence of proof that the 3-D model provided to us all by NASA official narrative as a means to publicly, falsely, and deceptively debunk the Face, is indeed in reality a fake!

We do have additional image data provided by NASA for this 2007 final official hit piece showing a video animation of which the color image of the Face from MSSS, on behalf of NASA, had provided to us is seen circling around and around allowing us to see perspective views of the shaping of the Face.

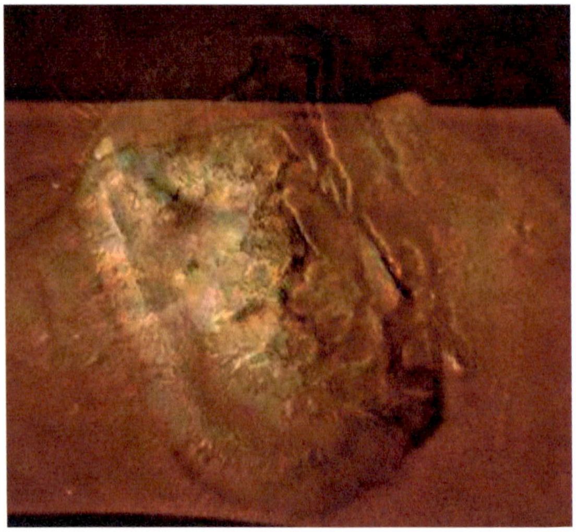

The video of the Face spinning around in movie form, would also have been a nice touch if Mars Express perspective views given unto us all, could also have been done in a video like what NASA provided. The video is showing lower resolution and less quality compared to the beginning of the video, where we see the full on color view of the Face we just studied into from NASA. After the full on Face view is seen, the video starts its circling perspective view by showing the Face at a side angle. But the point here I believe is that NASA wanted to be the big dog and limit ESA's thunder as it were, despite the obvious discrepancies of the shaping of the Face seen in comparison, by showing them up with an animated perspective view. Turning around and around allowing us to see what NASA is trying hard to make us believe that their view of the Face, is the truth.

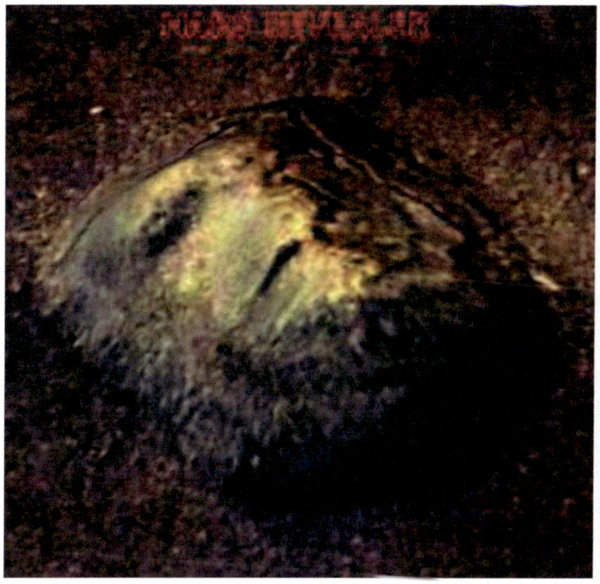

Even though ESA did not provide us with animation data showing the perspective views for the Face as they had presented of their Mars Express image data, we see clearly enough what the truth of the facts of which the image data shows. It is NOT looking the same, and NASA's video 'SPIN JOB' video does NOT take away validity into the ESA Mars Express image data facts presented and compared with the NASA inconsistent views of the Face. In fact again, it leaves us asking a question of whose views are correct. Or are they both incorrect?

Also, I would like us make a side note here, to compare this side image view of the Face from NASA video, to Dr. Mark Carlotto's perspective view showing the Face at a side angle from Viking Orbiter 35A72. Even here shows inconsistencies of the shape of the Face, between NASA's MOLA data, and Dr. Carlotto's **shape from shading** data samples, especially seen of this side perspective view, and seen at the top of the Face. It just not matches!

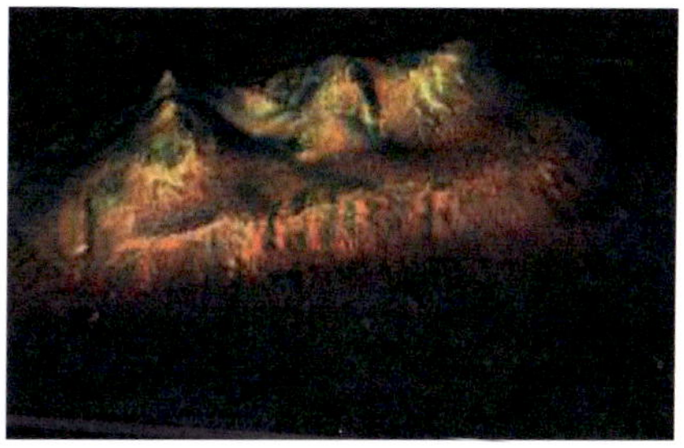

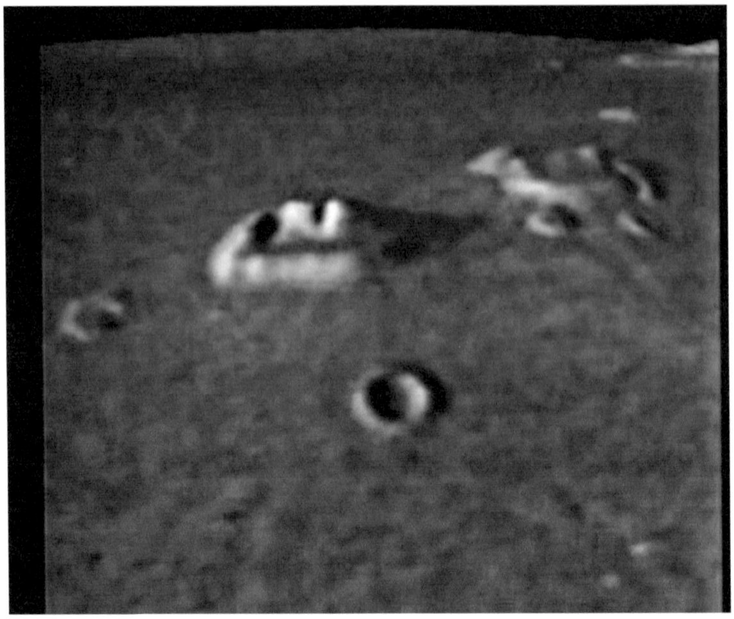

WHY? Because of NASA's manipulations of the MOLA image data to try to skew the reality making a faux 3-D model and misinform the public in shaping their perceptions of the Face's true shaping and design. That is my opinion based upon what the image data shows to consider this! The shaping of public perception based upon of a false view and image obfuscations for the famous Face of Cydonia. Why? To continue to hide the truth and its reality of a really incredible and intelligent techno-signature! SETI (Search for Extraterrestrial Intelligence) is achieved and ignored! Again, in the eyes of the public.

We will look at some views from NASA's 'Spin Job' video from 2007. The video animation plays for about one minute in length. Here are some views taken from along its spin path showing what they want us to see and believe that the Face is really shaped like. If anything, I look at this video as more of a comedy form of entertainment, rather than a true scientific or even a serious perspective view due to the lack of NASA's photographic candor.

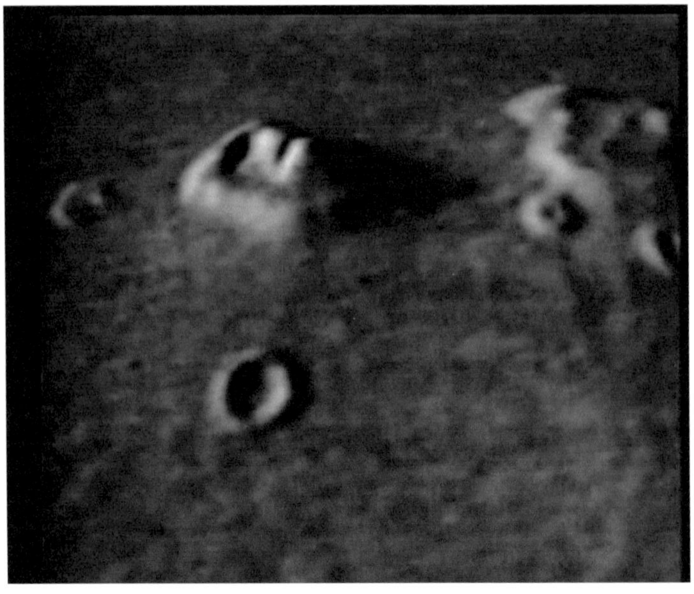

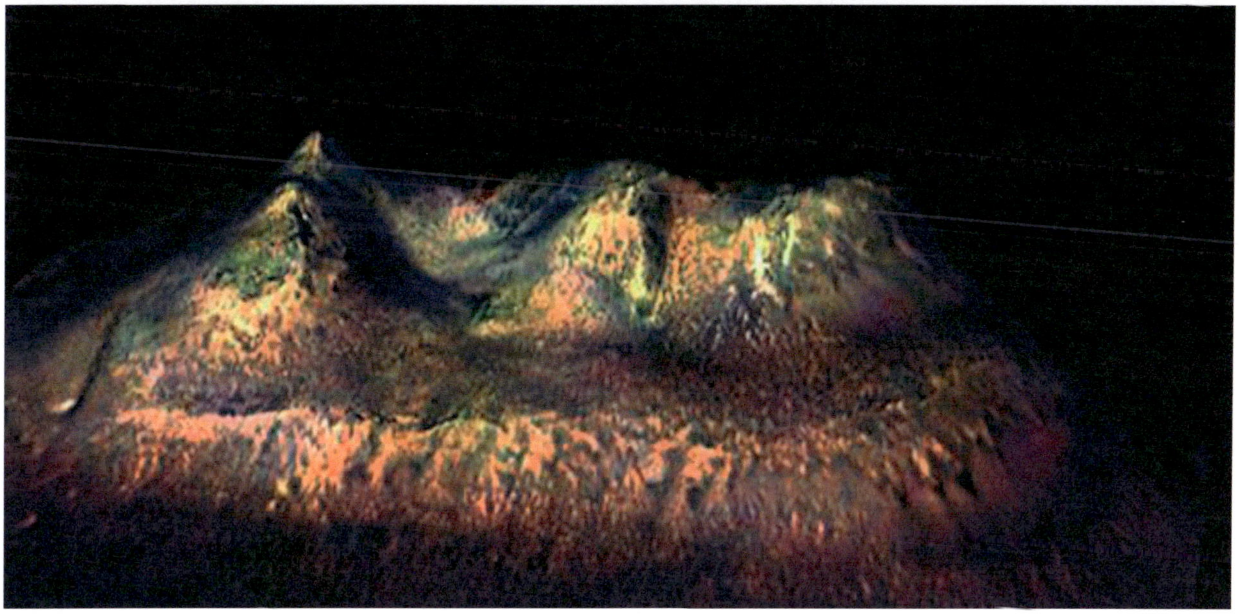

It is important to reflect upon what NASA presents as so called truth of the Face, and going out of their way to literally stop public inquiry and demand of further and closer study. Ignoring the Face's surface relief details, they try to make a smooth natural like feature, and also trying to make it look like an eroding dust heap of rocks, by changing its true shaping to try to make it look as un-face-like as possible. Making it look as ridiculous as can only really be discerned when seen and studied from image data that happens to reveal its truth right in front of your eyes. In comparison, the truth is revealed! As usual, NASA remains silent about it by ignoring that reality.

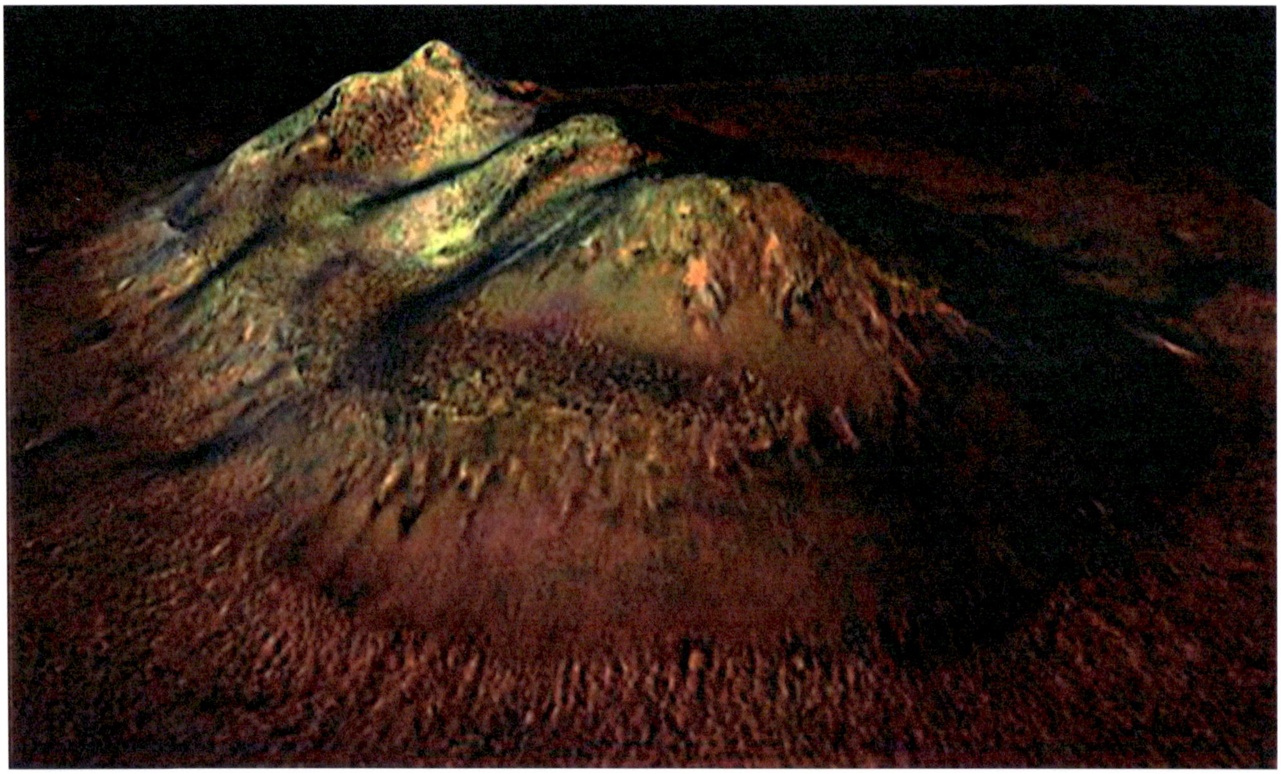

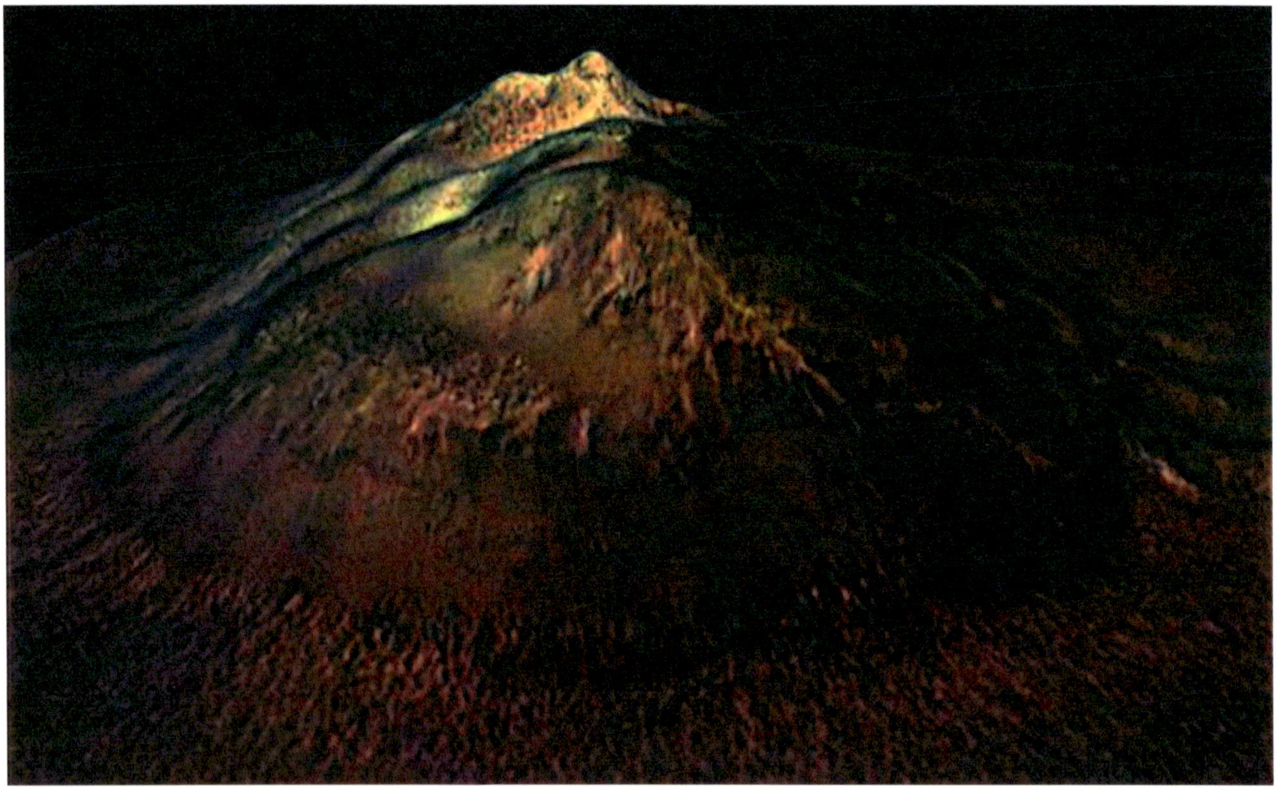

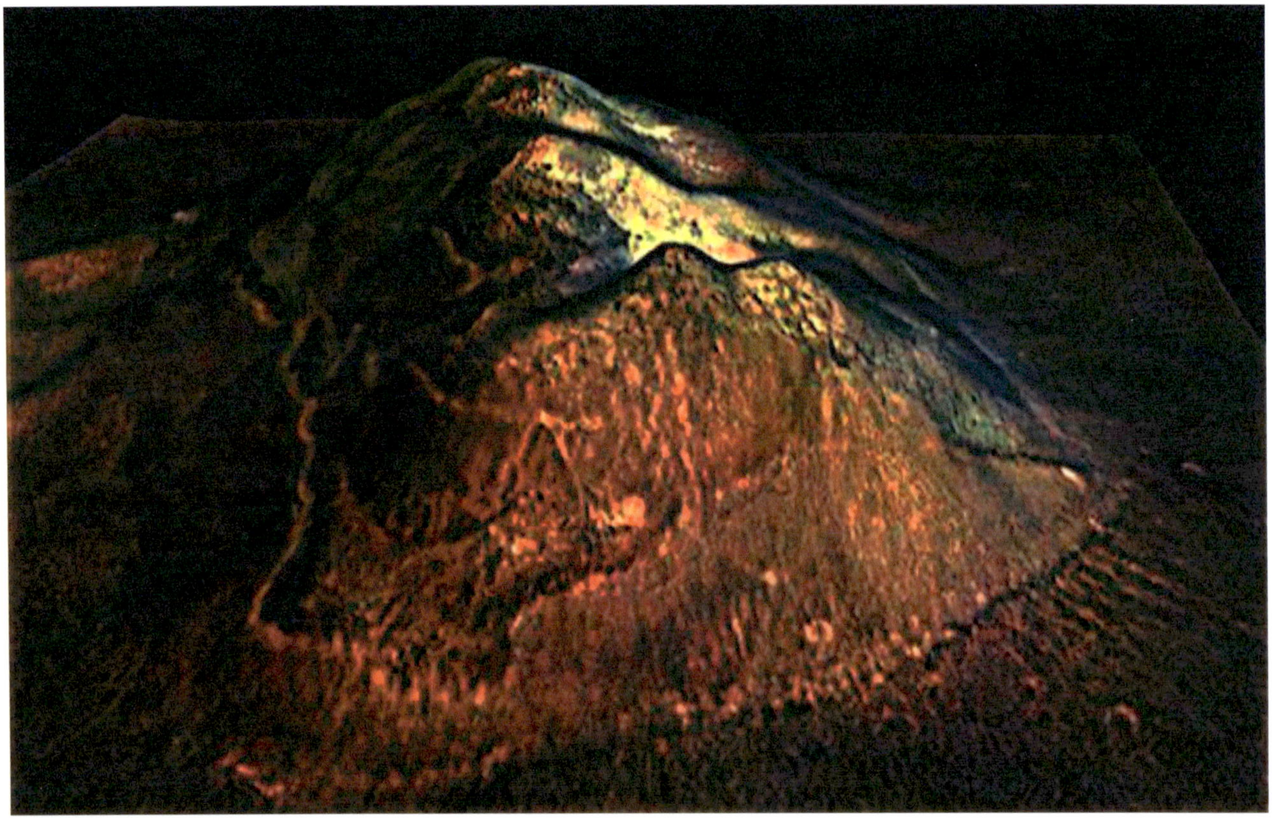

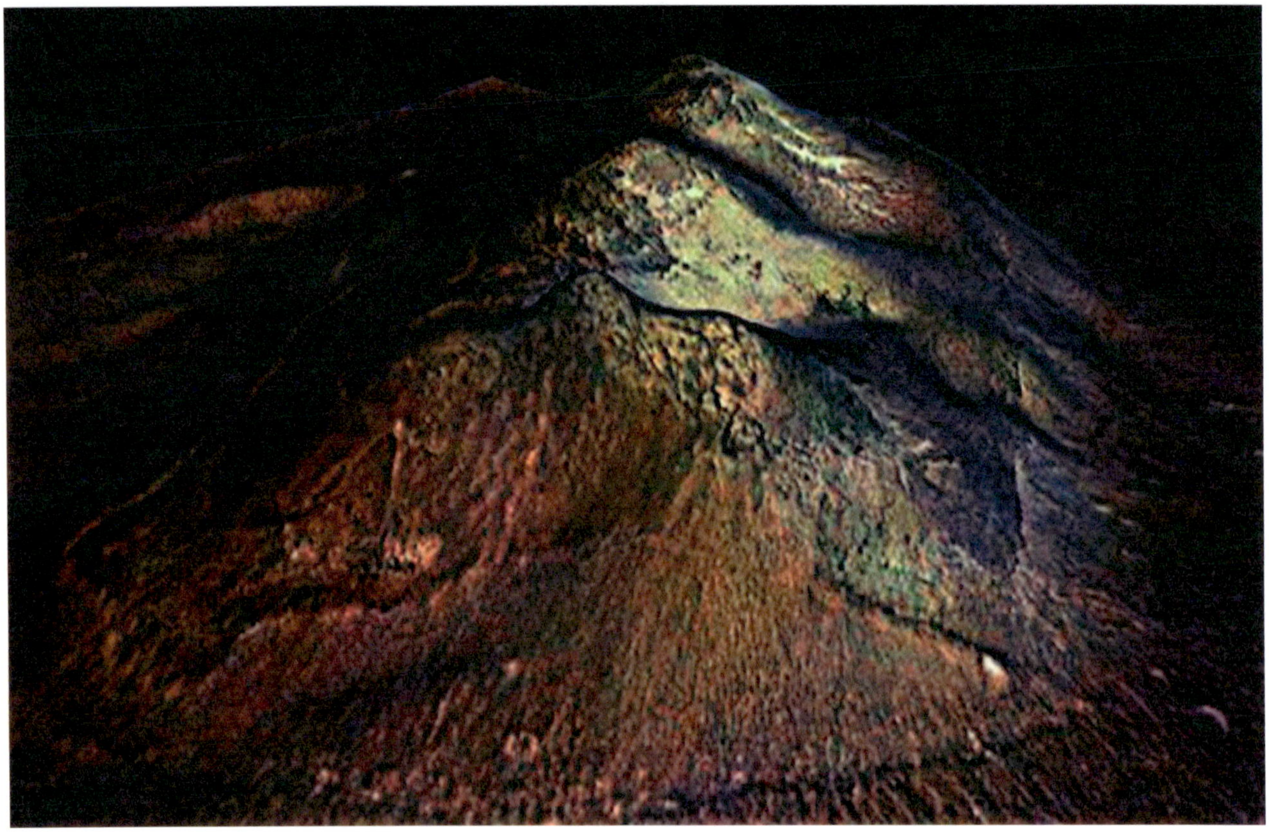

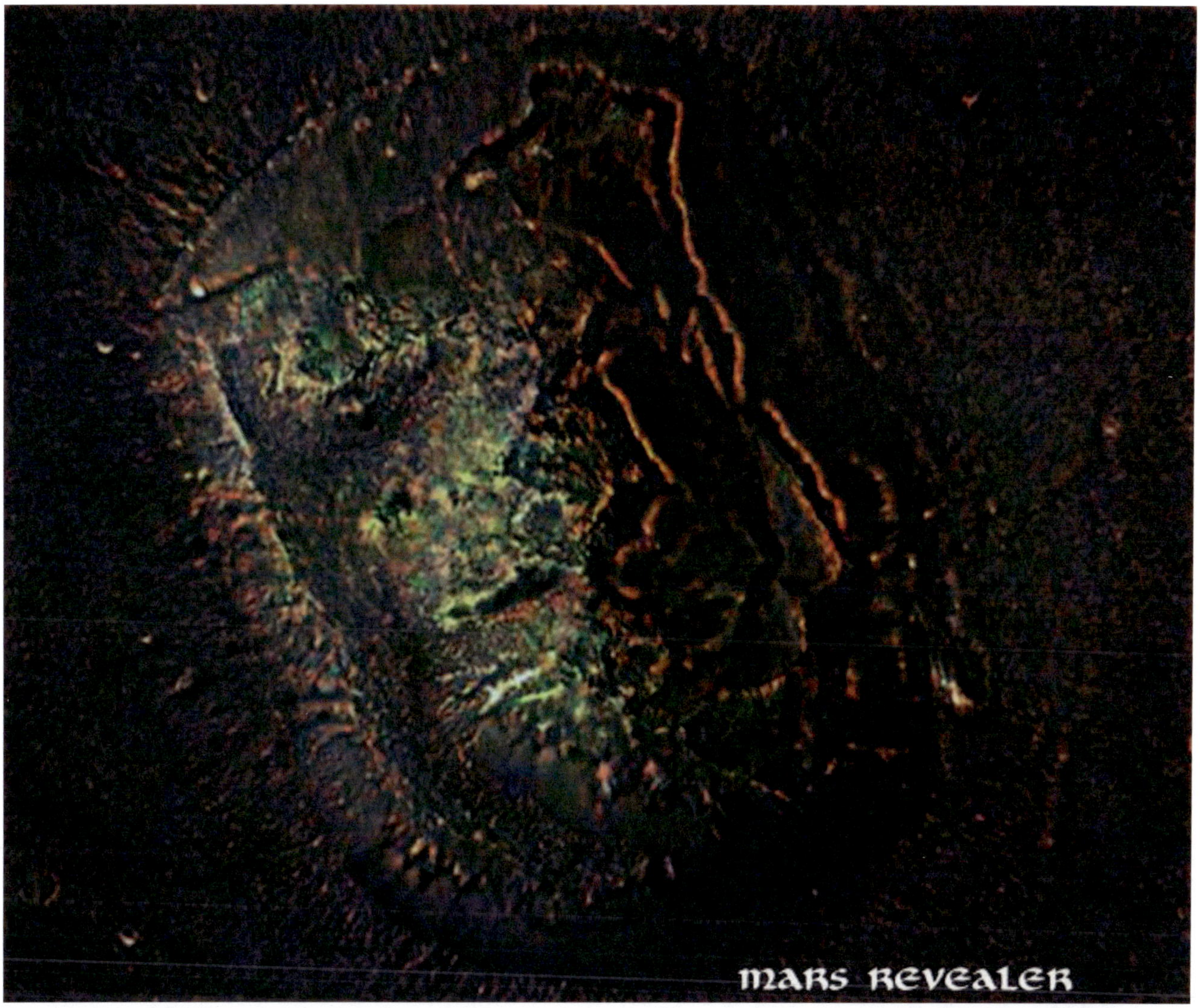

This is where a different yet more scientifically and realistic accurate procedure is needed to be done to put NASA' official model to the test, and all we need is just one image that has the right sun lighting angle and position which is needed to allow us to render an accurate 3-D model view from a two-dimensional image. That more accurate procedure is called shape from shading! We will look at this needed accurate procedure soon. But first, there are also a couple other ways to attempt to get a 3-D view of the Face in its limited form, by viewing it from above only, with no surrounding structured 3-D model.

The first of these couple methods is anaglyphs! **Anaglyph 3D** is the stereoscopic 3D effect achieved by means of encoding each eye's image using filters of different (usually chromatically opposite) colors, typically red and cyan. Anaglyph 3D images contain two differently filtered colored images, one for each eye. When viewed through the "color-coded" "anaglyph glasses", each of the two images reaches the eye it's intended for, revealing an integrated stereoscopic image The visual cortex of the brain fuses this into the perception of a three-dimensional scene or composition.

Anaglyph from Stereo Pairs. A stereo pair is a pair of images from slightly different perspectives at the same time. Objects closer to the camera(s) have greater differences in appearance and position within the image frames than objects further from the camera. Historically cameras captured two color filtered images from the perspective of the left and right eyes which were projected or printed together as a single image, one side through a red filter and the other side through a contrasting color such as blue or green or mixed cyan.

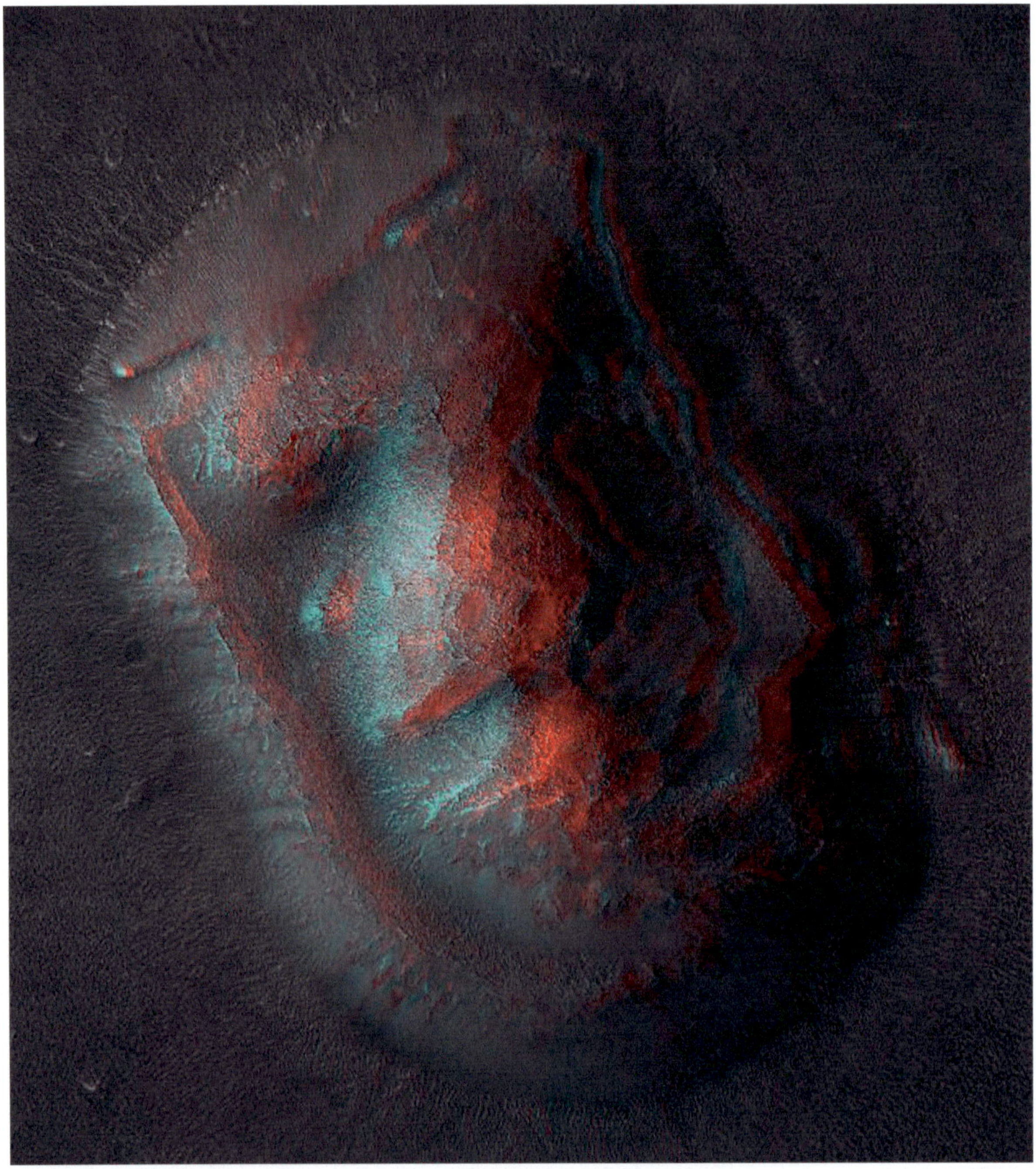

(HI-RISE IMAGES ESP_080535_2210 and ESP_080970_2210 ANALGLYPH STEREO PAIR BY NASA)

USE RED AND BLUE GLASSES TO SEE THE 3-D LIKE VIEW

https://www.uahirise.org/anaglyph/ESP_080535_2210_ESP_080970_2210_RED

Even with the Hi-Rise 3-D anaglyph, it still leaves much to our imaginations to provide an accurate opinion of its true and accurate shaping. In my opinion, this helps, and can be tested for comparison if and when we get a 3-D surrounding perspective views to better understand what we are seeing in the stereo pair anaglyph of only an above point of view of the Face. We need a data model.

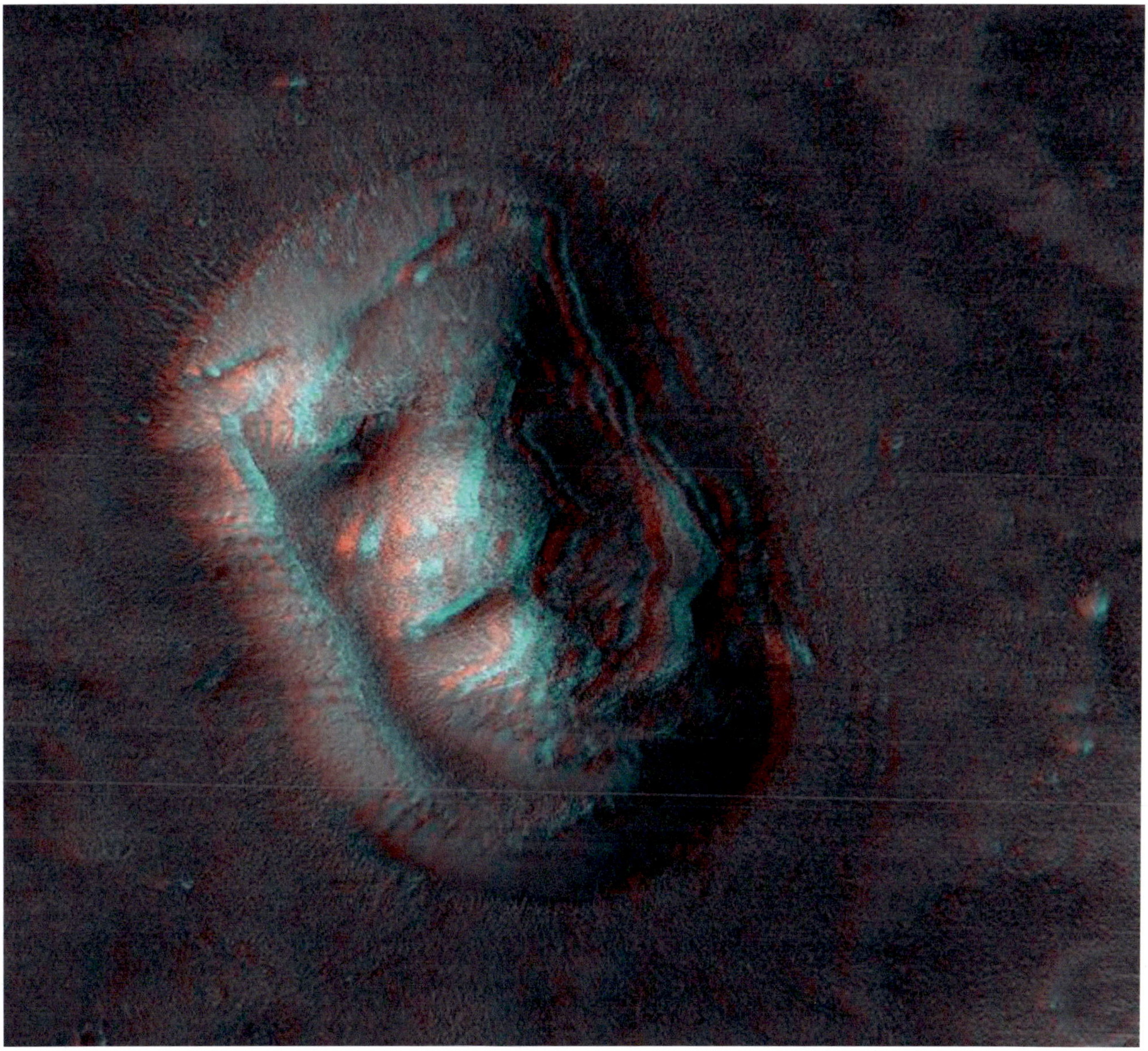

(HI-RISE IMAGES ESP_080535_2210 and ESP_080970_2210 ANALGLYPH STEREO PAIR BY DAVID GANNETT)

My friend **David Gannett of Mars X3-D,** had provided us with two anaglyph views also of **ESP_080535_2210 and ESP_080970_2210,** and **ESP_080970_2210 and ESP_080535_2210.** David, is a lifelong professional 3D photographer. David is A 3-D pioneer in this realm, he presents his discoveries in stereoscopic X3D on his YouTube channel, **"MarsX3D - "THE EVIDENCE".**

https://www.youtube.com/@MarsX3DTHEEVIDENCE/videos

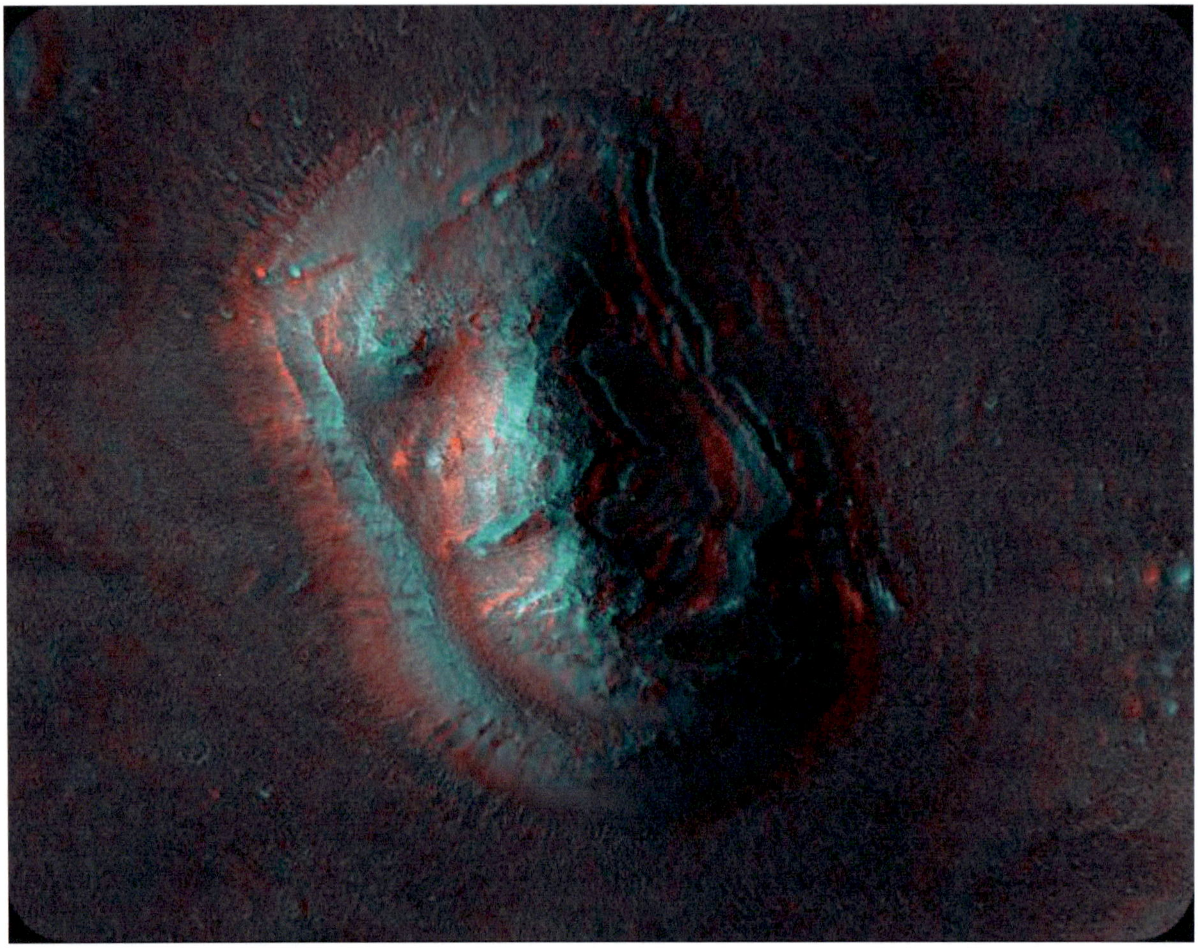

(HI-RISE IMAGES ESP_080970_2210 and ESP_080535_2210 ANALGLYPH STEREO PAIR BY DAVID GANNETT)

In and after viewing David's Mars X3D YouTube channel, you will see that David has pioneered a way to observe 3-D imagery without the use of any special 3-D glasses, and uses the natural use and easy **safe** exercising of the human eyes to see spectacular imagery that comes to 3-D life. **By using for a guide to see the 3-D images, you see a white dot seen above the two images to create the third image you will see by slightly crossing your eyes, that renders the third 3-D image you will see in the middle. This middle third image is created by using the 2 white dots seen above each of the 2 images, and slightly crossing your eyes not very much and will not cause any pain, to make the 2 white dots become one for the third middle image seen in 3-D. Once you have the 2 dots become on, seeing the third image in the middle, juts keep your eyes at that white spot in the middle and look down at the third image and you will see a stereoscopic 3-D image view.**

This is an easy eye exercise for your eye muscles as well and is a great way to view images come to life. This is especially great to use and to see what is being found in images on Mars via the rovers as David has shown many examples of. He has provided us for use in this book of a Mars X3-D view of an orbital view of the Face at Cydonia. On the next page, turn the book sideways, and see a 3-D view of the Face. David had used my processed version of MRO CTX image **G02_019057_2195.** Interesting and also yet at the end of the day, we need a model!

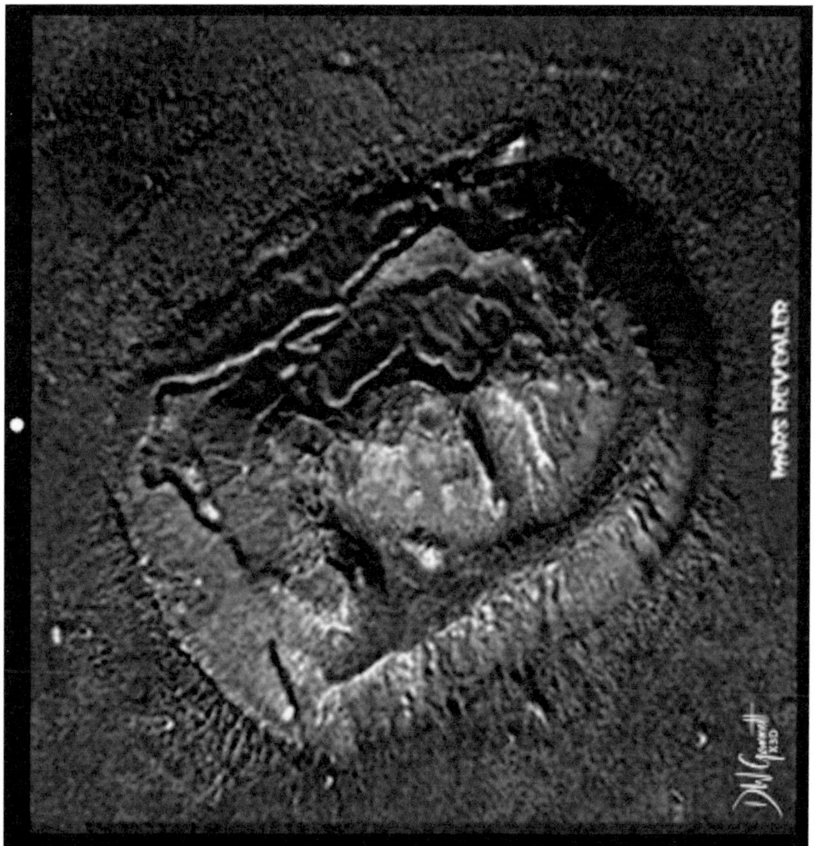
MARS REVEALER

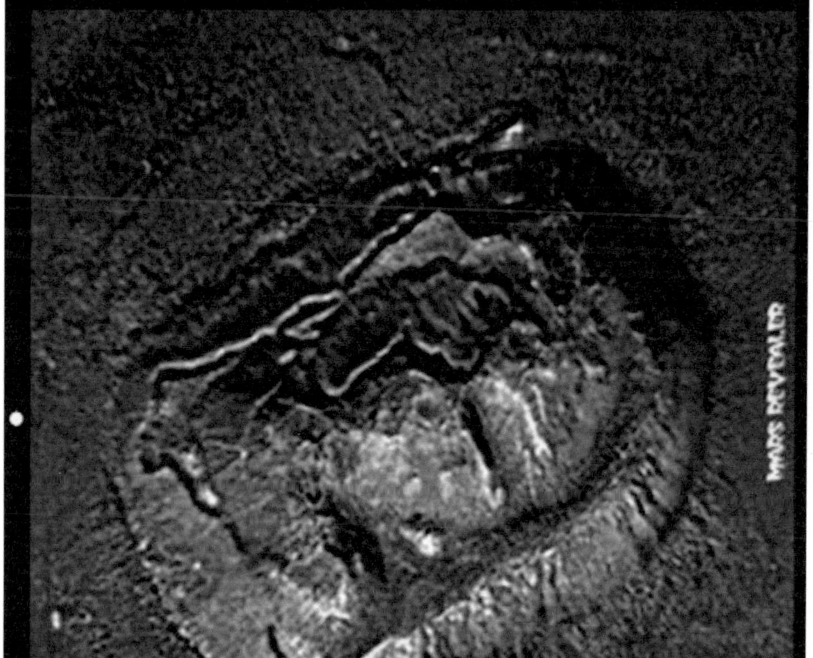
MARS REVEALER

The best other form of method in determining the accurate model needed and true shape for the Face at Cydonia is **Photoclinometry**, or also known as **shape-from-shading**, SFS, is a method used in computer science for obtaining accurate reconstructions of objects, especially faces. SFS is the process by which a 2-dimensional image of a surface is transformed into a surface map that represents different levels of elevation. It uses the shadows and light direction as reference points. It is used mostly to depict the surface of sculptures, to give an idea of how it would look in 3-dimensions. The techniques depend on very specific conditions, **especially light direction**.

Some artists use **photoclinometry** to digitize a 3-dimensional representation of a sculpture. Geologists and those that study planetary science use it to get an idea of how the surface of a planet looks like, and generate topographic maps and digital elevation models. Planetary applications began with investigations of lunar topography as early as 1951 and has since been applied to the Martian surface. To obtain accurate 3-D reconstruction of a face object that has very smooth regions and subtle shading, a traditional stereo or structure from motion algorithms may not be effective. Shape from shading is an effective approach complementary to stereo and SfM.

The visual information that can be extracted from the shading of an object can be usefully employed for the recovery of the shape of the viewed surface. Continuous surfaces are common in 3-D visualization, and the shape of their bumps, ridges, and indentations can contain important information. Light direction is very important to the quality of a photo clinometric image. Light that comes from directly over the surface (behind the camera) makes it hard to distinguish the shadows. Since it destroys important shadows required for the algorithms to work properly to get a realistic shaping model, let alone an accurate one.

Even using this amazing procedure, it musts be reiterated that the image **must still have an accurate sun lighting position and angle to get an accurate 3-D view,** even if the image itself is off nadir (taken at an angle), or if it is taken nadir (directly overhead). What matters to get an accurate 3-D shape view of the Face in this case, is having the sunlight position and sun angle in a correct spot when the image was taken, so that the shape from shading procedure can produce an accurate and realistic view.

This obviously is not such an easy task, especially if person or program using the camera taking the images is not aware of this invisible important data that can provide us all with the realistic results that is needed to make an accurate 3-D view using the shape from shading procedure. If this is not known, or is taken by chance, we would have to only hope and wait for many images to be taken at various times, and then hopefully having an image **eventually** to be taken, that is in the right spot. This can take many images to be taken and tested with shape from shading until this correct lighting conditions in an image is acquired and spotted, and tested and rendered.

On the other hand, if someone in control of the camera picture taking process is aware of this invisible goldilocks zone as it were, to get an accurate rendering for a 3-D shaping of the Face, or any feature really, then that person or program taking the images can then purposefully avoid taking images at the perfect sun position times and correct sun angles that can give us the accurate and realistic 3-D views of the Face being imaged that we can successfully render with the shape from shading procedure. This again, as seen in other areas of this book we went over regards to the Face, would add to the list of the continued deliberate obfuscations on NASA's part, of what the Face really is.

Yet however by chance from a lack of knowing, or even **possibly on purpose,** such an image had been taken and released to us all 12 years ago, in the year of famous and awaited apocalypse, in 2012, and also had been discovered and tested, and revealed many years later **by me!**

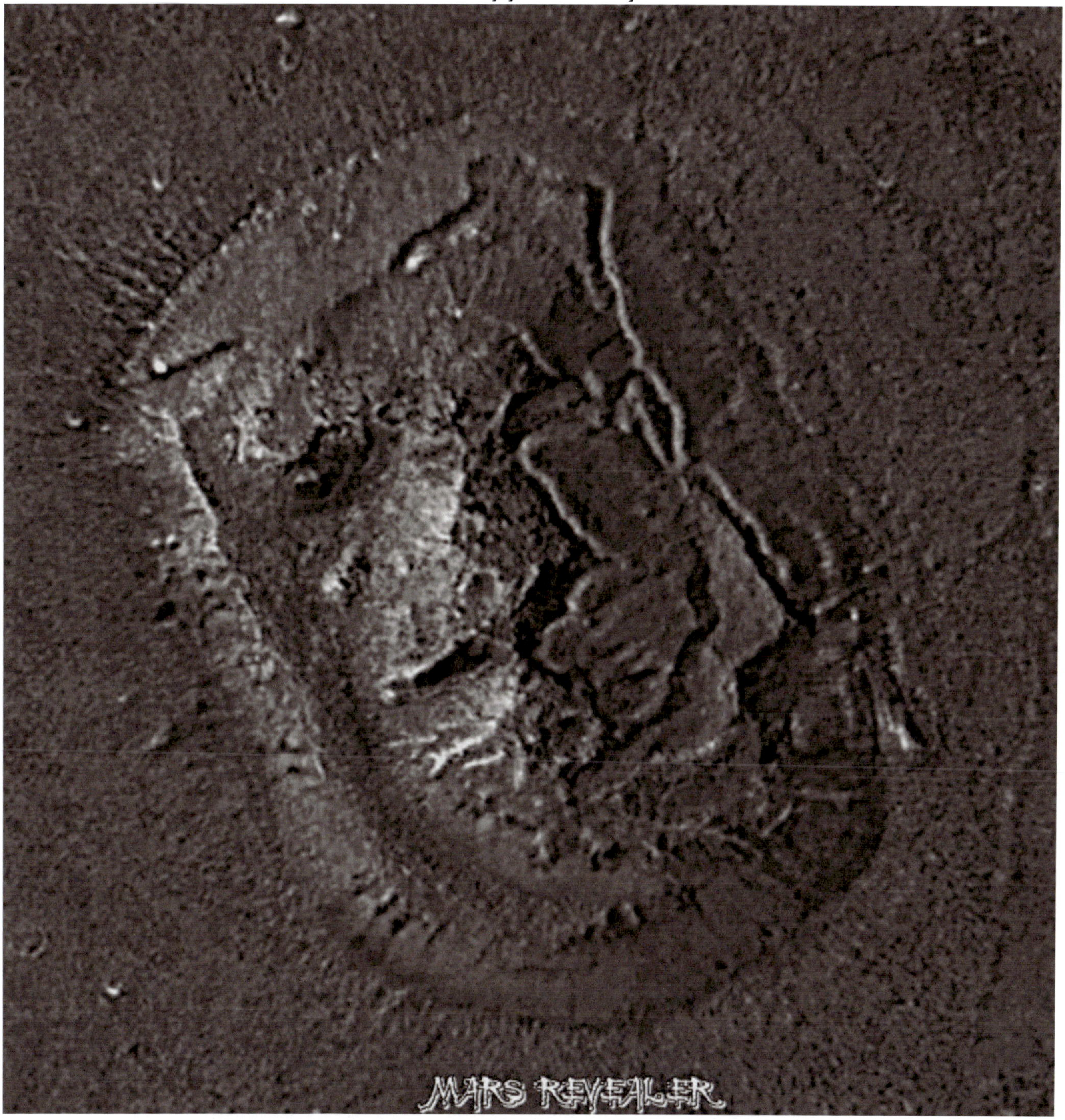

(MY PROCESSED RAW CTX IMAGE: D01_027417_2195 **IMAGE TAKEN:** **JUNE 1st, 2012**)

This is such a historic image. It is again, the only image of the Face from any Mars orbiting image mission, of various resolution that has been publicly released, to which we can get an accurate and true 3-

D shaping of its structure from an animated multi-perspective view. It is incredible of how its surface relief features are standing out making a viewing of how it will look as if we are flying around it. Talk about bringing the Face to Life! Pun intended!

Let us see what this now looks like. An unprecedented view of the Face seen as never seen before. An unprecedented view which even makes more sense and more realistic than anything NASA has given us of what they want us to think and believe the reality is. Yet that is NOT the reality. As we will compare and look at these major differences and discrepancies that is pretty much self-explanatory when comparing side by side views, or one above the other comparative views. It has nothing for me to explain really, as what counts in what I am saying, is seeing the actual image data for yourself to compare and decide for yourselves which side do you believe? The case I provide, or the deceptive obfuscated facts and manipulated image data and public narrative from NASA?

Let us see some comparisons showing each side's perspective views of the Face, to present and make my strong case right here and now.

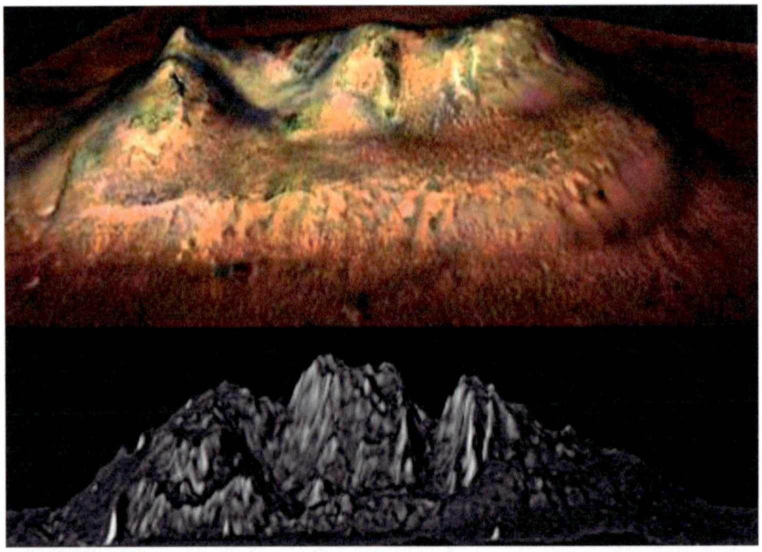

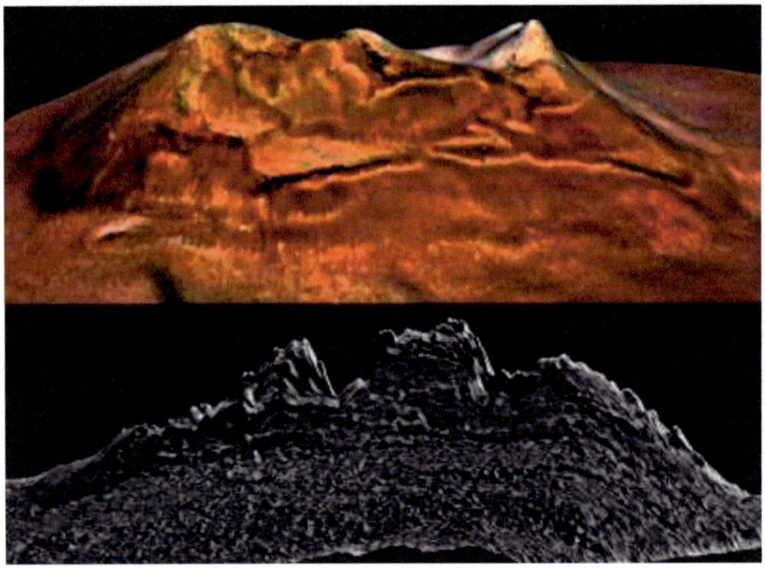

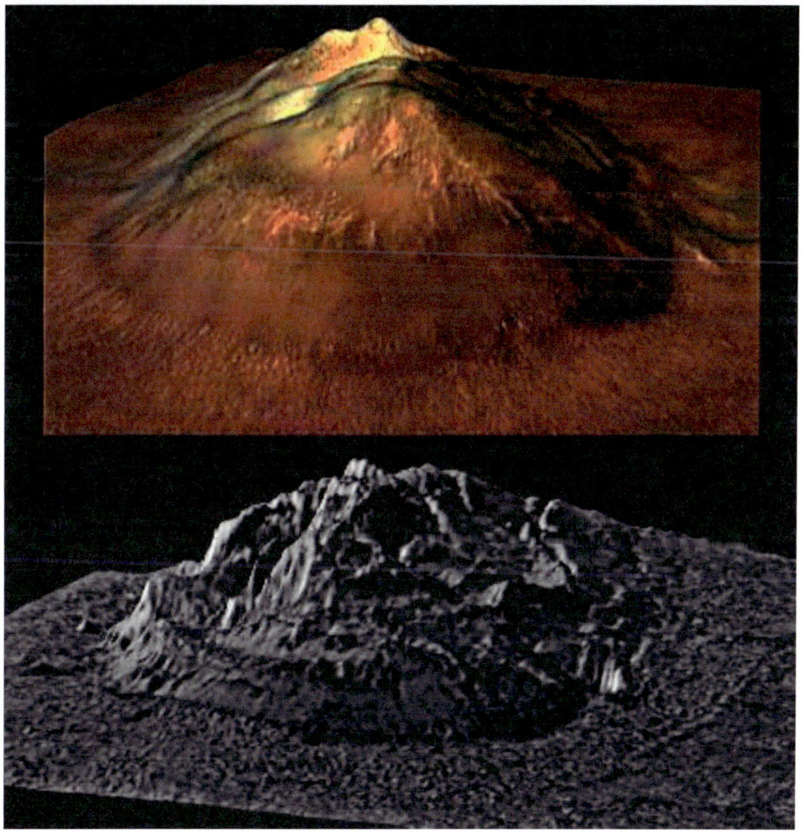

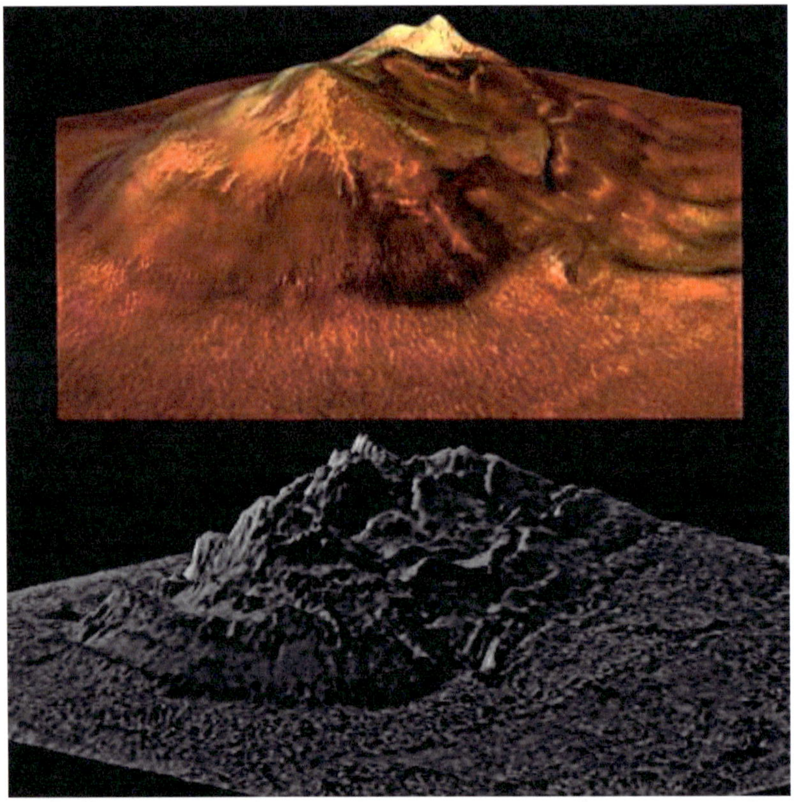

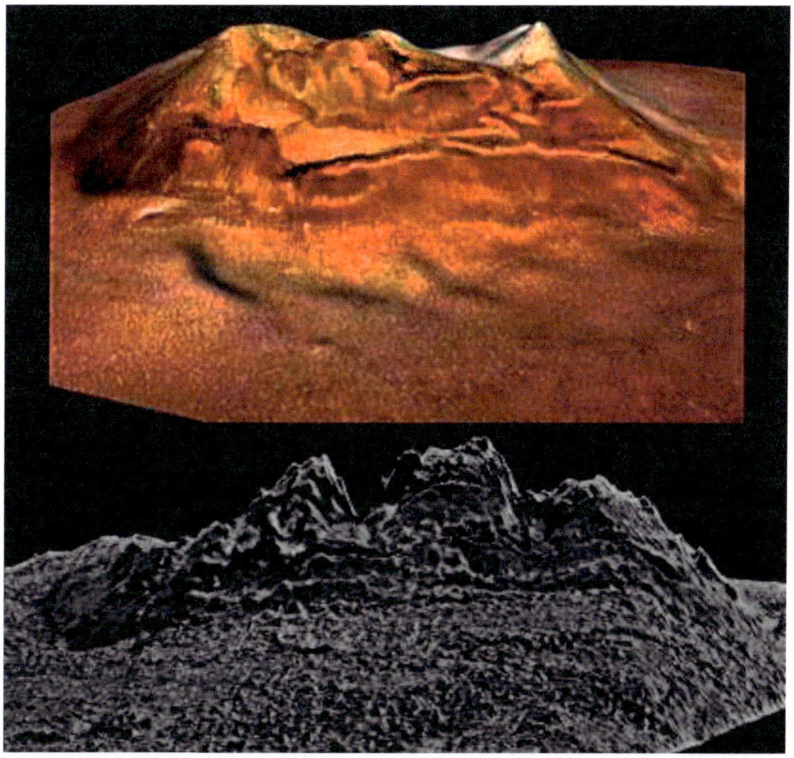

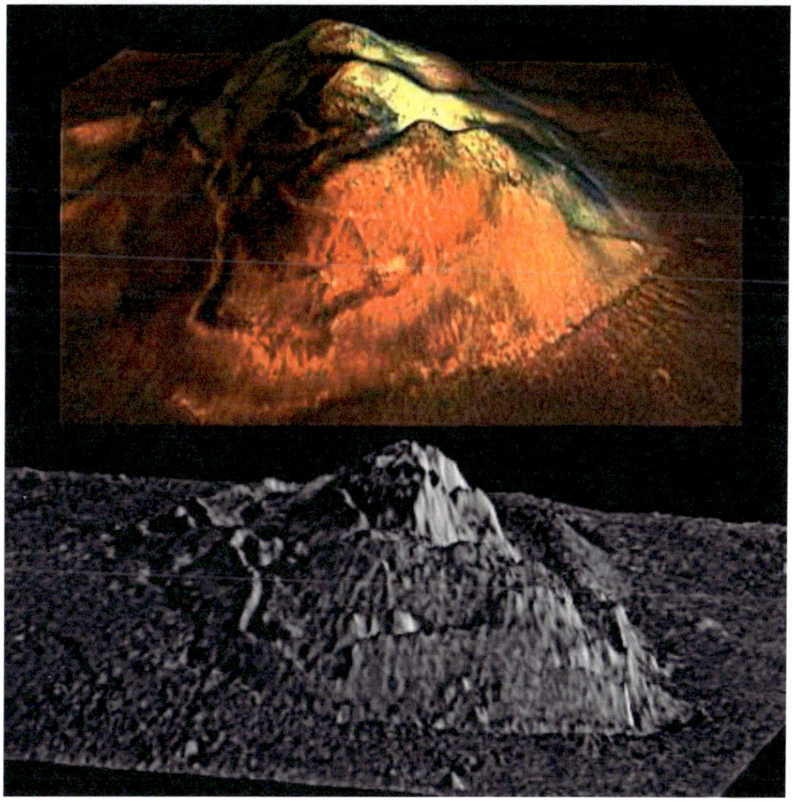

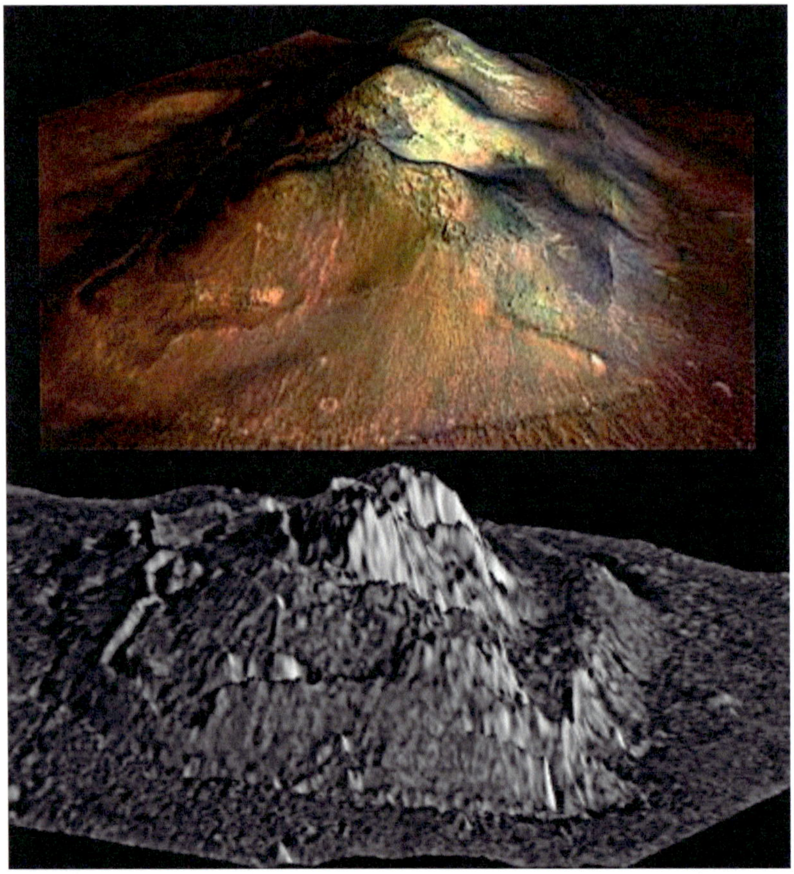

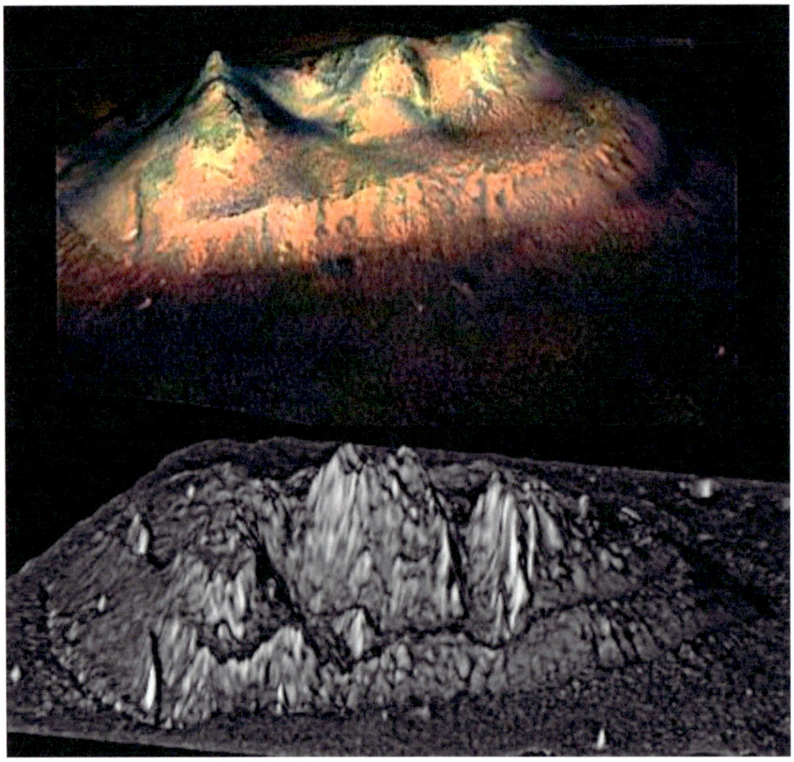

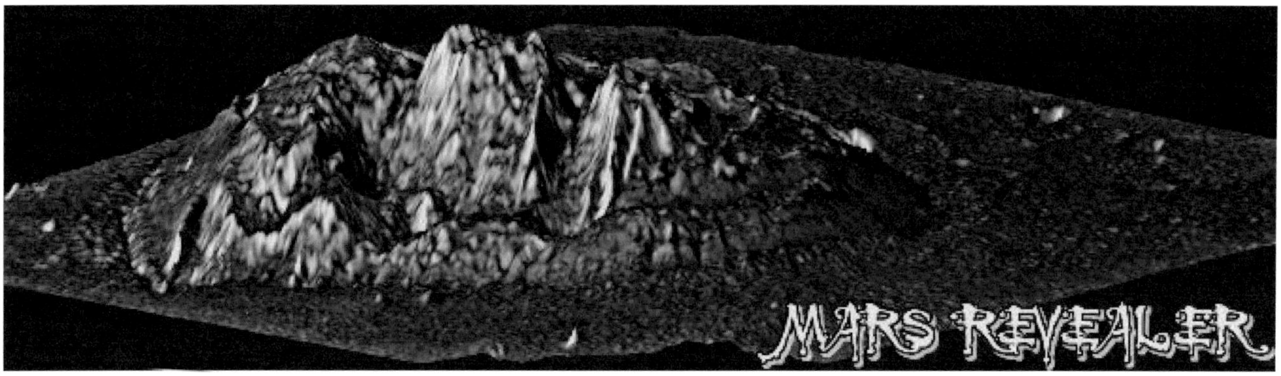

As you can see my words about this obvious NASA deception upon the world appears to be true. "**It has nothing for me to explain really, as what counts in what I am saying, is seeing the actual image data for yourself to compare and decide for yourselves which side do you believe? The case I provide, or the deceptive obfuscated facts and manipulated image data and public narrative from NASA?**"

After the shifting of scientific and accurate perspective, as well as challenging many people's cognitive dissonance, the informed '**new reality**' of the true shaping of the Face now seen for the world to get to finally see in fact reinforces what is seen from above and actually makes far more sense than what NASA releases to the world as being officially their so called scientifically proven truth which never made any sense to begin with! This is not hyperbole as this is in fact the truth. This entire book shows this from the onset of looking closely at all of their public official positions, lies, deceptions and possible coded information seen behind the surface that has been given and we have seen and you had to shift your thinking in how to possibly see it in how I was seeing it, and understanding it all leads us to ultimately to connect to the truth of the Face by finally being able to reveal and see its truth. By seeing its true shaping! This all pieced together as we have went over in this book to consider in what I believe is right and true, that **(it seems), this image view is what has been released to us all by NASA in the apocalypse (revelation) year of 2012, to connect all these dots laid out down through time to be able to one day seriously be looked at, and pieced together, and REVEALED! Yet, in an un-official way?!**

NASA certainly will have nothing to say in any admitting of this being possible or correct in what I said or presented of course by me here, other than call me insane or I am on drugs or something. LOL! **Maybe you think so also?** I not mind or really care to be honest. I did the homework, and have laid it out to you using multiple examples of image data of the Face **showing details you would never see from NASA in their public releases of imagery so easily.** I laid it out here for you all to see and read and try to follow of **all of their official statements**, and dared to make you think too deeply on all that was publicly positioned by NASA, and we were able to see **a possible behind the scenes coded story trying to be given and told to us all between the lines, behind the scenes**, that in what we went over, certainly matches the image data we are now finally **ALLOWED** to be able to see?! Truth is being given to us all secretly, or rather it has been given to us all in hiding, and slowly, behind their official position of the Face for many years for us **to hopefully one day be able to see this hidden secret code, and someone brave enough, or crazy enough to see it, and be able to try to present it?** Looking back in time, this is possible.

A special image which provides information, accurate unseen data to the human eye within image data that we can actually have comparative analysis from the image data, in comparison to what all other images data shows and lacks for this important and needed photoclinometry view. The only image ever released for the Face at Cydonia, where we are '**lucky**' by coincidence, or was released by deliberate silent purposes, to where **we get the correct lighting angle and sun position needed to render a correct shape from shading 3-D rendering that makes visual sense.**

We are **finally** able to see **the unprecedented acccurate views** of the side profiles of these **two faces,** as if we were standing on the surface on each side of the Face. Each side's facial profile likeness is obviously different than the other, and both sides profile Faces, appears to have open mouths. We now have a view as if these two 'species' represessentatives are looking at each other speaking or yelling at themelves. What are they saying?

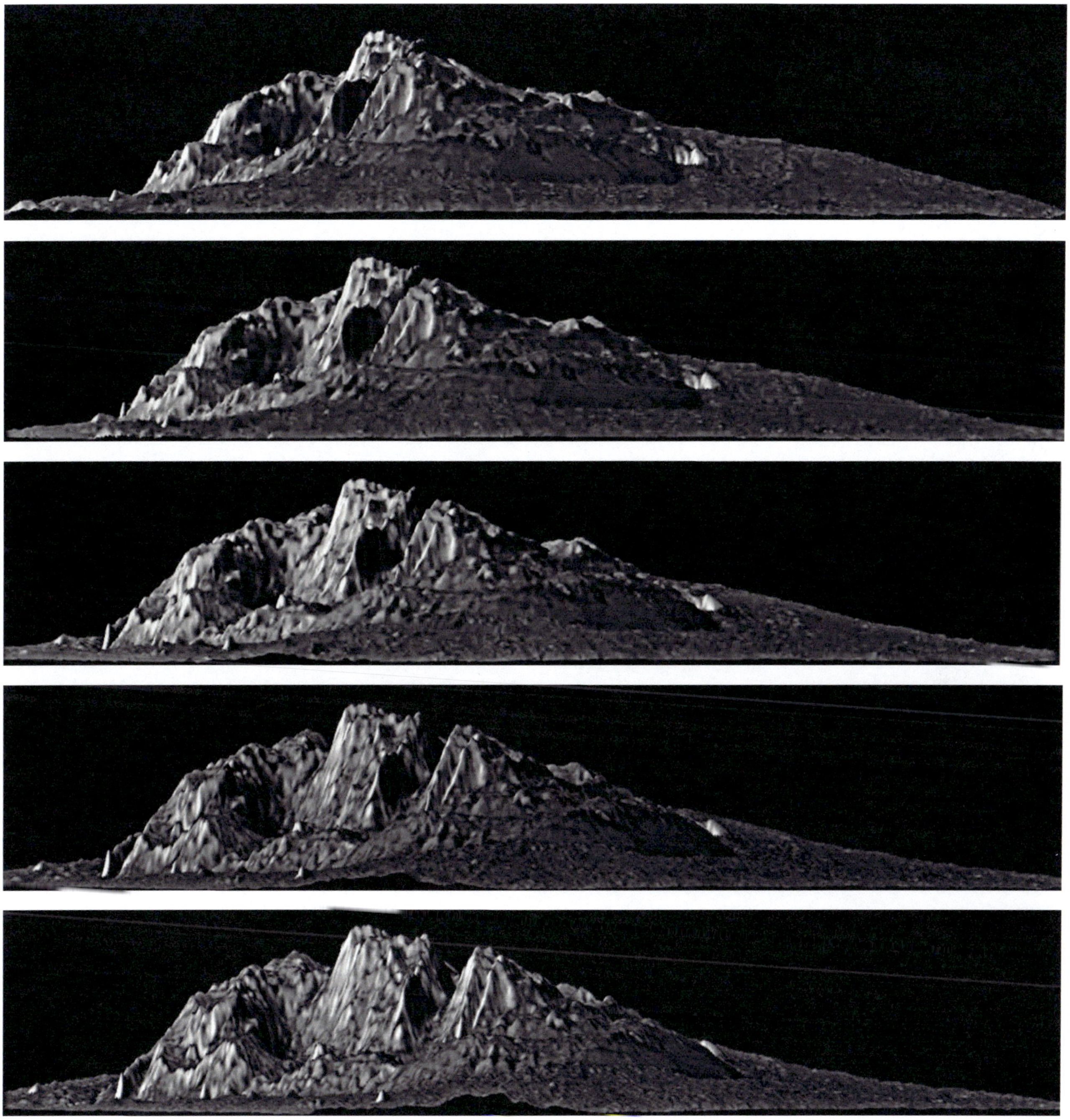

Watch a video showing these 3D perspective views here:
https://www.bitchute.com/video/8wXu7ilplJaL

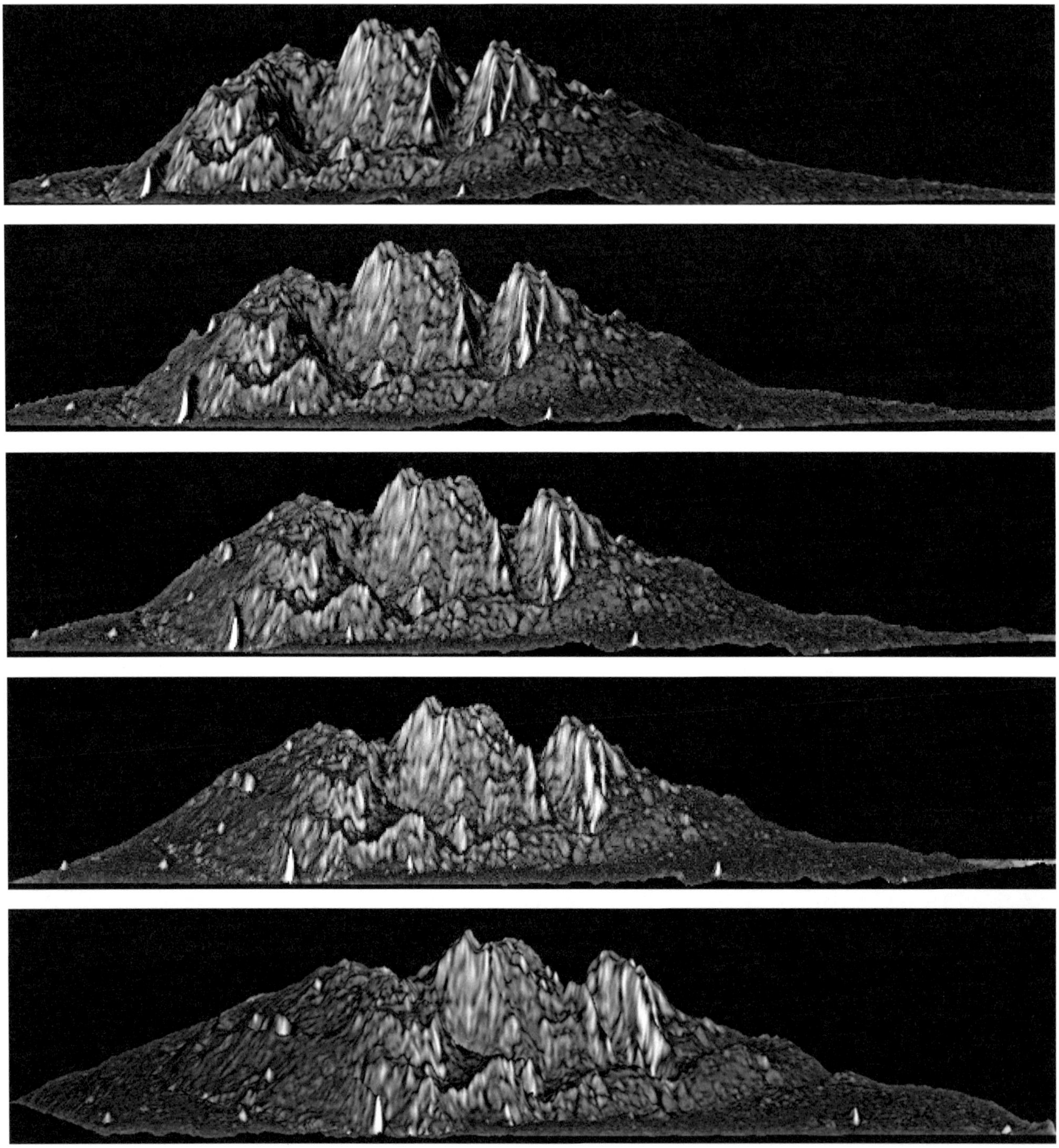

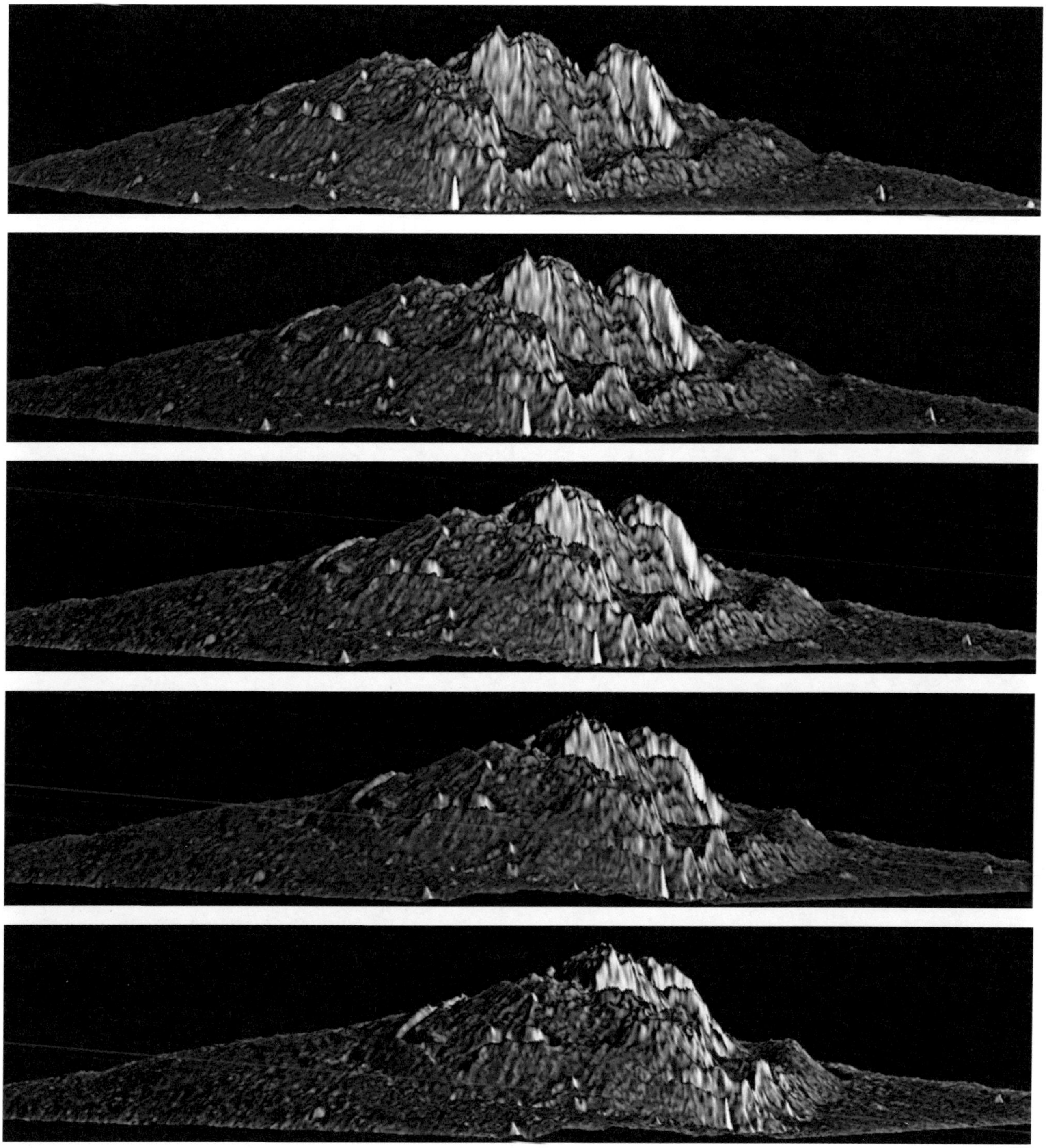

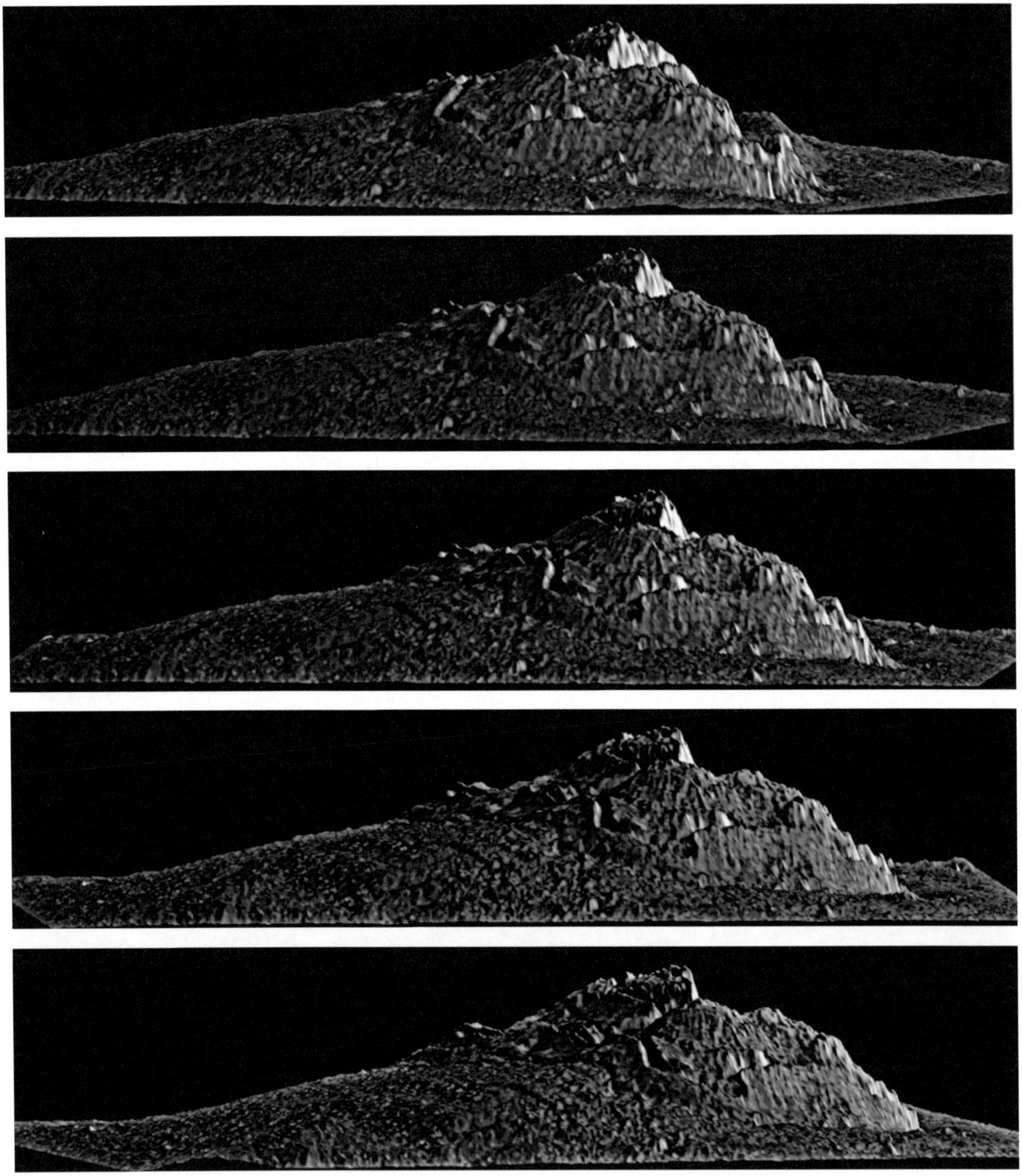

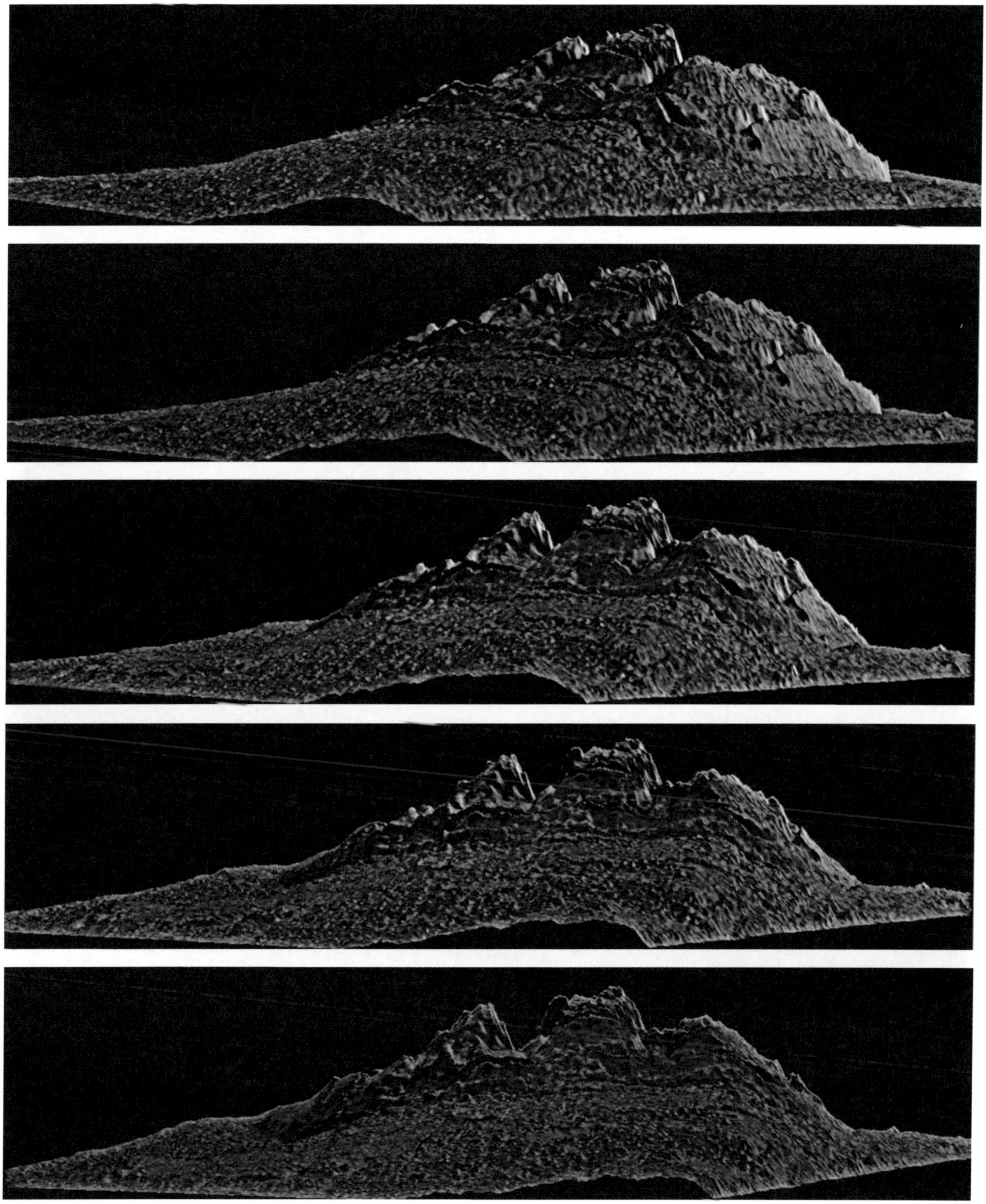

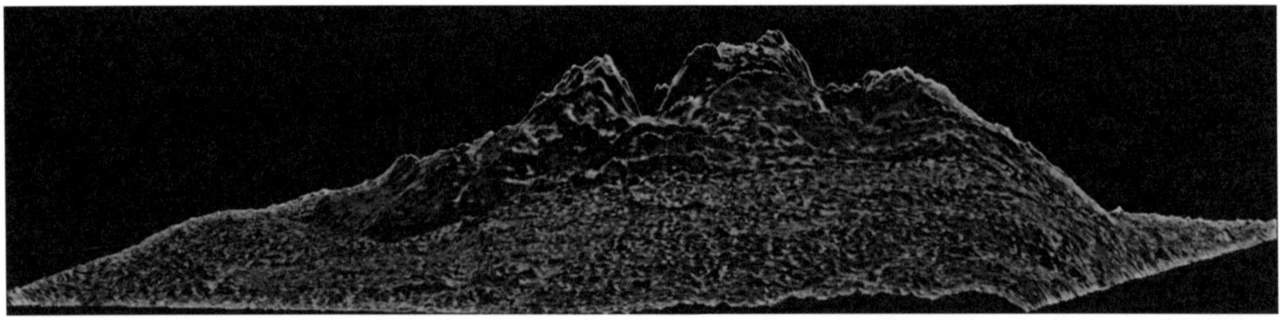

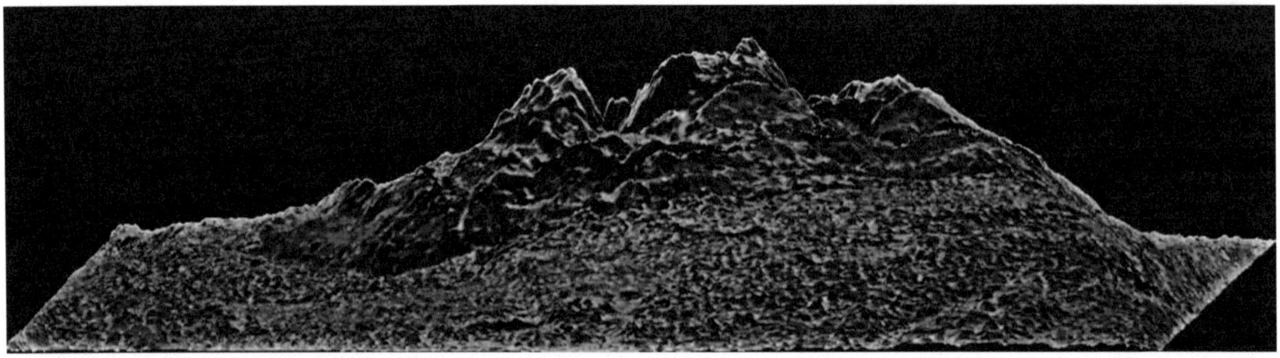

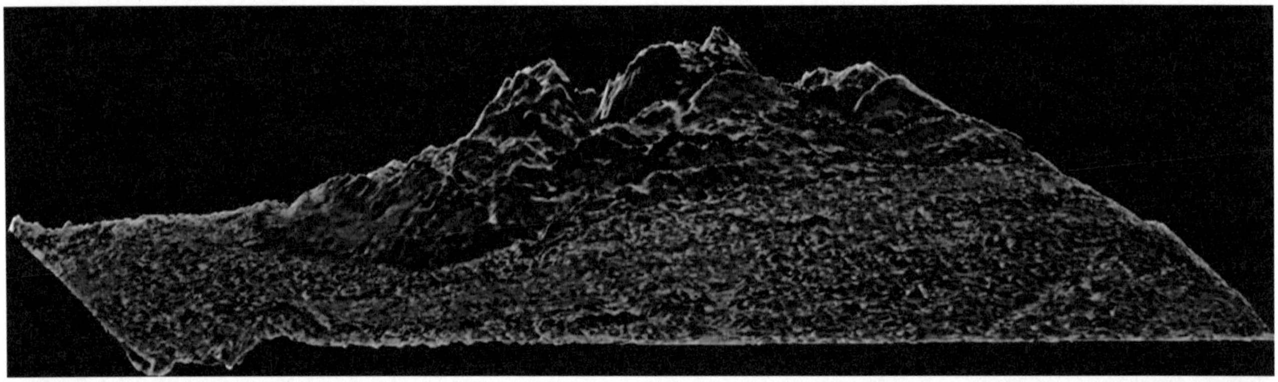

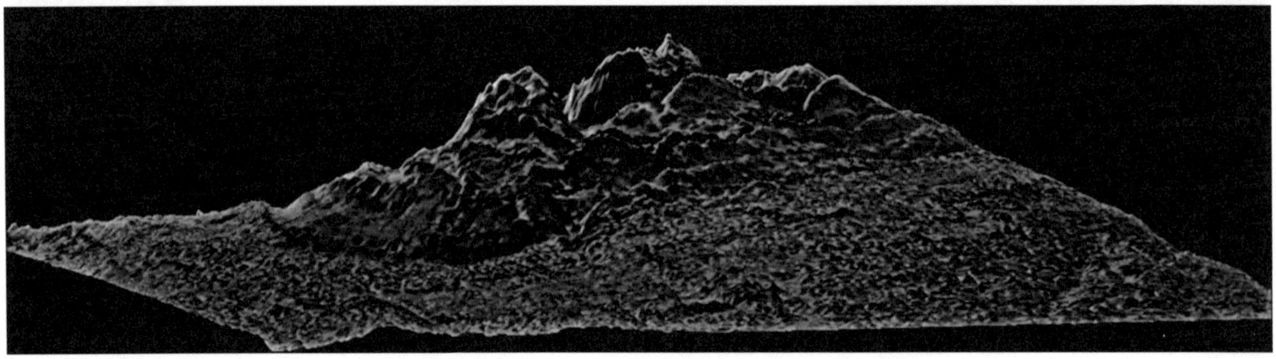

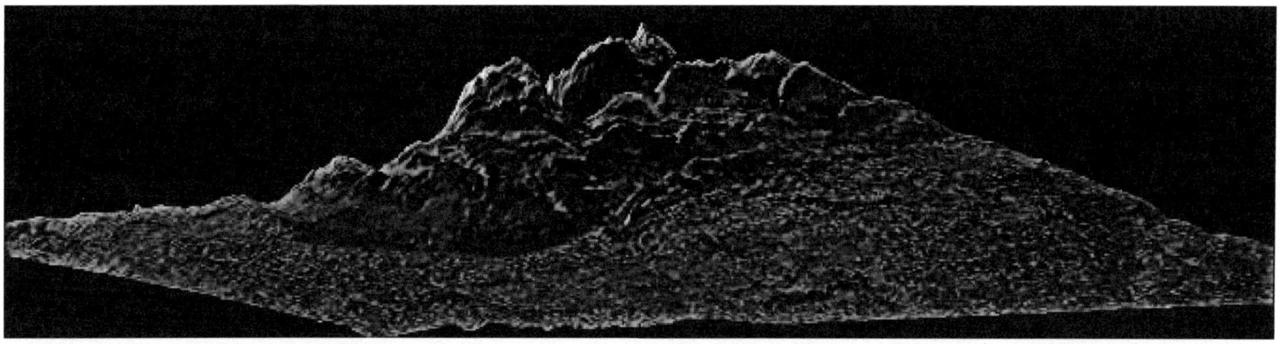

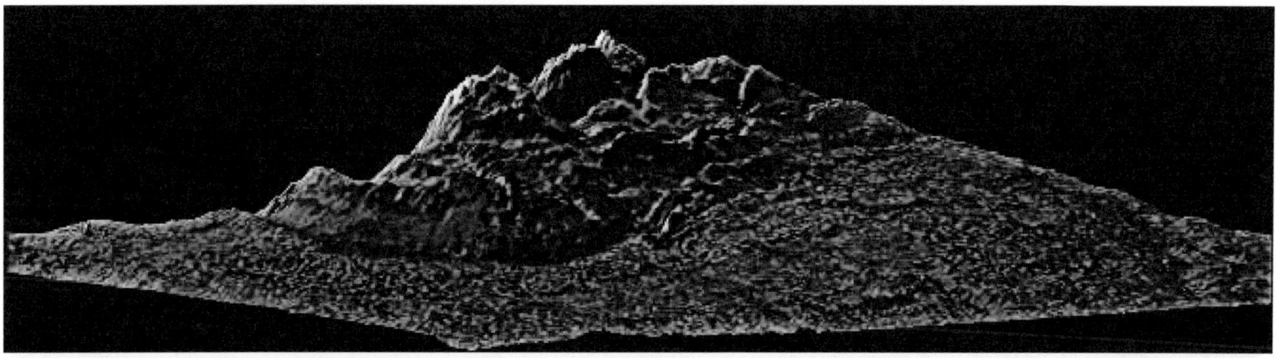

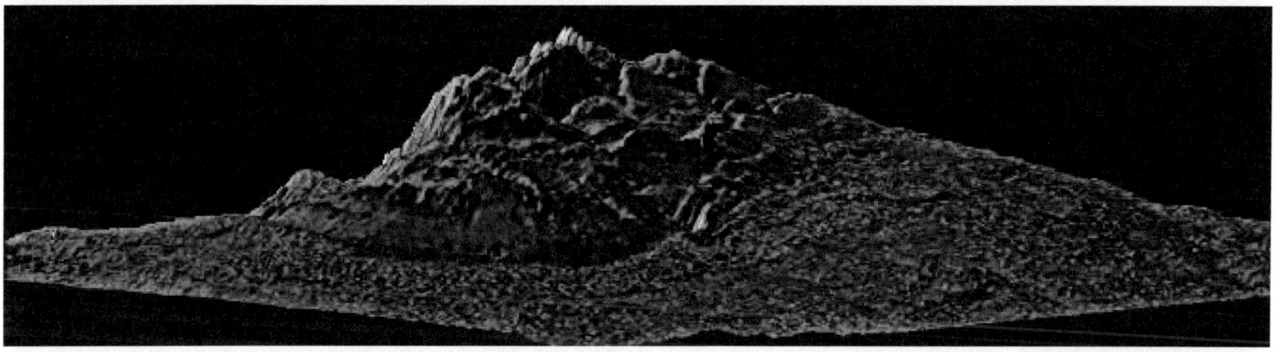

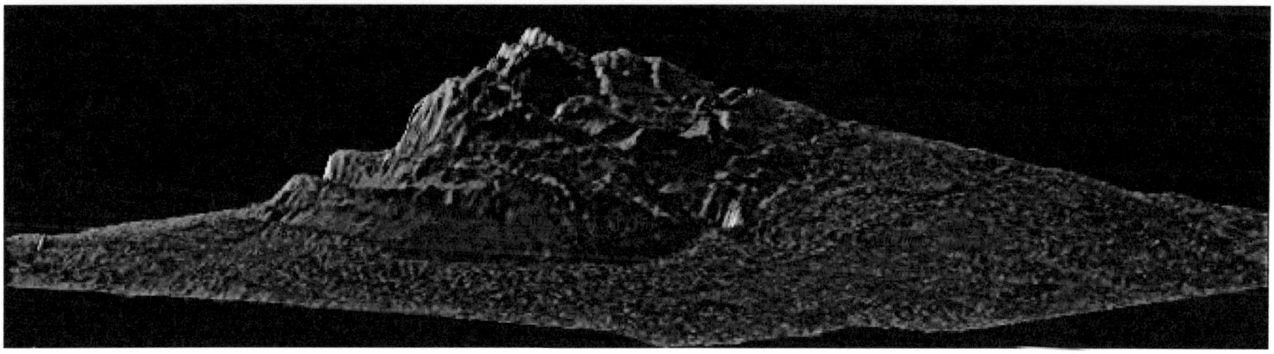

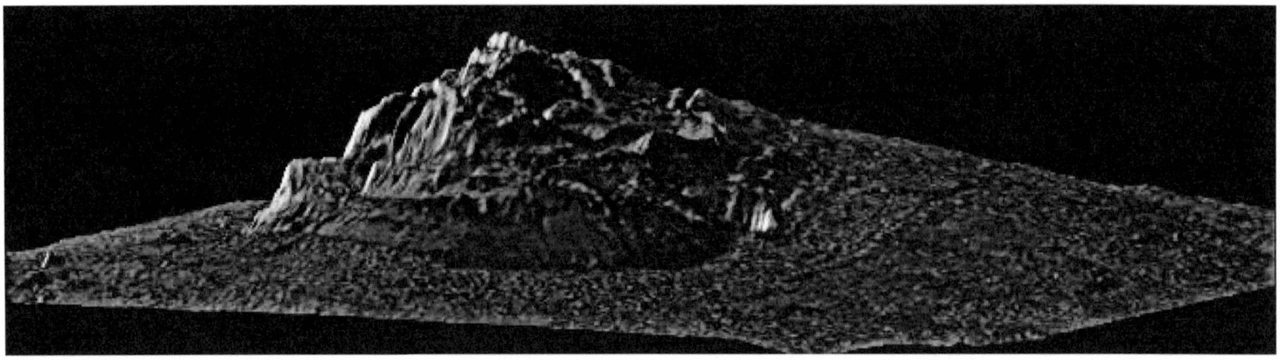

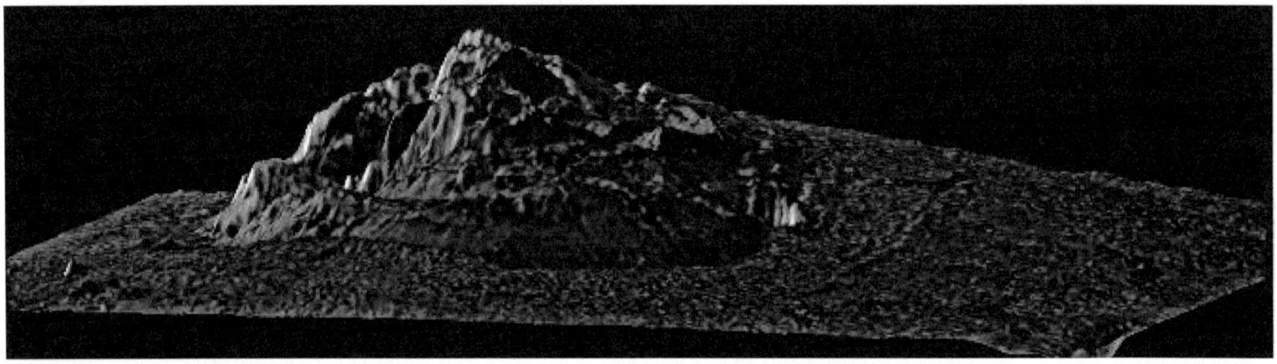

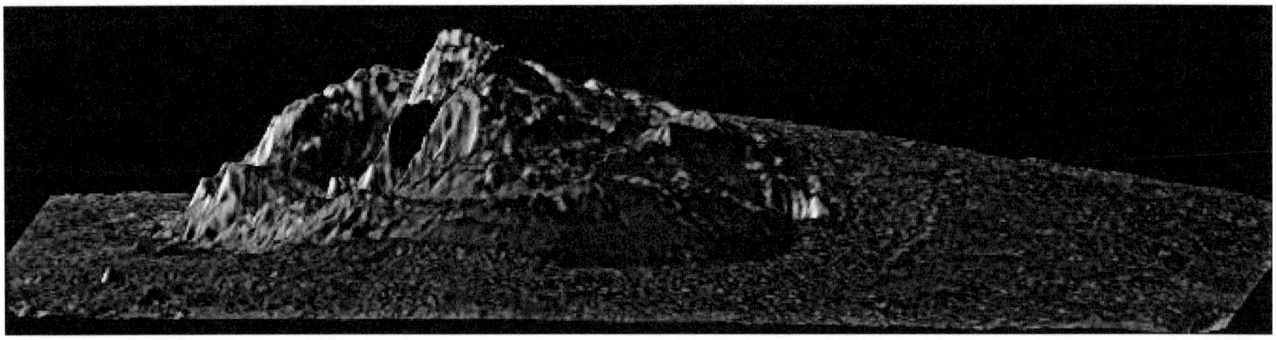

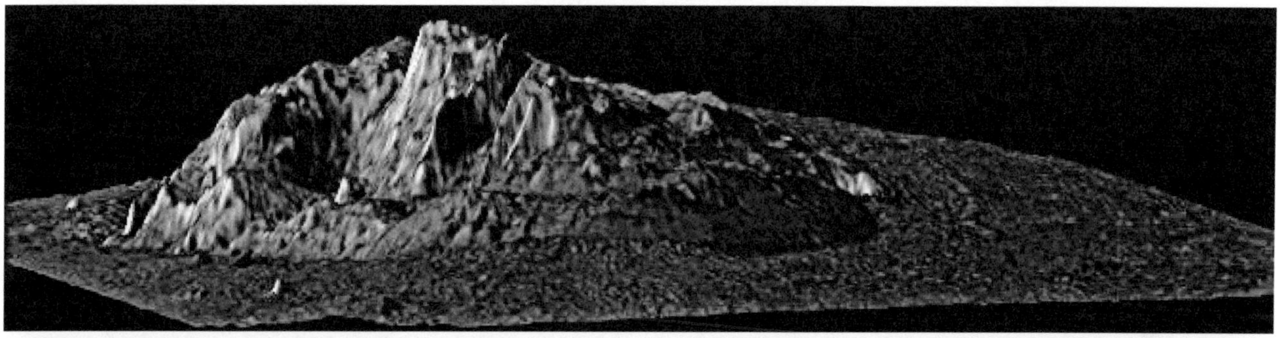

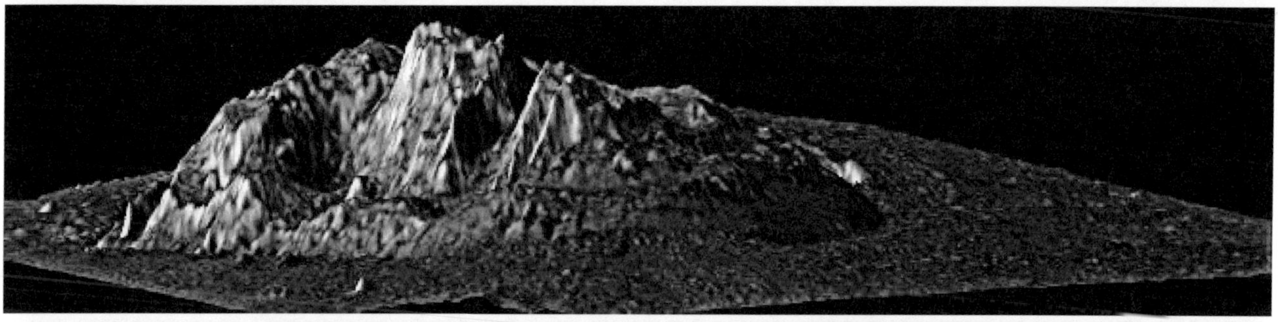

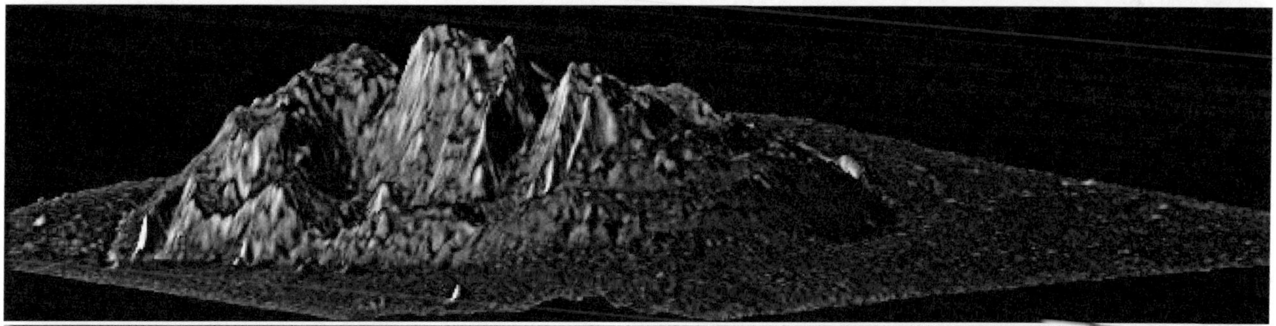

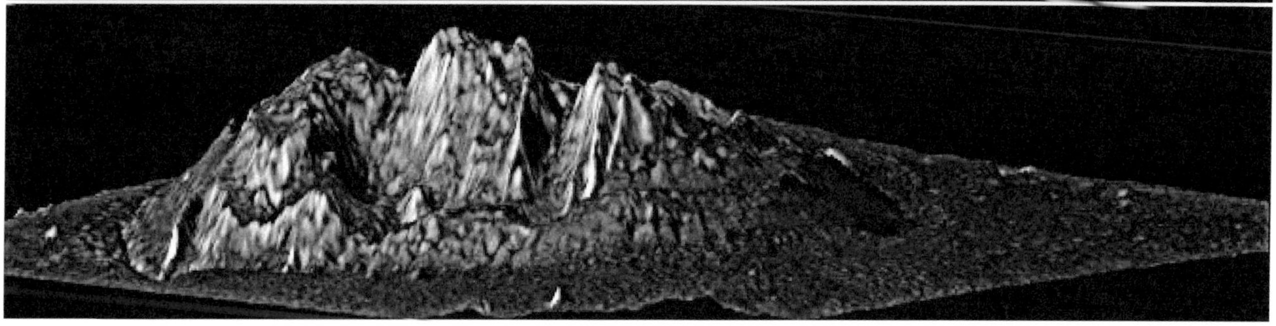

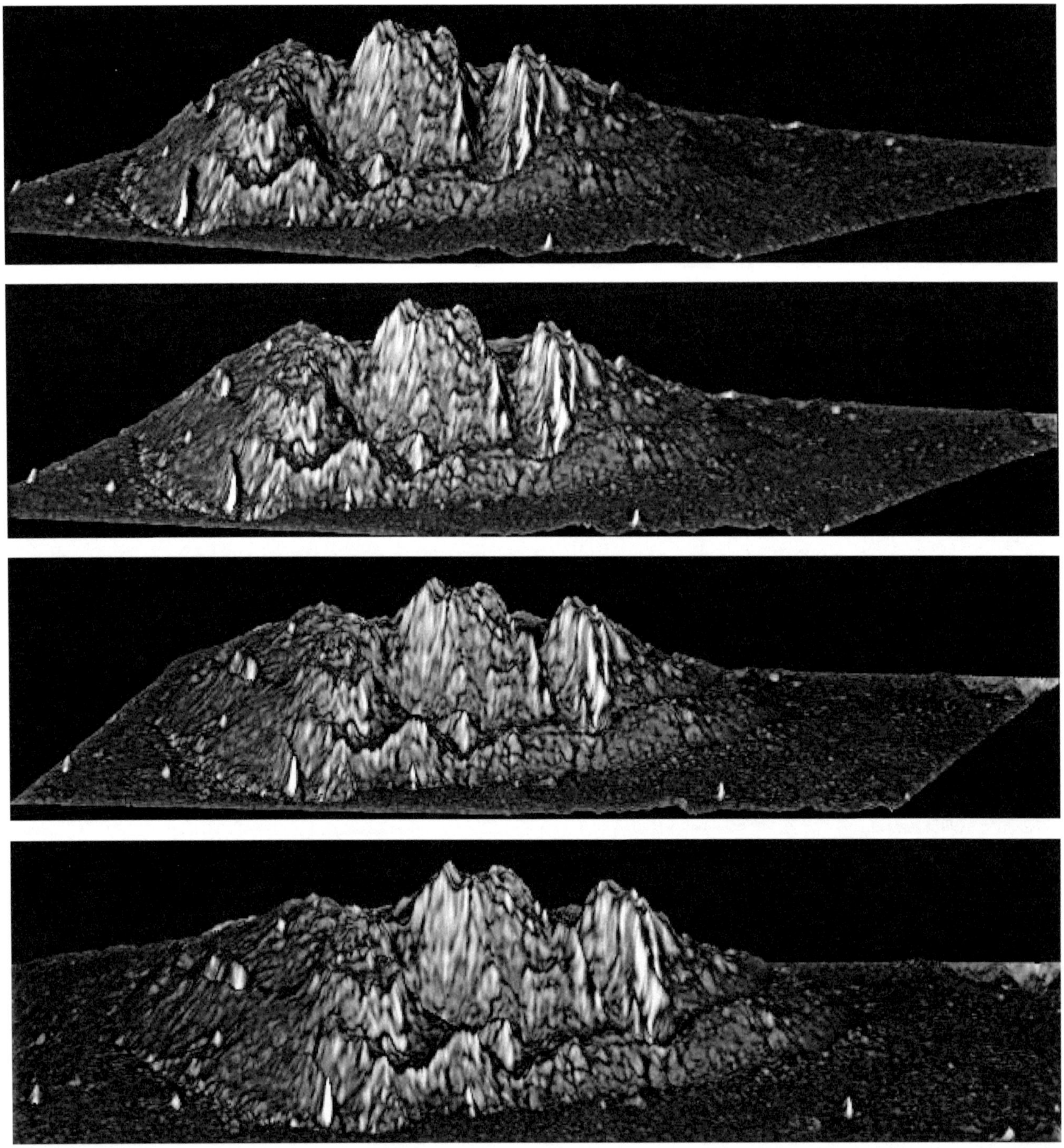

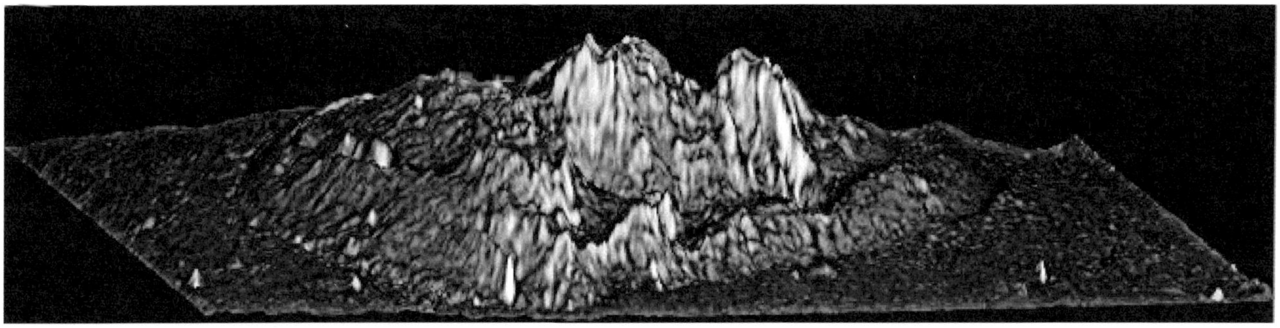

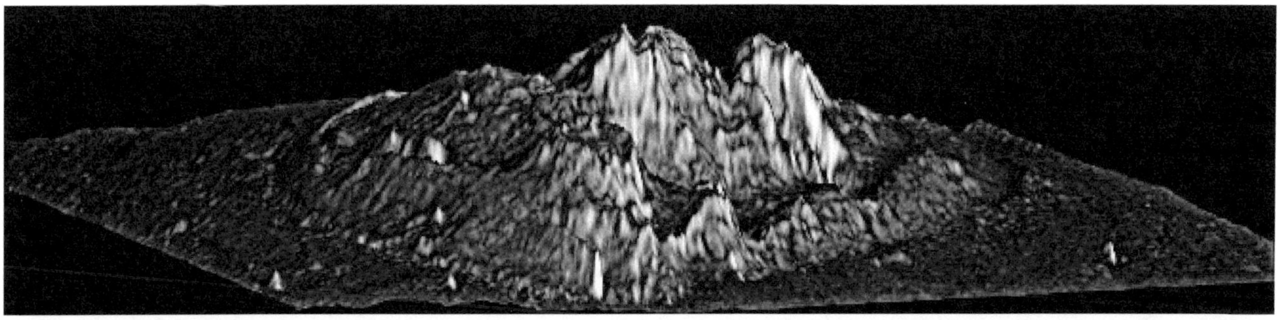

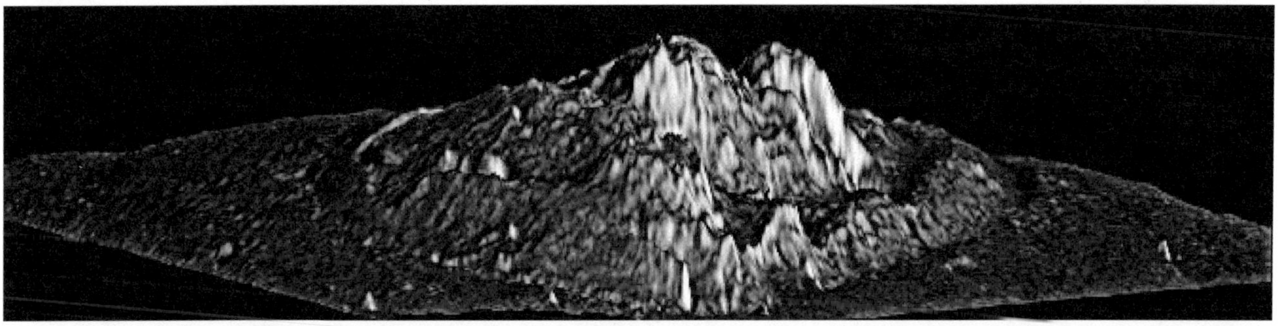

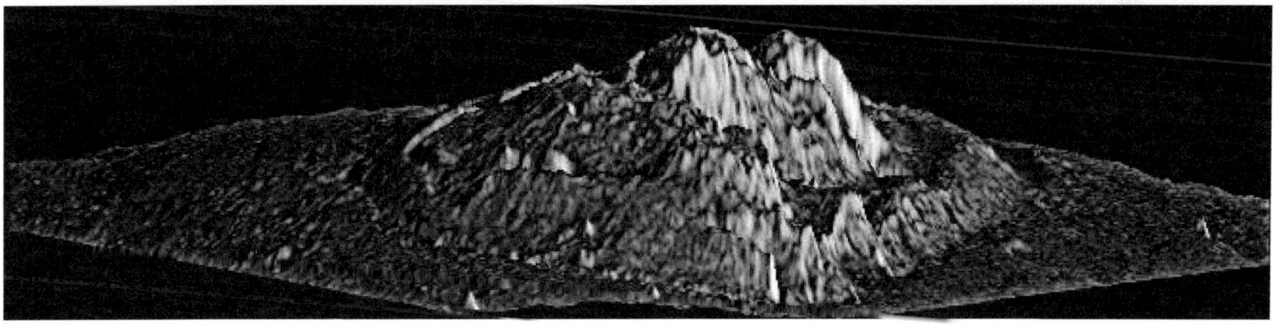

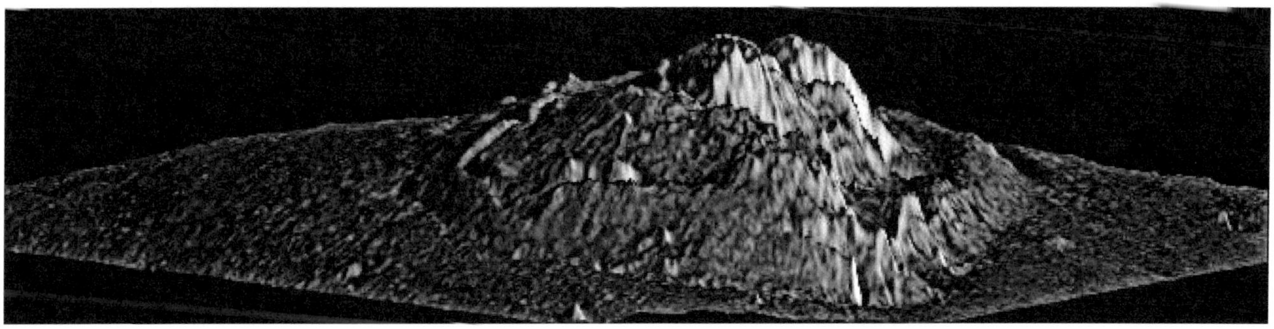

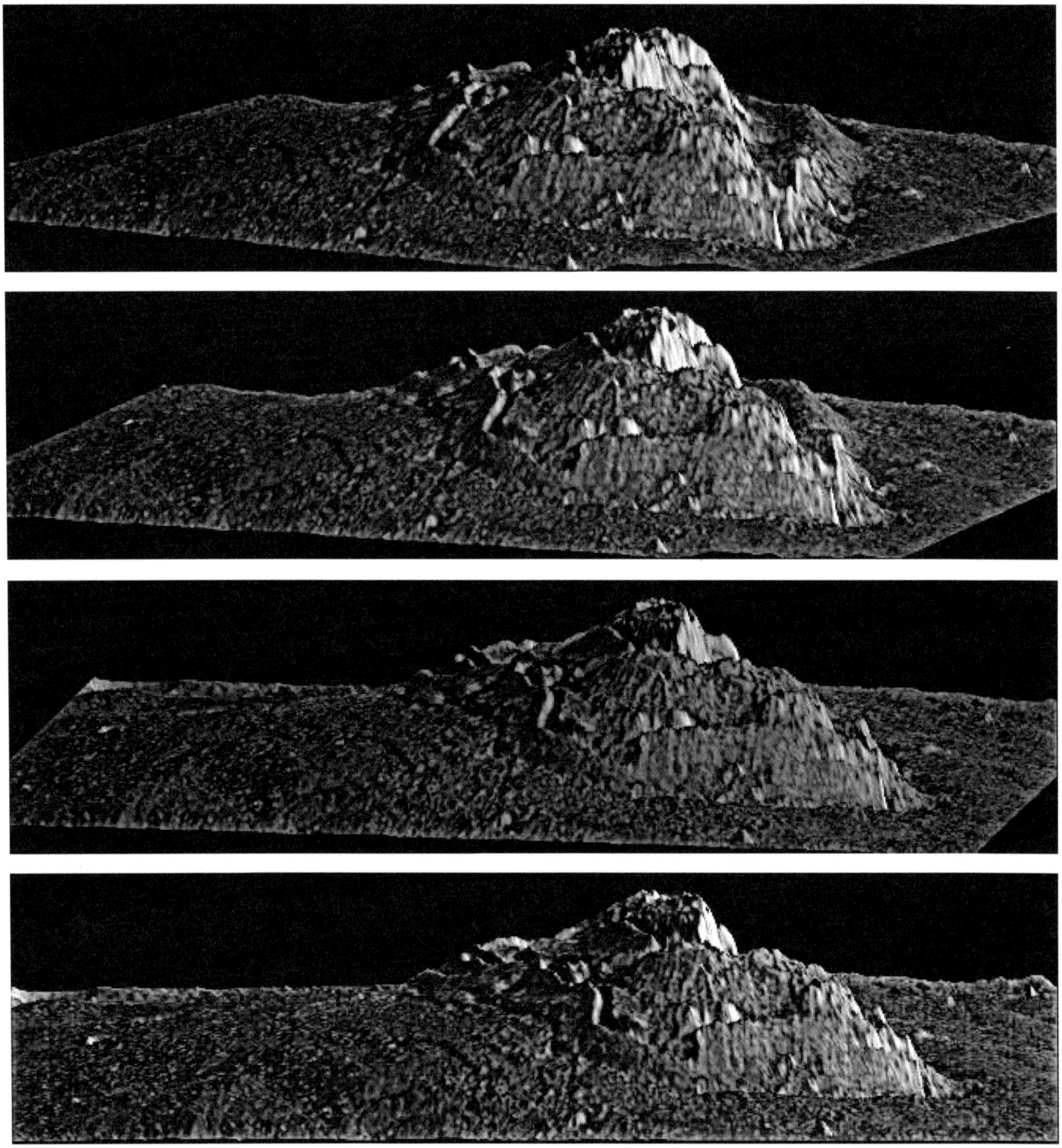

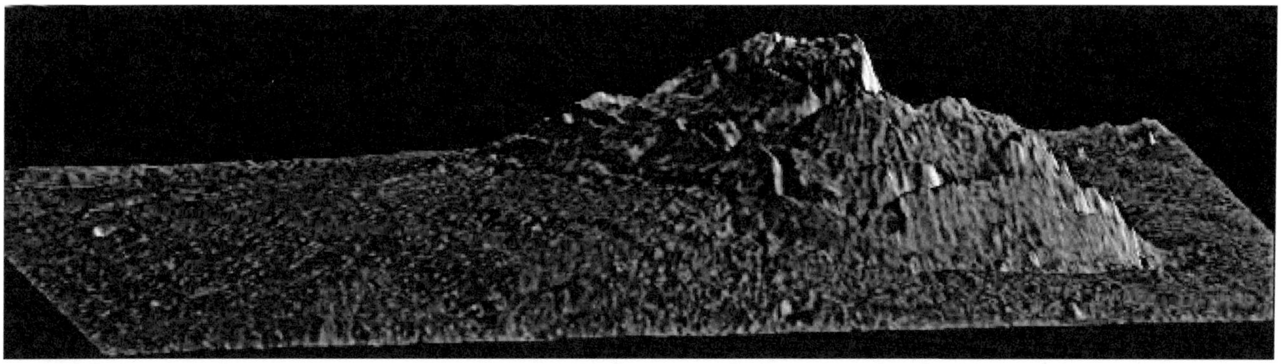

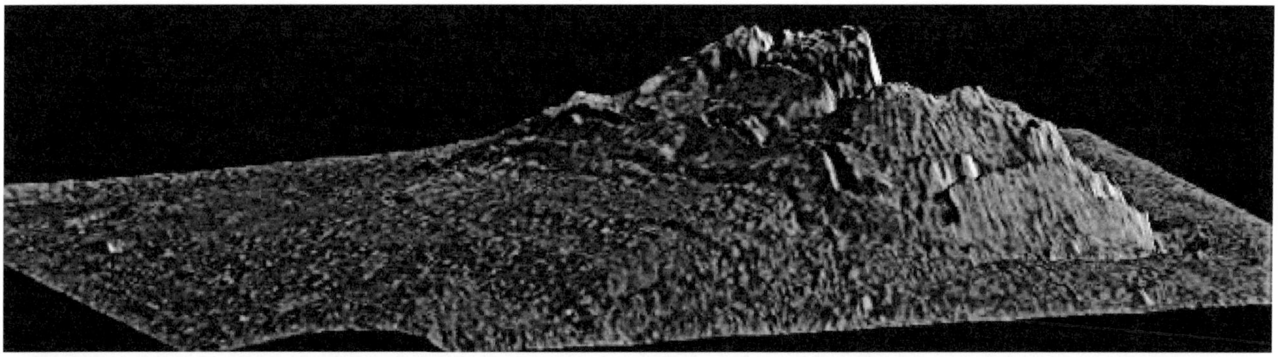

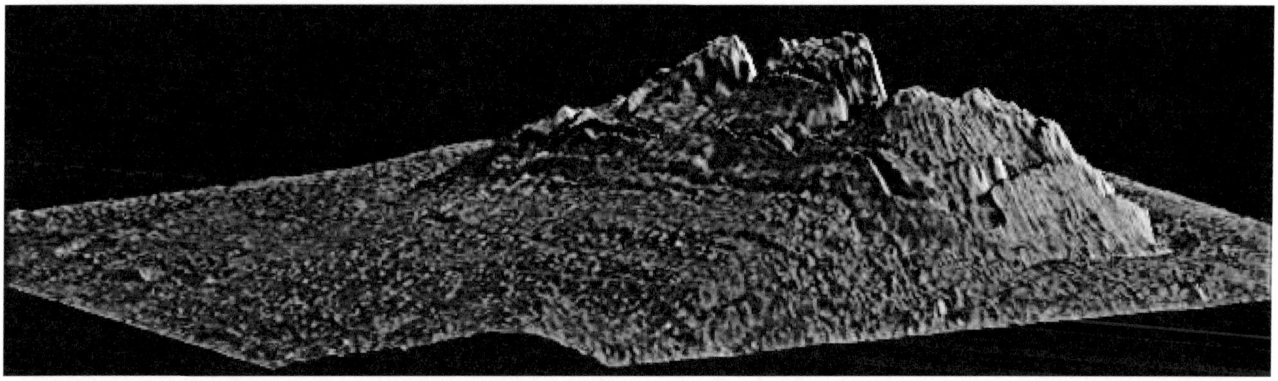

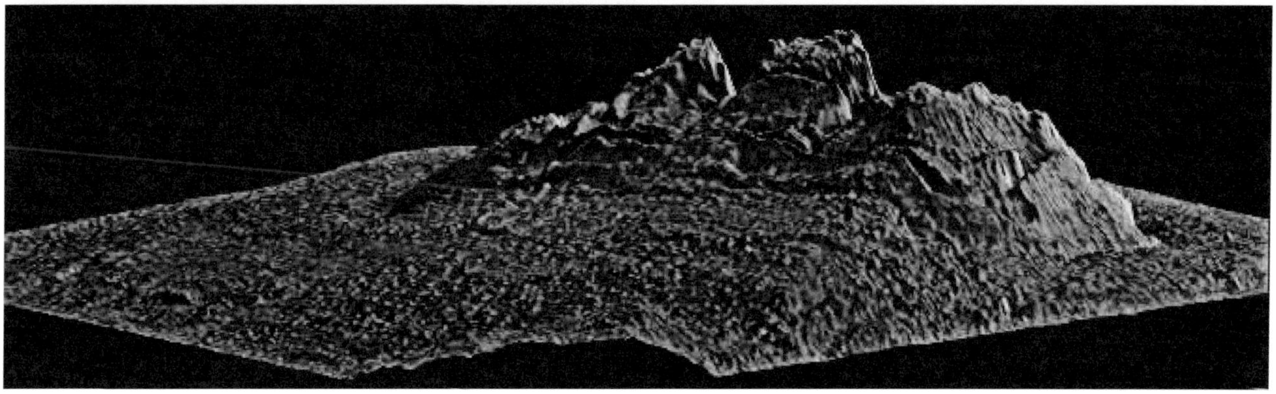

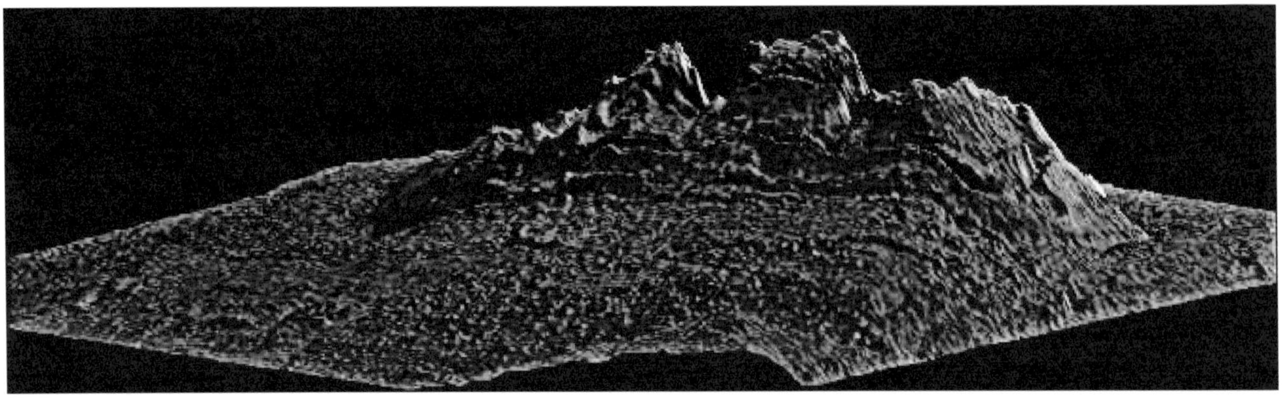

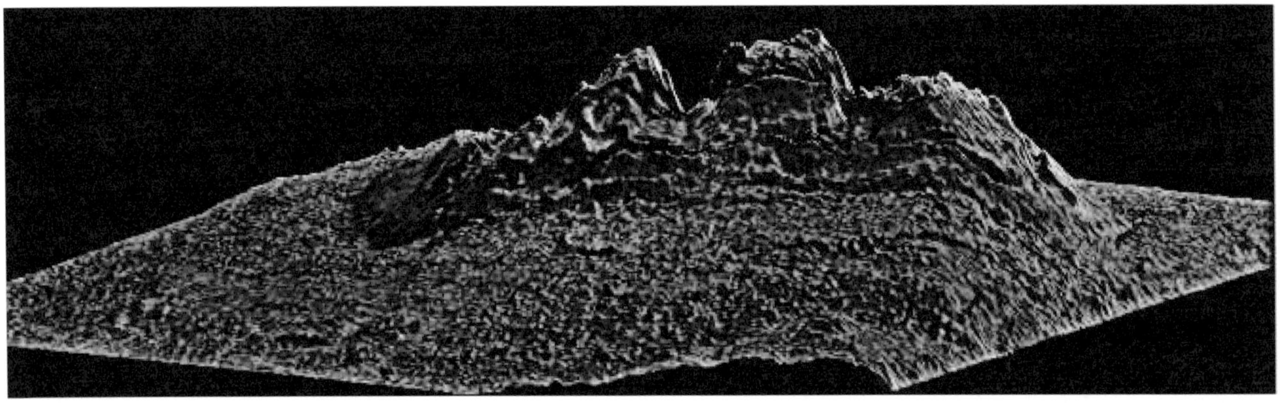

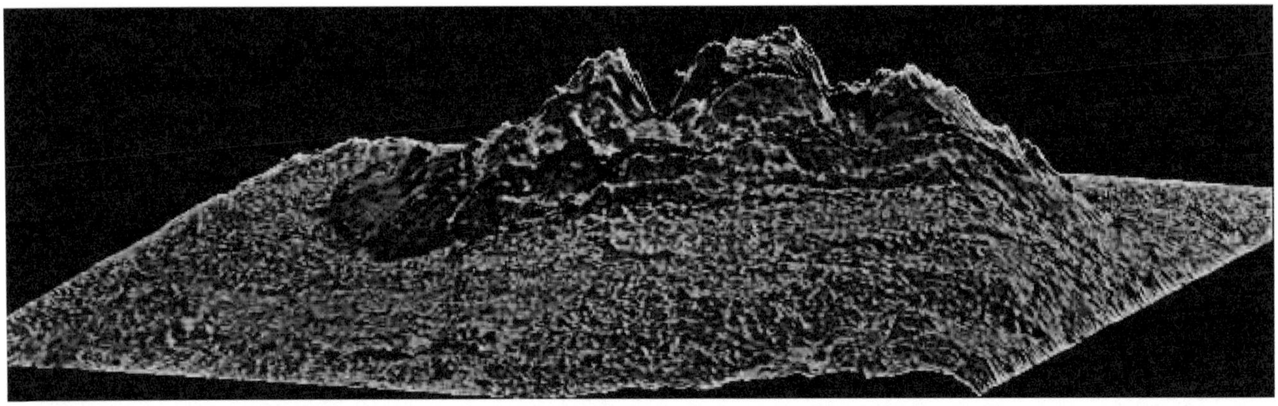

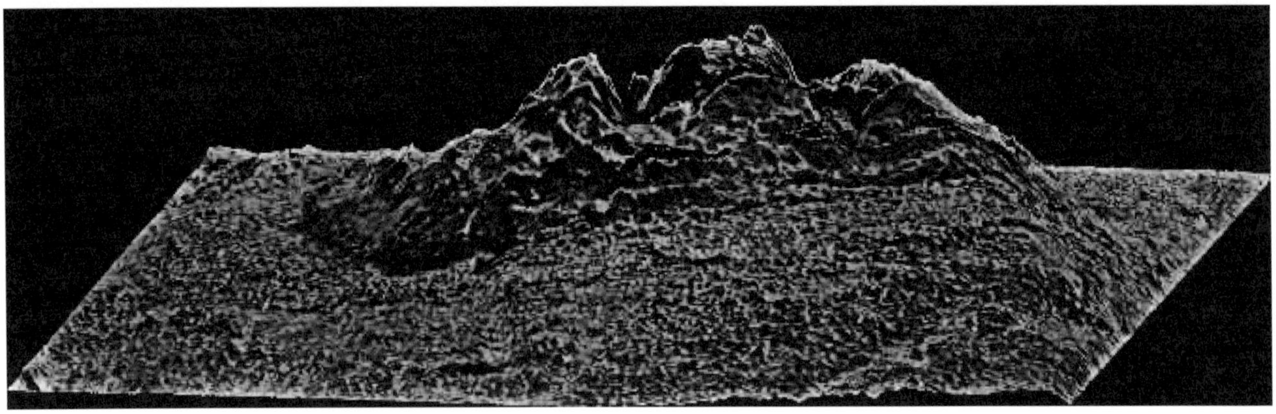

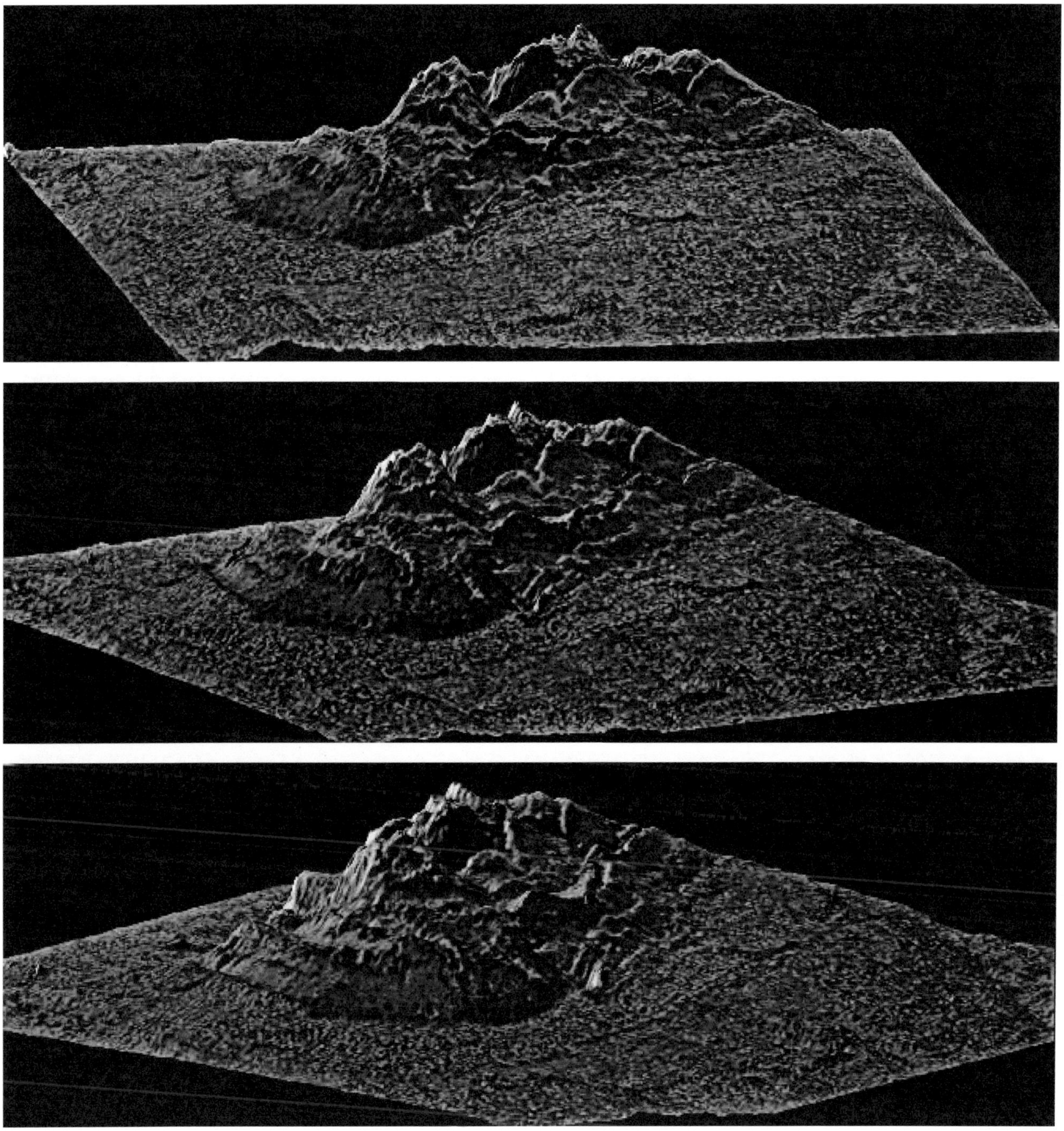

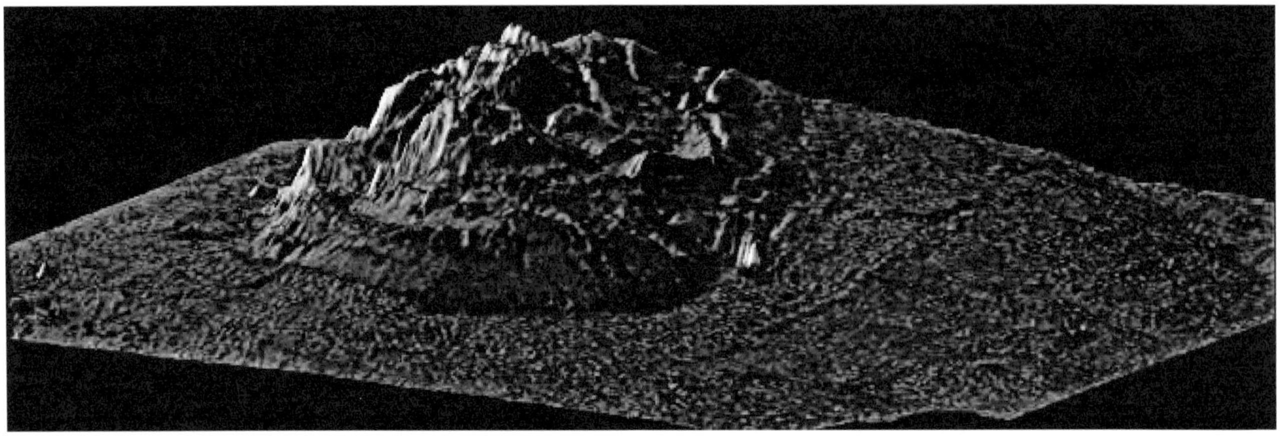

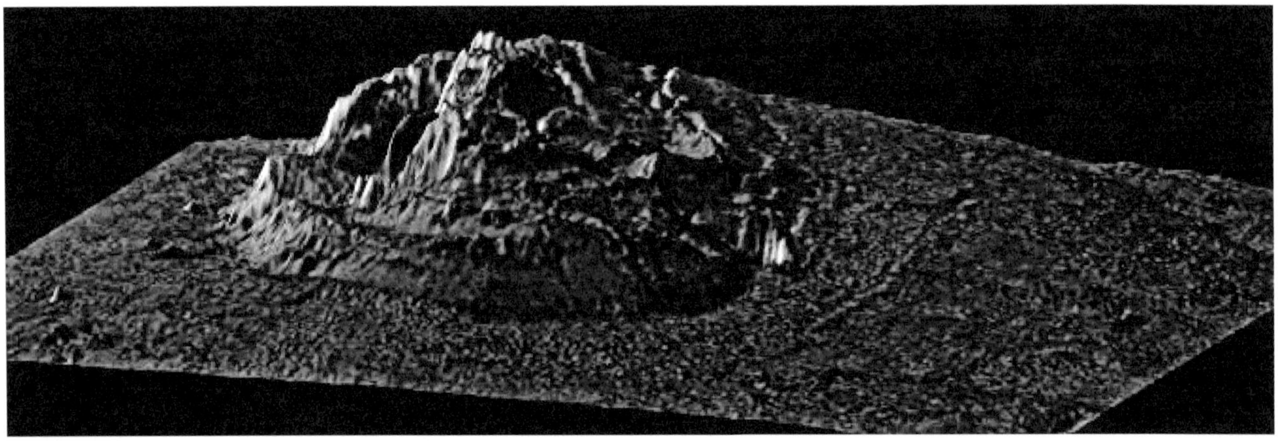

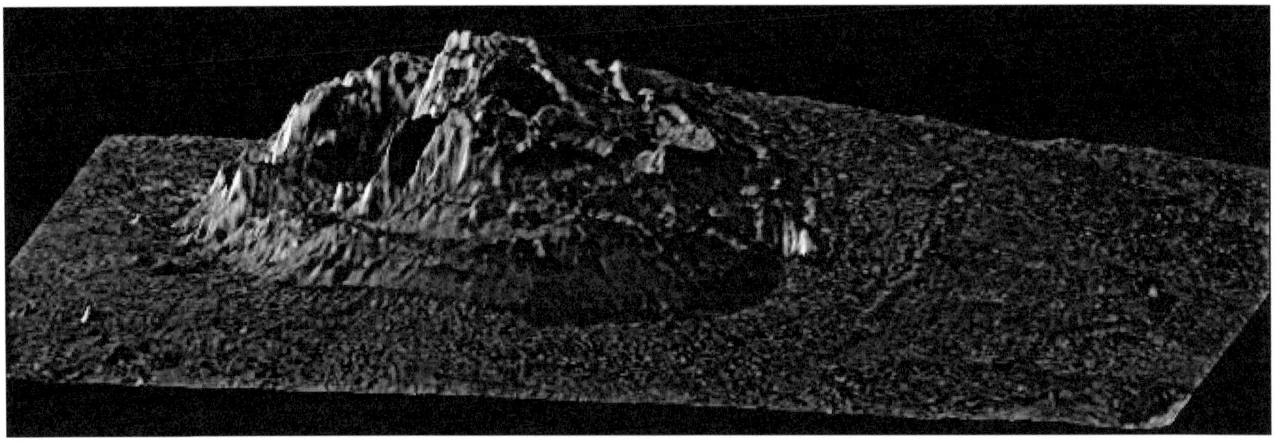

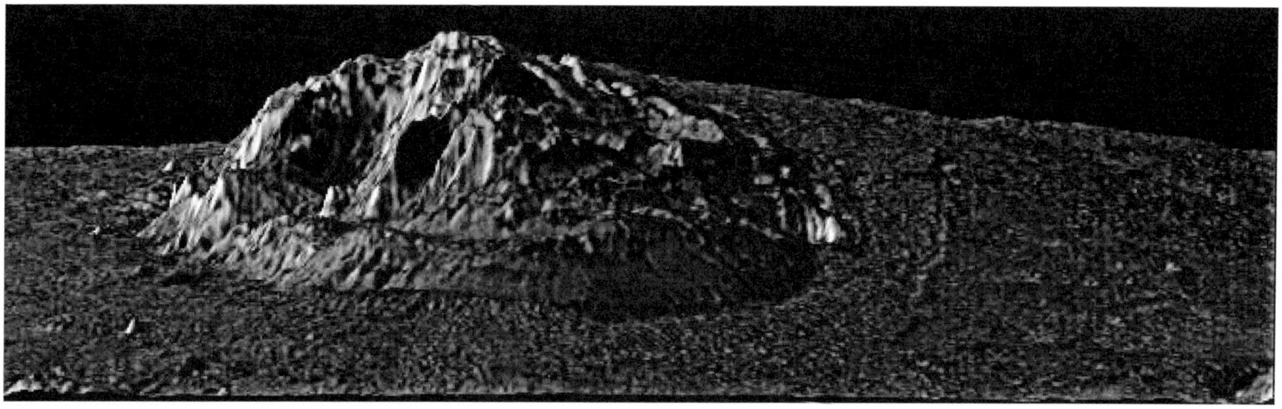

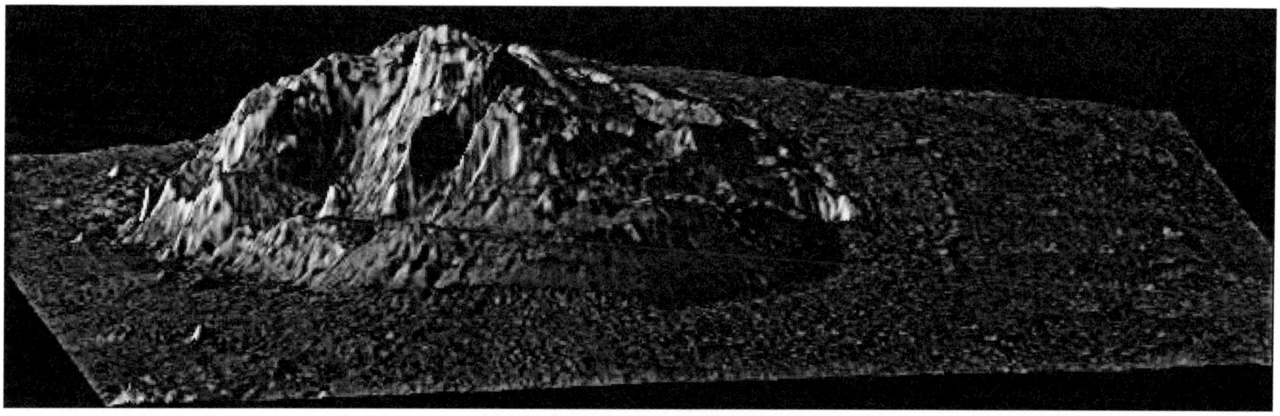

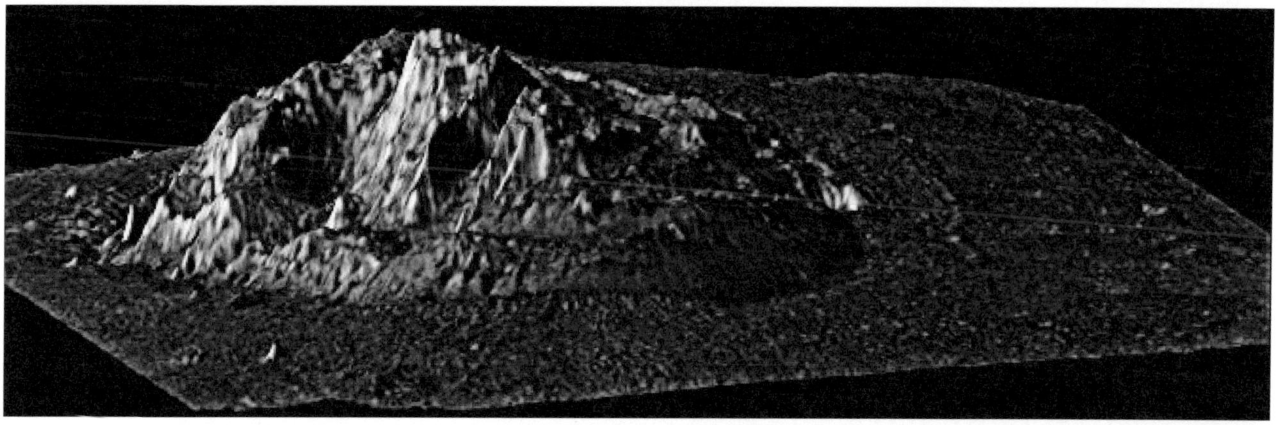

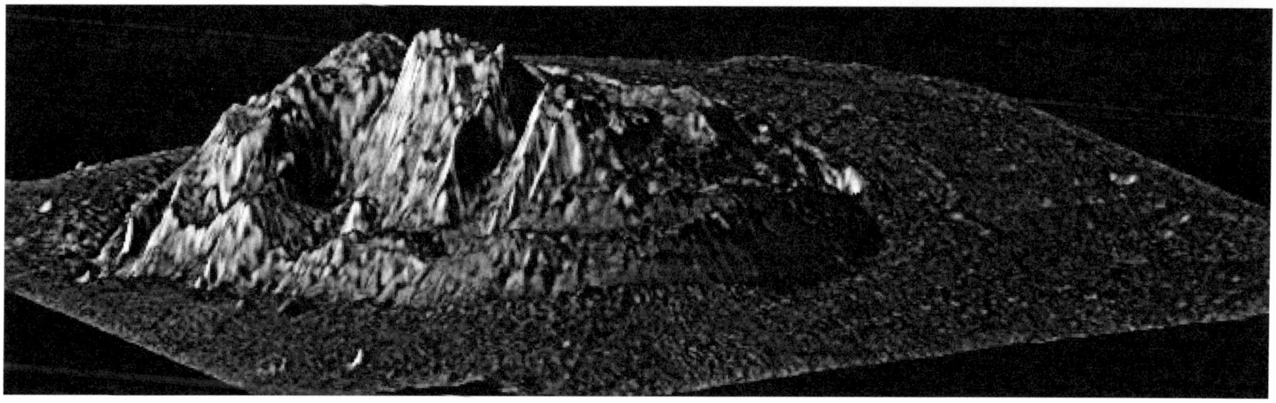

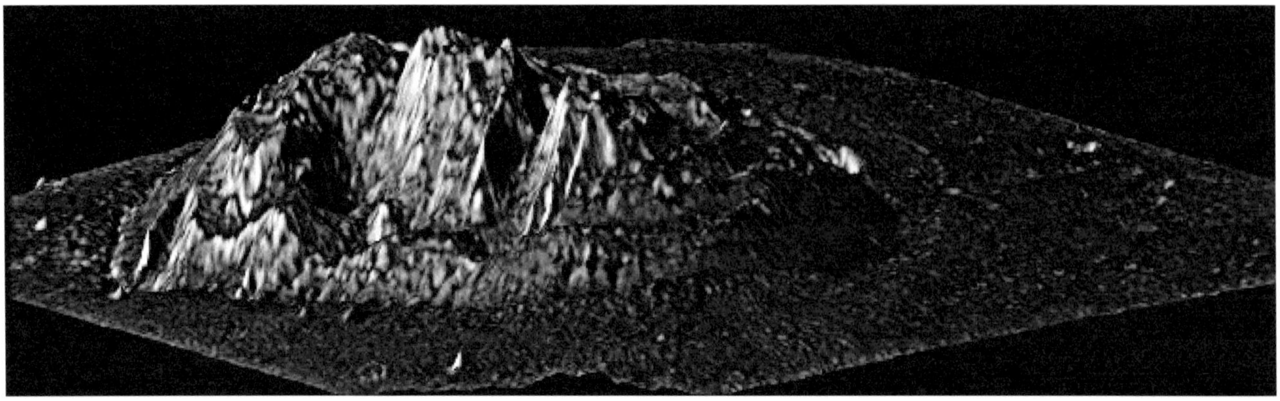

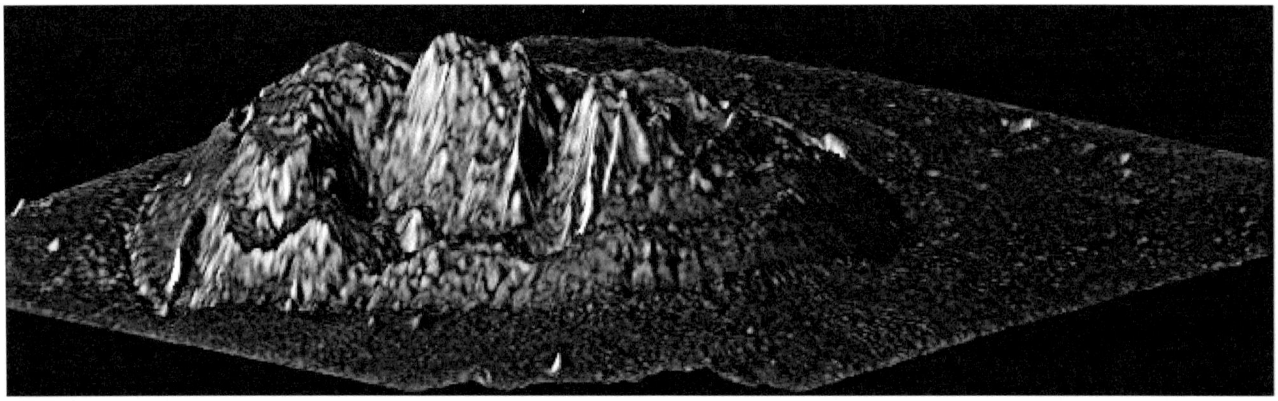

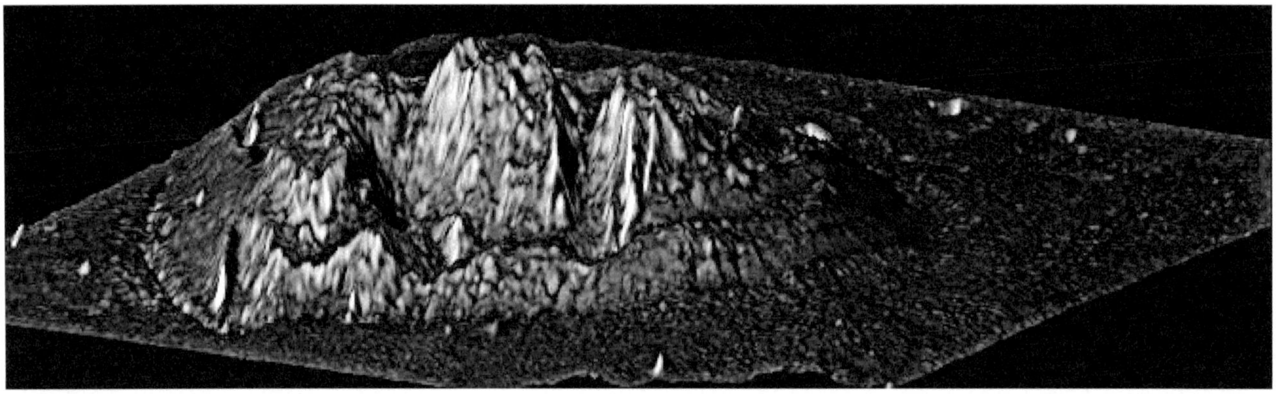

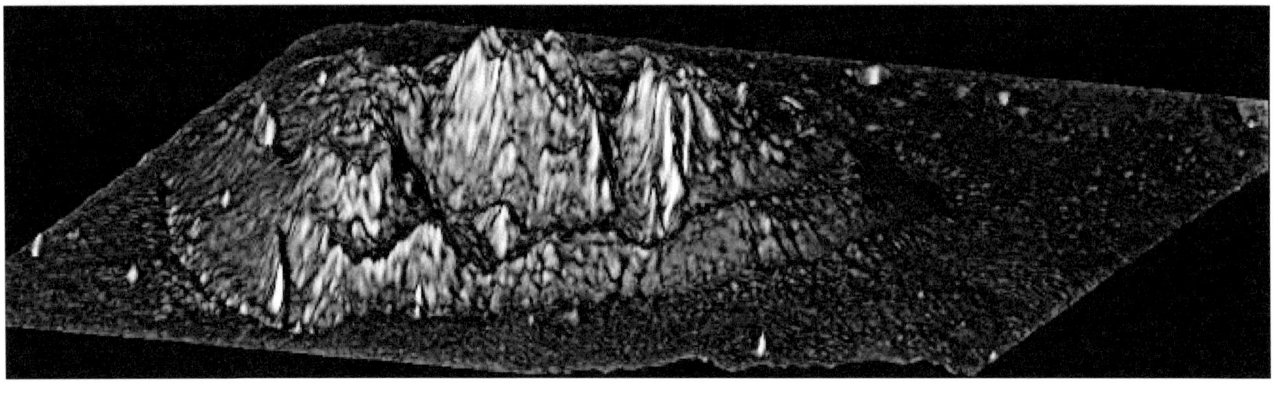

After almost five decades, we finally get to see and experience the true shaping and details of the Face. I can say with a 100% accuracy, that the '**true view**' of the Face at Cydonia is totally different and so unlike how NASA portrays the Face in false image presentations to the public and their official un-scientific wording of what we are actually seeing here. The old famous comment from the astronauts of the Apollo mission radioing in to mission control keeps coming to my mind, **"HOUSTON, WE HAVE A PROBLEM!"**

It is a problem that needs to be seen, and addressed by NASA, and by mainstream and independent Mars researchers alike, because it obviously raises questions that need to be answered, and raises many eyebrows, as never before has the Face ever been seen presented as shown here in any venue from any mainstream or independent Mars researchers. It is something new to chew on as it were. To see another perspective than the one that has been indoctrinated and falsely shown to us all by official narrative and policy, it creates a shock or even disbelief by some who are not so very receptive to this truth, because of what they thought they knew to be true. Some in seeing this truth, makes them wish and want to remain deceived to NASA's false truths. The reasoning is because as controversial as the Face was and still is, it is and was easier for some to feel better about what is seen or acceptable in their mind as to what they thought was true, (by NASA's outright obfuscation and deliberate lies in how they presented the Face) because of '**fear**' of what is now seen and being presented that changes everything.

The Face is now brought to life as it were. Making far more visual sense even for the Face's artistically sculpted surface relief being seen and what it actually makes the Face's likenesses and its shaping stand out as it was originally created to do and be, when designed and carved by the capable intelligence and abilities of its artistic creator. NASA obviously seen and still sees this truth, and still to this very days in 2025, they are still pushing the policy of obfuscation and deception using incorrect and manipulated image data to continue to force their false perspective of the Face to keep being drilled into all our heads and be accepted by the masses.

Sunrise to Sunset on the Martian Mimetolith

https://www.uahirise.org/hipod/ESP_080970_2210

To make my point, on April 25th, 2025, the Hi-Rise Picture of the Day (Hi-POD), released some data for the Face at Cydonia. In fact, in their very wording for it, it was just a doubling and tripling down of their official unscientific denial of the Face being a techno-signature creation of intelligence, and reinforcing that it is only a natural feature. I will like to add here another note of some importance, that this was released on Richard C. Hoagland's birthday! Richard Hoagland has famously and scientifically presented a case to NASA and the world regarding the artificiality of not just the Face, but of surrounding area features of Cydonia as well, that shows intelligent and artificial surface designing. I believe they provided us all on April 25th, for this Hi-POD of the Face, to not honor him, but to mock Hoagland!

Yet adding insult to further injury, in the wording provided, it shows another desperate attempt, to re-classify the Face with a human condition that is with them trying to continue to convince us all, that what we are seeing, is really not what we are really seeing. Adding a new term to apply to a faulty human condition that deceives us all of what we are well able now, to clearly see! The truth of the Face!

It is also worthy to note, that I have been showing images and videos on the internet since 2024 of the 'True Shape' of the Face as you all seen in this book. What you see and have seen here in this book is also no secret to NASA, as I explicitly had sent NASA and Hi-Rise team links and questions and comments about it. So I also am taking this April 25th, 2025 Hi-Pod release as another attempt to triple down on their official public position in reply to what I have been showing publicly also. Here is a set of images they also put forth within the Hi-POD release, and once again, using incorrect 3D perspective views and false shaping of the Face they still want us all to see and to think it really appears as.

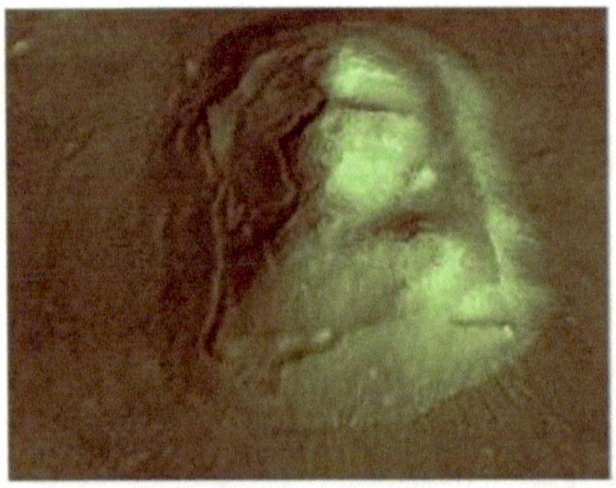
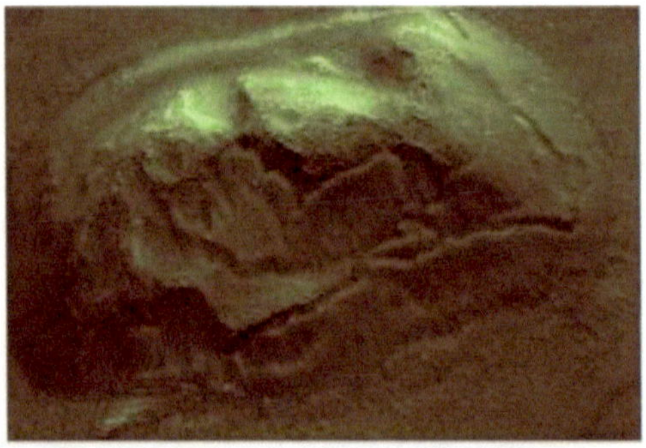
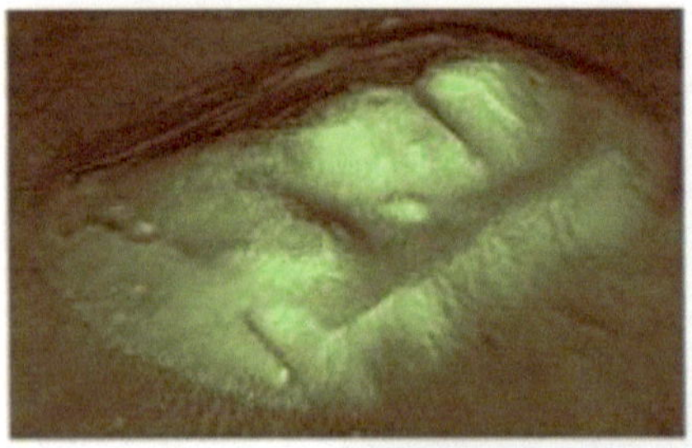

https://static.uahirise.org/images/2025/details/cut/ESP_080970_2210-4.jpg

So after seeing their presentation, I decided to also make my own presentation image with using the same 3D perspective image position views of the Face, but showing its true shaping. **Once again, what I reveal actually makes more visual sense when comparing to what NASA presented in their continued attempts to mislead and obfuscate the truth from us all in their 3D perspective views.**

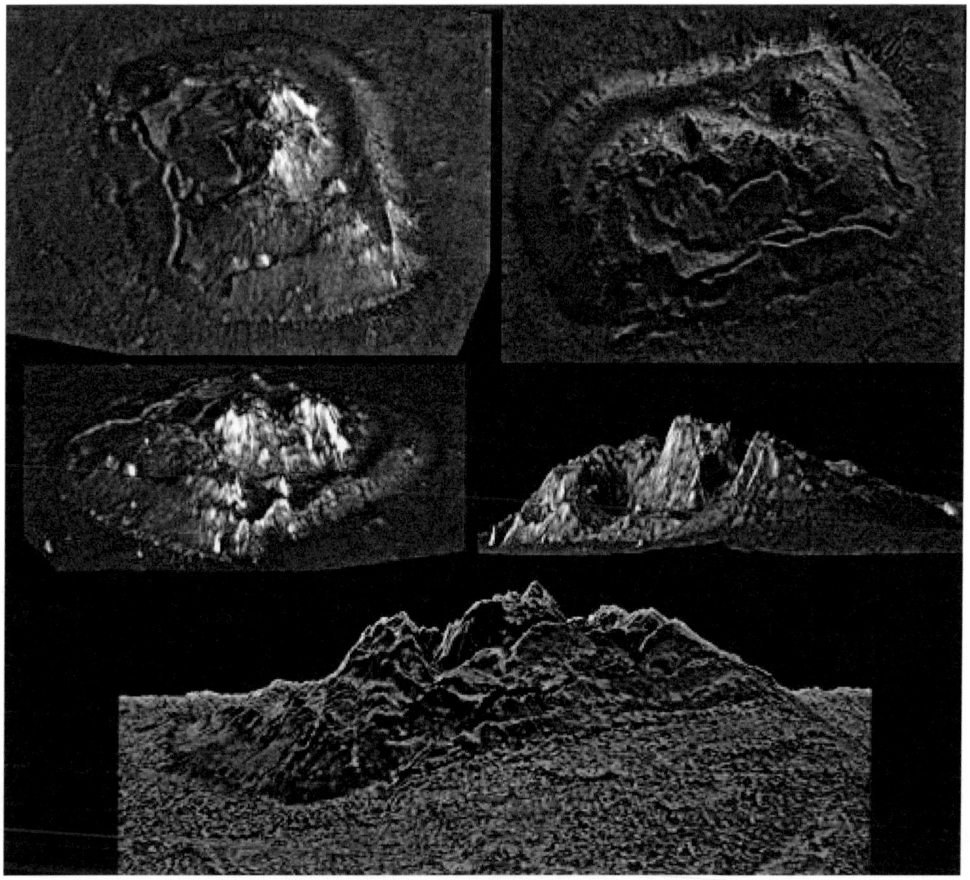

With this in mind, seeing what they written as official statement for this Hi-POD release, shows their unscientific hypocrisy, to again, add insult to injury to blame us all for seeing this amazing intelligent techno-signature for what it really is, and trying to force us all to accept their un-scientific psychological diagnosis speculations as the reason for what we are really seeing. As you all seen and learned of in this book, NASA has demoted once again the Face from being known and called '**The Face**', to then being called '**a popular landform in Cydonia**', then going to being called '**terrain sample in Cydonia**', is now receiving a new title. The new title being attributed to the Face is "**mimetolith!**" Let's look at their official wording for the April 25th, 2025 Hi-POD of the Face;

"This particular mesa acquired a certain notoriety when an image was captured by the Viking 1 orbiter on 25 July 1976. The shadows cast across the feature created a pareidolic perception of a human face. This feature became one of the most famous examples of a "mimetolith," a pattern created by rocks that may come to mimic recognizable forms through the random processes of formation, weathering and erosion."

It appears to me that NASA is desperate now to continue to try to reinforce their official policy position. It appears to me also that now NASA is finally addressing the Face's surface relief itself! Just simply and falsely stating for years that it was only '**tricks of light and shadow**' that made the Cydonia feature look like a Face. The **surface relief itself, was never addressed as even being an issue before** that helped the feature appear as being a face!

It is in my opinion of course, that once NASA seen what I was revealing to the world on the internet and what I have been sending to them directly to see, they seen and know that what is seen is only because of what the surface relief of the Face actually shows to reveal in the shape from shading evidence examples. We know they are addressing now finally the surface relief because of their own words, ***'a pattern created by rocks that may come to mimic recognizable forms through the random processes of formation, weathering and erosion."***

It is kind of funny actually because they address a pattern! That was **CREATED** (the obvious correct term that is kept silent by steering away from intelligence), by **'random natural processes of formation'!** NASA subtly admitting that the Face is indeed patterned, and processed (created) as a formation! Created by what? **ROCKS!** Rocks that create patterns to resemble or mimic recognizable forms, by using **'random natural processes such as weathering and erosion'.**

So we are now being told that the rocks are to blame, for creating recognizable forms. Also being told to us all that **the rocks,** is using random processes of formation (which they cannot describe legibly) and with weathering and erosion as being the tools that **the rocks** used, as being the artist, to create mimicking recognizable forms! Such as faces and other geometric shapes! Smart rocks! This excuse is so unscientific and absurd and is comical. It is a continued avoidance on NASA's part to show and tell the truth!

This is all just new information being given to us all to add to the rest of the preponderance of evidence of NASA's obfuscations and deceptions as we have seen laid out in this book. I am not complaining however, as we are getting more information and data to add to this book, making this book literally providing you all the most up to date information and images for the Face, to help make my case even stronger to help validate even more what I am revealing to you of the Face.

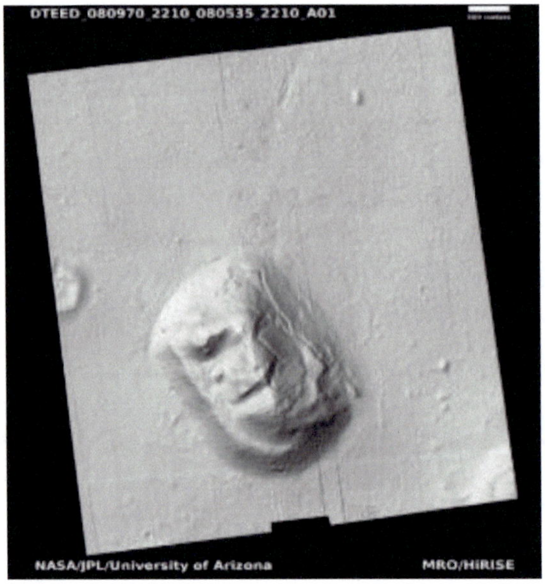

https://www.uahirise.org/dtm/ESP_080970_2210

Along with the Hi-POD April 25, 2025 release, we are given a Stereo Digital Terrain model (DTM). A stereo pair of this feature was taken by Hi-RISE and was turned into a digital terrain model. Hi-RISE DTM's are made from two images of the same area on the ground, taken from different angles, on different orbits. Hi-RISE DTM's provide valuable data, not only for researchers, but also for visualizing the terrain.

But what we have below, is the stereo pair **shaded relief** DTM model of the 2 latest Hi-Rise Cydonia Face images given to us in November 2023 and December of 2023.

The shaded relief is a form of photoclinometry, or shape-from-shading, is the process by which a 2-dimensional image of a surface is transformed into a surface map that represents different levels of elevation. It uses the shadows and light direction as reference points. It is used mostly to depict the surface of sculptures, to give an idea of how it would look in 3-dimensions.

The techniques depend on very specific conditions, especially light direction, as Shaded relief maps are particularly useful when it comes to representing the three-dimensional surface of the Earth, or even of Mars! By using light and shadow to mimic the effect of sunlight on the terrain, these maps provide a realistic representation of the land surface. Shaded relief, also known as hill-shading, which adds a touch of realism to map representations, by showcasing the shape of the terrain as it would appear when illuminated by a point light source.

Shaded relief DTM model overlaid with Hi-Rise grayscale image of the Face, to help accentuate relief details.

Another treat given to us all from the Hi-POD April 25th, 2025 release is the colorized elevation map created by the shaded relief DTM. DTM's are very powerful research tools. It allows us to better understand what we are looking at regards to its height, when looking down at the surface features being looked at. This is done by using a spectrum of colors representing highs and lows, as seen by the shaded relief image model, with white representing its highest point levels, and purple representing its lowest point levels. The colors in between represents obviously the height levels in between. This crucial and needed information also helps us to determine the Face's accurate and true shaping.

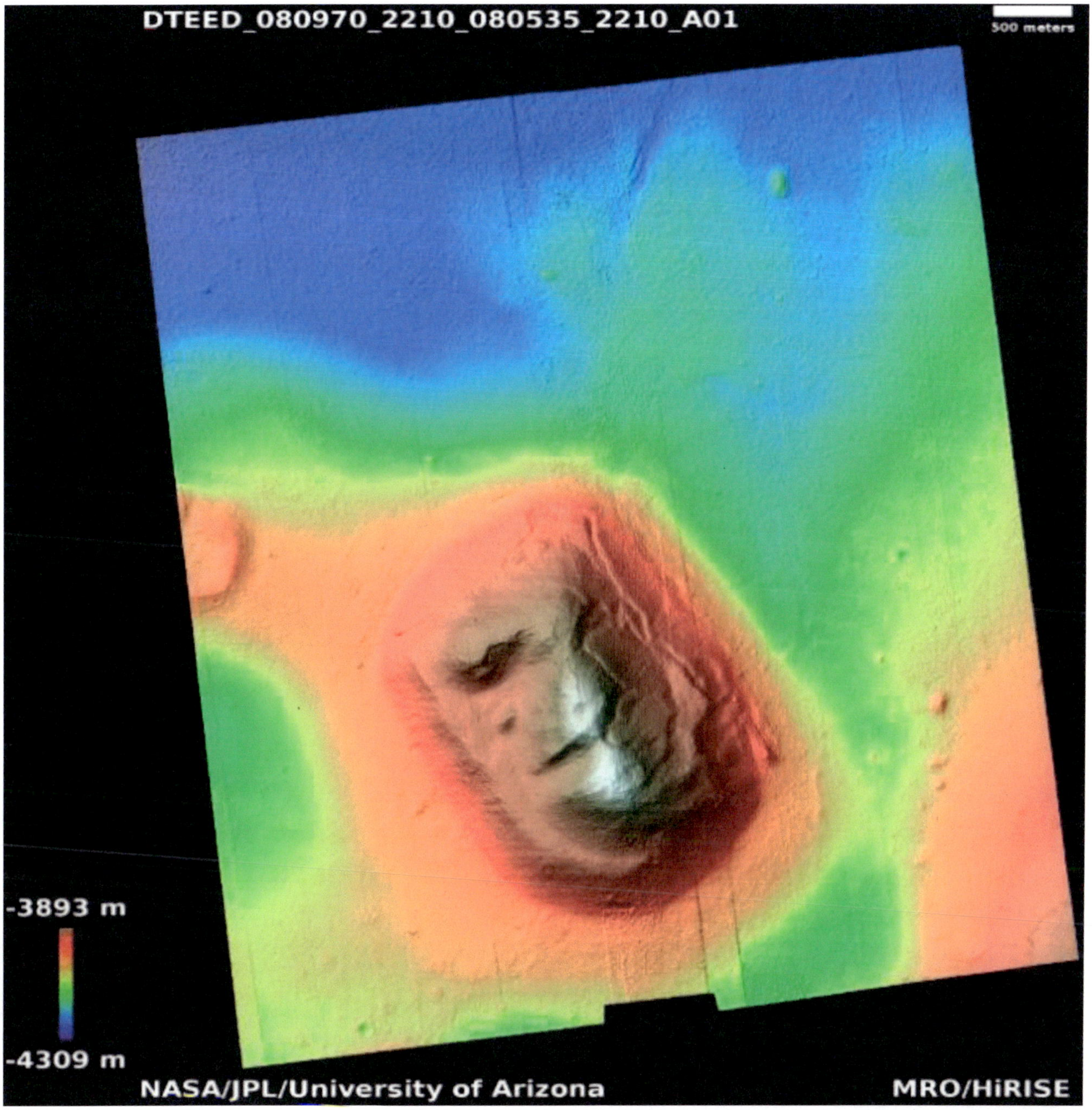

According to the colorized elevation map, it shows us all that the top of the Face is reaching up to 602 meters in height, which in American standards means that the Face is 1975.066 feet tall from a lowest bottom surface color level. The Face sits upon a mound that its orange and yellow sits upon the green level. So the Face from the green level to the white level is 502 meters in height, which means the mound the Face sits upon is in American standards, means that the Face mound and the Face that sits upon it, is 1646.982 feet in height.

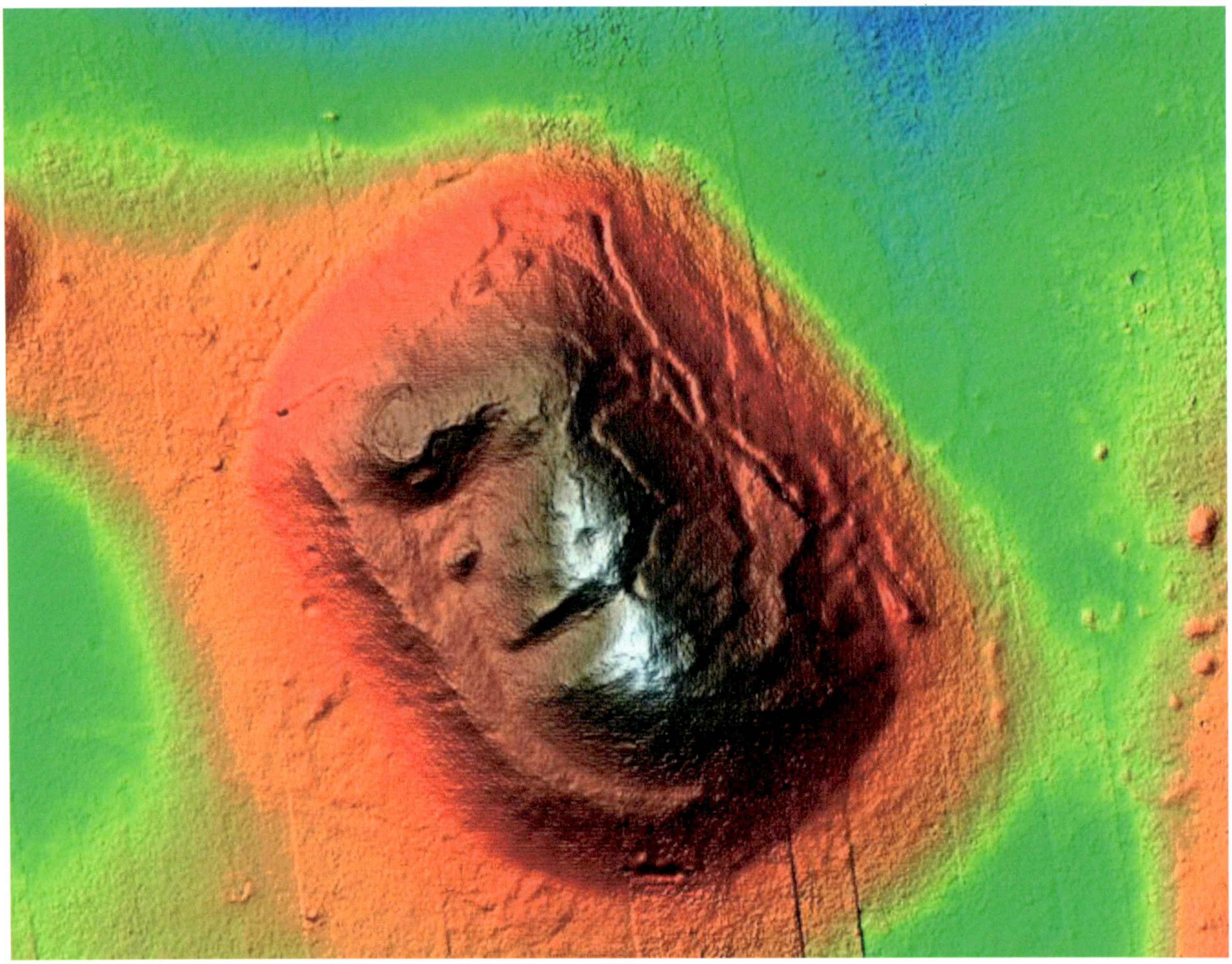

Therefore the base of the Face itself, seen in the red up to the white, is showing us all that the Face is **1500 feet in height!**

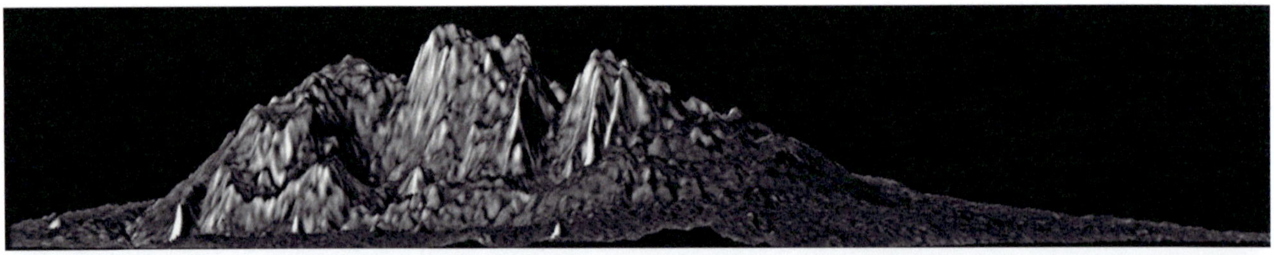

Added also to the Hi-POD from April 25th, 2025, was given to us all an interesting treat. A model was provided to us all of the DTM (Digital Terrain Model) of the shaded relief image, that allows the added data to help me **validate my claims** I made in this book, **and what I presented as my evidence** as to the true shaping of the Face. In NASA's official wording on the Hi-POD release page states: ***"For example, with the DTM we were able to recreate the shadows on the day of the original 1976 Viking image.***

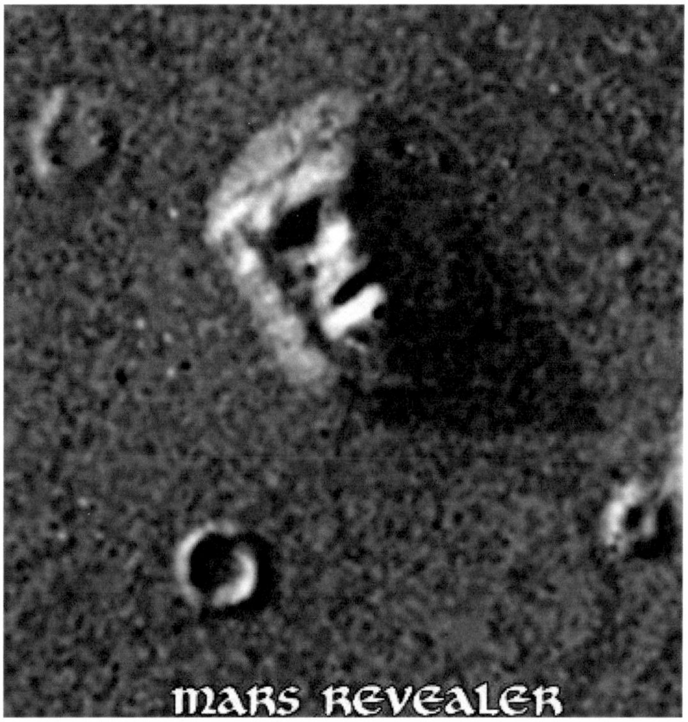

"By modeling the sun angle as it would have appeared on 25 July 1976 throughout the day, the features of the 'mimetolith', only take shape when the lighting is lowest just before sunset. Visualizations such as these provide an accurate view of what could be seen by a ground-based or aerial explorer.

Although it is a recreation, it is a good way to test the data collectively, to see if it is consistent to the information and especially what the unprecedented image views I have presented showing the true shaping of the Face at Cydonia is revealing. The fact is, **it is very consistent!**

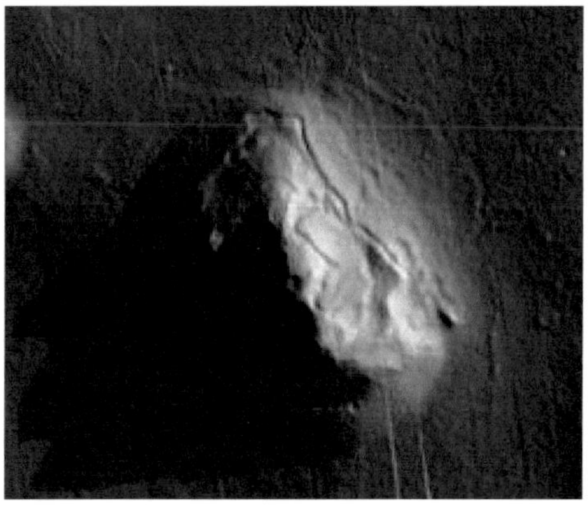

GIF VIDEO LINK FOR SHADED RELIEF MODEL WITH 'RECREATED' SHADOWING:

https://static.uahirise.org/images/2025/details/cut/ESP_080970_2210-2.gif

In my opinion, **why conduct recreations** to mimic the shadowing of the face from sun up to sun down, when we can simply have visible imagery released publicly showing the shadows positions on the Face, and around it using actual image data itself?! That way we can compare that multiple image data examples with the results of NASA's '**recreations**'? It is a valid point and also a very valid question to ask. After so many years since the famous 35A72 image was taken in 1976, we have yet to receive any modern imagery from any sources that recreates what was first imaged by Viking Orbiter in July of '76.

Well, I cannot say that is entirely true, as the European Space Agency (ESA) and their Mars Express mission, did in fact take an image in orbit above Cydonia, when **we are being given a view of the Face when the shadowing was seen in about the same position as seen in the Viking Orbiter image.** The problem is that the resolution of the image and its quality is much lower than that of the Viking Orbiter image 35A72 itself.

MARS EXPRESS IMAGE H_2872_0000_ND3

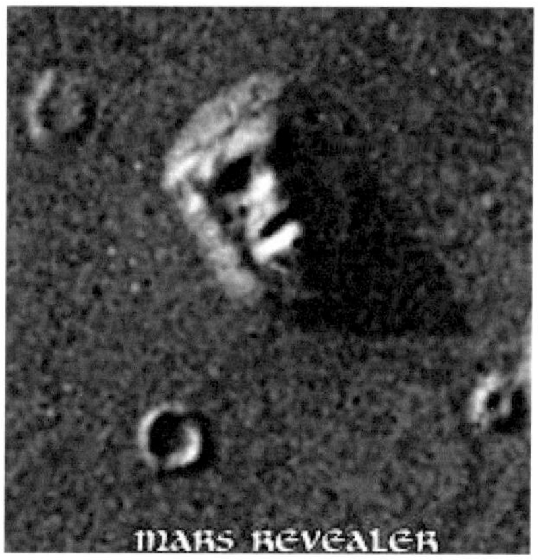

With this only Mars Express image **H_2872_0000_ND3,** is what we have to see to be able to compare. No imagery from any other space agency's Mars orbiting missions have provided us all with these incredible views. It is a joke we only get the low quality view from Mars Express which has an incredible imaging system, capable of much higher resolution, only provided us all with anything of its shadowing nearly recreating the 35A72 image, so it can be done.

In fact, NASA's Mars Odyssey orbiter's THEMIS visible imager has shown that there should be no issues acquiring such images of the Face showing its shadowing positions. NASA reveals that fact because they provided several images of the Face being seen before sunrise, or near sunrise, or after sunrise, **(but after its shadowing being cast away from the Face)** that show what the Hi-POD April 25th shaded relief model of the Face shows when we see the **'recreated'** shadowing on the Face that NASA applied to it. That Mars Odyssey image number for the best detailed image of this is early morning view is: **V96459003.**

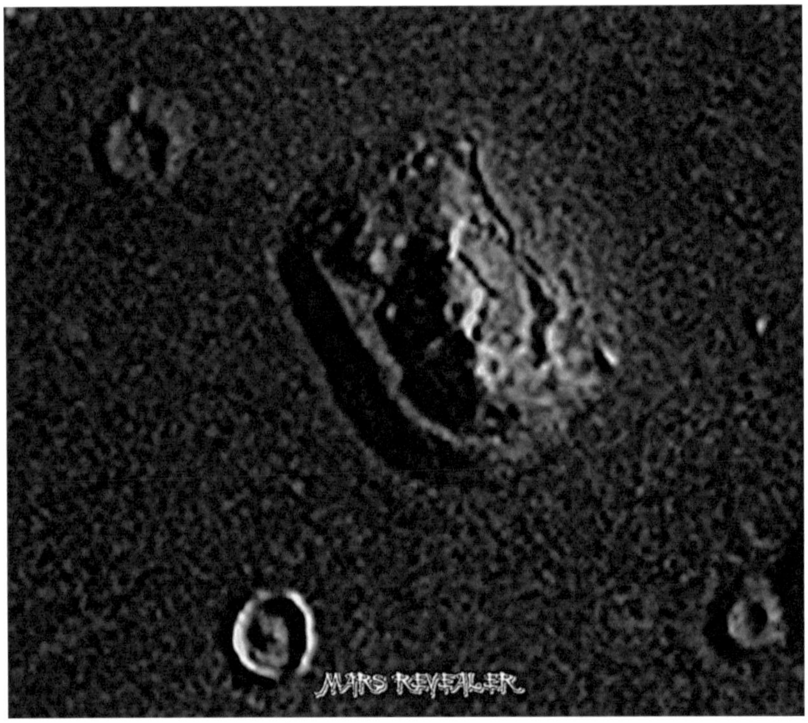

MARS ODYSSEY IMAGE: V96459003

This is a great image, especially out of all the early morning views of the Face from Mars Odyssey. These views although lower resolution than modern high resolution imagers, these views add much validating support of my claims and evidence in showing and revealing the true shaping of the Face. It also helps prove my claims about NASA's obfuscations and official claims for the Face is also is also validated as even this Mars Odyssey image view is even helping to contradict NASA's own claims in their official statement positions, and also contradicts NASA in the **false shaping** images of the Face they provide and claim is seen in 3D and is actually how it really looks in reality as they present it, **(as we have seen examples of here in this book).**

This Mars Odyssey image **V96459003** itself alone, **'DEBUNKS'** NASA's official positions. **Why?** Simple. Because of its **'Tricks of Light and Shadow'**, that they keep trying to hide, in order to help hide **the true**

shaping of the Face! This is exactly also why we never get to see high resolution images of the Face seen in various shadowing positions to see its relief contours in various lighting conditions that exposes its shaping.

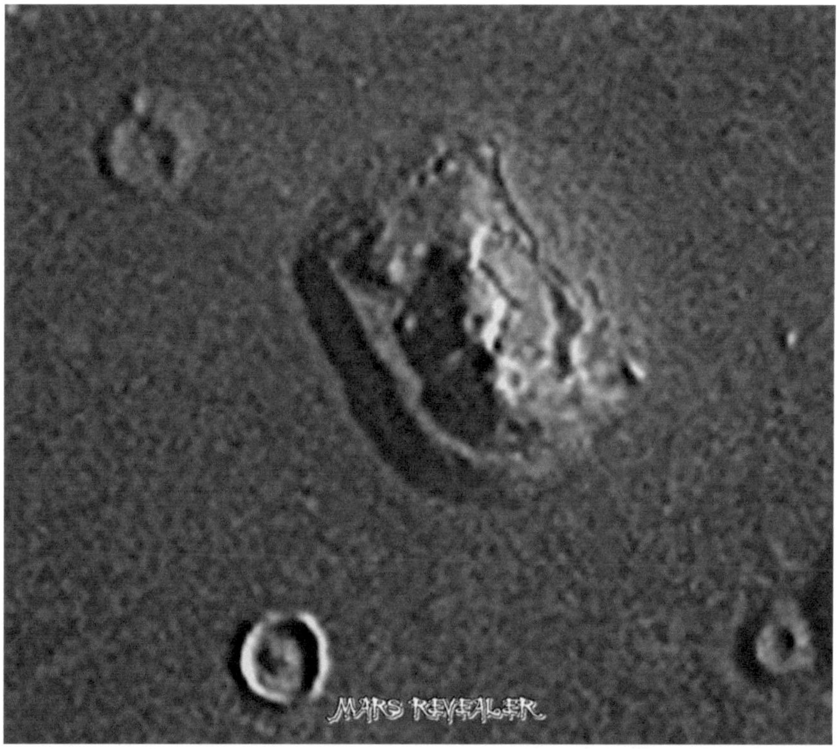

Mars Odyssey image V96459003 overlaid with the Shaded relief DTM shadow model.

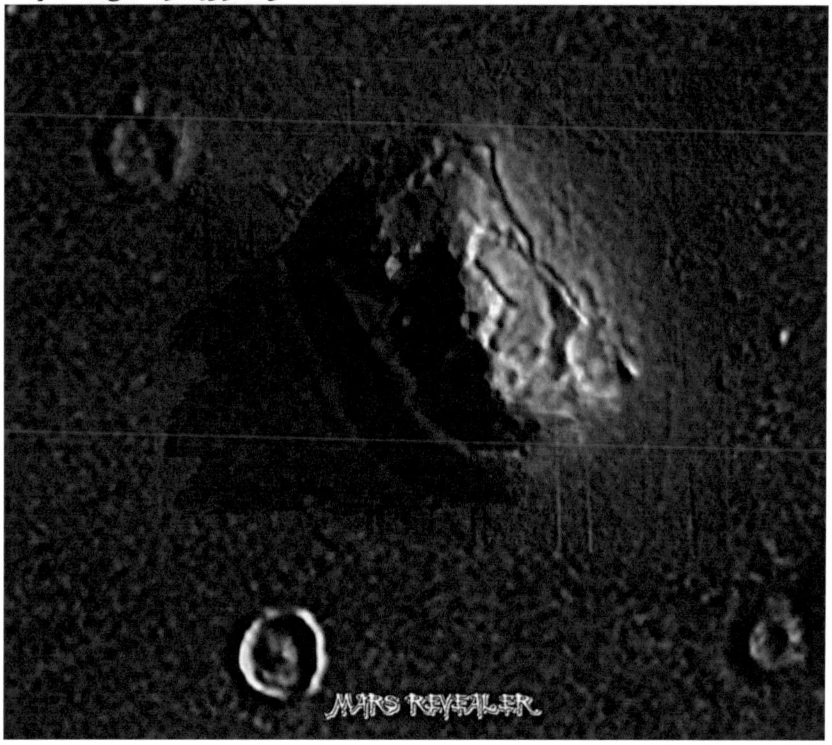

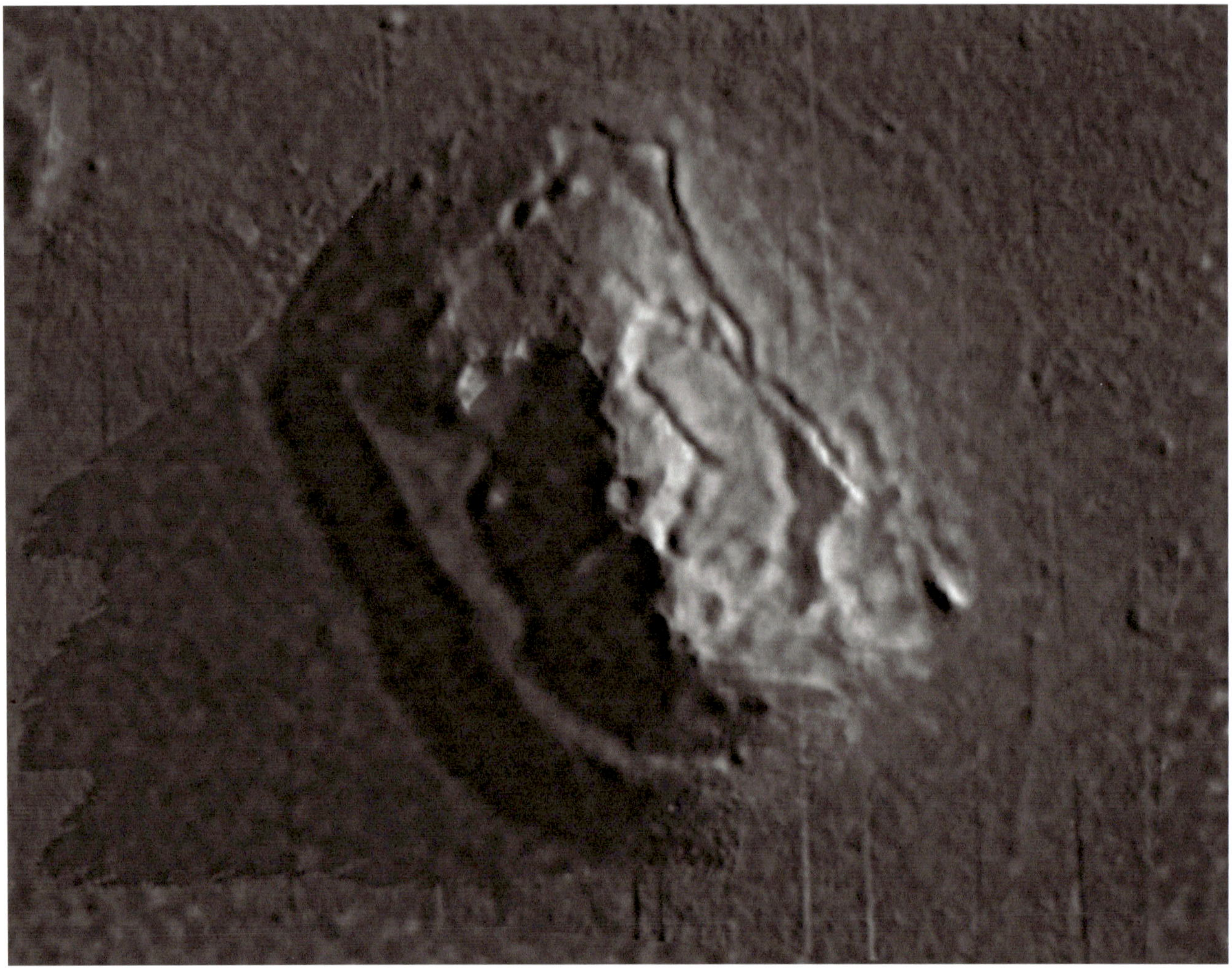

Mars Odyssey image V96459003 overlaid with the Shaded relief DTM shadow model.

As you watch the gif video start, you can see that the Face at Cydonia is casting its shadow to the left. Look at that shadow '**carefully**'**!** That shadow does NOT look like anything that would be put forth by any of the so called 3D perspective views NASA puts forth over the years **(that you noticed never has the same shaping)!**

How can the Face as seen in those '**FALSE**' so called true shaping views from NASA, put forth a shadow that resembles, a **HOMINID?!** To be specific, it's casting a shadow that resembles a gorilla-like side profile as it were!

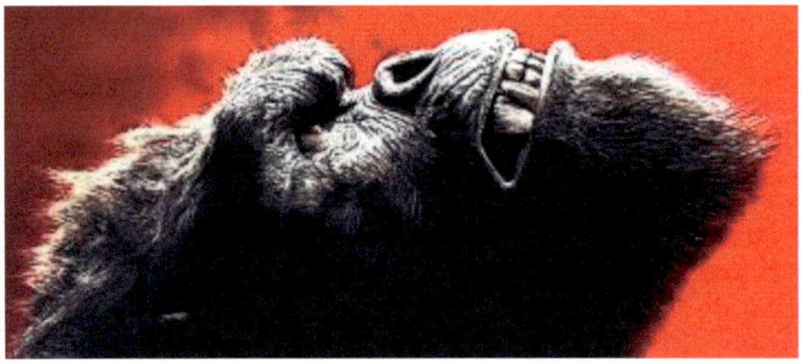

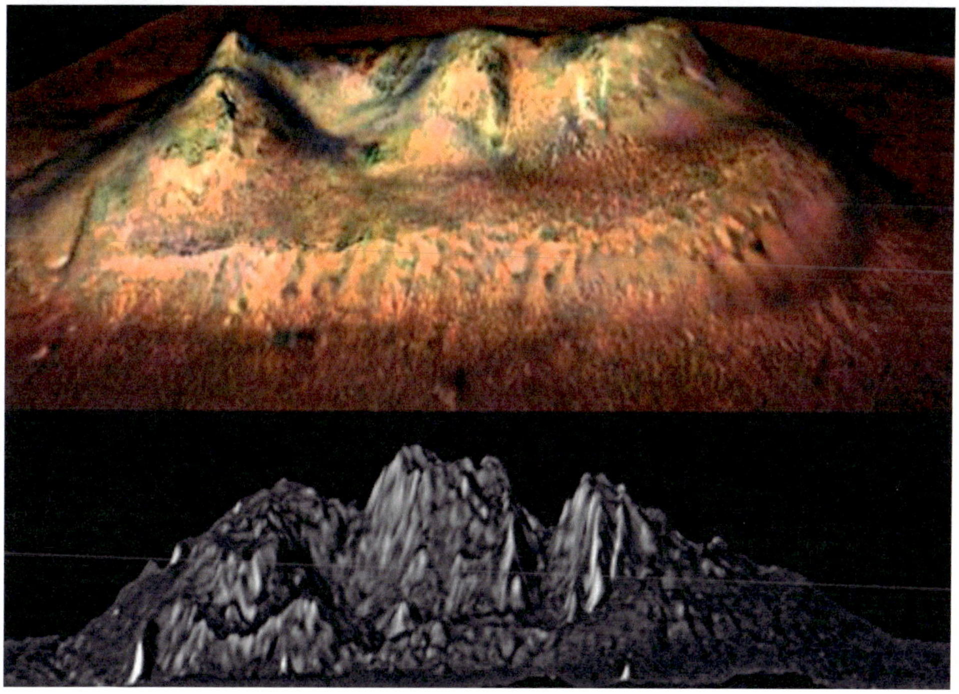

The brow is uncanny running into its forehead. The space between the brow line above the eye and the maw of its **open mouth**, looks very consistent to the **true shaping** of the Face which I revealed for all the world, to finally see after nearly 50 years!

I have provided the side profile views of both sides of the Face showing its true shaping. Notice again the deep open mouth seen on both sides that is far much more consistent to the shadow being cast. Again also notice that NASA itself, is showing it to be by their own data presented, of how the Face really appears **in reality** as seen in the mimic by the **'recreated'** shadowing. The shaded relief shadow model also exposes the deception! **NASA contradicted and contradicts themselves and their own official claims, and can easily be seen and proven of that fact of reality, just by studying the image data which they themselves have provided to us all.**

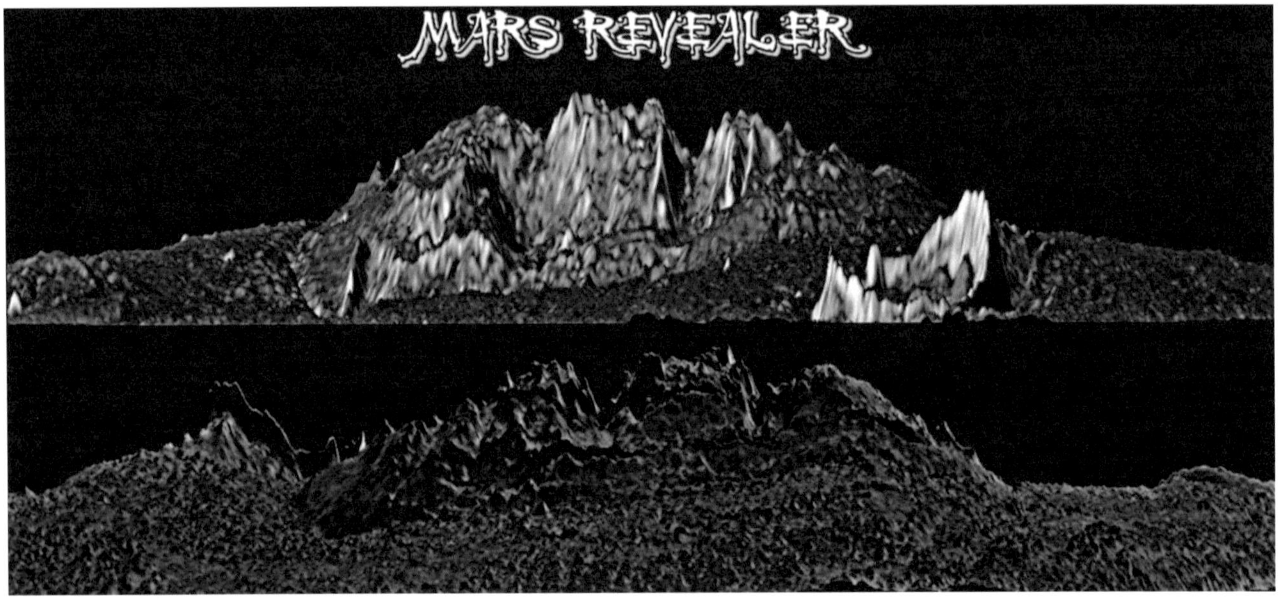

The Face is **not wispy smooth, nor is it flat.** It is not merely a naturally eroding heap of sand, dirt and rocks that NASA not even knows what to call it, a mountain, a massif, a hill, a butte, or a mound that apparently magically changes it's shaping over time with each new 3D image perspective view that NASA releases to the public.

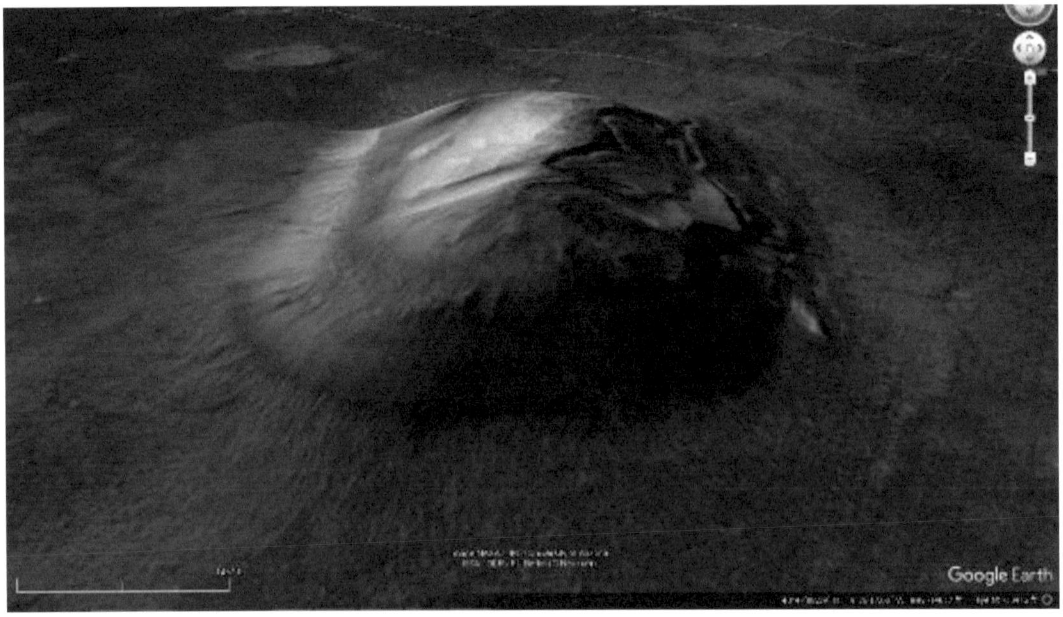

Again, look at the perspective views. **IT NOT MATCHES THE SHADOWS CAST!** Look at the contours and carved details of the shaded relief model, and see the consistency to the unprecedented evidence I reveal and present to you all in this book.

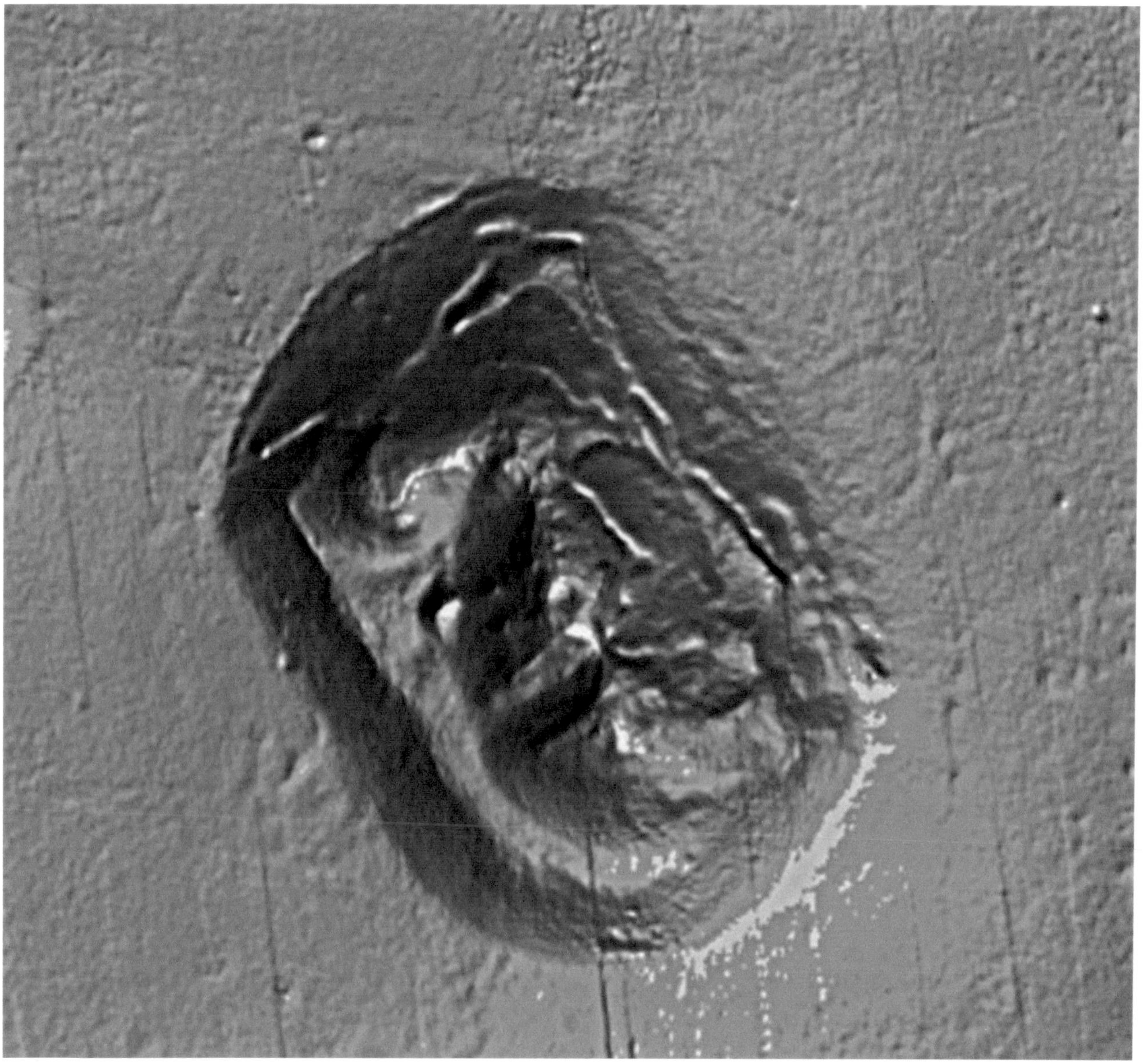

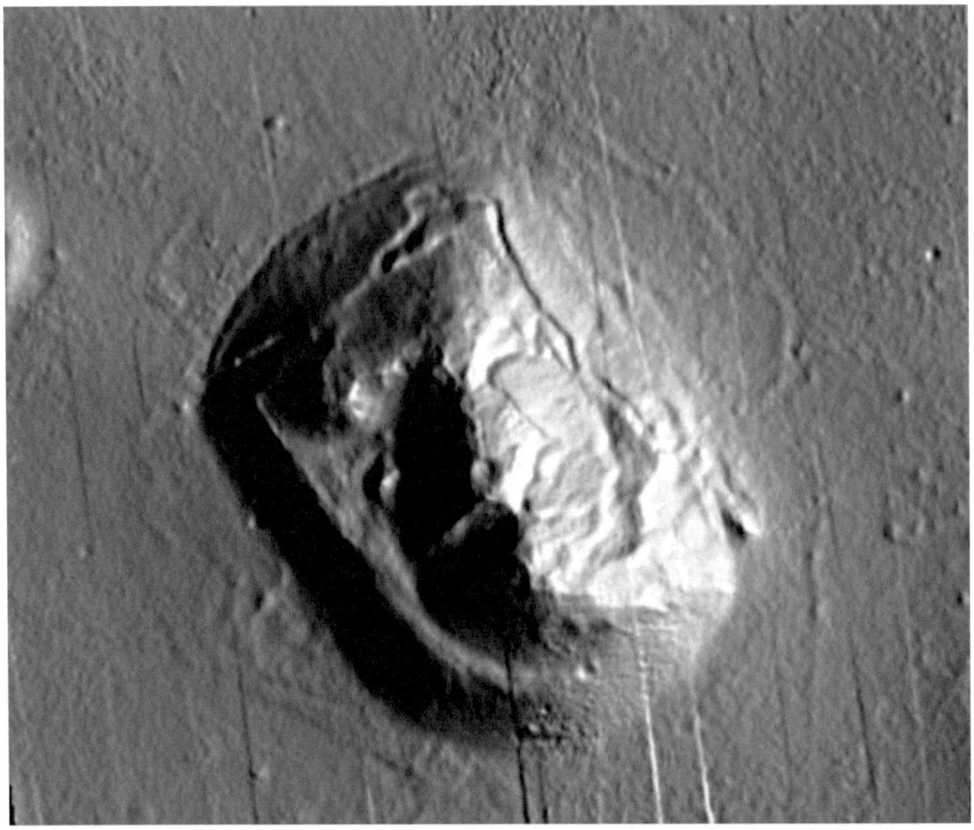

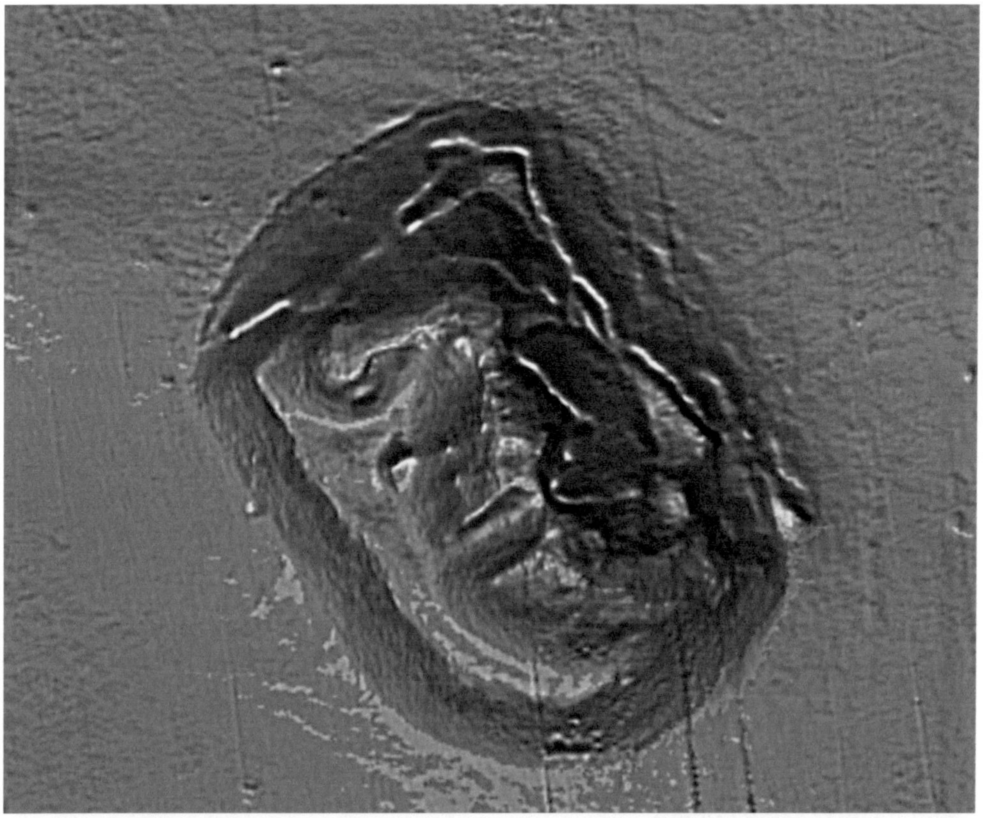

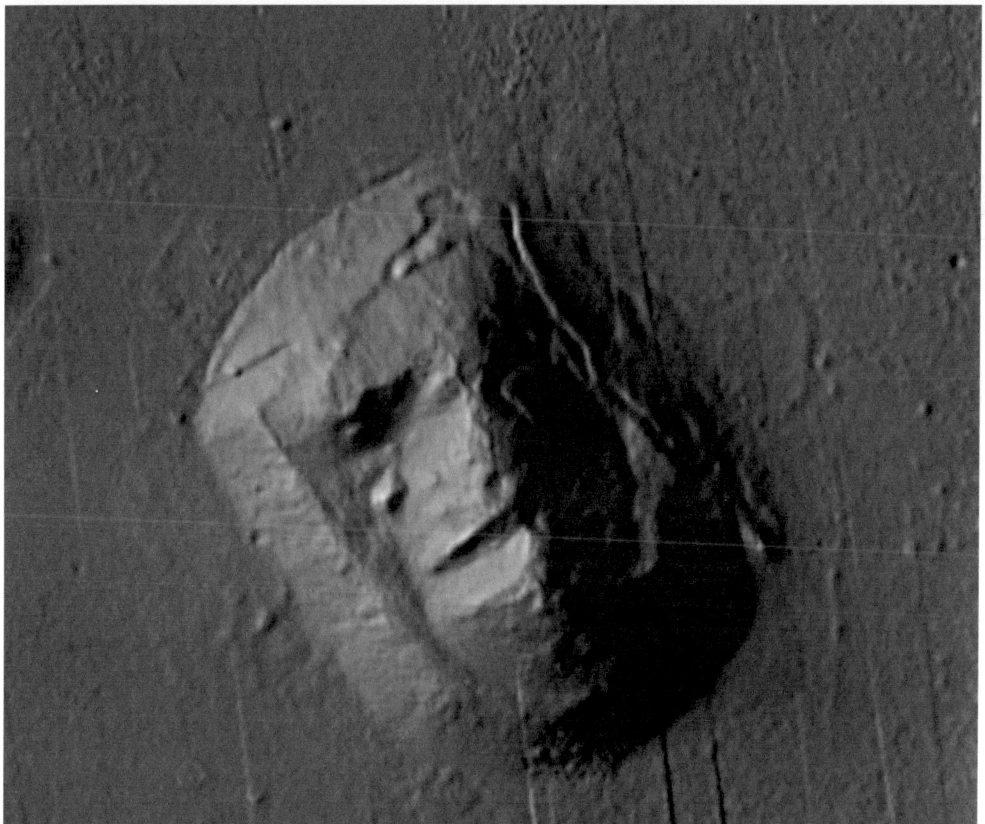

With that being said and **most importantly shown to you all, there are however other independent Mars researchers who seemingly do not want to see or accept the truth as seen and presented here.** Another known Mars researcher, (name withheld), was on a phone call with me on April 30th 2025, five days after the April 25th 2025 Hi-POD release for the Face. Obviously the Face was our subject of discussion. This 'un-named' independent Mars researcher and I were speaking about this very book as it was almost finished. I was asking him about advice for publishing this book, as he is a published author with regards to the subject of Mars.

This is where and when he became triggered for some reason, and went on a tirade with me about what I had revealed as the 'true shaping' of the Face by the shape from shadow renderings that I have been placing up on the internet for about a year before the completion of this book, and with me passing it around the internet and making sure that it was being seen by the researchers of Mars and those interested who like to watch and see what people put forth of their findings etc. I especially made it available to be seen by NASA. It was at this time in speaking with him about the publishing this book, where he wanted to scathe me about the true shaping of the Face and wanted to outright claim that what I was and am presenting and revealing of the Face is not real.

This 'un-named' Mars researcher wanted to be adamant that what I am presenting, revealing, and publishing in this book about the true shaping of the Face is false and a fault of my own for making it appear as it does. Insinuating I had done something wrong with the shape from shading program to render it to get that true shaping, and being adamant that what is seen, was not accurate. Insinuating also that the shape from shading program was not accurately doing what it was created to do and creating a false shaping and incorrect relief detail.

In essence, he was arguing against my claims, my work, and what I had to show of it all. He verbally and vehemently was trying to debunk all of that, without being able to present any evidence to the contrary.

Now mind you, I provide the image data for everyone to see, basing all of my claims on the image data. I also provide my correct processed views of the majority of the images I provided from the raw format, revealing all its details that NASA's public official releases of those same images is obfuscating/hiding the majority of all its surface relief details! I reminded him that I also have provided multiple images showing and proving that fact in this book as well. So I provide the correct processed image data, and I provide links to the majority of them to compare, and I also dared him to debunk what is seen in the correct processed imagery and also for the shape from shadow renderings.

This 'un-named' Mars researcher, seemingly annoyed with my work and presentation of the evidences I provide and reveal for all to see has NO IDEA how to actually scientifically or otherwise present anything to me or obviously to anyone at all, that can back up his claims that what I am revealing is false. Now consider this to weigh in on the 'credibility' of the annoyed 'un-named' Mars researcher who was suddenly accosting me of what my claims are of the Face and what I reveal of the Face, **does not even have the skills or even knows how to process any of the raw images** to reveal the details not so easily seen or even not seen at all in NASA's public releases of the same images.

I told him that **while not meaning any disrespect,** (as we know each other for many years and are friends and colleagues in this research) that he not knows what he was talking about in regards to the true shaping of the Face as I reveal it to really be in order for him to debunk it, again without providing any image data evidences or even any knowledge of the rendering process, that would show and prove to the contrary. He had and still has no valid argument to successfully debunk what I reveal.

I told him also that I have been studying the Face so very closely since 1998 **(which he knows as to that fact),** and in order for me to really study it and to learn about it, I had to learn how to reveal what NASA obfuscates to allow me to reveal that very so important surface relief detail of the Face and its surrounding terrain. I have found no one also doing the same, **including himself**, to be able to reveal these so important details.

So I told him respectively, that he and I cannot even speak on the same level of understanding because of **his ignorance to the details of the Face itself, let alone its true shaping.** All he and the majority can only do is to make '**enhancements**' to images which can help some details be seen better than originally presented in NASA's public releases of imagery. These '**enhancements**' usually do no justice to actually being able to view and correctly process the raw image data that can be revealed of its surface relief details without enhancing anything.

I also reminded him also that as for many years as we known each other, that he claims falsely as NASA does, that the Face is eroded, and eroding and is made of mere rocks and sand and dirt, that was sitting upon also a plain of mere desert sand and rocks and dirt. The evidence I provide, because of the so important surface relief of the Face and surrounding terrain, shows otherwise! So how can we speak on the same level about the veracity and validity or lack thereof, in my claims and presented evidences and images, if he not even knows what he is looking at regards to the Face to be able to simply debunk my claims and my work for all these years without doing his own homework to challenge what I present and claim? That is how science works!

I reminded him that since because of his ignorance of the surface relief details of the Face, as well as its surrounding terrain features, that he was also only studying and misunderstanding what is seen in the images because of the fact the images are in 2D, NOT in 3D! I again respectively told him that he not knew what he was talking about or even what he was looking at to determine the Face's truths. **He does not know how to replicate or even test the image data I present, or let alone how to render the 2D imagery into 3D imagery using the shape from shading program used to reveal the truth of the 'true shaping' of the Face at Cydonia.** It is indeed valid and REAL!

This is where the 'un-named' Mars researcher scolded me and claimed that he was a sculptor, and that it was and is a valid proof to debunk my claims and evidences presented regards to the surface relief and the true shaping of the Face and surrounding terrain. So once again he rudely stated that I was incorrect of my claims and work and evidences presented. I reminded him that I also am an artist, and that has nothing at all to do with how to reveal this truth. Being an artist or a sculptor is a helpful tool to try to determine what we are seeing in 2D imagery, but is not a valid evidence or proof without being able to present and show that evidence and proof!

He then told me that **he was going to also be writing a book** for the Face at Cydonia. I told him **"GOOD FOR YOU"!** (Now I was not sounding very respectful as it was obvious he was all of a sudden after so many years of my claims, and also seeing my evidences shown over the past year of the true shaping for the Face he was deciding to confront me and sadly attempt to debunk me and my work with only words and feelings).

He then went on to tell me that he does not agree, nor will he ever agree or '**believe**', that what I was claiming and most importantly, what I was '**showing**', is anywhere near accurate or real. He was acting or pretending, to be an authority above me as like I was asking for or even needed his approval to reveal the truth of the Face, as evidenced and shown to you all here in this book.

Instead of making any statements to me about his disapproval of the evidences and the images and videos I posted online for at least one year revealing the true shaping of the Face, he waited until him and I speak on the phone with me asking for his advice on publishing this book, to say all this to me? I was offended and he shown me no respect as to my work and claims, as it is not my ego speaking when I say I am his superior in this research in terms of ability and skills to process images and to reveal what is seen in them, as well in diligence of looking at and studying and finding and revealing features and structures on the Martian surface than he himself is.

I reminded him that he was not an authority, especially above me in this research, to merely make claims that my work and claims are false! Again without any scientific or any other methods to prove me wrong! This was not professional of him to do and say to me, nor was it scientific! He was simply triggered, for some reason. Later as I thought about the conversation that took place, I remembered how he all of a sudden out of nowhere told me he was going to be writing a book for the Face at Cydonia.

I then understood then what really the issue was that must have triggered him, which was me daring to publish a book about the Face where obviously it will **contradict** how he was hoping his book for the Face will turn out as well as **conflict** with what he will be making and presenting as a case for regards to the Face inside his future book. He cannot compete with it and certainly will not be able to obviously be as convincing. So after me telling him that he was not an authority on this subject, I stated to him I however am an '**expert**' on the subject of the Face (which I am), and I not needed or wanted his approval of my claims or what my work shows. I just wanted his advice to help publish this book. This is my work, my case, my presentation, using and being based upon the image data, not my feelings or my beliefs!

He mockingly replied to me in a voice of sounding as if he was an authority above me regards to this subject and **telling me that he not wanted to see me embarrass myself!** To this **repulsive and insulting** statement from him, I could not believe he said this to me and I grew angry and I tried then to explain to him the process used to show the truth regards to revealing the Face's true shaping. He simply not understood anything of what I was trying to explain to him in a professional manner providing details, and he simply said '**You need to check your math**'! Check my math? What math? I told him there is no math to do in order to reveal the true shaping for the Face. He just ignored my reply and scoffed at me and said **'I'm sorry Gary, I do not approve of it, nor will I ever agree with you that it is valid at all. It all looks spikey'. The Face does not really look like what you reveal, and the shaping is not accurate at all.'**

Once again, I replied back to the 'un-named' Mars researcher that he has no idea what he is talking about or even seeing regards to the Face, and this was not a debate to be held between he and myself as he was not and is not a peer enough to be able to debate it or prove anything to refute my claims and my work and what I reveal of the Face that no one else ever has. **Period!** I told him **I am not the one who will be embarrassed!** I done my homework, he should have done his.

Before I allowed him to reply to my statements, I immediately asked him outright; "On a scale of 1 to 100 per cent, if he believes that the 'so called' 3D perspective view images that NASA releases of the Face that show its false shaping is actually being accurate and the truth?" **He replies to me that he believes in the image data from NASA showing how they present the Face and want to brainwash us all that is how the Face really appears 90%.** I could not help but to literally laugh at him, as I was surprised, yet not so very surprised at the same time, of his ignorance of the details of the Face. I thought he may have had at least a bit more understanding regards to the Face, as **a 'serious' Mars researcher should.** He replied to my laughing at him stating to me in an unapprovingly tone of voice saying; **"The way you are presenting and revealing the Face to appear as, is in fact 'SCARY'!"**

I laughed even harder at him after he telling me this ridiculously funny statement, so mockingly I replied to him saying; **"So that is what your real problem is about my work, claims and image data I reveal? Telling me it is not real, and that you will never accept it, because it is too scary looking for you?** (I laughed even more and even harder at him) **"Therefore by the Face as it really appears scares you and you choose to ignore the science, to ignore the data, nay-say and deny with verbally debunking it all without evidence and proof to the contrary shows that you see and understand now why NASA keeps this truth from us officially, and therefore you are in full support of their policy of obfuscation regards to the Face in terms of what the people of the world get to see and to know about? That being the case, you should just quit being a so called serious Mars researcher, because you are too scared of the truth and support the people of Earth continuing to be lied to and denied the truth by NASA. Showing us all that they and you also, have the right to determine what we the people of Earth get to see and know about it? That**

would and does then, make you a FRAUD! I no longer have respect for you in this important research and work!"

It was at this point he hung up the phone on me. Ha ha ha ha ha! Need I tell you that he did **NOT** give me any helpful advice in publishing this book?

The reason why I went into such great details of my conversation with this now exposed '**non-serious Mars researcher**' here in the pages of this book, is because of an issue I arose earlier on page 316, about **FEAR!** The same fear that the '**non-serious Mars researcher**' and NASA share, if and when the world's people see and learn the truth of the Face at Cydonia. The fact that it scares them, does not mean that this all deserves to be obfuscated and lied about. Once again, science is NOT about feelings. Science is about discovery and adapting to reality for which progresses our growth and knowledge as a human species on this planet, as well as on Mars, and implicating and understanding the reality of the universe for which we are embarking upon our future presence within it.

What NASA and the '**non-serious Mars researcher**' are doing is trying to steer you to see and to believe something else that they want you to think and believe, rather than seeing and knowing the truth and questioning them about it which is against the policy and dictates by those who think that we all are not worthy enough to know the truth. It's because of their own fears about what the reality really means for us all, as well as regards to our real place in the universe. The point is that the '**FEAR**' of the reality of what the Face really is and implicates, allows them to think that they have the rights to dictate for the whole world what that truth is without science, without real data, and without a basis in truth at all. Therefore acting as gate keepers of the reality and just continue lying to us all indefinitely with useless wordings based upon beliefs and feelings and deceptions, obfuscating surface relief details in images, and falsifying accurate shaping of the Face. Plausible deniability is what this is called.

They also have the '**FEAR**' of being exposed because of their official statements, presentations, so called mainstream scientific articles, books, TV shows and radio interviews that has helped push people's perceptions and beliefs upon which their work has influenced the public as to what the reality is. **Obviously what I am claiming, showing and revealing, contradicts their work and their so called serious Mars research, and feel threatened of their public perception and positions of themselves to be lessened, or being embarrassed. So some will not allow themselves to agree or to believe that what I am revealing and showing of the Face is indeed real. It was not them who revealed it or who was abled enough to seriously research it to be able to reveal or debunk it all themselves!**

That is their problem, not mine! I done my homework all these many years, and I did not deserve, nor do not deserve how this '**non-serious Mars researcher**' treated me, pretending and acting to me as if he was somehow granted to be above me and better than me in this work. Yet this is far from the truth!

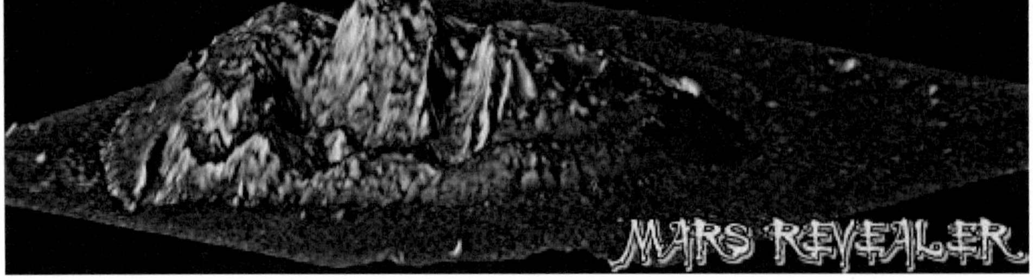

Tell me anyone and everyone, who is it that should feel embarrassed? **Is it me?**

Just like George Haas of *The Cydonia Institute* '**BRAVELY**' and openly stated publicly, that NASA '**reveals**' that the shaded relief model of the 'recreation' of the shadowing for the Cydonia Face, casts a 'spiky' shadow in the likeness of the side profile, of an ape! George Haas is seen '**BRAVELY**' admitting that that what I reveal as the '**True Shaping**' of the Face is indeed really consistent to the data NASA puts out, that contradicts their own 3D shaping views and certainly debunks all that NASA has had to say about the Face. George Haas again, is correct when he **'BRAVELY'** States publicly that **'Someone at NASA, just gave ME a 'spiky' gift of Light and Shadow!**

Please **'anyone'?** Dare to debunk me and my work and all I revealed in this book? **PROVE ME WRONG!**

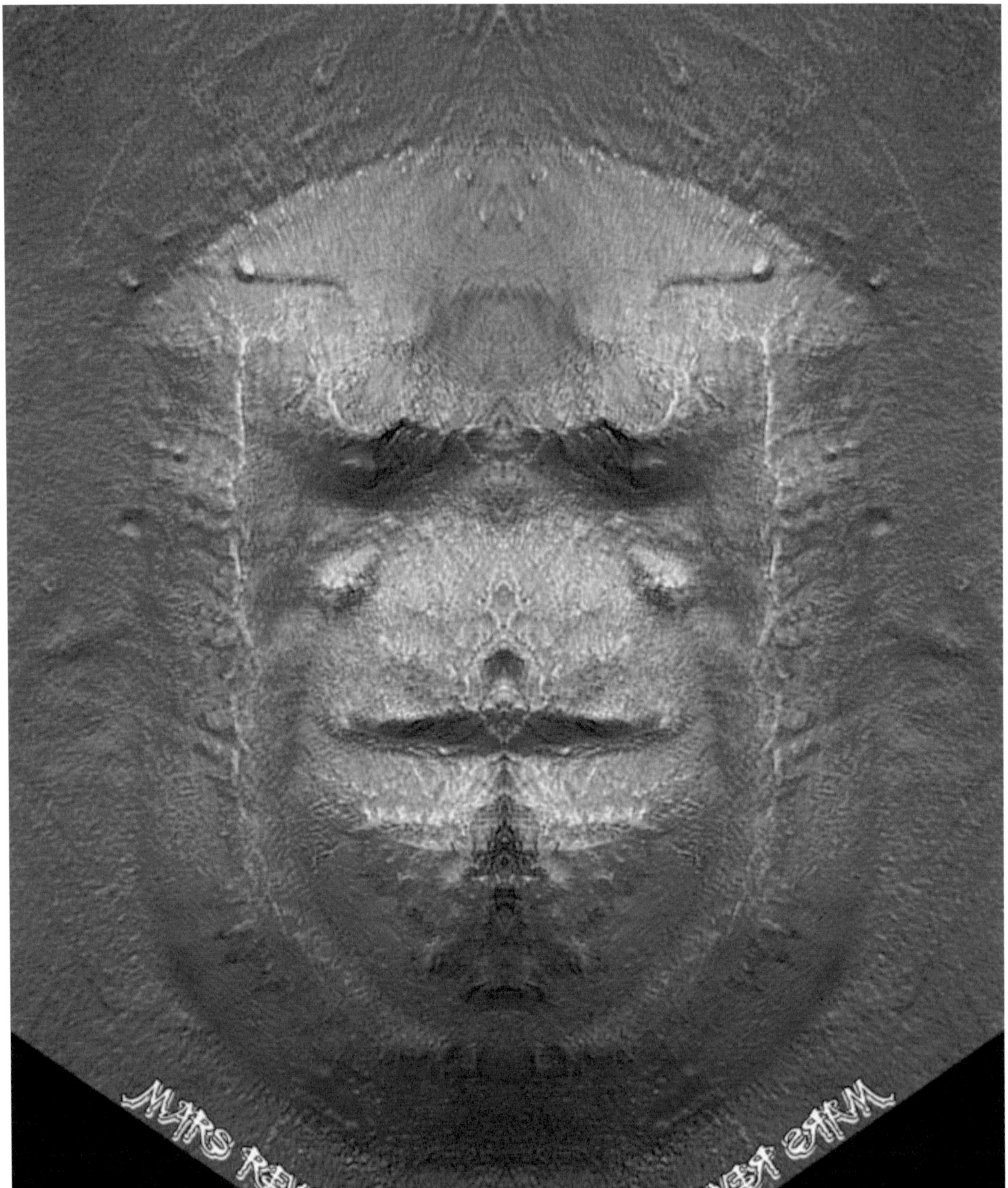

CHAPTER 12:

THE FACE OF CYDONIA: WHOSE FACE IS THIS?

Many people over the many years since the Face was first imaged and people's imaginations ran wild, the assumption of what was being seen or speculated was that the Face looks human, or was a human looking face. This is because of the shadowing of the Face that allowed the hominid side of the Face to be standing out. The data has been acquired over the years (as we have seen) and the reality shows us all that it does not represent humanity showing a human face, but it does in fact have the look of appearing as being a possible unknown kind of hominid species, such as a kind of ape.

There are people who believed that the Face was representing a human face and when the image data does not show this, but looks instead like a type of ape species, belief in seeing any type of face what so ever was diminished or was totally destroyed. For many of these people, and scientists, it is a mind freak to see and hard to accept that a type of hominid ape face as being seen instead of a human one, because humans have the tendency to think and believe that humanity is the highest form of an intelligent species amongst hominids who the only ones who creates structures and civilizations. They feel put out of their faslsely perceived place of greatness in the universe and feel most offended and they consciously cannot accept that a different hominid species being characterized and considered known to be beneath them, would be represented in such a way, whether it is seen on Earth or especially that is being seen on Mars.

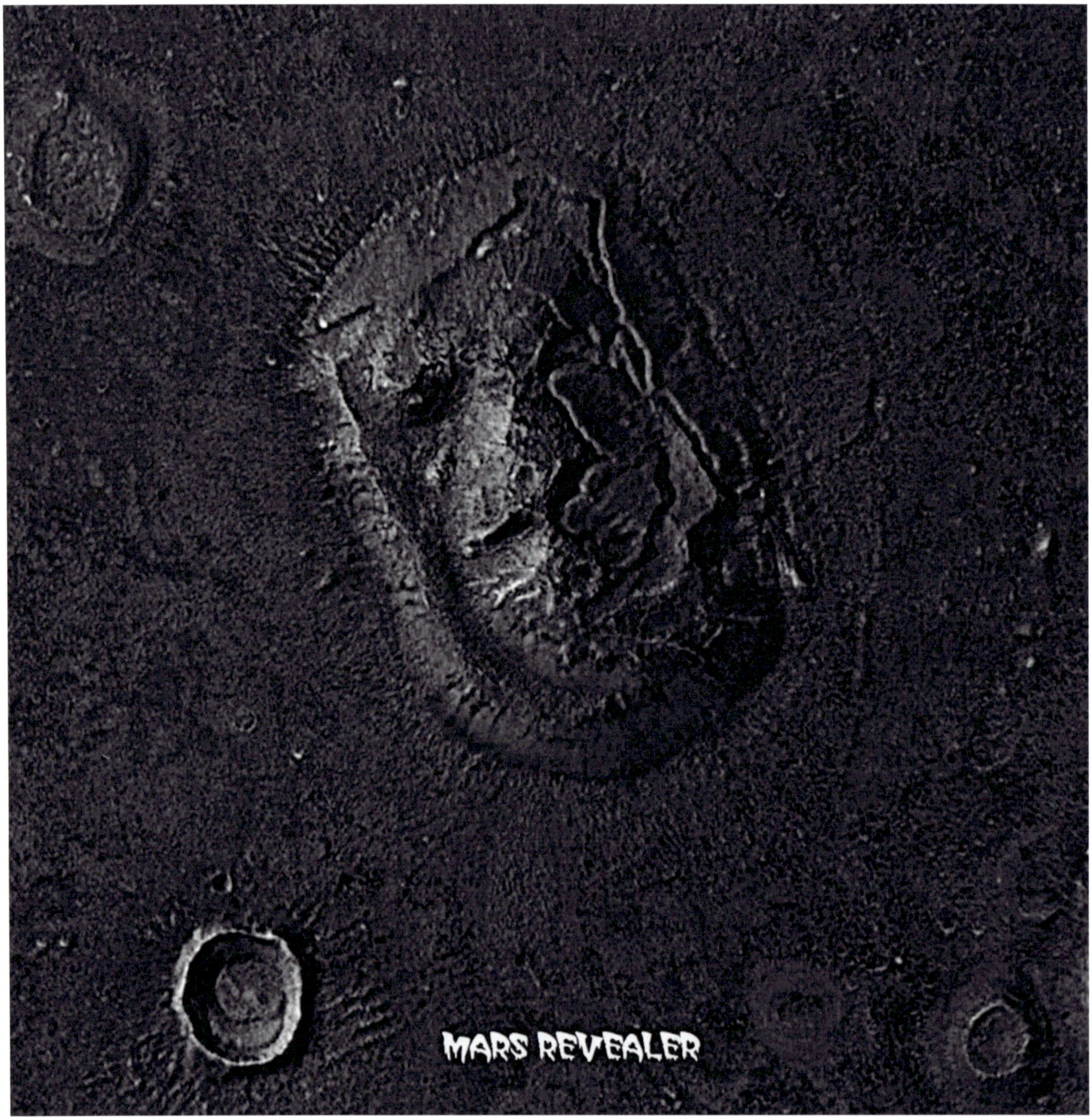

(MY CORRECTLY PROCESSED RAW CTX IMAGE: D01_027417_2195 **IMAGE TAKEN: JUNE 1st, 2012)**

One can only speculate about what type of hominid being or ape like species which is being represented, on Mars! That is an issue that is needed to be thought about hard. The main issue to me obviously seen however, is that the Face at Cydonia is **NOT** representing humanity as we know and accept it, therefore the Face, cannot technically also be called **'man-made'!**

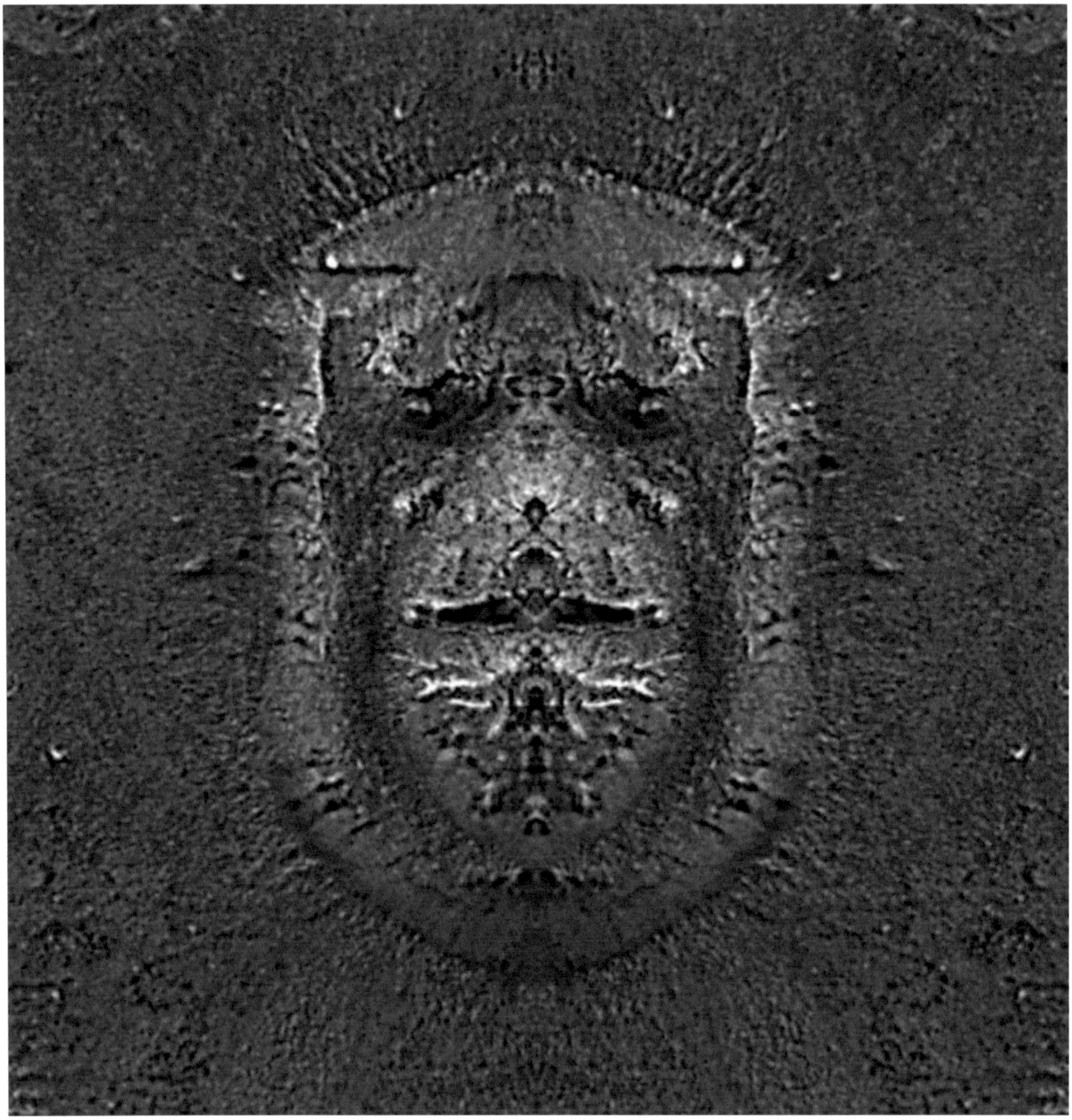

It is not withoiut merits however to consider and study the obvious anthropologic implications of being apart of this amazing trechno-signature. Just because this is being seen and implicated on Mars, does not take away the importance and the need to study the science of anthropology relating to the Face at Cydonia. To those who not know, or need a reminder, anthropology is the science of human beings. Especially the study of human beings **and other ancestors 'through time and space'**, and in relation to **'physical character'**, environmental and **'social relations'**, and culture.

The Face itself represents a **'linguistic'** anthropology, as the Face appears to be consistent with a linguistic duality for the Face which represents or communicates not just one identity as it were, or rather meaning does not represent only one face. Instead the Face at Cydonia represents two sides, or two Faces, where each side is communicating a message where each side has something to say.

Not only does the main attraction to the Face at Cydonia lead us to see the hominid nature that is the first thing most people see when looking at it, but there is also yet another **physical character** likeness that we as humans, hominids, are also seemingly quite familiar with and recognizable to us being seen here on Earth. That likeness of the physical character of the second face apparently is represented by a Lion!

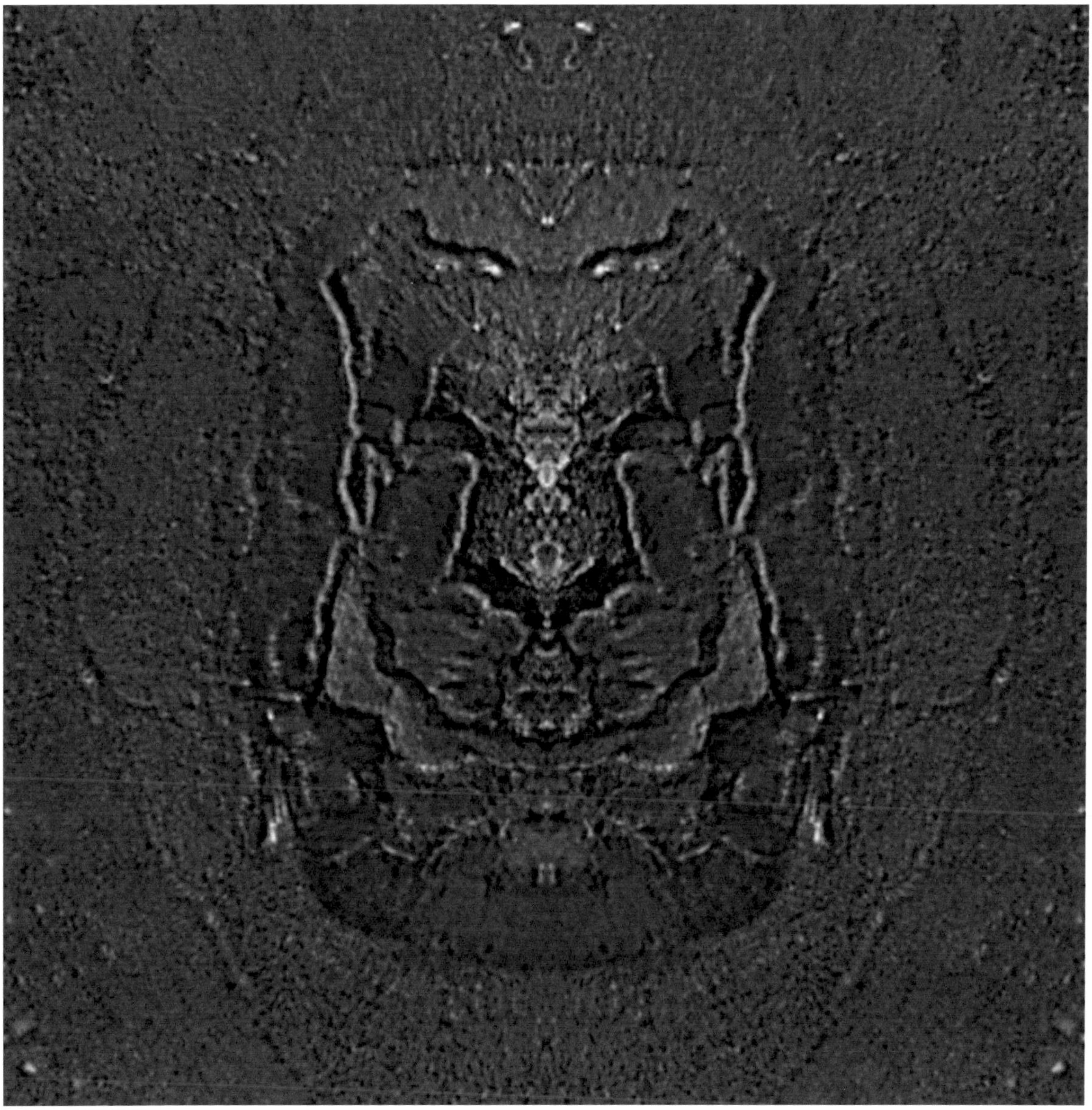

An amazing anthropolical question to ask here to me is, what is the **social relation** to hominids and felines? The answer must lay in the linguistic anthropology, we just need to find a way to understand and speak its language in the iconography of this linguistic heiroglyph **carved** and left on the surface of Mars in the deep ages of time and space staring up into the cosmos declaring itself of its magnificent presence, where it resides forevermore.

Hominid and Lion?

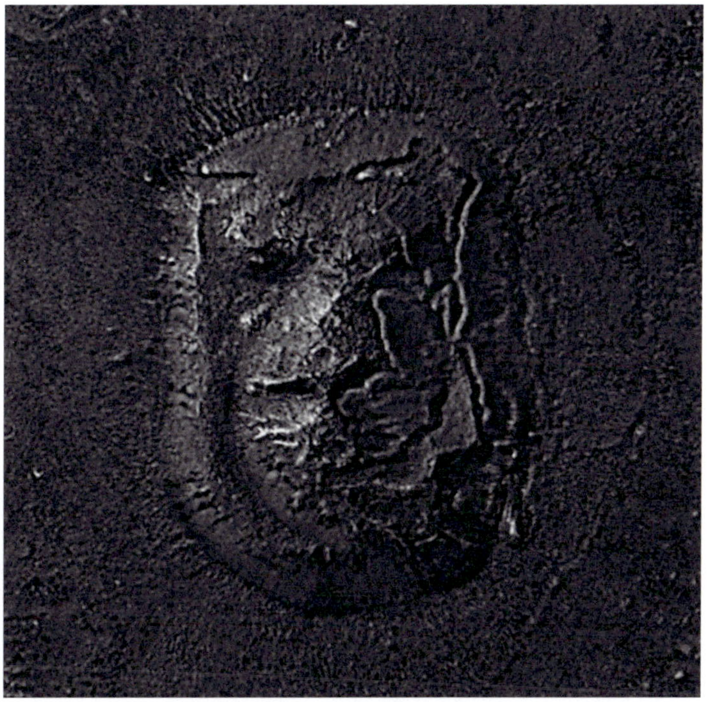

Or.....Hominid and Lyran?

Or, Saurian and Lyran?

The anthropology of humankind as we know it may in fact be far more complex and diverse than what we think and know, and its answers do not reside here simply on Earth, but in the cosmos! Humankind is also considered a young species as modern humans, known as Homo sapiens, are believed to have emerged around 300,000 years ago in Africa, and are the most common and widespread species of primate, and the last surviving species of the genus *Homo* and the broader *Australopithecine*. However, the evolutionary lineage leading to modern humans dates back millions of years, with these early ancestors appearing as far back as six million years ago. They are the great apes characterized by their hairlessness, bipedalism, and high intelligence. Humans have large brains, enabling more advanced cognitive skills that enable them to thrive and to adapt in varied environments, develop highly complex tools, and form complex social structures and '**civilizations**'.

***Australopithecus* was an early ancestor of modern humans.**

A **civilization** is any complex society characterized by the development of the state, social stratification, urbanization, **and symbolic systems of communication** beyond signed or spoken languages (namely, **writing systems** and **graphic arts**).

Symbolic communication of human and feline split face ancient Mesoamerican artifact.

Symbolic communication of human and feline split face ancient Mesoamerican artifact.

This symbolic communication in and of itself is even the greatest techno-signature applied to the Face at Cydonia in terms of anthropology of ancestor hominids and humans here on Earth in terms of ancient human civilizations where this same linguistic symbolic communication and iconography of ancient artifacts has also been discovered and evidenced using the same symbolic characters, the hominid and the feline. It is greater astonishment for some however that aside from the linguistic communication symbolism iconography being seen as the Face at Cydonia, it is more astonishing that we even see a characteristic of a hominid and of a feline at all being seen on Mars.

Hominids and felines we are aware of and know about their realities of these likenesses of these characters because these very same characters still apply to our human knowledge and exist still in our human experience today, as well as in our ancient past. These are characters of life as we know it here on Earth, therefore it is not a surprise to see hominid and feline ancient artifacts from past ancient civilizations depicting this ancient techno-signature of symbolic communication being applied.

It would be and is indeed a surprise, to find this same techno-signature of symbolic communication and characters of an anthropologic origin and history, being seen on Mars where there should not be at all showing any hints or likenesses of hominids and felines, or any other known character of life as we know it here on Earth in our known human experience and knowledge. It should be impossible! How would this exist at all?

This linguistic techno-signature seen as the Face at Cydonia is something of which NASA will never address as a serious issue let alone even hint of it as being a scientific issue. If they were to do so, they will never have been able to have the plausible deniability and obfuscation efforts to succeed to mislead the world as to the true nature of the Face at Cydonia. If they were to acknowledge the symbolic communication iconography of the Face, they would have to admit an anthropological science and archeological science is needed to be applied in studying the Face. That fact alone would admit it is an obvious un-natural intelligent techno-signature demanding study.

The study into those sciences of the Face also would push far beyond the scope of knowledge and beliefs of anthropology for the time line for the Face's creation most likely goes far back in time, far beyond what is believed about the origins of ancient ancestors and modern humans. In other words, it directly challenges and destroys their false belief and false teaching of Darwinian evolution that they force upon us as being real science, but never was able to use science to prove that it even exists.

It has been determined that the Face at Cydonia most likely was created at least 150 million to 500 million years ago when Mars had a more Earth-like atmosphere and Earth-like conditions where Mars was much wetter and warmer than it is today. Mars was alive. **Mars is still alive!** Yet modern Mars is only a vestige of her former glory. It is needed to be brought up here also that is the time frame upon which Mars was severely decimated by an un-natural means, as its techno-signature has also been discovered at Mars also in 1976. That techno-signature was also a signature of war, as it was the direct scientific techno-signature of a weapon having been used at Mars to destroy its atmosphere and to destroy and to decimate its life and civilization. That weapon techno-signature type which was discovered and determined, was xenon 129, which only can exist un-naturally, as it is the isotope made directly from nuclear weapon detonations.

These nuclear weapon detonations seemed to have been detonated at least at two places above Mars, one detonated above Mare Acidalium, and also above Utopia Planitia, in the region of Utopia Planitia called Galaxias Chaos. The radioactive isotope itself has been found, studied and determined of its age of creation from its detonations, again, from 150 million to 500 million years ago. You can read all about this from my friend and who is a nuclear physicist and Mars researcher, Dr. John Brandenburg in his book:

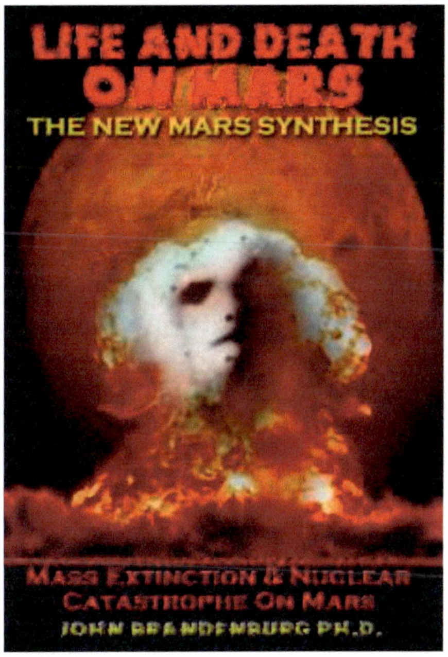

(Everything you have been taught about Mars is wrong. The terrible truth: Mars was actually Earthlike for most of its geologic history. Mars held a massive and evolving biosphere. Mars was the wracked by a mysterious and astonishing nuclear catastrophe. We are, biologically and culturally, the Children of Mars.)

https://www.amazon.com/Life-Death-Mars-New-Synthesis-ebook/dp/B004XJ4YT2/ref=sr_1_1

With that being said, it is noteworthy to notice that one of the detonation spots, Mare Acidalium, is also where Cydonia and the Face reside! This amazing fact is also no coincidence, as it shows a specific targeting area showing purpose and intent in direct connection with Cydonia and the Face. That being the case, it allows us to assume also that the face at Cydonia is also at least the same age of which the nuclear detonations had occurred. The Face can of course be far older existing at Cydonia long before those nuclear detonations.

The odds are greater the Face was existing during the timing of the nuclear detonations rather than the Face being created after the nuclear detonations, which would have been seriously irradiated for millions of years. Herein lies the problem, as once again, the Face's techno-signature of symbolic communication of hominid anthropology as well as felines again, also dates back hundreds of millions of years in the least, not on Earth, but on Mars.

So much further back in time than the mere six million years of the proposed age of our human ancestors' somehow first coming into existence upon the Earth. This also implicates the common sense approach to this problem that hominids and felines, are not simply unique and only limited to life as we know it upon Earth, but are a much older presence and origin of species existing already in the cosmos.

That raises the question then, of how ancient hominids and felines have really come into existence upon the Earth? Was it by so called natural means of Darwinian evolution that is universal in creating hominids and felines out of nothing, to just come into existence naturally on all worlds suitable for life? That idea in and of itself sounds far more ridiculous to consider than considering the symbolic communications of hominids and felines on Mars hundreds of millions of years ago, also being found and depicted here on Earth's ancient civilizations linguistics.

Or, was it more likely that somehow, these species came upon the Earth, by being brought to the Earth by an extraterrestrial intelligence to proliferate upon the Earth? Was this delivery of species brought to the Earth by ancient surviving Martians to somehow replenish the types of life lost on what had once flourished on Mars? In my opinion, this makes the most sense to consider, and therefore these species are also not limited to Mars or Earth.

It is funny however to notice that by looking at NASA's 2007 hit piece false 3D shaping views of the Face, as we looked at in previous chapters, showing of their misleading tampered with image data to try to make the Face look natural, yet when viewed from its side view of the 3D perspective, it does kind of look like the **Australopithecus**!

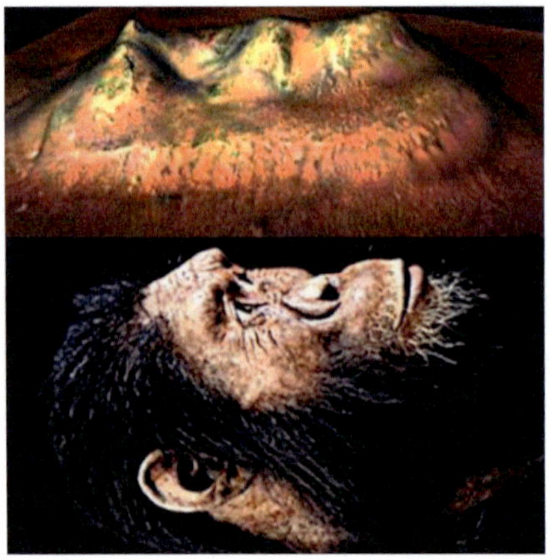

In going back to the idea of life of various forms and species being brought to Earth, there is another false view being attributed to the Face at Cydonia as regards to its origins. This false view of the origin of the Face at Cydonia has to deal with an alien race called the Anunnaki.

The Anunnaki are an extraterrestrial race of beings as has been discovered about and detailed about through ancient linguistics from the known first human civilizations from six thousand years ago written in cuneiform on clay tablets, such as mostly have been found by the thousands in the ancient ruins of Nineveh in ancient Mesopotamia, or rather in the lands of modern Iraq. It is attributed in these tablets telling the stories about these Anunnaki who came to Earth less than 500,000 years ago on a mission to extract Gold from this planet Earth in order to fix a deteriorating ozone and atmosphere for their home world which was an existential threat to the planets survival and also to the survival of the Anunnaki themselves.

So no, they were not the Martians or even the ancient Martians, but they are connected to Mars regards to their history of their presence here on Earth in after coming to Earth. A famous translator for this ancient cuneiform language named Zechariah Sitchin has written many books in detailing the information provided within the ancient historical contexts of the Anunnaki detailing their mission coming to Earth and the history of them living here on Earth and toiling to extract the much needed gold to bring back to their home world Nibiru to be processed and lifted aloft to fill in with fine gold dust and fix their ozone and therefore fixing the decimation of Nibiru's atmosphere. You can read all about this from Sitchin's books of the 12th Planet series.

Now is a time once again to focus on anthropology studies of humans, as also detailed in their historical record as ascribed and described by the Sumerians/Mesopotamians found detailed on these cuneiform tablets, there is a direct connection of these Anunnaki and modern humans so therefore, this must be considered in the study of anthropology itself because it gives the historic story and scientific process and description and deliberate actions on part of the Anunnaki in creating modern humans. This is very important to consider here as remember again, the Anunnaki had come to our planet between 450,000 to 500,000 years ago. Remember also, that modern humans have seemed to come into existence **according to science** around 300,000 years ago.

It is also important to note that the lives of these Anunnaki are lengths spanning into the thousands, tens of thousands, or even in hundreds of thousands of Earth years. Modern humans live to a maximum of 120 Earth years if lucky. Most never near that age though, though in some cases it is known. It is told to us that we were created by them to alleviate the Anunnaki workers toiling in mines and other places on and inside the Earth, whereas they were burdened with the work, and wound up rebelling against the work.

The solution was to create a being fashioned in their likeness, to understand, communicate, have intelligence to follow commands and to use and work with tools, and to do the toilsome burdensome work to mine the gold from the Earth to satisfy the Anunnaki workers and to satisfy the demands for the gold to be brought to Nibiru for fixing their atmosphere. In regards to our creation in making a terrestrial being, created in their likeness, the Anunnaki learned that the being to help to create the new worker beings created in their image, had already been existing here on earth when they decided and planned to create this new primitive worker.

Our so called ancestors! The hominid ape species, where fossilized remnants shows obviously various species with differences between each other. It is with and in modern humans however, that these differences and changes are dramatic and complex, and is very difficult if at all, can there be any evolutionary changes of these 'human ancestors' on such a short period of time to have been able to have occurred naturally. Changes in ancestor species of traits such as high intelligence (in comparison to primitive hominid apes), brain sizes, skeletal differences, especially regards the DNA of modern humans compared to hominids and primitive apes. The so called '**missing link**' from our 'ancestors' that explains modern humans as a species, apparently are the Anunnaki themselves.

So to learn about and discover about all that if you are not already aware yet about the Annunaki, I strongly suggest that you read the entire 12th Planet series, by Zechariah Sitchin. I do know enough about the subject to write a book about them, but that is not for this book. I am merely making important points that need to be focused upon, especially when it comes to the Face at Cydonia and its origin, and especially where it comes to **the false assertion** by many people and independent researchers that the Face at Cydonia is the evidence and proof that the Anunnaki are the ones responsible for the Face of Cydonia to be created and that it is in fact an Anunnaki creation. It is **NOT**!

The idea for this false assertion comes from a lack of understanding of what is being described in one of Zechariah Sitchin's books titled, '**The Lost Book of Enki**'. I shall detail a summary of the information told in it regards to the Anunnaki, Mars, and even creating a Face on Mars. It is a very compelling and a very important story being detailed by Sitchin that if accurate in its historical context, I would agree that this is adding a preponderance of evidence to the truth of the existence of the Anunnaki coming to Earth, as well as to adding preponderance of evidence of their truth of the circumstances of our creation as well.

As it was in fact the Anunnaki and the creation of a Face on Mars connects to a time before modern humans were even contemplated about being created by them. To better understand what this is all about, and where this false belief stems forth from, we need to look at the circumstances that brought about the creation of a Face on Mars' surface by them.

As a summary of the history of the Anunnaki coming to Earth, This story begins actually on Nibiru before discovering of the Earth. As this matter occurred over challenging for kingship of the throne of Nibiru by Nibiru's then King Alalu, and the challenger Anu. I will not get into the politics and reasons for this challenge for the throne as that needs to be read in Sitchin's books to be understood fully. It is an incredible story and worth reading all about it.

So Anu challenged Alalu to a fight of wrestling to determine who shall be king of all Nibiru. Alalu lost and his kingship was deposed, with Anu becoming king of all Nibiru. Alalu became scared for his life and stolen a space ship which had incredible 'weapons of terror' aboard. Alalu knew the stories and information about the solar system and the prehistoric formation of the solar system of which they were also a part of. He set a course on a dangerous path through space from Nibiru, to travel inward to the sun in the attempts to discover and come to Earth to discover the gold to acquire to save Nibiru, which had been going through a catastrophic calamity of a diminishing atmosphere, and it was determined the only solution was gold to acquire to solve the issue.

The Anunnaki had no resources to locate or acquire any gold on their world or any other, accept through a legend. A legend of the inner solar system where gold was hinted at being in abundance on a world that sat between Mars and Jupiter, being a large watery world with ran with rivers of Gold. Then somehow that large planet was destroyed by blowing up or by being blown apart, which then created the asteroid belt between Mars and Saturn. The Anunnaki had determined that it was too dangerous to go try to have the asteroids mined for its gold in the debris of the former planet. They have sent several missions to try doing that, but none never returned.

Alalu was taught about the former makeup of the solar system before the destruction of that world. Our earth, was a chunk of that old former world. Alalu figured he would escape Nibiru to head toward the sun in an attempt to discover the segment remnant planet of the old world it was once a part of. He figured that he alone, can discover that new world and also discover the much needed gold to attain in order to have hope to regain the kingship of Nibiru by being its salvation by discovering and acquiring the desperately needed golden material. If he failed to do so, certain death he still faced despite escaping from Nibiru for fear of his life because of his challenged fight and the loss of the fight and of his kingship, with Anu triumphantly taking the throne from Alalu.

Alalu thought of his potential for success and his planned desire to regain kingship of Nibiru by doing what no other Anunnaki, including Anu, which could not attain that much needed gold for Nibiru's salvation. Alalu used the 'weapons of terror', to blast his way opening up spaces to pass through the asteroid belt between Mars and Jupiter, which he had successfully done and the way to Earth was open for him to discover, which he did. Alalu landed on Earth successfully and the ship's sensors detected the much needed gold. Alalu sent a broadcasted transmission to Nibiru, letting the Anunnaki all know that he, and he alone, has discovered the gold of salvation for Nibiru. Telling them he had found the salvation on the remnant segment of the destroyed planet between Mars and Jupiter, and that its orbiting position was much closer to the sun that it once was before the destruction of that large planet.

From the Lost Book of Enki, we will see what information is on the subject of Alalu and a face being constructed on Mars surface. Let us read:

"The fate of Nibiru is in my hands; to my conditions you must give heed!"

Those were the words of Alalu, from dark-hued Earth to Nibiru they were by the Speaker beamed.

When the words of Alalu to Anu, the king, were conveyed, Anu astounded was; astounded were the counselors, amazed were the sages. "Alalu is not dead?" (they to each other had asked). "Could indeed he on another world be living?" (they with disbelief were saying). "Was he not on Nibiru hiding, in the chariot to a place of concealment gone?" The commanders of chariots were summoned, savants the beamed words considered. "The words from Nibiru did not come; from beyond the Hammered Bracelet (asteroid belt) were they spoken). This was their finding, this to Anu the king they reported. Stunned was Anu; the happening he pondered.

"Let words of acknowledgment to Alalu be sent", to the assembled he was saying.

"Anu the king, to you his greetings sends; of your well-being to learn he is pleased; For your departing from Nibiru there was no reason, enmity is not in Anu's heart; If gold for salvation you have indeed discovered, let Nibiru be saved! The words of Anu Alalu's chariot (space ship) did reach; Alalu to them quickly answered: "If your savior I am to be, your lives to save, Convene the princes to assembly, my ancestry declare supreme! Let the commanders make me their leader, bow to my command! Let the council pronounce me king, on the throne Anu to replace!"

(After speaking and detailing messages and answers asking for proof of Alalu's claims, it was determined he was telling the truth. We are told an answer regards to making Alalu being claimed king of Nibiru):

"Let Alalu be king on Earth," a verdict of the sages awaiting: "If Nibiru it will save, let there be a second wrestling; who shall Nibiru rule let it determine!"

"Let it so be!" Anu announced. Alalu a second time I shall then wrestle, let the winner be on Nibiru king!" The words of decision to Alalu were conveyed; He pondered them and agreed: "Let Ea, my son by marriage, to Earth come! Let gold from the waters be obtained, let it for salvation on Nibiru be tested; Let a second wrestling kingship by me or Anu settle!"

"So be it!" Anu in the assembly decreed.

(So as time went on, the Anunnaki's mission to Earth to acquire the gold was under way and Earth was reached. Landing parties set up performing their various tasks and areas and positions of control and powers was being divided and decisions made for who leads what and where. Tasks and mission objectives are not going as exactly planned and was no easy task to acquire the needed gold on a fast enough pace as desired. Anu sends his other son Enlil, to Earth to which Anu will follow afterward to arrive on Earth after Enlil's arrival. After arguing about tasks and where to set up control areas on Earth, Anu's two sons EA and Enlil, argued about who shall be in command. Anu had them drew lots to determine who will control and rule on Earth.)

"Let Ea forever Eridu as his home retain!" Anu to Enlil was saying, "Let his being the first to splash down forever be remembered, Let Ea as Earth's master be known; Enki, Earth's Master, let his title be!"

His father's words Enlil with a bow accepted; to his brother he thus said: "Enki, Earth's Master, your title name shall henceforth be; I Lord of the Command shall be known."

To the heroes in assembly Anu, Enki, and Enlil the decisions announced. The tasks are assigned, success is in the offing! Anu to them was saying.

"Now farewell I can bid you, to Nibiru with quiet heart I can return!"

Forward toward Anu Alalu stepped. "A grave matter has been forgotten!" he shouted. "The mastery of Earth to me was allotted; that was the promise when the gold finds to Nibiru I announced!"

(Alalu was not satisfied with how things were turning out, as he was made to be Earth's ruler, yet Anu's sons were the ones whom it seemed was being put in control of all mission objectives and the apparent ruling of the Earth.)

"Nor have I the claim to Nibiru's throne forsaken, By Anu to share all with his sons, it is a grave abomination"!

Thus did Alalu to Anu and the decision challenge. Without words was Anu in the beginning, then with anger he spoke up:

"By a second wrestling let our dispute be decided, let us the wrestling do here, let us do it now!"

With disdain Alalu took off his clothing; likewise did Anu unrobe. In nakedness did the two royals begin to grapple, a mighty struggle it was.

Alalu bent his knee, to the ground Alalu fell; Anu on the chest of Alalu with his foot pressed down, victory in the wrestling thereby declaring. By wrestling the decision was made;

"I am the king, to Nibiru Alalu shall not return!" So was Anu saying as he removed his foot from the fallen Alalu. Up as a lightning Alalu from the ground arose. By the legs Anu he pulled down. His mouth was wide open, swiftly he the malehood of Anu bit off. The malehood of Anu did Alalu swallow!

In pained agony did Anu a cry to the heavens shout; to the ground wounded he fell. Enki to the fallen Anu rushed, Enlil the laughing Alalu captive held. Heroes Anu to his but carried, words of accursation against Alalu he uttered.

"Let justice be done!" Enlil to his lieutenant shouted. "With your beam-weapon let Alalu be killed!"

"No! No!" Enki fiercely shouted. "Justice is within him, in his innards poison has entered!"

They took Alalu to a reed hut, his hands and feet as a prisoner they bound.

Now this is the account of the judging of Alalu, And of the happenings thereafter on Earth and on Lahmu (Mars). In his reed hut Anu was hurting, in the reed hut to him Enki applied the healing.

Before these Seven Who judge Alalu was brought; his hands and feet were untied. Enlil was first to speak: "In fairness a wrestling match was held, Alalu the kingship to Anu forfeited!"

"What say you, Alalu? Enki him this question asked.

"In fairness the wrestling match was held, the kingship I forfeited!" Alalu said. Having been vanquished,

"Alalu an abominable crime performed, the malehood of Anu he bit and swallowed!" Thus did Enlil the accusation of the crime make. "Death is the punishment!" Enlil was saying.

In the silence Alalu slowly began to speak: "On Nibiru I was king, by right of succession I was reigning; Anu was my cupbearer. The princes he aroused, to a wrestling he me challenged, For nine counted circuits (orbits around the sun) I was king on Nibiru, to my seed kingship was belonging. On my throne seat Anu himself sat, to escape death to distant Earth I made a dangerous journey. Salvation for Nibiru I, Alalu, on the alien planet discovered! Return to Nibiru I was promised, in fairness the throne to regain! Then to Earth came Ea; the one by compromise the next to reign Nibiru he was designated. Then came Enlil, the succession from Anu to himself claiming. Then Anu came, by lots he tricked Ea; Enki, the Lord of Earth, he was proclaimed, Of Earth, not of Nibiru, to be the master. Then to Enlil command was granted, Enki to the distant Abzu was delegated. My heart of all that was aching, my chest from shame and anger was bursting; Then Anu his foot upon my chest placed, upon my aching heart he was treading!"

In the silence Anu spoke up: "By royal seed and law, by fair wrestling did I gain the throne. My male hood you bit off and swallowed, my offspring line to discontinue!"

Enlil spoke up: "To the crime the accused admitted, let the judgment come, Let death the punishment be!"

"Death to Alalu by itself will be coming, what he had swallowed in his innards death will bring!" Enki was saying. "Let Alalu for the rest of his days on Earth be in prison!" Anzu was saying. Their words Anu was contemplating; anger and pity both him engulfed. "To die in exile, let this be the judgment!" Anu was saying. In amazement the judges at each other glanced. What Anu was saying, they wondered.

"Neither on Earth nor on Nibiru shall the exiling be!" Anu was saying. "On the way there is the Lahmu planet, (Mars) **with waters and an atmosphere it is endowed.**

Enki, as Ea, thereon made a pause; "of it as a way station have I been thinking. Its netforce (gravity) is less than that of Earth forceful, an advantage in wisdom to be considered;

Anu spoke: "In the celestial chariot Alalu shall be taken, on my departing from Earth he with me shall make the journey. Around the planet Lahmu (Mars) we shall make circuits (orbits), to Alalu a sky chamber we shall provide, to the planet Lahmu (Mars) in it he will be descended. **Alone on a strange planet** an exile he shall be, His days to his last day by himself to count!" Thus did Anu words of judgment utter, in solemnity were the words intended.

By unanimity was this judgment upon Alalu imposed, in the presence of the heroes it was announced.

"Let Nungal be my pilot to Nibiru, therefrom chariots bearing heroes again to Earth to pilot. Let Anzu join for the journey, of the descent to Lahmu (Mars) take charge! So did Anu commandments utter.

On the morrow departing was readied; all who depart by boats to the chariot were ferried. "A place for landings on firm soil (on Earth) you must prepare!" Anu to Enlil was saying. "How Lahmu (Mars) **as a way station to utilize,** plans you should be making!"

Farewells there were, both joy and sorrow. Limping did Anu on the chariot embark, with his hands tied did Alalu the chariot enter. Then to the heavens the chariot soared up, and the royal visit had ended. They around the Moon made a circuit; Anu by the sight was enchanted.

Toward red-hued Lahmu (Mars) they journeyed, twice about it they circled. Lower toward **the strange planet** they came, mountains sky-high and tears in the surface they noticed. Where Ea's chariot had once landed they observed**; by a lakeside** it was located. Slowed by Lahmu's (Mars') netpower (gravity), in the chariot the sky chamber they readied.

Anzu, its pilot, then unexpected words to Anu was saying: "With Alalu to the firm soil of Lahmu (Mars) I shall descend, With the sky chamber to the chariot to return I wish not! With Alalu on **the strange planet** I shall stay; until he dies I shall protect him. When he dies of his innards' poison, as befits a king him I shall bury! As for me, I shall have made my name; Anzu, they will say, against all odds to a king in exile a companion was, He saw things by others unseen, **on a strange planet** he faced **unknown things!"**

There were tears in the eyes of Alalu, there was amazement in the heart of Anu.

"Your wish shall be honored", to Anzu Anu said. "Hereby let a promise by me to you be made, By my raised hand to you I this swear: On the next journey a chariot by Lahmu (Mars) shall circuit (orbit), its skyship to you shall descend. If alive it shall find you, the master of Lahmu (Mars) you shall be proclaimed; **When a way station on Lahmu (Mars) shall be established,** its commander you shall be!"

Anzu bowed his head. "So be it!" to Anu he said. Into the sky chamber Alalu and Anzu were ushered, With Eagles' helmets and Fishes' suits they were provided, with food and tools they were supplied. From the circling chariot the skyship departed, from the chariot its descent was observed. Then from view it disappeared, and the chariot to Nibiru continued. For nine Shars (shar is an orbit of Nibiru around our sun, its year equals 3600 Earth years) was Alalu king on Nibiru, for eight Shars, Eridu he commanded. In the ninth Shar, to die in exile on Lahmu (Mars) was his fate.

(After Anu and crew leaving the orbit of Mars back to Nibiru, an unmentioned amount of time has passed and a return to Mars was finally done upon which returning to Mars, they found Anzu also deceased. They found him by a lakeshore of which they found him from a location beacon operating inside his helmet. Using technology and the food of life, they revived Anzu.)

About Alalu they him then inquired; of Alalu's death, Anzu them told. He led them to a great rock, from the plain heavenward protruding. There to them what had happened he was telling: "Alalu soon after the landing from unremitting pain to scream began. From his mouth his innards he was spitting; in agony he peered over the wall!" Thus was Anzu to them saying.

He led them to a great rock, like a mountain from the plain heavenward rising. "In the great rock a cave I found, Alalu's corpse therein I hid, Its entrance with stones I covered." So was Anzu to them saying.

They followed him to the rock, the stones they removed, the cave they entered. Inside what of Alalu remained they found; "He who once on Nibiru a king, was now a pile of bones in a cave now lying! For the First time in our annals, a king not on Nibiru has died, not on Nibiru was he buried!" So did Ninmah say. "Let him in peace for eternity rest!" she was saying. They the cave's entrance again with stones covered;

The image of Alalu upon the great rock mountain with beams they carved. They showed him wearing an Eagle's helmet; his face they made uncovered. Let the image of Alalu forever gaze toward Nibiru that he ruled, toward the Earth whose gold he discovered!

So that is the story of Alalu and the creation of **his face** on Mars' surface. It was needed to go over that story as detailed as possible to us all in order to understand the details of it, and to understand the details of the face, **his face,** which was created on Mars' surface by beams from above to carve it.

Many people after reading that book of The Lost Book of Enki, felt for certain that the Face at Cydonia was and is the face of Alalu. Those people are the ones who not know the Face at Cydonia, and this can be shown easily because they also ignore the very details of the story of Alalu's face being carved on Mars. This is very important to notice and consider, as details matter. Let us go over these important details to show by what we are told, and by comparing to the image data we have seen from Mars as presented in this book, the Face at Cydonia cannot be, and is **not** the face of Alalu!

We learned that the Anunnaki came to Earth 450,000 to 500,000 years ago. It was at the beginning arrivals to Earth, when Alalu was sentenced to exile and death on Mars. That is half of a million years ago.

We learned that at that time of the Anunnaki arriving to Earth, and after observing Mars, we are told that Mars was **with waters and an atmosphere it is endowed.** We were told that they found Anzu by **a lakeshore.** This tells us all that 500,000 years ago, Mars obviously had a stronger atmosphere than it does today, as well as having water being seen and able to be accessed by the Anunnaki on the surface. It is interesting to also note however, that it seems they had to wear helmets. Nothing is said about the helmet use was for purposes of breathing, but it is probable.

It was decided that a way station between Earth and Nibiru would be established on Mars, with Anzu as its master. That effort and work began immediately. Anzu was also described as having **saw things by others unseen, on a strange planet he faced unknown things!**

This is telling and allows interesting speculation. It is possible the **unknown things he faced,** was its alien markings, that sculpted landscapes on Mars surfaces, with bizarre structures, interesting features and Faces, **by others unseen, who are not there any longer?** The remnants of the **unknown** civilization's techno-signatures from the ancient past being seen on Mars surface, which we are told was not a normal planet, but being called **a strange planet?** It is interesting to speculate about.

It is worthy to note also, that nothing was being said of the radioactivity remnants of what we scientifically know now for certain today, of the techno-signature of nuclear weapons having been used on Mars, and the evidence discovered of the Xenon 129 isotope proving that historical fact. I am certain however that the Anunnaki must have noticed that fact, as the great 'weapons of terror' Alalu used to open a pathway through the asteroid belt were of nuclear weapons, for him to be able to find and come down to the Earth's surface.

We also were described how Alalu's space ship had sensors that penetrates the surfaces and also checks and tests the atmospheres. This tells me they knew of the radio activity on Mars, but at the time of 500,000 years ago, it obviously was deemed safe and livable and adaptable enough to establish a way station on Mars **with an atmosphere and with water being endowed!**

However the most important information and the point of telling all of this to begin with, is the Face! The Face at Cydonia, if it attributed to what information we hear from the Anunnaki perspective, would mean that we have a minimum age for the Face's creation at almost 500,000 years ago. According to Anzu, the attributed Face of Alalu was described by Anzu claiming he led them **to a great rock, like a mountain from the plain heavenward rising.** It seems we are to assume that it must be close by to the lakeshore upon which Anzu was found and revived from death.

We are told that Anzu stated:**"In the great rock, a cave I found, Alalu's corpse therein l hid, Its entrance with stones I covered."**

We were then described of the purposeful manipulation of the **"great rock"**, which was described as **"Like a mountain, from the plain heavenward rising,"** was modified to resemble: **"The image of Alalu upon the great rock mountain, with beams they carved. They showed him wearing an Eagle's helmet; his face they made uncovered."**

It is obvious we are being told that it was the space craft that used its **beams,** which had **carved Alalu's face, wearing an Eagles helmet, and they made his face UNCOVERED!** It is at this exact description being described that tells us and shows us, that the Face at Cydonia cannot be and is **NOT** Alalu's face!

The Face of Cydonia does **NOT** share these characteristics at all. It does not show a symmetrical Face, of a whole face which was uncovered, while wearing an eagle's helmet. It also does NOT show a human looking face. Why is this important to keep in mind?

According to the Anunnaki, in the historical records in cuneiform, it is they, who had created modern humans, from merging and manipulating the DNA from the Earth's primitive ape hominids, who walked on its two hind legs, and arms swaying as they walked, with their Anunnaki DNA to create a hybrid being. Again, **creating us in their image. Meaning we look like them!**

The Face at Cydonia does **NOT** look like, or represent a human looking face! As we have seen, the Face at Cydonia is a split face symbolic linguistic communication iconography.

There also is nothing that looks like an eagle's helmet. The eagle's helmet also tells us another interesting fact, that the eagle is known also to the Anunnaki before they ever went to Earth! There is nothing of the Face at Cydonia that even resembles an eagle, let alone an eagle helmet!

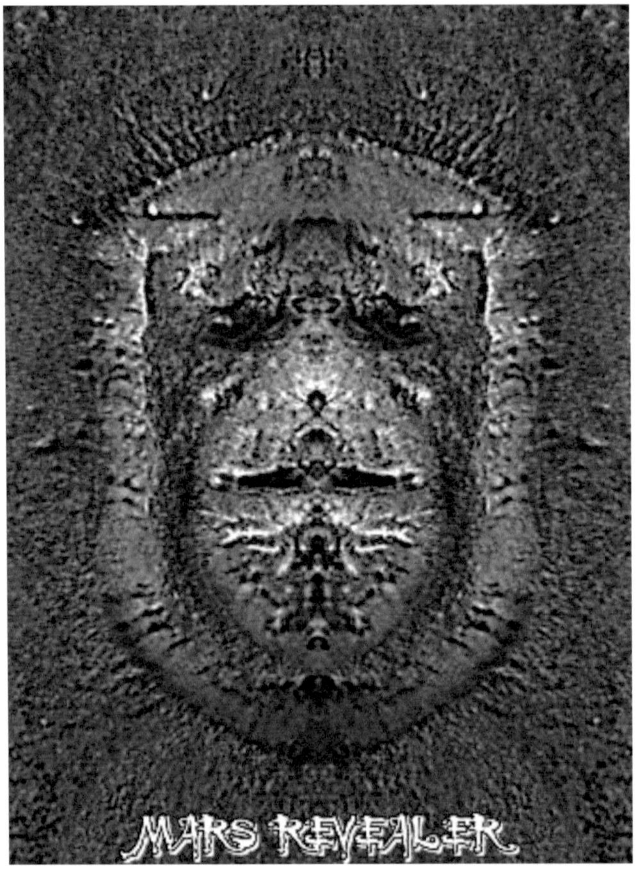
MARS REVEALER

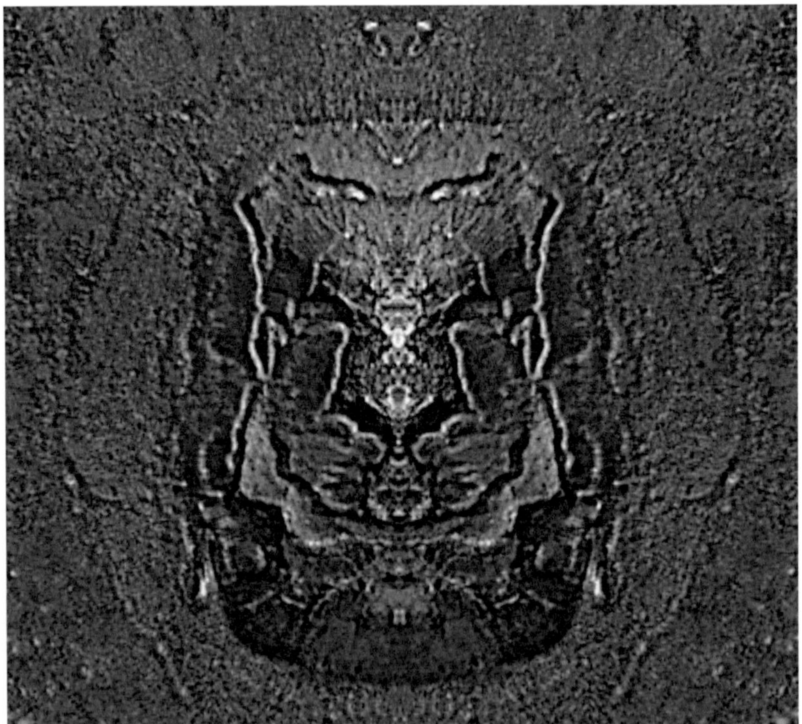
MARS REVEALER

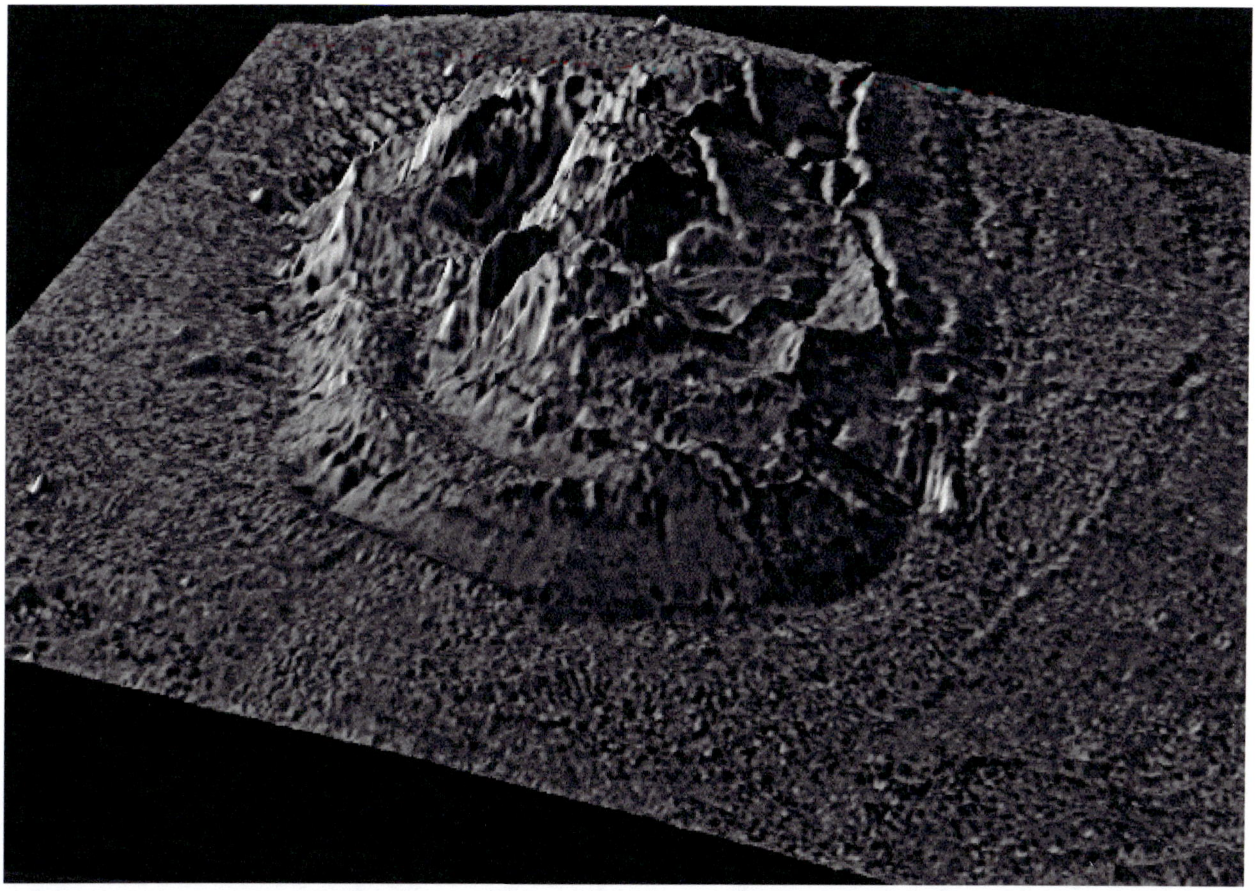

Now what this tells us of the Anunnaki story, is that if is true, it is an ancient link alluding to and cluing us all down through time, of there being a face on Mars created by an intelligent race of alien beings. Another interesting aspect of that information is the detailing to us all of how they created the face of Alalu. Again also notice that we were not told that they created a split face symbolic communication. It was quite clear how it was carved and descriptive of its likeness. **"The image of Alalu upon the great rock mountain, with beams they carved. They showed him wearing an Eagle's helmet; his face they made uncovered."**

So how can this be? Yes, you may agree with me, that the Face at Cydonia absolutely does not look like a human face being seen whole, uncovered, and while wearing an eagles helmet. It just not matches up to the description given. Despite this fact there are those who still claim it is Alalu's face, because it just has to be.

In their defense, this is understandable frustrating. We have information left to us from an extraterrestrial race here on Earth, giving details of **a mountain sized rock raising heavenward having been artistically modified by beams that carve (lasers),** to depict a face of one of their own in remembrance. Giving us the idea long ago in time that there was a mountain sized face to be found on Mars. If the Anunnaki were just myths of fiction, how then can there be such a consistent connection of a face on Mars? Again, a preponderance of evidence to frighteningly validate the Anunnaki story.

To now having received images in modern times showing what appears to be a mountain sized face looking up into the heavens on the surface of Mars, the odds are beyond astronomical for this to be a mere coincidence. I do feel and know however that what we are seeing on Mars as the Face in Cydonia, is not representing Alalu's face. The problem this brings to people's minds is mind boggling to me as to why they cannot figure it out the problem here. The fact we are having ancient extraterrestrials telling in records about a Face on Mars, is again clarifying and validating the concept itself, as well as the evidence for it in modern images.

The problem we all face here is that it does not match up. So what does that mean to you or what does that tell you? We see a different mountain sized face than the one being described by the Anunnaki, and the Face we are relating it to of being Alalu's face. The simplicity of what this then should tell us all is that obviously, if the Face at Cydonia is not Alalu's face, then someplace else on Mars, Alalu's face as described, **must exist!**

However, most who know anything of the subject or who study Mars, **it is publicly known that there has not been any other mountain sized carved faces structures found in any images from Mars.** Yet the fact that this is the only face mountain in the public knowledge, it therefore has to be Alalu's face, because there are no others, and the Anunnaki left a record to us about it thousands of years ago. (*I will make a conclusion statement more about this point soon.*)

There is also another amazing point which was made about Alalu's face and direct connection also to my research of the Face of Cydonia and what **I have proposed about the Face since 1998, which is that the Face of Cydonia was artistically sculpted from above!** It would be far easier to use **beams that carve** from above, who can easily and with precision, be technologically abled to create what is seen rather quickly, as compared to hundreds or thousands of 'Martians' walking all over it working by laser sculpting the face. It is ridiculous to think it was created by foot workers on the surface. The architects of the Face at Cydonia had to have a programmed 3D perspective model in knowing how and where to carve for its multidimensional designing that also consists of geometry, and master artistry, and their knowledge of how light effects these multidimensional aspects in its artistic designing of its surface relief that brings this created feature to life. (*Pun intended*). Remember, the Face of Cydonia is one and a half miles in length, by one mile in width, and 1500 feet in height; **'Like A Mountain'!**

As we seen already proven in this book, the Face is not smooth, nor is it an eroding dust heap of sand and rocks that changes over time by erosion, as it is not a trick of natural light and shadow. Yet it's very artistic carved designing uses tricks of light and shadow to show its elaborate handiwork by positions of daily sun positions, as well as various seasonal oribital positions and planetary shifting of precessional positions. That is my opinion of what my study of the face tells and shows me. This is a proven concept already, as again, we have enough data of the Face at various times of day and seasons that we see various details that are not so easily seen at other times.

To understand and believe that the Face of Cydonia, and obviously that of Alalu's elusive Face mountain structure is easier to use beams that carve from above, to artistically sculpt the entire mountain with precision and skill, is to consider that technology which we have here and currently use on Earth today that also uses beams that carve to create excellent artistic precision and very ornate 3D models of the programmed design; Laser etching!

(I am at the time as I write this, am trying to prove this point by recreating the Face of Cydonia by finding and using a laser etching service skilled enough to help create a table top 3D scale model. I hope to have it at the time of publishing this book).

To myself, having myself stated openly since 1998, that lasers had sculpted the Face from above, and when some years later, learning about the Anunnaki information, excited me and convinced me in the least that I was right. It made most sense. The Anunnaki confirmed and validated my theory that the Face, at least to myself, who came up with that theory that the Face was laser carved from above to begin with. I never made that theory a secret and I was very public about it.

I will also say once again, that it is **NOT** and was **NOT JUST THE FACE**, which was artistically sculpted alone, but the area around the Face, as well as its '**civilization**' techno-signatures on the plain it all sits upon!

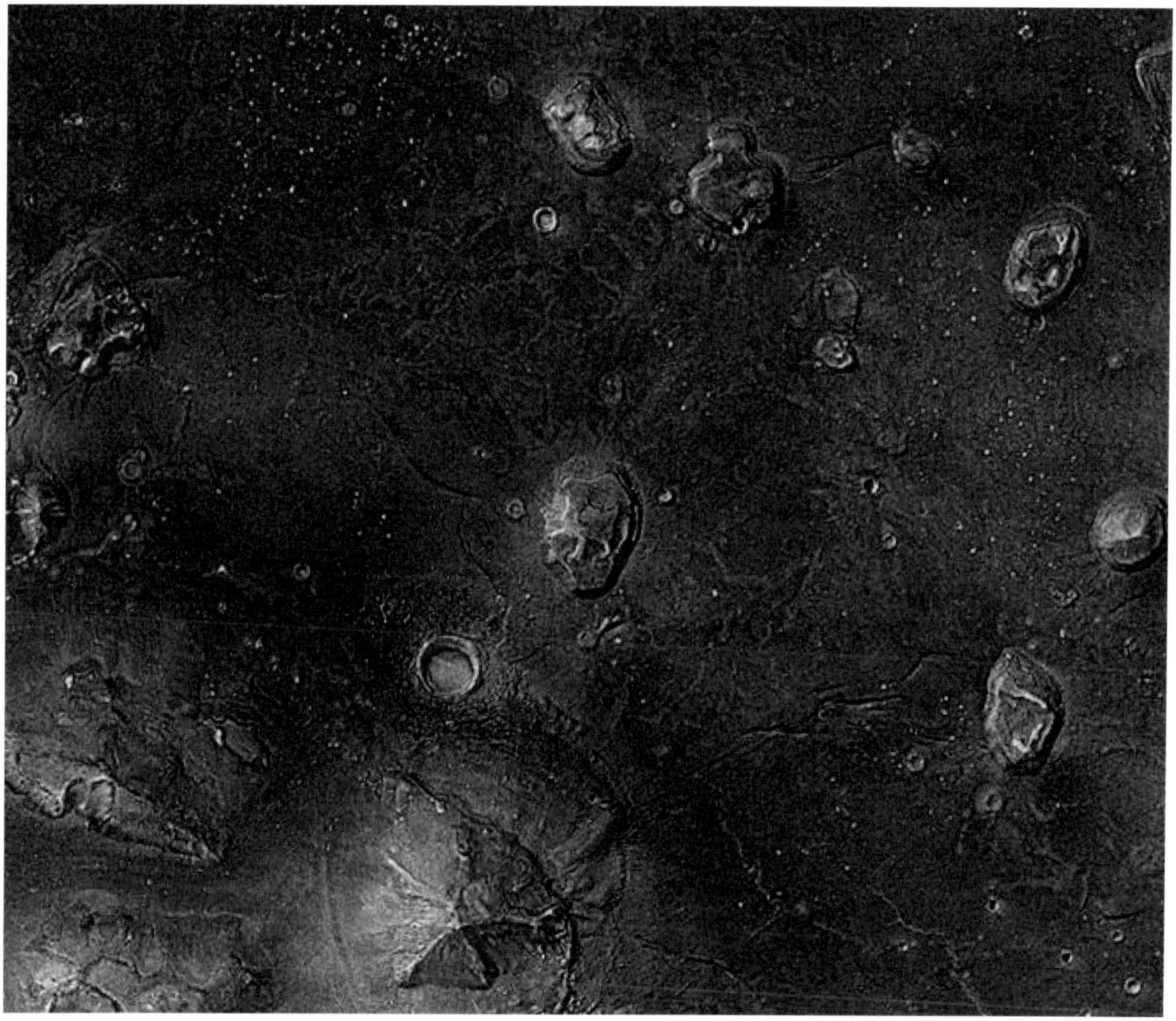

If one goes by the Anunnaki story of creating a face on Mars, and the fact that a way station was immediately set up there on Mars by the Anunnaki, again, at most, its age would be up to 500,000 years ago! Is this the age of all we see there at Cydonia? Are the Anunnaki responsible for all this planetary landscape 3D laser etched modifications on such a massive scale?

Or, did the Anunnaki's idea for creating a mountain sized laser etched sculpturing of a face, was based on an idea that they '**saw things by others unseen,** (*these 3D lasered art-scapes and carved faces and structures created, by others*), **on a strange planet they faced unknown things?**

Maybe the Anunnaki were already familiar with the concept obviously of using **beams that carve**, to artistically sculpt faces out of mountains, and use the same to artistically carve up the planetary surfaces they encounter?

What may have been unknown to them however, are the creators of what unknown beings or ancient civilization whose strange world they had come to see, and to reside upon as a way station? These are serious things to consider, but we must remember that between 150 million to 500 million years ago, the area of Cydonia resides upon, was deliberately blasted with a nuclear weapon, that according to Dr. John Brandenburg, judging by its techno-signatures, its weapon size was that of the Empire State building. At most, the true age and creation for the Face and all of the laser sculpted region of Cydonia itself was older than the nuclear weapon assault.

Another note to add also, is that the Anunnaki did also have nuclear weaponry (weapons of terror), and used them on their world, in space, and even on our Earth during their time here. It was not the Anunnaki who created all that on Mars which was targeted by also an ancient unknown race of intelligent beings with weaponry to destroy the ancient civilization of another unknown race of intelligent beings on Mars that **DID** create the Face and Cydonia, up to over 500 million years ago! The Anunnaki were not around then, nor were human kind as we know it.

The thing we need to do, is to locate and differentiate between the Anunnaki's modifications of surface features, or their refurbishments of ancient Martian sites and landscapes that they created at most up to 500,000 years ago, to that which has been there for hundreds of millions of years. This is a fascinating field of research and study, as well as its archeological prospects.

Regardless if you think or believe that the Anunnaki were ever at Mars, which is not the point. The point is, that someone was there in the past, which also had an ancient intelligent civilization established there, regardless whether it was an indigenous race of intelligent beings, or one that was spacefaring and which had come to Mars and had thrived there. Much more input and research needs to be done to establish those sets of facts and solve those mysteries, but by all means, never ask NASA or consider that they will be a part of doing that publicly and openly.

Now it is time to make my couple of points in conclusion of this book. I would like to again focus on the Anunnaki and the creation of Alalu's face, and how people directly attribute the Face at Cydonia as evidence that the Anunnaki story is real and true, and that the Face is Alalu's face. As we went over and looked at the image data, it is safe to assume and to say that the face at Cydonia does not match to the description of what the Anunnaki story testifies about.

When pointing out the evidence and facts to people to show and tell them that they are incorrect, they mostly get defensive because they cannot answer about why it not matches to the Anunnaki's description of Alalu's face, and seeing the obvious discrepancies for what the Face at Cydonia details actually shows. When defending their research and belief of the Anunnaki story, they all reacted by questioning the obvious saying: **"What else can it be? The Anunnaki claimed on the record thousands of years ago, that they created the Face on Mars!"**

After that question and statement, the next obvious logical and understanding statement would be said: **"There are no other faces found there, or we would have seen it! So, it has to be Alalu's face told to us by the Anunnaki, even though it obviously not matches the Anunnaki's description of Alalu's likeness"**

My reply:

"REALLY? THERE ARE NO OTHER FACES FOUND THERE ON MARS???

WOULD YOU HAVE SEEN IT???

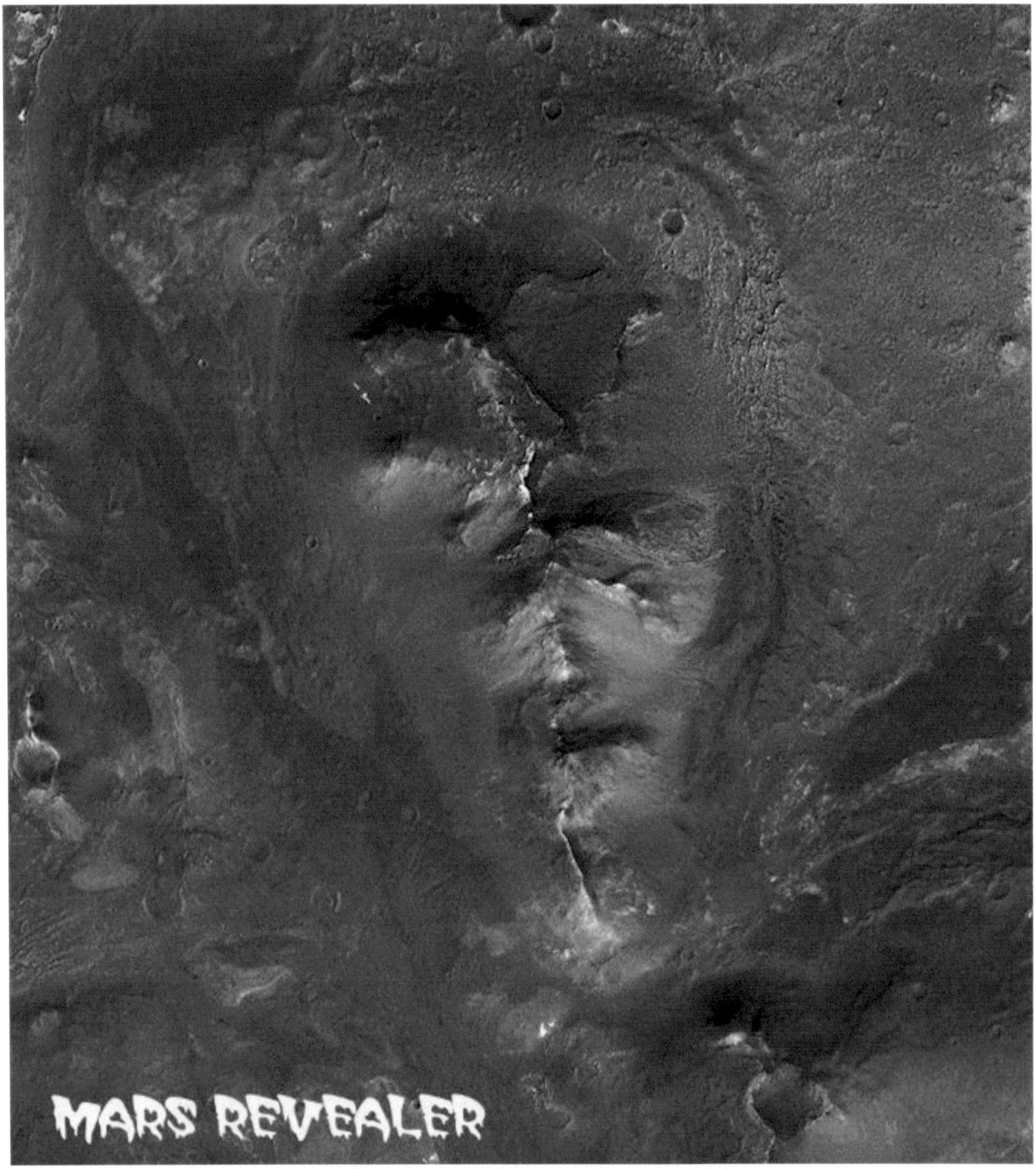

To be continued…….

Final Thoughts...

Regardless of its implications and possibilities, the Face of Cydonia is not a **mere** natural object, mountain, mesa, hill, massif, or butte as which all of these natural terms has all been falsely categorized by NASA for what this greatest amazing intelligently designed techno-signature found in the solar system this Face really is.

In fact to leave you with my final statement about another unintended experiment that occurred around 2018, I went to post the image seen below on Twitter. At around that time, twitter ran an A.I. (artificial intelligence) algorithm in connection to images people would post on their pages. It would ask 'questions' about the image you were trying to upload and post. It is funny how A.I. on Twitter seen and recognized and also confirmed the reality.

The question I received from the A.I., I assume so it can tag someone before finalizing the post on Twitter was:

"**WHOSE FACE IS THIS**?"

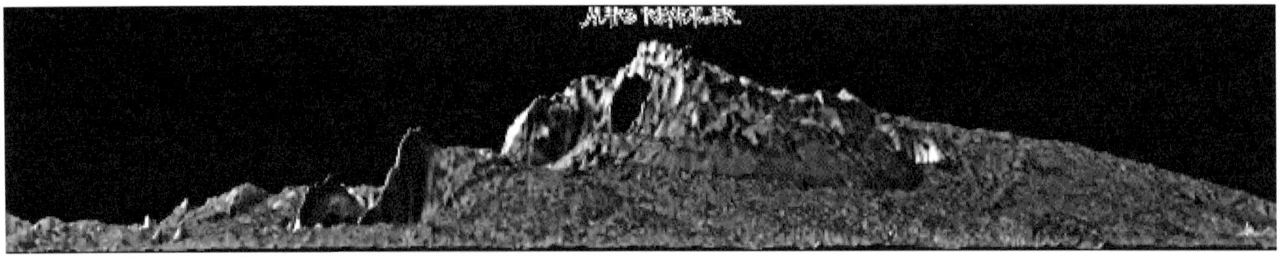
MARS REVEALER

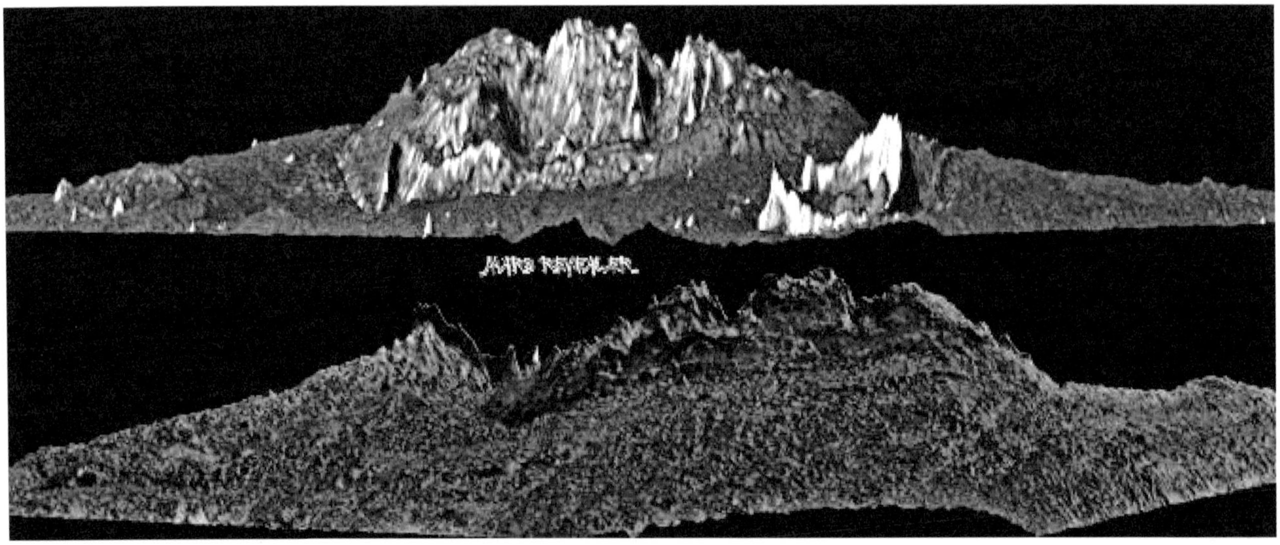
MARS REVEALER

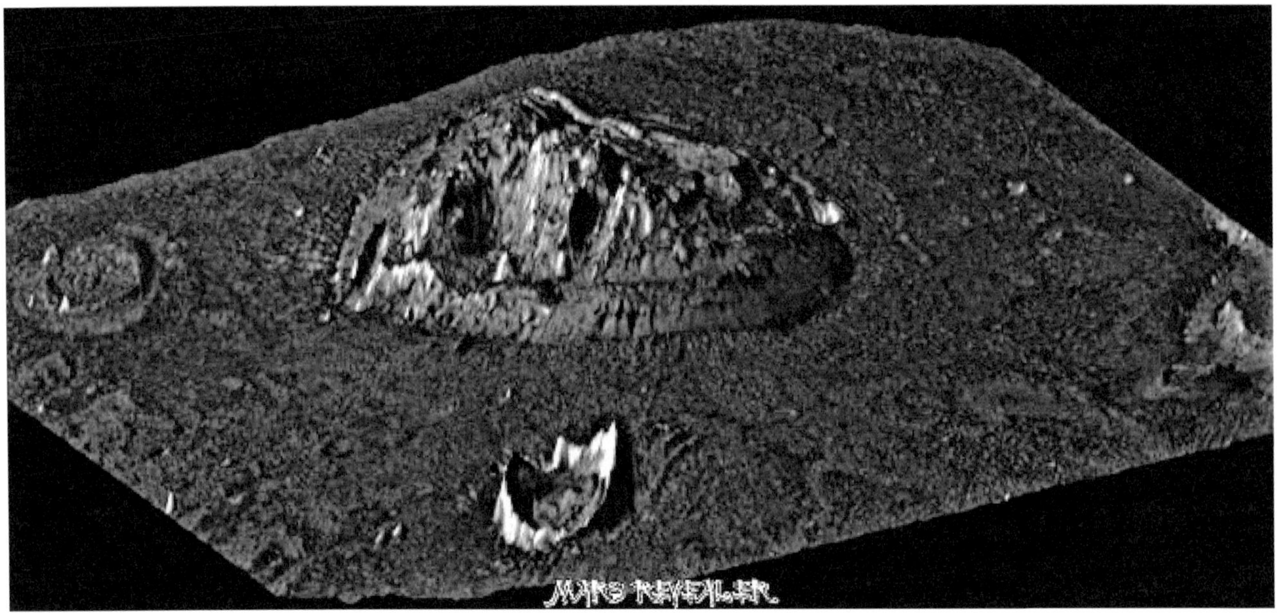
MARS REVEALER

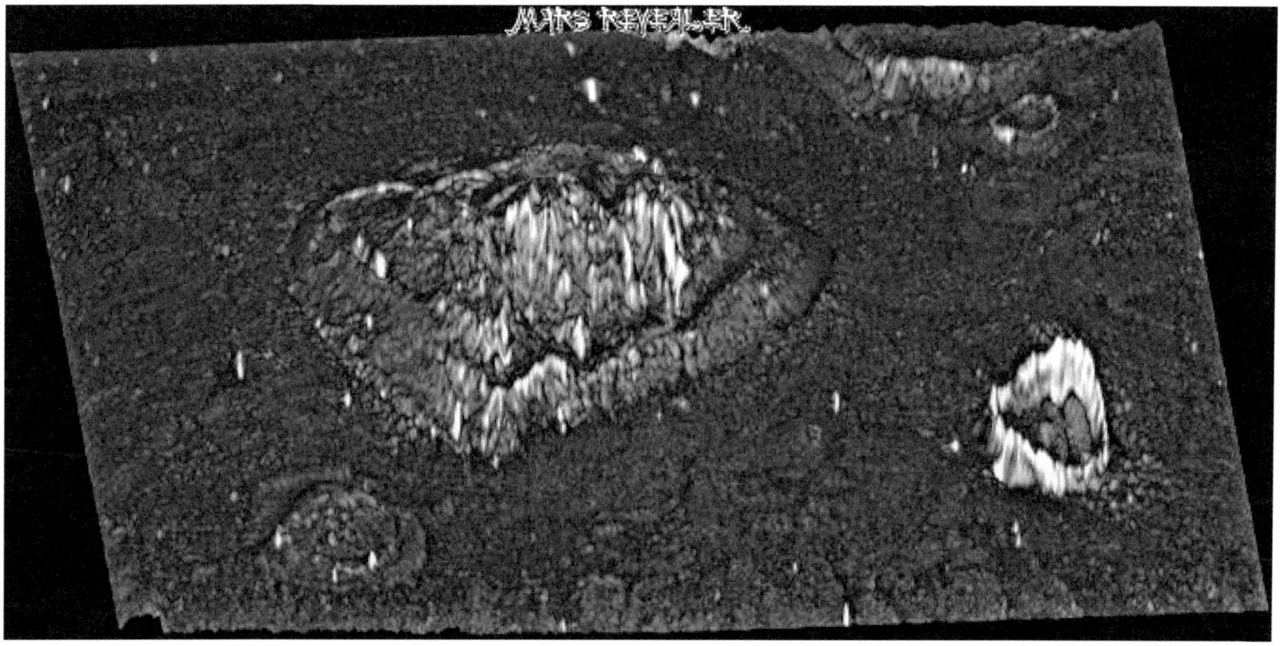
MARS REVEALER

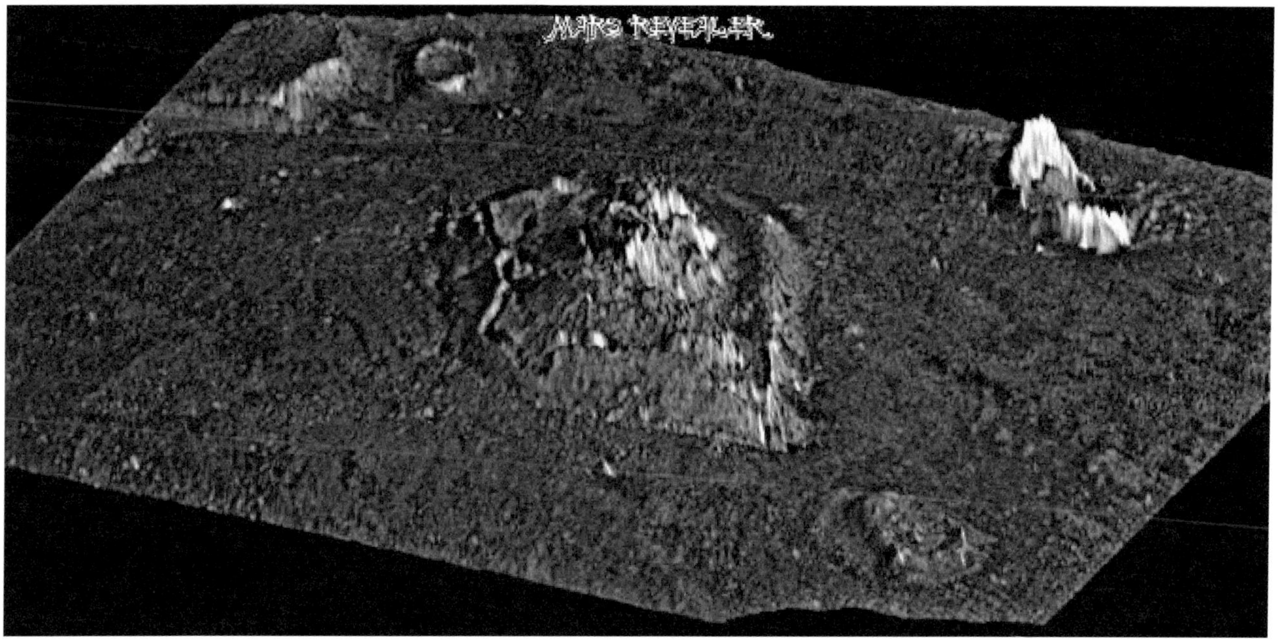
MARS REVEALER

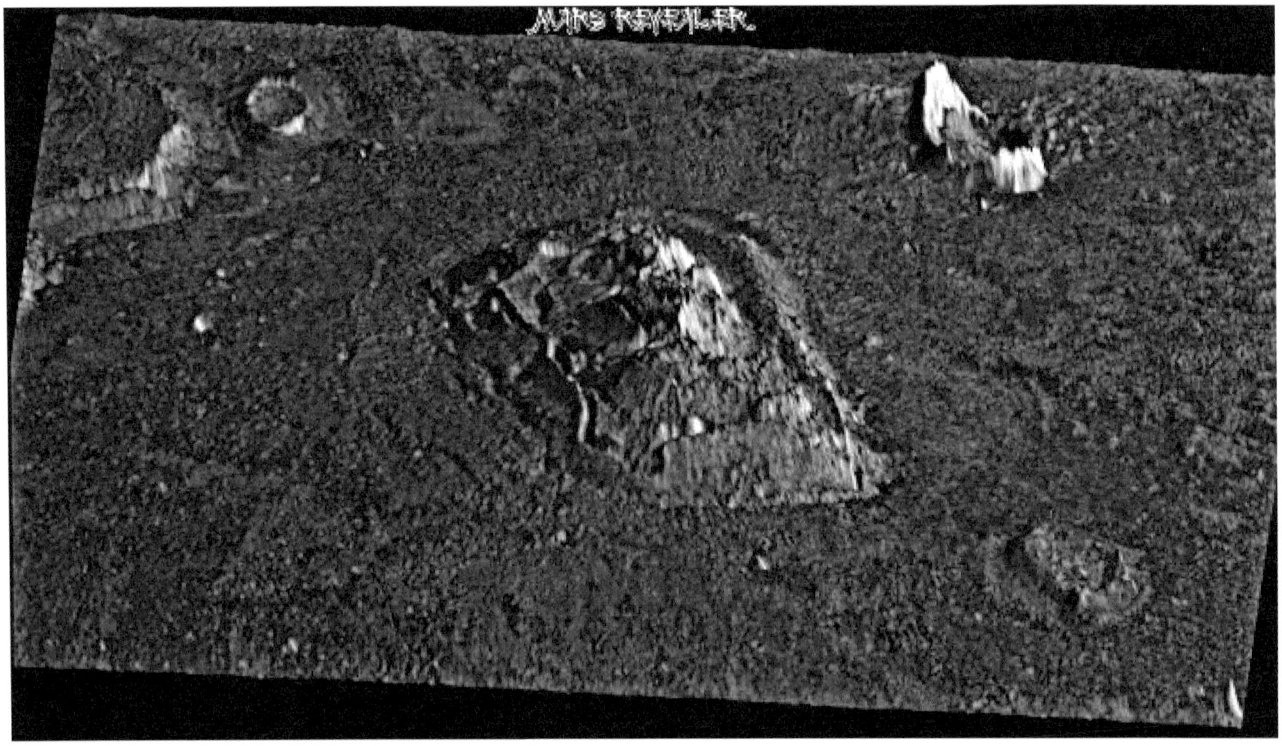
MARS REVEALER

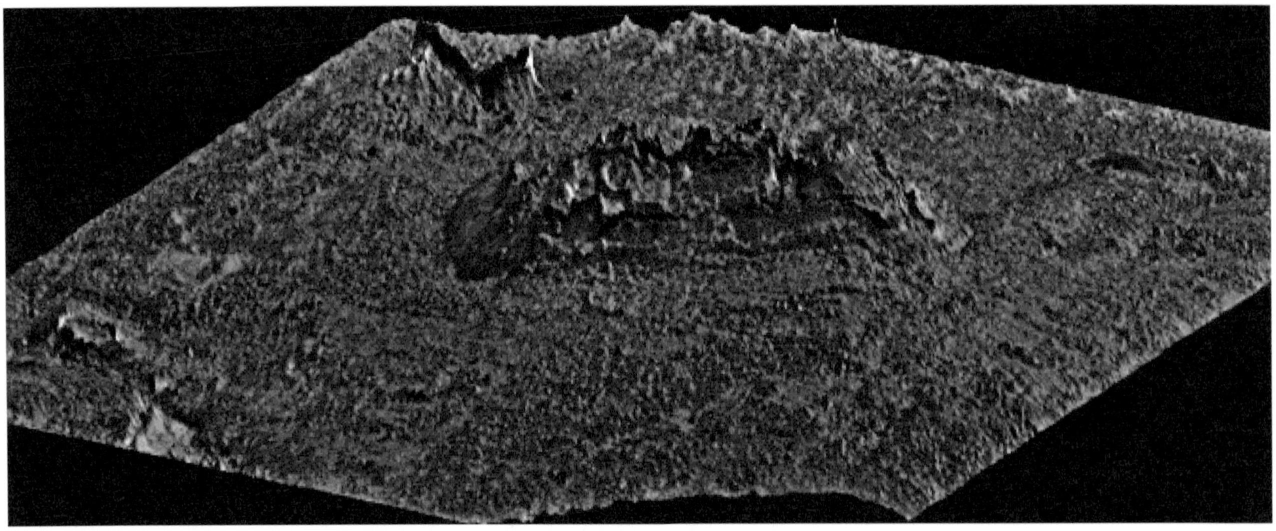

Made in the USA
Monee, IL
29 May 2026

51163468R00231